MECHANICAL KINESIOLOGY

MECHANICAL KINESIOLOGY

Jerry N. Barham, Ed.D.

Professor, Department of Physical Education,
University of Northern Colorado
Greeley, Colorado

with 507 *illustrations*

The C. V. Mosby Company

Saint Louis   1978

The C. V. Mosby Company
11830 Westline Industrial Drive, St. Louis, Missouri 63141

Library of Congress Cataloging in Publication Data

Barham, Jerry N
    Mechanical kinesiology.

    Bibliography: p.
    Includes index.
    1. Human mechanics. 2. Kinesiology. I. Title.
[DNLM: 1. Biomechanics. 2. Kinetics. 3. Movement.
WE103 B251m]
QP303.B37   612'.76   77-23969
ISBN 0-8016-0476-1

CB/CB/B 9 8 7 6 5 4 3 2 1

Preface

Mechanical Kinesiology is an introduction to the mechanical foundations of human movement. However, the text is not intended as a total treatment of these foundations. My experience is that a text which tries to cover all the facets of a subject matter simultaneously is often more confusing to the beginning student than enlightening. For this reason, I have delimited this text largely to the basic principles of human mechanics, with applications being made to movement problems most often found in typical physical education programs. The specific movements studied therefore are not all inclusive but are given only as examples of the kinesiological applications of the mechanical principles involved.

I have divided each lesson into three parts: a part on the text proper; a part entitled study guidelines, which includes behavioral objectives, summaries, and applications of the concepts and principles presented in the text proper; and a self-evaluation test, which students can use to measure their progress toward meeting the lesson objectives.

To me education is a problem-solving process which promotes changes in the behavior of students. Therefore the desired student behavior changes are stated for each section of each lesson in the form of specific behavioral objectives. A behavioral objective is one that is stated in terms of what is expected of students in the way of their personal performances, and not in terms of teacher performances and/or the performances of textbook writers. The self-evaluation section at the end of each lesson allows students to determine their own individual achievement of the stated objectives.

The applications and problem solving section of the study guidelines is included in each lesson to serve the two ends of (1) helping the student develop skill in analytical thought and numerical calculation, both of which are required for the successful study of human movement, and (2) promoting the mastery of the text proper by actually putting its ideas into practice. Perhaps the exhortation of James, i, 22, is appropriate here: "But be ye doers of the word, and not hearers only deceiving your own selves."

No more than the simplest mathematics is used in this text, and no prior knowledge of physics is assumed. A review of the elementary mathematics required for a thorough understanding of the topics covered is presented in Appendix D.

The progression of concepts is from the more basic to the more applied, and from the more general to the more specific. The basic materials are presented with the idea of allowing the student to grasp the total concept before proceeding to the more specific movement applications involved.

The treatment concentrates, as mentioned before, on those mechanical principles primary to an understanding of human movement performances. To make such an emphasis possible, I have omitted peripheral material that might be of value from a general education standpoint since, whatever its merits, including such material would have necessitated delving into ideas of marginal concern to the students of human movement at the possible expense of key kinesiological concepts.

Whenever possible I have tried to present derivations of important results throughout the text, thereby exhibiting rather than merely describing the inductive and deductive methods of science. I have also attempted throughout the text to encourage in the student a feeling of participation that I believe is essential for learning.

However, to completely eliminate the role of students as mere passive receivers of information, I strongly recommend that this text be accompanied by appropriate laboratory work and by supplementary readings.

Special thanks are due to Jerry Krause, for his assistance with the study guideline sections; Stu Horsefall and Jack Tandy, for their art work; Bob Waters, for his photography; Lindy Minihan, Nelson Ng, Dan

Caster, Bart Smith, Jeff Broida, Holly Summerlot, Mike Dunafon, Cliff Harris, Dehaven Hill, Patty Sanchez, Jim Goldstone, Rick Marquez, Fred Thompson, Brad Campbell, Larry Martinez, Dave Marrufo, and Keith Anderson, for appearing in photographs; Gertie Fellinger, Marlene Krieger, Joan Lehr, and Debbie Marks, for typing the manuscript; the hundreds of students who worked with and helped to improve the mimeographed editions of the manuscript; Lyle Knudson and Nelson Ng, for helping to proofread the manuscript and to prepare the teacher's guide; and Dr. Nancy Van Anne, for her many helpful suggestions.

Jerry Krause and I have prepared a programmed textbook to facilitate the learning of basic concepts: Kraus, J. V., and Barham, J. N.: *The Mechanical Foundations of Human Motion: a Programmed Text,* St. Louis, 1975, The C. V. Mosby Company.

Jerry N. Barham

Contents

PART I

INTRODUCTORY CONCEPTS

Concept of mechanical kinesiology

CONCEPT OF KINESIOLOGY

In a world where emerging disciplines are staking claims on words as though they were gold mines (which they are), perhaps the meaning of the word "kinesiology" as used in this text is not evident to every student. Kinesiology is defined as the study of human movement in all its ramifications. Therefore kinesiology is concerned with one of the most complex of all phenomena associated with the most complex of all living organisms—the movement behavior of the human being.

Study of human movement

The movements of humans through their various environments can be studied from three basic points of view. First, we can study human movements from the standpoint of the perceptions and motivations which prompt the movements and the neuronal mechanisms which control them. Second, we can study human movements from the standpoint of the biochemical processes that initiate and sustain them. Third, we can study the movements of people from the standpoint of the time, distance, and force relationships involved. The first approach is identified, for the sake of

internal consistency, as being *psychological kinesiology,* the second as *physiological kinesiology,* and the third as *mechanical kinesiology*. Human movement, of course, requires all three of these approaches in order to be comprehended in its entirety (Fig. 1-1).

Human movement defined

For the purposes of this text, human movement is defined, without considering the psychological and physiological components involved, as the change in position of the body or body segments in space and time through the application of varying amounts of force.

Since no two objects can occupy the same space at the same time, movement involves a consideration of space-time relationships. It does not matter whether the change in space and time is one of location, volume, or shape. Therefore human movement can take many forms such as running (change in location), expanding the chest (change in volume), or bending an arm (change in shape). However, all these movements resolve themselves into only two fundamental types, *linear* and *angular,* and thus all motion can be described in either linear or angular terms.

3

KINESIOLOGY

Study of
human movement

MECHANICAL	PHYSIOLOGICAL	PSYCHOLOGICAL
KINESIOLOGY	KINESIOLOGY	KINESIOLOGY

Study of
spatial, temporal, and
force variables in
human movement

Study of
biological and
biochemical variables
in human movement

Study of
behavioral and
neurological variables
in human movement

Fig. 1-1. Conceptual model of kinesiology.

Linear and curvilinear paths of motion

The two basic pathways that a moving object can follow are (1) a straight-line path and (2) a curved path. The movements which follow these pathways are known as linear and curvilinear, respectively. These two classifications of movement and the subdivisions of primary importance for the curvilinear classification are defined as follows:

1. *Linear movement.* Motion that follows a straight-line path. This type of motion is also called translatory because all parts of the moving body travel exactly the same distance, in the same direction, and at the same time.

2. *Curvilinear movement.* Motion that does not follow a straight-line path.
   a. *Angular or circular movement* (Fig. 1-2). Motion that follows a circular path, i.e., that remains a constant distance from a fixed point. The constant distance from a fixed point is, of course, the radius of a circle. This type of movement can also be called rotatory. The three terms *angular, circular,* and *rotatory* are synonymous from the standpoint of mechanics and can be used interchangeably in the description of movements.
   b. *Parabolic movement* (Fig. 1-3). Motion

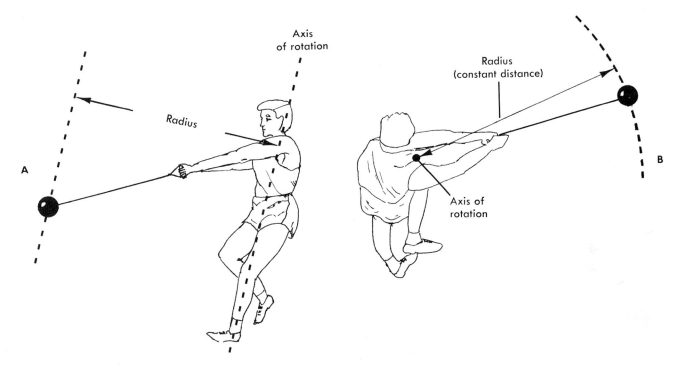

Fig. 1-2. Circular motion. **A,** Side (sagittal) view. **B,** Top (transverse) view.

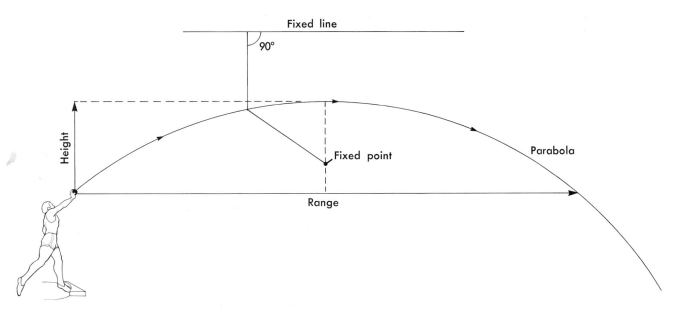

Fig. 1-3. Parabolic motion.

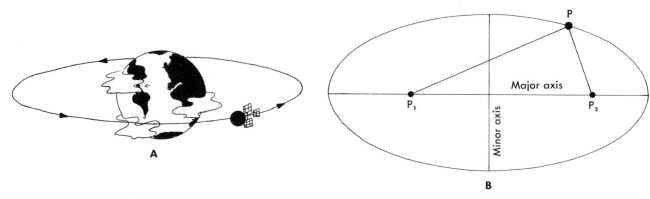

Fig. 1-4. Elliptical motion. **A,** Orbiting satellite. **B,** P_1 and P_2 are two fixed points.

that follows a path which is always an equal distance from a fixed point and a fixed line. This is the pathway followed by a projectile in flight in the earth's atmosphere when it is moving at an oblique angle to the horizontal and the effect of air resistance is ignored.

NOTE: The *elliptical pathway,* which is the path of a point the sum of whose distances from two fixed points is constant, is not listed here because it is not a normal pathway followed by moving objects within the earth's atmosphere. It will not therefore be studied in this text. An elliptical path is illustrated by a satellite in orbit around the earth (Fig. 1-4).

Perhaps it should be pointed out here that we are concerned with a classification of motion pathways in this discussion, not of motion directions. Confusion can be avoided by remembering that the direction of *all* motion, including the various types of curvilinear motion, at a given *moment* in time, is always linear. This concept will be developed later. The direction of curvilinear motion during a given *interval* of time, however, is classified as being either clockwise or counterclockwise.

The two types of motion found in *most movements* of the human body are linear and angular. For example, the linear movement of the body as a whole (e.g., running a 100-yard dash) is produced by the angular movements of the limbs. However, the parabolic path is commonly seen in the motion of projectiles. The parabolic pathway of an object projected through the air is unique to aerial activities. The trajectory followed by the body in the air during a long jump in track and field, for instance, is a parabola.

Even though the human body at one time or another performs all three types of movement—linear, angular, and parabolic—angular movements are the most typical of the musculoskeletal system; that is, the angular movement of a body lever rotating at a joint because of the force imparted to it by a muscle contraction is the type of motion typically found in the musculoskeletal system. The interrelationship of the primary movement types associated with human movement is depicted in Fig. 1-5.

Sometimes linear and angular movements repeat themselves back and forth over the same general path. This special motion is called *vibratory* or *harmonic* motion. The terms vibratory and harmonic are synonyms and refer to any to-and-fro motion which generally follows the same repetitive path. Two other terms often used to describe harmonic motion are *periodic* and *oscillatory.* Any sort of motion that repeats itself is called periodic. If the motion is back and forth over the same path, it is called oscillatory. These types of motion will be discussed in Lessons 19 and 24.

Concept of motor behavior

Human motor behavior is the movement of the human body resulting from muscle forces acting

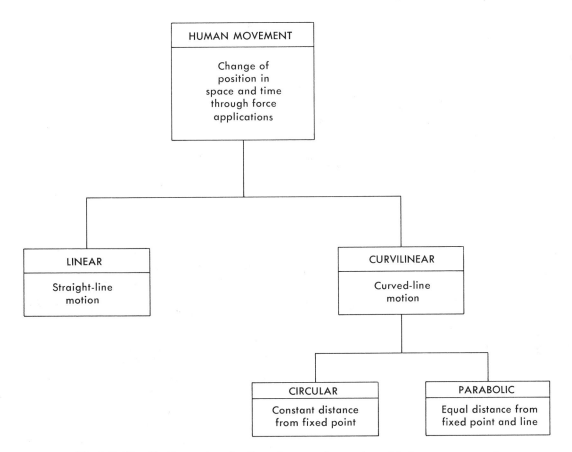

Fig. 1-5. Classification system for the primary pathways traced in human movement.

through the muscles, tendons, bones, and joints of the body. The function of muscles in producing human motion can be likened to the function of an electric motor, water turbine, or gasoline engine in other machine systems; that is, muscles, through the conversion of chemical energy into mechanical energy, serve as the movers of the body. Therefore muscles may be conceptualized as being the motors of the human body.

In identifying muscles as motors, the word *motor* is used as a noun but it also can be used as an adjective. As an adjective the word means "causing or imparting motion." Following are some of the ways the word motor is used as an adjective:

> **motor activity or behavior** Movements of the body produced and controlled through the contraction of skeletal muscles; example, the act of running
> **motor task** A specific type of goal-oriented motor activity; example, the act of running a 100-yard dash in track and field
> **motor performance** The actual execution of a motor task by an individual

CONCEPT OF MECHANICAL KINESIOLOGY

Since kinesiology is defined as the study of human movement, mechanical kinesiology is defined as that branch of kinesiology concerned with studying the mechanical factors affecting human movement, i.e., with applying the physical laws of mechanics, a branch of physics, to the study of human motion. Mechanical kinesiology can also be defined as that branch of kinesiology concerned with the "mechanical analysis" of human motor behavior.

The mechanical analysis of human motor behavior requires that we separate or break down movement performances into their mechanical components. This process of separating or breaking down any whole into its parts is called *analysis*.

The two types of kinesiological analysis are identified as *kinematic* and *kinetic*. A kinematic analysis is concerned with the scientific observation and description of movement variables. A kinetic analysis is concerned with the study of these variables as interacting factors that cause movements and movement changes. The disciplines concerned with these descriptive and causal types of analysis are called kinematics and kinetics, respectively.

Relationship of mechanical kinesiology to biomechanics

The two foundational sciences upon which all life sciences are based are physics and chemistry, and the application of these sciences in the study of life processes is called biophysics or biochemistry, respectively. Each of these sciences can be further subclassified. Biophysics, for example, can be further broken down into biomechanics, biothermogenics, bioptics, bioelectrics, and so forth. Thus biomechanics is a subdivision of biophysics and is concerned with the study of *all* biological movements. Such diverse movements as those performed by lower animals, plants, and the visceral organs of mammals are all within the academic domain of biomechanics.

Since movement is a characteristic of all biological functions, biomechanics is just as much a concern of the physiological kinesiologist as of the mechanical kinesiologist. The heart and blood vessels, for example, must be studied from a biomechanical standpoint if their physiological functions are to be fully understood. Mechanical kinesiology, however, is concerned only with the study of such gross human motor behaviors as are involved in specific sport, dance, and work performances. Thus mechanical kinesiology can be conceptualized as being a subdivision of both biomechanics and kinesiology, for both these disciplines are considerably broader in scope.

Delimitations of this text

Since movement behavior is generally studied from the three basic approaches previously identified as physiological, psychological, and mechanical, there are three basic approaches to the descriptive and causal types of analysis. These are identified as the kinematic and kinetic divisions of physiological, psychological, and mechanical kinesiology. As also mentioned previously, we will not be concerned in this text with the physiological and psychological types of analysis but instead will direct our attention solely toward mechanics, mechanical kinesiology, and the mechanical analysis of movement variables.

Because of the enormous amount of information that pertains to the mechanical functioning of humans within their many environments, the subject is divided here into three parts. The first part, which is the focus of this text, is confined to the mechanical foundations of human motion and is primarily concerned with the fundamental methods and principles of mechanics involved in the kinematic and kinetic analysis of human motion. The second part, sometimes called *anatomical* or *structural kinesiology*, pertains to a study of the structures and mechanical functions of the specific muscles, bones, and joints that form the movement or motor apparatus of the human body. The third part, primarily concerned with the detailed analysis of specific motor skills, deals with how the different body segments and joint actions are linked together in the proper sequence for the most effective and efficient production of total body movements.

Perhaps it should be emphasized that we are dealing here with the arbitrary division of a total subject matter into three parts. Even though the mechanical foundations of human motion are the primary concern of this text, numerous examples of the mechanical analysis of motor apparatus structures and functions and of motor skills will be given throughout the various lessons.

Perhaps it should also be pointed out here that we will not be concerned in this text with the total discipline of biomechanics, which, as already discussed, applies the principles of mechanics to the study of life processes. In other words, we are not concerned in mechanical kinesiology with the study of all conceivable forms of movement associated with human beings and with other forms of life. For instance, the biomechanical functions of the heart, lungs, and other visceral organs are relegated to physiological kinesiology and will not be studied here. Likewise, the movement of impulses through the nervous system is relegated to psychological kinesiology. The movements of plants and lower animals are omitted entirely.

ACADEMIC DIMENSIONS OF MECHANICAL KINESIOLOGY

The mechanical kinesiologist, like all scientists, is concerned not only with the solution of practical problems but also with the advancement of man's knowledge of human movement as a natural phenomenon. He has, in other words, an academic as well as a practical interest in his subject.

Academically a mechanical kinesiologist is primarily interested in identifying and understanding the mechanical events involved in the successful performance of motor tasks. This interest normally leads to original research—which can be observational, theoretical, or experimental in nature.

The *observational* researcher in mechanical kinesiology simply observes the movements of different performers and identifies those which consistently produce the best results. The researcher might observe, for instance, that all successful major league baseball batters have certain movement patterns in common. Thus he or she might infer that these patterns are necessary for successful batting performances. This, of course, may or may not be true but it does lead to the next logical step, which is to develop a theoretical explanation for the observations.

The *theoretical* study of movement patterns involves special application of the mechanical laws and principles of classical physics. The assumption here is that if a movement pattern is essential for successful performance then a reason for this fact can be hypothesized from a knowledge of classical mechanics. Not all theoretical hypotheses, of course, are generated to explain current practices. They may, in fact, be outgrowths of creative attempts to develop better and previously untried techniques.

The formation of a theoretical hypothesis, however, usually leads to some type of *experimental* research through which the tenability of the hypothesis is tested.

Through the procedures of observational, theoretical, and experimental research, we are able to develop movement models for the various types of performance. Thus we can speak of the "proper" or "model" techniques of performing such skills as throwing a baseball, shooting a basketball, or clearing a high hurdle in track and field. The words "technique" and "mechanics" will be used as synonyms in this text when they refer to the process of movement. Therefore, when we speak of the mechanics of a performance, we mean the specific movements used by a performer in the execution of a motor task.

In summary, it should now be evident that the study of movement techniques and the development of movement models are what comprise the academic subject matter of mechanical kinesiology.

PROFESSIONAL APPLICATIONS OF MECHANICAL KINESIOLOGY

The professional applications of mechanical kinesiology are chiefly focused in the area of physical education instruction. It is assumed here that the physical educator is primarily a teacher of motor skills and that the improvement of motor performance is one of his or her main professional responsibilities.

It is also assumed that the teaching competence of a physical educator is largely determined by his or her ability to analyze a student's movement performances in terms of cause-and-effect relationships; that is, a competent teacher of motor skills should be able to analyze a performance and answer such questions as (1) what is right about the performance (using a performance model), (2) what is wrong, (3) why it is wrong, and (4) what must be done for improvement. This skill in analysis should, of course, be combined with certain communication skills and with certain leadership skills. Communication skills are necessary for communicating the results of the analysis to the student in a positive manner. Leadership skills are also necessary for motivating the student to use the results of the analysis in his or her practice sessions.

The background for a movement analysis can come from two basic approaches. The first approach is that of the lay participant, whether as a performer or a spectator. Golf can be studied, for example, by the weekend golfer (performer), and football can be studied by a housewife (spectator). Each of these approaches leads to a certain amount of insight into movement problems;

but it is also possible that neither the performer nor the spectator can recognize the reasons *why* these movement problems exist, and it is most unlikely that either knows *how* these problems might best be solved. Thus there is a need for a professional individual who *is* capable of analyzing a movement in terms of cause-and-effect relationships. The golfer will probably go to a competent golf teacher for assistance with his swing, and the housewife will probably be content to leave the development of football players to their coaches.

Much can be learned about a motor skill by being a performer of that skill and by watching others perform, but even more can be learned by combining these experiences with the academic approach of mechanical kinesiology. In other words, doing and watching combined with a study of the principles will give the student much more insight into the nature of a skill and develop much greater analytical ability than will either experience alone.

STUDY GUIDELINES
Concept of kinesiology
Specific student objectives

1. Describe kinesiology.
2. Identify the three approaches involved in the study of human movement.
3. Differentiate between psychological kinesiology, physiological kinesiology, and mechanical kinesiology.
4. Define human movement.
5. Differentiate between linear and curvilinear types of motion.
6. Differentiate between angular and parabolic types of motion.
7. Define human motor behavior.
8. Differentiate between a motor activity, motor task, and motor performance.

Summary

Kinesiology is the complete study of human movement. It can be approached from three points of view: psychologically (the study of human motor behavior and its neurological components), physiologically (the study of the biological and biochemical components of motor behavior), and mechanically (the study of the

physical laws of mechanics applied to human motion).

The change of position of the body or its segments in space and time through force applications is termed movement. This motion traces the pathways of linear (straight-line) motion and curvilinear (non–straight-line) motion. The most common types of curvilinear motion found in human movement are angular and parabolic. Angular movement occurs at the body's joints, while parabolic motion is the characteristic motion of projectiles. All movements of the human body can be analyzed into linear or angular components. Motor behavior is movement of the body caused by the contraction of skeletal muscles acting through their tendons, bones, and joints. A specific type of goal-oriented motor activity is called a motor task while a motor performance refers to a specific individual's execution of a motor task.

Performance tasks

1. Consult a dictionary or other reference for a description of kinesiology.
2. Scan through this section of the lesson and write down your listing of the most important points.
3. Write down dictionary definitions for psychology, physiology, and mechanics.
4. Complete Fig. 1-6 with examples of each of the most common pathways followed in human movement.
5. List physical education examples of a motor activity, motor task, and motor performance.

Concept of mechanical kinesiology
Specific student objectives

1. Define mechanical kinesiology.
2. Develop a conceptual model for mechanical kinesiology.
3. Define the term "analysis."
4. Differentiate between kinematics and kinetics.
5. Differentiate between mechanical kinesiology and biomechanics.
6. Describe the delimitations of this textbook.

Summary

Mechanical kinesiology can be thought of as the mechanical analysis (breaking down) of hu-

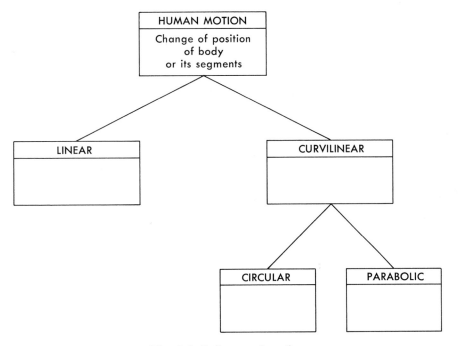

Fig. 1-6. Pathways of motion.

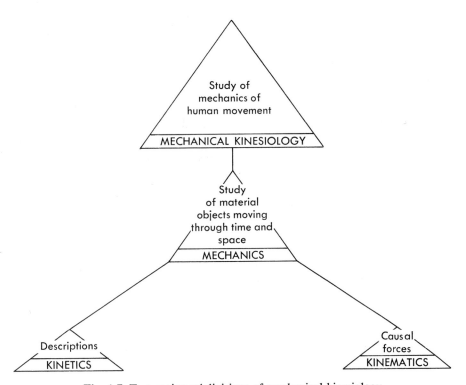

Fig. 1-7. Two major subdivisions of mechanical kinesiology.

man motor behavior. Mechanics is the study of material objects and their movement through time and space. The two major branches of mechanics are kinematics and kinetics. A complete conceptual model of the interrelationships involved in the study of human movement is shown in Fig. 1-7.

The total subject matter dealing with the mechanics of human motion is divided into three parts. The first part, which is the primary concern of this text, contains the mechanical foundations of human motion. The second part, sometimes entitled structural kinesiology, is centered on the anatomy and mechanical function of specific joint actions. The third part focuses on the overall motion effectiveness and efficiency of the body in the performance of specific motor tasks.

Performance tasks

1. Consult a dictionary for definitions of physics, mechanics, and kinesiology. Compare them to the descriptions of these terms given in this text.
2. Identify the two areas of mechanical analysis used to study human motion.
3. In your own words, describe mechanical kinesiology and compare it to the description on p. 7.

Academic dimensions of mechanical kinesiology
Specific student objectives

1. Distinguish between theoretical and practical study of a knowledge area.
2. Identify academic areas of interest in mechanical kinesiology.

Summary

Mechanical kinesiology is concerned with theory and practice. Academically a scholar is interested in studying the subject matter from a standpoint of pure curiosity, i.e., to advance man's knowledge of human movement with the knowledge being an end in itself.

This would include such tasks as identifying and understanding mechanical events involved in a motor task performance. It could also lead to original research designed to produce theoretical models of performance and the experimental verification of these models.

Performance tasks

1. State reasons for the inclusion of both theoretical and applied material in the study of an area of knowledge.
2. List several scholarly tasks that could be undertaken in mechanical kinesiology.
3. The study of movement techniques and the development of movement models are (academic/practical) _____ tasks.

Professional applications of mechanical kinesiology
Specific student objective

1. Identify professional application tasks in mechanical kinesiology.

Summary

Mechanical kinesiologists are concerned with practical applications as well as academic theory. The professional applications of mechanical kinesiology are based on the assumption that the physical educator is primarily a teacher of motor skills. It is also assumed that the teaching competence of a physical educator depends upon the ability to analyze a student's movement performances in terms of cause-and-effect relationships. The professional physical educator should be able to analyze a performance and decide (1) what is right (using the performance model), (2) what is wrong, (3) why it is wrong, and (4) what must be done for improvement. Analysis skills need to be combined with communication and leadership skills to respectively communicate the results of the analysis to the student and motivate the student to use the results in his or her practice sessions.

Performance tasks

1. Discuss the difference between knowing how to perform a motor skill and knowing *why* it should be performed that way.
2. Discuss the statement "A good player has the best background to become a good coach."
3. Would you add any other skills to the three mentioned in the statement "A physical educator should be able to teach (analyze), communicate, and lead"?

SELF-EVALUATION
Students should use no reference materials for this progress test, and they can check their answers by referring to Appendix A.

1. Define kinesiology.
2. The three kinesiological approaches to the study of human motion are
   a. behavioral, functional, and mechanical
   b. mechanical, physiological, and psychological
   c. both a and b
   d. neither a nor b
3. In your own words, explain the meaning of mechanical, physiological, and psychological kinesiology.
4. Define "human movement."
5. The basic motion pathway(s) is (are)
   a. linear or straight-line
   b. curvilinear or non–straight line
   c. both a and b
   d. neither a nor b
6. The two major branches of mechanics are
   a. dynamics and kinetics
   b. kinematics and kinetics
   c. dynamics and statics
   d. dynamics and kinematics
7. The two most common types of curvilinear pathways are
   a. angular and circular
   b. angular and parabolic
   c. circular and rotatory
   d. circular and linear
8. The description of motion is called
   a. dynamics   c. kinetics
   b. kinematics   d. statics
9. The casual analysis of motion is called
   a. dynamics
   b. kinematics
   c. kinetics
   d. statics
10. Structural kinesiology focuses attention on
    a. anatomy
    b. mechanics of joints
    c. both a and b
    d. neither a nor b
11. Mechanical kinesiology is primarily concerned with
    a. mechanical laws
    b. motor skills
    c. both a and b
    d. neither a nor b
12. Motion along a "straight-line" path is termed
    a. angular
    b. linear
    c. rotatory
    d. parabolic
13. Motion along a "circular" path is called
    a. angular
    b. linear
    c. both a and b
    d. neither a nor b
14. Diagram a conceptual model for mechanical kinesiology below; include definitions.
15. Since this text focuses on mechanical kinesiology, it must be limited to exclude
    a. kinetics and kinematics

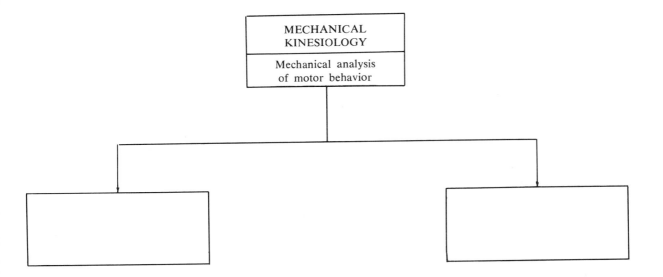

b. psychology and physiology
c. both a and b
d. neither a nor b
16. Academically a mechanical kinesiologist would be interested in the following aspect(s) of a motor performance:
a. identifying and understanding the mechanical event
b. research in the area

c. both a and b
d. neither a nor b
17. A professional application of mechanical kinesiology by the physical educator would be
a. movement analysis
b. experimental research
c. both a and b
d. neither a nor b

Concept of mechanical analysis

Body of lesson
  Qualitative and quantitative types of kinematic analysis
    Qualitative analysis of human motion
    Quantitative analysis of human motion
  Divisions of kinetics
  Introduction to the deductive and inductive types of mechanical analysis
  Overview of the kinetic and kinematic analysis of human motion
Study guidelines
Self-evaluation

QUALITATIVE AND QUANTITATIVE TYPES OF KINEMATIC ANALYSIS

In the last lesson we defined mechanical kinesiology as being concerned with the "mechanical analysis" of human motor behavior. Key to this definition is the word *analysis,* which is a noun that denotes the process of separating or breaking down any whole into its parts so as to determine its nature and to determine the proportions, functions, and/or relationships of its parts. The word also denotes the results of such a breakdown. The two basic types of mechanical analysis were identified in the last lesson as kinematic or descriptive and kinetic or causal.

The purpose of this section is to further delineate the subject matter of kinematics. In the next section the subject of kinetics will be taken up. A conceptual model of mechanical kinesiology and its kinematic and kinetic subdivisions is presented in Fig. 2-1.

The two types of kinematic or descriptive analysis are here identified as qualitative and quantitative. A qualitative analysis consists of (1) identifying the components of a whole (naming) and (2) evaluating these components in terms of some value system (comparing and judging). A quantitative analysis, on the other hand, deals with the accurate determination of the amount or percentage of the various components of a thing

(measuring or counting). Thus a quality and a quantity consist of the following:

1. Quality
   a. A characteristic that constitutes the basic nature of a thing or is one of its distinguishing features
   b. The degree of excellence which a thing possesses
2. Quantity
   a. The property of anything which can be determined through either the process of counting or the process of measuring
   b. The exact amount of a particular thing

Qualitative analysis of human motion

The two types of qualitative analysis are identified as nominal and evaluative. A *nominal* analysis is concerned with the identifying and naming of movement components. An *evaluative* analysis is concerned with determining the relative value of each movement component.

The most basic operation involved in any kind of qualitative analysis is the operation of differentiating or "telling apart." We can differentiate people according to sex, for example, and we can call all those of one kind "men" and those of the other kind "women." This qualitative process of naming things is called a *nominal*

15

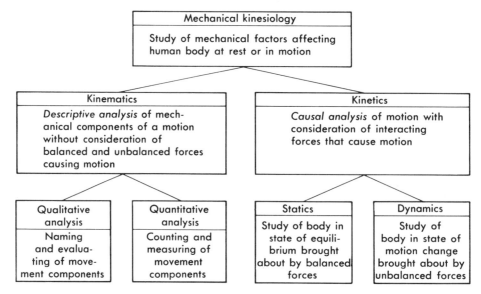

Fig. 2-1. Conceptual model of mechanical kinesiology. (From Krause, J. V., and Barham, J. N.: The mechanical foundations of human motion: a programmed text, St. Louis, 1975, The C. V. Mosby Co.)

analysis and it usually results in the establishment of a system of names called a nomenclature.

A nominal analysis also involves the employment of an already established nomenclature in the recognition and identification of things as they exist in nature. This type of analysis is made possible through the process known as classification, where the characteristics of a thing that lead to its recognition are used also to classify or establish its identity. Thus the nominal analysis of a movement performance would consist of recognizing and identifying its component characteristics. Several qualitative terms have already been used in this and the preceding lesson—such as *linear, angular, parabolic, motor task, motor performance, kinematics,* and *kinetics*. These terms point out the importance of the observer, as well as the motion information observed, in making a qualitative analysis.

Often we can decide not only that two things are different but also that one of them has more or less of some quality than the other. Thus we say that a qualitative analysis consists of both a nominal analysis (naming) and an evaluative analysis. The process of ranking or ordering things according to a value system is called an *evaluative analysis*. Our language is rich in comparative

words—bigger, smarter, better, taller, hotter, and so on. Because we can make comparative judgments, these words are useful. This type of analysis is necessary in the expert judging of athletic competition to determine the relative quality of performance. The judging of gymnastics, of figure skating, and of diving are examples of evaluative analysis.

If after we have classified our objects through a nominal analysis we can then order the classes in such a way that each class has more of the quality upon which the classification is based than has any of the preceding classes, we have an evaluative scale.

An example of an evaluative scale in common use is the scale of running ability used in track and field. In a 100-yard dash, for example, running ability is often determined by the *order* in which the different participants finish the race. The basic rule by which numerals are assigned to classes on an evaluative scale is this: "The order inherent in the formal system of the numerals shall correspond to the order empirically established among the classes."

Numbers have a conventional order: 1, 2, 3 . . . etc.; and letters have a conventional order: A, B, C . . . Z. If we use numbers or letters to identify our runners, the conventional order of the letters or numbers must correspond to the order in which the runners have been placed by the operation of having them all run in the same race and observing the order of their finish. Starting with the first runner to cross the finish line, we may assign the numerals 1, 2, 3, 4, 5, 6, and 7 to the seven runners or just as logically A, B, C, D, E, F, and G.

The numerals assigned to the classes of an evaluative scale behave like numbers only insofar as they have the conventional order of numbers. In other words, the ordered relation of the numerals is important, but their absolute values—the differences between the numerals assigned to adjacent classes—are not. This is exemplified in a track event in which the order is important but the difference between first and second or second and third is not.

Quantitative analysis of human motion

In addition to making qualitative comparisons, we sometimes are able to quantify these comparisons. Sometimes we can say not only that John is faster in a 100-yard dash than George but also that John is 3 seconds faster. This boy is not only stronger than another—he is 30 pounds stronger. In analyzing a track event, for example, we are sometimes interested in more than a qualitative analysis, i.e., who finishes first, second, etc. We are sometimes interested in a quantification of the differences between the finishers.

The process of determining the magnitudes of motion variables is called a quantitative analysis. Human movement variables are quantified through either the operation of counting or the operation of measuring. Whenever numerals are assigned by counting, the variable to which they are assigned is said to be *discrete*. When they cannot be assigned by counting, the variable is said to be *continuous*.

The difference between discrete and continuous variables can be illustrated by thinking of the difference between the number of peas and the amount (volume) of pea soup. Peas are quantified by counting; pea soup is quantified through measurement. Peas therefore are discrete variables, while pea soup is a continuous variable. It should also be noted that continuous variables can be subdivided into smaller and smaller units

while discrete variables must retain their identities as whole objects and cannot be subdivided.

The quantitative analysis of continuous variables obviously involves measurement. Thus this type of analysis is subject to the usual limitations concerned with assumptions, validity, reliability, and so forth.

The fundamental quantities of human motion, as will be further discussed in Lesson 7, are time, space, and force. From these basic quantities we are able to derive such measurable quantities as speed, velocity, acceleration, energy, and so forth. Knowledge of motion quantities, both basic and derived, makes the motion more understandable and the description of the motion more precise.

Summary

The different types of qualitative and quantitative analysis involved in the kinematic analysis of human movement can be summarized by the following list:

> *Qualitative analysis*
>   Nominal analysis: that performed through the operation of *naming*
>   Evaluative analysis: that performed through the operation of *comparing* and *judging*
> *Quantitative analysis*
>   Discrete variable analysis: that performed through the operation of *counting*
>   Continuous variable analysis: that performed through the operation of *measuring*

Organization of the kinematic analysis discussions

Part II of this text is devoted to an examination of the methods and principles of kinematics as found in the study of human motion.

Section 1, entitled "Kinematic Analysis of Fundamental Qualities," is mostly concerned with identifying and naming the fundamental segmental movements of the body. These segmental movements are defined as the basic qualities of human motion. Thus this section deals mostly with the nominal type of analysis.

Sections 2, 3, and 4 are devoted, respectively, to the kinematic analysis of (a) the fundamental quantities of space, time, and force, (b) the spatial and temporal dimensions of linear, parabolic, and angular types of motion, and (c) the spatial, temporal, and force dimensions that are combined in various ways to describe the torque, work, and power productions of the body's machinery. Thus these sections deal mostly with the quantitative type of analysis.

From the foregoing discussion it can be seen that Part II is mostly concerned with the nominal and quantitative types of analysis. Very little specific attention is given to the evaluative type of qualitative analysis since this type is unique to the study of specific motor tasks and skills. Numerous examples of qualitative analysis, however, are given throughout the text.

DIVISIONS OF KINETICS

The kinetics of human motion is divided into two parts, statics and dynamics. While *statics* deals with bodies in a state of equilibrium, a condition brought about by balanced forces, *dynamics* deals with changes in motion brought about by one or more unbalanced forces. Thus force, which is any push or pull tending to cause motion, can be either balanced or unbalanced. If it is unbalanced, a motion change occurs, and when it is balanced, no motion change occurs. In other words, during static states the body either remains at rest or, if in motion, continues to move with a constant speed in a straight line.

The core consideration in kinetics is, of course, an analysis of the causes of motion states governed by Newton's three laws of motion. All the sections in Part III of this text, dealing with the mechanical principles of kinetics, revolve around these three laws. The laws themselves are presented in Lesson 17, and then several lessons are devoted to their applications in the study of linear and angular motions and in the analysis of external forces.

INTRODUCTION TO THE DEDUCTIVE AND INDUCTIVE TYPES OF MECHANICAL ANALYSIS

A mechanical analysis can be employed either deductively or inductively.

A *deductive* analysis is characterized by reasoning that proceeds from the known to the unknown, from the general to the specific, or from a premise to a logical conclusion. For example, Newton's laws inform us that all motion changes of a body are the result of unbalanced forces acting upon it. Thus, whenever we observe a change occurring in the motion of a body, we know

logically that an unbalanced force is acting in the situation. Another example of this type of analysis is the application of the kinematic conclusion that the linear velocity at the end of a lever undergoing rotation is directly proportional to its length and to its angular speed. We can therefore conclude deductively that a tennis server who serves with a straight elbow will impart greater velocity to a tennis ball than is possible with a bent elbow as long as no loss of angular speed is involved. We can also see that a deductive analysis starts with a specific motor performance situation in which the analyst identifies the mechanical characteristics and then evaluates the situation with respect to criteria (mechanical principles) derived from Newton's laws.

An *inductive* analysis, on the other hand, is characterized by reasoning which proceeds from particular facts or individual cases to a general conclusion. For example, it might be observed that all successful runners in track and field have certain specific movements that they perform consistently alike. Therefore it might be postulated that these movements are essentials for success. Should these movements later prove to be explainable through the application of Newton's laws and be verifiable through experimental procedures as indeed being essential for success, they will be established as valid mechanical kinesiology principles which then can be used in the deductive analysis of future specific running performances.

A principle is considered here as a comprehensive and fundamental rule, law, or assumption. Thus the identification and verification of principles is the result of an inductive process, while the application of these principles in the solution of specific movement problems is a deductive process.

The formulation and application of mechanical principles in the analysis of sport and dance skills through inductive and deductive methods will be further discussed throughout this text.

OVERVIEW OF THE KINETIC AND KINEMATIC ANALYSES OF HUMAN MOTION

It should now be evident that the mechanical analysis of human motion is concerned with developing and verifying the theoretical mechanical models that can, hopefully, be used in the evaluation of motor task performances and motor skills. As mentioned in Lesson 1 in the discussion of the academic dimensions of mechanical kinesiology, the three approaches that can be used in the study of motor tasks and motor skills are identified as observational, theoretical, and experimental.

The observational approach to the study of motor tasks and skills is that of kinematics, in which the objective is a clear description of the motion qualities and quantities involved. The theoretical approach is that of inductive and deductive logic, in which the objective is to develop a theoretical hypothesis that can be used to explain the observations. In other words, the objective of the theoretical approach is to "make sense out of experience." The experimental approach is concerned with the verification of the theoretical hypothesis, and the objective here is to subject the hypothesis that grows out of the theoretical analysis to the "test of experience."

It should also be obvious to the student now that a kinetic analysis of cause-and-effect relationships is based upon the kinematic description of the motor tasks and motor skills to be analyzed and that the reliability and validity of the kinetic analysis are dependent upon the reliability and validity of the underlying kinematic observations. Also it should be evident that the qualitative evaluation of actual motor performances is based upon the development and verification of kinetic performance models and that the reliability and validity of the evaluation are dependent upon the reliability and validity of the underlying kinetic models.

STUDY GUIDELINES
Qualitative and quantitative types of kinematic analysis
Specific student objectives

1. Describe mechanical kinesiology.
2. Describe analysis.
3. Identify and describe the two types of mechanical analysis.
4. Identify and describe the two types of kinematic analysis.
5. Identify and describe the two types of quantitative analysis.
6. Differentiate between a quantity and a quality.
7. Identify and describe the two types of qualitative analysis.

8. Identify examples of nominal and evaluative analyses.
9. Identify examples of continuous and discrete variables.

Summary

Mechanical kinesiology has been described as the mechanical analysis of human motor behavior. Separating or breaking down any whole into its parts or the results of this process are known as analysis. The two basic types of mechanical analysis are kinematic or descriptive and kinetic or causal.

The two types of kinematic analysis are qualitative and quantitative. A qualitative analysis consists of (1) identifying the components of a whole (naming) and (2) evaluating these components in terms of some value system (comparing and judging). A quantitative analysis deals with the measurement of the amount or percentage of the various components of a thing (counting or measuring).

A *quality* is (1) a characteristic that constitutes the basic nature of a thing or is one of its distinguishing features and (2) the degree of excellence a thing possesses. A *quantity* is (1) the property of anything which can be determined by counting or measuring and (2) the exact amount of a particular thing.

Qualitative analysis consists of the operation of differentiating, or "telling apart," called a nominal analysis. A nominal analysis (naming) usually results in a system of names, called a nomenclature, and a classification system.

Quantitative analysis, which deals with determining the amount of motion components, is carried out through counting (discrete variables) and measuring (continuous variables). The basic measurable quantities of human motion are time, space, and force.

Performance tasks

1. Consult a dictionary for definitions of kinesiology, mechanics, and analysis.
2. Mechanical kinesiology has been described as the mechanical (analysis/synthesis) _____ of human motor behavior.
3. The two basic types of mechanical analysis are
   a. kinematic and descriptive

b. kinetic and causal
c. both a and b
d. neither a nor b

4. Kinematic analysis of human movement is concerned with
   a. causal forces
   b. motion description
   c. both a and b
   d. neither a nor b

5. Develop a written list of descriptions for
   a. mechanical kinesiology
   b. analysis
   c. kinematics
   d. kinetics

6. Write out an organizational outline for the kinematic analysis of human motion.

7. The two types of kinematic analysis are
   a. causal and descriptive
   b. counting and measuring
   c. both a and b
   d. neither a nor b

8. A qualitative analysis consists of
   a. naming
   b. evaluating
   c. both a and b
   d. neither a nor b

9. Quantitative analysis centers on
   a. comparing and judging
   b. counting and measuring
   c. both a and b
   d. neither a nor b

10. The property which can be determined by counting or measuring, i.e., the exact amount of something is a
    a. causal analysis
    b. description
    c. quality
    d. quantity

11. Developing a sex classification would be a type of (qualitative/quantitative)_____ analysis.

12. Determining a "third place" finish in a track event is a type of qualitative analysis known as (evaluation/nominal)_____.

13. Measuring a 4-minute mile would be a (qualitative/quantitative)_____ analysis.

14. Write out a list of descriptions for
    a. kinematic analysis
    b. qualitative and quantitative analysis
    c. quality and quantity

d. nominal and evaluative analysis

15. Give examples of each classification listed in the preceding task.

Divisions of kinetics

Specific student objectives

1. Describe kinetics.
2. Identify and describe the two divisions of kinetics.
3. Describe a force.

Summary

Kinetics is the causal analysis of motion. It is divided into statics (bodies in a state of equilibrium resulting from balanced forces) and dynamics (objects in a state of changing motion brought about by unbalanced forces). Force is any push or pull tending to cause motion. Static states result when an object remains at rest or, if in motion, continues to move with a constant speed in a straight line. This static state is caused by balanced net forces. Unbalanced net forces produce motion changes in speed or direction.

Performance tasks

1. The two divisions of kinetics are
   a. statics and dynamics
   b. mechanics and kinematics
   c. both a and b
   d. neither a nor b

2. Kinetics is concerned with (causes/description)_____ of motion.

3. The analysis of objects in a state of equilibrium would be (dynamics/statics)_____.

4. Objects in a state of equilibrium are produced by (balanced/unbalanced)_____ forces.

5. An object at rest or moving with constant speed in a straight line is considered to be in a (dynamic/static)_____ state.

6. Identify three illustrations of motor tasks that exemplify both dynamic and static situations.

Introduction to the deductive and inductive types of mechanical analysis

Specific student objectives

1. Describe deductive analysis and inductive analysis.
2. Illustrate deductive and inductive types of analysis as involved in the study of human motor performance.

3. Describe a principle.
4. Identify an example of a motion principle.

Summary

A mechanical analysis can be employed either deductively or inductively. A deductive analysis proceeds from the *general* to the *specific,* from a premise to a logical conclusion, or from a known principle to an unknown. A principle is considered as a comprehensive and fundamental rule, law, or assumption. A deductive analysis starts with a specific motor performance situation in which the analyst identifies the mechanical characteristics and then evaluates the situation with respect to mechanical principles derived from Newton's laws.

Inductive analysis is characterized by reasoning which proceeds from *particular facts* or individual cases to a *general conclusion.* The identification and verification of principles are the result of an inductive process. Application of general principles in the solution of specific movement problems is a deductive process.

Performance tasks

1. Reasoning from the general to the specific is (deductive/inductive)_____reasoning.
2. An analysis that starts with a specific motor performance in which the analyst identifies the mechanical characteristics and then evaluates the situation with respect to principles derived from Newton's laws is (deductive/inductive) _____.
3. A principle derived from observation of a number of motor performances would result from a (deductive/inductive) _____ analysis.
4. Write out descriptions of
   a. principle
   b. kinetics
   c. deductive analysis
   d. inductive analysis
5. Illustrate deductive and inductive analyses with two specific motion analysis situations.

Overview of the kinetic and kinematic analyses of human motion
Specific student objectives

1. List and describe the three approaches used in the study of motor tasks and motor skills.

2. Describe kinetic and kinematic analysis.
3. Illustrate the mutual dependence of kinetics and kinematics.

Summary

Mechanical kinesiology is concerned with the development and verification of theoretical mechanical models that can be used in the evaluation of motor task performances and motor skills.

The three approaches used in mechanical kinesiology to study motor tasks and motor skills and for the development and verification of mechanical models are identified as observation, theoretical, and experimental.

The observational approach to the study of motor tasks and skills is that of kinematics: a clear description of the motion qualities and quantities involved.

The theoretical approach uses inductive and deductive analyses to develop a theoretical hypothesis that can be used to explain the observations, i.e., to make sense out of experience, and to construct a mechanical model.

The experimental approach is concerned with the verification of the theoretical hypothesis and/or model, i.e., to subject the results of the theoretical analysis to the test of experience.

Kinetic analysis is based upon the kinematic description of the motor tasks and motor skills to be analyzed. The qualitative evaluation of actual motor performances is based upon the validity and reliability of the underlying kinetic models.

Performance tasks

1. Develop a written list of descriptions for
   a. kinematic analysis
   b. kinetic analysis
   c. observational, theoretical, and experimental approaches to the study of motor tasks and motor skills
2. Kinematics is primarily used in the approach called
   a. experimental   c. theoretical
   b. observational   d. all of the above
3. State how kinetics and kinematics are mutually dependent in human motion analysis.
4. Subjecting a hypothesis to the "test of experience" would be done primarily in the approach termed
   a. experimental

b. observational
c. theoretical
d. all of the above

SELF-EVALUATION

Students should use no reference materials for this progress test, and they can check their answers by referring to Appendix A.

1. The mechanical analysis of human motor behavior is known as
   a. mechanical kinesiology
   b. motor performance
   c. both a and b
   d. neither a nor b
2. Separating or breaking down a whole into parts is termed (analysis/synthesis)_____.
3. The two types of mechanical analysis are called
   a. nominal and evaluative
   b. qualitative and quantitative
   c. both a and b
   d. neither a nor b
4. Kinematic analysis focuses upon (cause/description)_____.
5. Qualitative and quantitative analyses are the types of (kinematics/kinetics)_____.
6. Quantitative analysis deals with
   a. counting
   b. measuring
   c. both a and b
   d. neither a nor b
7. Qualitative analysis centers upon
   a. naming
   b. comparing
   c. both a and b
   d. neither a nor b
8. A characteristic that constitutes the basic nature of a thing and the degree of excellence which a thing possesses is a (quality/quantity)_____.

9. A discrete variable is a quantity that can be (counted/measured)_____.
10. The "time" of a motor performance would be a (continuous/discrete)_____ variable.
11. Describe kinetics and kinematics.
12. The two divisions of kinetics are
    a. dynamics and statics
    b. mechanics and kinematics
    c. both a and b
    d. neither a nor b
13. Objects in a state of equilibrium are produced by (balanced/unbalanced)_____ forces.
14. Reasoning from a specific to the general is (deductive/inductive)_____ reasoning.
15. A principle derived from observation of a number of motor performances would be the result of (deductive/inductive)_____ analysis.
16. A comprehensive or fundamental rule, law, or assumption is known as a
    a. motor skill
    b. motor task
    c. both a and b
    d. neither a nor b
17. List and describe the three approaches to the study of motor tasks and motor skills.
18. The observational approach depends primarily upon
    a. kinematics
    b. kinetics
    c. both a and b
    d. neither a nor b
19. Identify one way in which kinematics and kinetics are mutually dependent.

LESSON 3

Introduction to mechanical analysis systems

INTRODUCTION

A motor performance can be analyzed in terms of its products or outcomes and in terms of its process. The two products which are analyzed in a motor performance are its effectiveness and its efficiency. Since motor performance is goal oriented, the *effectiveness* of the performance is evaluated by the extent that the goal is achieved. The *efficiency* of a performance, on the other hand, is evaluated by the ratio of the mechanical work accomplished to the total energy expended.

The terms effectiveness and efficiency refer to the outcomes or to the *products* of a performance. The performance process, however, is what leads to the performance products. The *process* of a motor performance is referred to as *motor skill*. The motor skill of a performer is evaluated in terms of his or her mechanical proficiency or the extent that his or her movements conform to proper movement techniques as established by the laws of mechanics.

It should now be evident that the kinematic analysis of motor performance parameters is really concerned with the analysis of motor effectiveness, motor efficiency, and motor skill.

When we analyze a movement performance, we either temporarily record the movement in the short-term memory circuits of our brains or we more permanently record it through the use of analytical instruments. We shall identify these two approaches as being (1) the analysis of live performances and (2) the analysis of recorded performances.

23

Analysis of live performances

The study of live performances is the way physical educators most often analyze the movements of their students; that is, they observe and analyze movements during an actual contest or in an actual gymnasium setting. They usually do this by what they call "eyeballing" the performance. Through experience and training teachers can acquire a visual model or picture to be used in evaluating the effectiveness, efficiency, and skill of their students. They probably also have a verbal checklist of (1) performance goals, (2) factors which influence performance, (3) recommended segmental positions and movements that constitute the performance, and (4) the most common errors usually found in the performance. In other words, they compare the actual movements observed with the visual and conceptual models of effectiveness, efficiency, and correct form (skill) they have stored in their brain. Thus the two analytical instruments indispensable to the analysis of human motion are the human eye and the human brain.

Analysis of recorded performances

The external instruments used for recording performances can be classified as either software or hardware. The software instruments consist of such devices as paper and pencil tests, charts, and rating scales. The directions for the administration of software instruments as well as the operation of hardware devices are called software routines. The hardware instruments are classified as mechanical, electronic, and optical (more commonly called photographic). When photography is used to record motion, it is called cinematography.

Cinematography involves the use of the camera to record motion for subsequent analyses. Cameras used for cinematographic analyses range from still instruments to sophisticated motor-driven high-speed motion picture devices.

Analyzing a performance by means of cinematography has a decided advantage over analysis with the unaided eye. The ability of cameras to record visible transient phenomena, to enlarge or reduce spatial relations, and to slow down or speed up action has made them highly useful in both teaching and research.

The study of motion frame by frame on a stop-action projector enables the physical educator to critically evaluate every aspect of the performance which may not have been observed with the unaided eye during the actual performance. In addition, cinematographic techniques allow the physical educator to make relatively accurate measurements of joint movements as well as velocity of the body and its moving parts directly from the projected image. Because of these advantages and others, the use of cinematographic techniques is essential in the teaching-coaching process.

The recording of movement quantities through the use of mechanical and electronic devices as well as through cinematography is becoming increasingly popular among physical educators. For example, mechanical and electronic devices for measuring joint angles, forces, and time are widely used today and the techniques involved will be discussed in detail in the kinematic chapters of this text.

Types of mechanical analysis systems

A mechanical analysis system is an assembly of units (e.g., hardware and software instruments and software routines) combined to work as a larger integrated unit having the capabilities of all the separate units. The two processes involved in this system are data acquisition and data reduction, which are defined as follows:

> **data acquisition** The gathering, measuring, and recording of information
> **data reduction** The transforming of raw information gathered by measuring or recording equipment into a more condensed, organized, or useful form

As we have already seen, the three hardware analysis systems classified according to the types of devices they use are mechanical, electronic, and photographic (optical). We will combine the mechanical devices with the software performance tests, charts, and rating scales commonly found in most physical education programs for the purpose of discussion in this lesson. The photographic and electronic systems will then be discussed separately.

INTRODUCTION TO MECHANICAL HARDWARE AND SOFTWARE ANALYSIS SYSTEMS

The mechanical devices used to record movement parameters include such instruments as mechanical stopwatches, measuring wheels, force dynamometers, and goniometers (*gōnia* is the

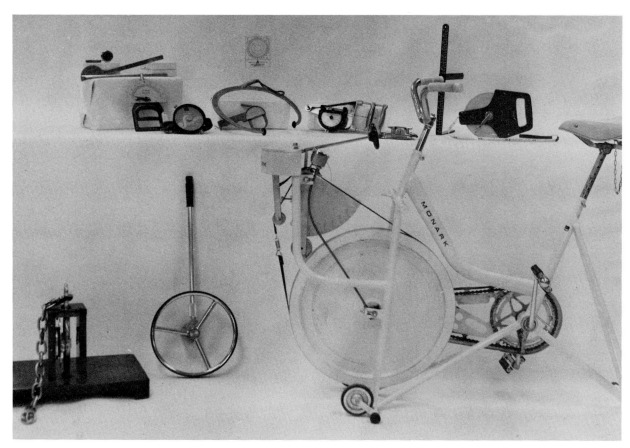

Fig. 3-1. Selected mechanical hardware instruments used in the measurement and evaluation of human variables. These devices are classified and named in the outline below.

Greek word for angle). The mechanical hardware instruments located in the kinesiology laboratory at the University of Northern Colorado (UNC) are listed and described below and are illustrated in Fig. 3-1. Since these devices are used in the administration of performance tests, they are discussed here in relation to these tests. The directions for the administration of the tests and for the use of paper and pencil tests, performance charts, and rating scales are called software routines. These analysis systems are used most often in the analysis of motor effectiveness.

Length-measuring devices
  Measuring wheel
  Tape measure (Gulick anthropometric)
  Yardstick
  Calipers
    Chest depth
    Shoulder breadth
    Skinfold
    Anthropometer

Angle-measuring devices
  Protractor
  Leighton flexometer
  Manual goniometer
Time-measuring device
  Stopwatch
Force-measuring devices
  Back and leg dynamometer
  Smedley dynamometer
  Preston grip dynamometer and adaptor
  Cable tensiometer
Work-measuring device
  Quinten-Monarch bicycle ergometer

Analysis of motor effectiveness: an overview

The analysis of motor effectiveness requires that the results of the performance be compared with the performance goals or purposes. The first step in this process is to measure the performance results, which can be accomplished through the use of either objective or subjective methods. The objective measures are obtained from performance tests and charts, while the subjective measures come from rating scales. The analysis of most performances requires both methods. In basketball, for example, there are objective *tests* of accuracy in shooting, of speed in passing, and of ability to jump as well as objective *charts* of ability to intercept passes, to block shots, and to secure rebounds. As yet, however, there are no objective means of measuring the ability of players to properly employ strategy and tactics. Thus subjective *ratings* of these abilities are made. Subjective ratings of performance with respect to some standard can be very useful, especially if made while the players are engaged in game play. The effectiveness of a football blocker, for example, in the accomplishment of his blocking assignment and in the employment of tactics could be rated from the viewing of game films.

The evaluative analysis of motor effectiveness can be done on an *absolute* basis by comparing an individual's performance measures with world, national, state, district, city, and/or school records or with some other norm. The evaluation can also be done on a *relative* basis by comparing the performance measures with the individual's own goals, interests, and aptitudes.

Analysis of motor effectiveness through the use of motor performance tests

The two basic purposes of performance are (1) the prevention of motion, in which the primary concern is with the static applications of force, and (2) the production of motion, in which the primary concern is with the dynamic applications of force. In static activities such as the holding and carrying of weights, in which there is a balancing of forces, the effectiveness of the performance is measured by either the amount of weight or the time that a weight can be held or carried. In dynamic activities, which are characterized by an unbalancing of forces, effectiveness is usually measured by either the speed or the accuracy of the movement. Some of the dynamic performance quantities which can be measured directly through the administration of performance tests are listed as follows:

Speed of performance
  Speed of human locomotion; examples: track and swimming events
  Speed of moving external objects; examples: a baseball pitch, a tennis serve

Distance covered by a performance
  Distance covered by a jump; examples: standing or running long jump, triple jump
  Height covered by a jump; example: standing or running high jump
  Distance covered in locomotion; example: by a runner during a time interval
  Distance that external objects can be projected; example: a javelin thrown, a baseball batted
Force employed in a performance; example: amount of weight that can be moved through a push, pull, or lift; force that can be registered on a dynamometer
Accuracy of performance
  Determined by counting
    Percentage scores; examples: basketball goal shooting, baseball batting
    Number of trials; example: golf
    Number of points; examples: archery, bowling, pistol or rifle shooting
  Determined by measuring
    Distance from target; examples: horseshoes, football punting out-of-bounds
Difficulty of performance
  Achievement progressions; examples: ranking of tumbling and diving stunts on a scale of difficulty so the effectiveness of the tumbler or diver is partly determined by the difficulty of the stunts he or she can perform
Competitive standing; examples: *percentage* scores (which represent the won-lost records following a round-robin tournament), *rankings* following such events as a track meet or a ladder tournament in tennis

Analysis of motor effectiveness through the use of motor performance charts

A performance chart is a form for recording specified incidents or occurrences. Following is a list of things that can be recorded on a performance chart:

1. Number of times a specified action occurs; example: rebounds by a basketball player
2. Location where a specified action occurs; example: area on the floor from which a basketball player makes his or her shots
3. Success of the specified action; example: percentage of shooting success by each basketball player from different areas of the court

The scorebook used in many sports is, to a certain extent, an example of a player performance chart, but it is quite limited in the information it yields. Physical educators often construct their own charts for recording the performances of their students.

Analysis of motor effectiveness through the use of rating scales

Subjective but systematic ratings of performance are used frequently in dance, gymnastics, diving, posture, and so forth. A rating represents a judgment made by the observer. The rating actually compares the performance of an individual with that of other individuals, with a theoretical model, or with some other standard in terms of a particular ability or factor. This method is frequently used by basketball and football coaches to determine who will make up the first and second teams. Sometimes ratings can be used as a substitute for objective tests when the latter are unavailable.

INTRODUCTION TO PHOTOGRAPHIC ANALYSIS SYSTEMS

The photographic devices used in the analysis of performance parameters can be subdivided into those concerned with data acquisition and those concerned with data reduction.

Data acquisition

A photographic data-acquisition system includes the hardware devices and software routines involved in the recording, processing, editing, and projecting of filmed data. Typical photographic data-acquisition systems are illustrated and described in Figs. 3-2 to 3-4. An outline of the photographic data-acquisition devices located in the kinesiology laboratory at the University of Northern Colorado (UNC) shows the three major types of photographic analysis systems:

Still photography system (Fig. 3-2)
  Recording equipment
    Cameras
      Honeywell Pentax, SP-IIA
      Polaroid, 145
      Polaroid, Graph Check sequence camera, model 300 (outdoor)
    Lens for Pentax
      Telephoto (Bushnell, 90 to 230 mm)
      Wide-angle (Bushnell, 28 mm)
      Macro (Bushnell, 50 mm, f 4.0)
    Attachments
      Honeywell 882 Strobonar flash
      Pentax slide copier, automatic bellows
      Tripod
    Lighting equipment
      Strobolume
      Strobotax
      Blinking lights for underwater use

  Processing and editing equipment
    Slide selector
    Processing kit
      Developing tank
      Thermometer
      Kodak Ektachrome K-4 kit
    Copy stand
  Projection equipment
    Kodak Carousel slide projector
    Opaque projector for Polaroid systems
    Overhead projector for transparencies
Motion picture system (Fig. 3-3)
  Recording equipment
    Cameras (16 mm)
      Photosonic (Actionmaster 500)
      Photosonic (P 16)
      Bell & Howell 70
    Lens
      Angenieux zoom, 12 to 120 mm (one for each Photosonic)
      Pan Cinor zoom, 17 to 70 mm (for Bell & Howell)
    Attachments
      Battery packs (one for each Photosonic camera)
      Battery for Bell & Howell
      Hercules tripods (one for each Photosonic camera
      Timing pulse generator
    Lighting equipment
      Two banks of Colortran Mini-Brute lights
  Editing equipment
    Viewer/editor (Maier-Hancock, 1600)
    Portable hot splicer (Maier-Hancock, 1635)
  Projection equipment
    Vanguard projector and stand
    Lafayette 16 mm analyst projector
    Kodak Ektagraphic 8 mm projector
Videotape and television system (Fig. 3-4)
  Recording equipment
    Camera
      Panasonic
    Lens
      Zoom
    Attachments
      Tripods
    Videotape recorder
  Editing equipment
    Special effects generator
  Monitoring and projection equipment
    Panasonic monitor
    Telebeam projector

Data reduction

A cinematographic data-reduction system includes the devices and routines involved in the digitizing and mathematical analysis of filmed data.

Digitizing systems. Digitizing systems, used to

Fig. 3-2. Selected still photography devices used in the measurement and evaluation of human movement variables. These devices are classified and named in the outline on p. 27.

Fig. 3-3. Selected motion picture photography devices used in the measurement and evaluation of human movement variables. These devices are classified and named in the outline on p. 27.

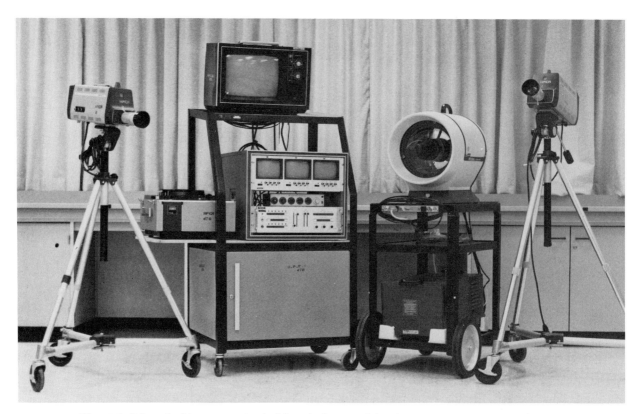

Fig. 3-4. Selected videotape and television devices used in the measurement and evaluation of human movement variables. These devices are classified and named in the outline on p. 27.

put visual images into digital or numerical form, can be classified as those that involve (1) paper and pencil procedures, (2) mechanical devices and procedures, and (3) electronic devices and procedures.

1. According to Logan and McKinney (1977) there are two general digitizing techniques that employ *paper and pencil* procedures: the "body contour technique" and the "point-and-line" technique. The two techniques can be used in conjunction with each other or separately.

The body contour technique involves tracing the contour of the body outline on paper. A workable image size should be projected on a sheet of paper 8½ by 11 inches and drawn. These drawings can also be done on transparencies approximately the same size for subsequent projection using an overhead projector. A good example of a body contour drawing is presented in Fig. 10-1.

The second digitizing procedure is called the point-and-line or stick figure technique. In this technique, points are used to indicate joint centers or other anatomic landmarks and the lines connecting these points indicate the body segments. This technique is illustrated in Fig. 12-11. The lengths of body segments as well as the angles formed between the various segments can be scaled directly from either the body contour drawings or the stick figure drawings to yield digital data for subsequent analysis.

2. A *mechanical digitizing* device located in the UNC kinesiology laboratory is the Vanguard motion analyzer (Fig. 3-5). This device also yields digital data that can later be mathematically analyzed.

3. The *electronic digitizing* device used in the UNC laboratory is the Graf pen system (Fig. 3-6). The advantage of this device is that it can be interfaced directly with a digital computer or

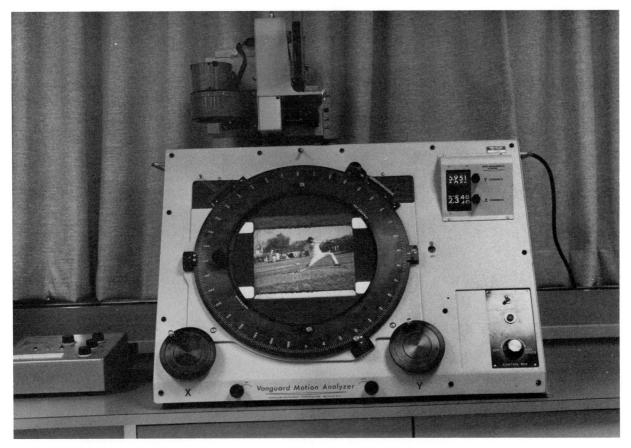

Fig. 3-5. Vanguard motion analyzer. This device is used to mechanically obtain data from film in the form of x-y coordinates and angles.

programmable calculator for subsequent data analysis. Fig. 3-6 shows the digitizer interfaced with the Hewlett-Packard 2100S microprogrammable systems computer located in the UNC kinesiology laboratory. The computer system is illustrated and described in Fig. 3-7.

Mathematical analysis systems. Once the analysis information has been placed in digital form, it must be further analyzed through the use of mathematical formulas. Most students using this text will perform their mathematical analyses with pencil and paper. A slide rule or hand calculator, however, can make the calculations much easier, and digital computers and programmable calculators which can be programmed to mathematically analyze data in an on-line situation and in real time have obvious advantages in terms of speed and efficiency.

INTRODUCTION TO ELECTRONIC ANALYSIS SYSTEMS

The electronic devices used in the analysis of performance parameters can be subdivided, as were the photographic systems, into those concerned with data acquisition and those concerned with data reduction.

Data acquisition

An electronic data-acquisition system consists of three major parts: (1) a transducer to convert physical phenomena into electrical signals, (2) a signal conditioner to modify the signals, and (3) a recorder to graphically display the electronic data. A typical electronic data acquisition system is illustrated and described in Fig. 3-8. As can be seen, the basic recorder used in the UNC kinesiology laboratory is the Norco Physio-

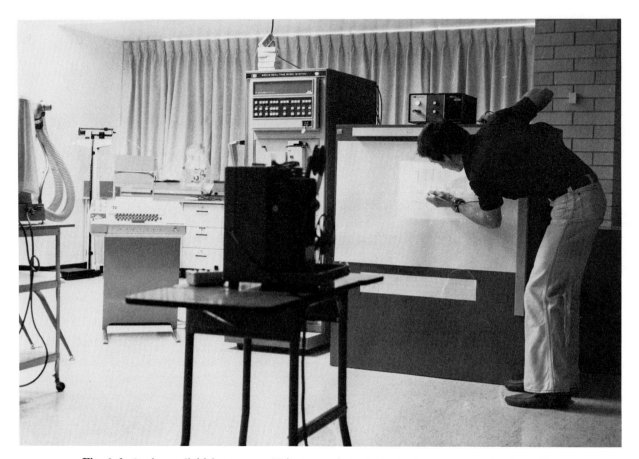

Fig. 3-6. Graf pen digitizing system. This system is used to obtain x-y coordinates from film. The digitizer is interfaced with a Hewlett-Packard 2100S microprogrammable systems computer which further analyzes the digital data.

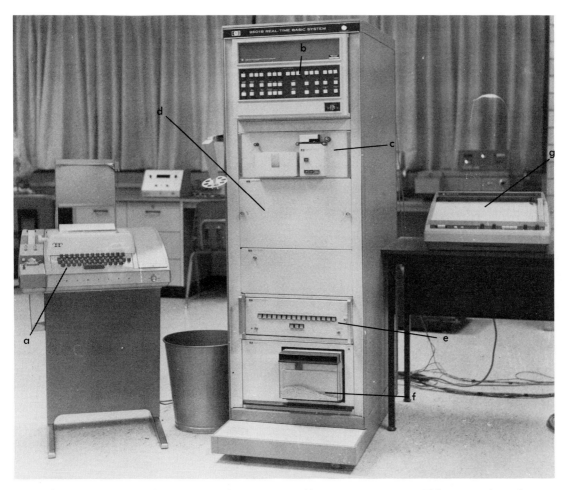

Fig. 3-7. Hewlett-Packard 2100S microprogrammable systems computer. This system is used for the mathematical analysis of data. It consists of *(a)* a teleprinter and interface, *(b)* a 2100S computer, *(c)* a paper tape reader, *(d)* an analog-to-digital interface subsystem, *(e)* a multiprogrammer for digital I/O, *(f)* a paper tape punch, and *(g)* a graphic plotter.

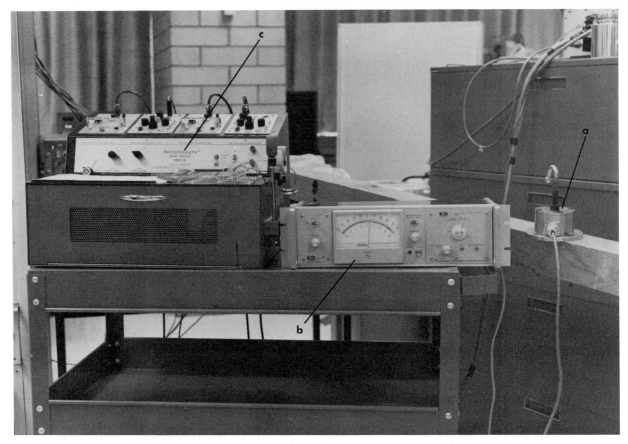

Fig. 3-8. Typical electronic data-acquisition system. Shown here are *(a)* the Daytronics LVDT force transducer, *(b)* the signal amplifier, and *(c)* a Norco Physiograph recorder.

graph, which is illustrated by itself in Fig. 3-9. The electronic data-acquisition equipment in the UNC kinesiology laboratory is classified as follows:

Electrogoniometry system (for angle measurement) (Fig. 3-10)

Elgons (transducers of the potentiometer type). These produce an electrical signal which is directly proportional to the angular displacement.

Signal conditioner. This is a module which calibrates the recorder and amplifies the electrical signals from the elgons.

Recorder. This can be either a Norco Physiograph or a digital computer (or both).

Electrodynamometry system (for force and torque measurement) (Figs. 3-8 and 3-11)

Transducers. There are two types.

Piezoelectric. These are pressure-sensitive transducers which measure forces, torques, and their rectangular components.

LVDT. This type of transducer *(linear variable differential transformer)* measures the force components acting parallel with the sensing probe.

Signal conditioner. This module amplifies the signal from the piezoelectric and LVDT transducers.

Recorder. This can be either a Norco Physiograph or a digital computer (or both).

Electrochronoscopy system (for time measurement) (Fig. 3-12)

Metronome

Electronic stopwatches

Dekan automatic performance analyzer

Gralab timers

Banked programmed timers

Electrocalorimetry system (for work and energy measurement) (Fig. 3-13)

Ergometer (to measure work)

Quinten-Monarch bicycle ergometer.

Quinten treadmill (remote control, high speed). This has a heart-rate monitor.

Harvard step benches.

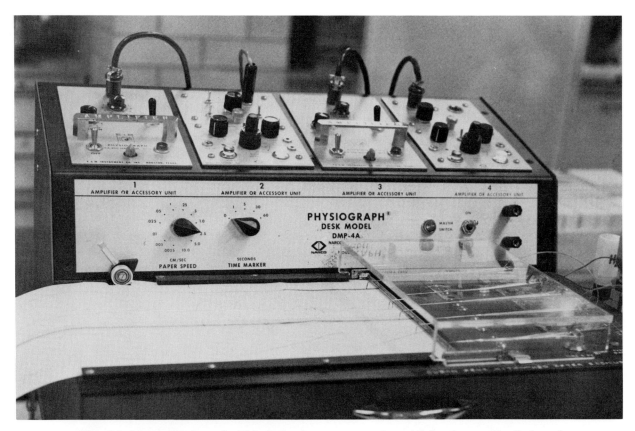

Fig. 3-9. Norco Physiograph. This device is an x-y recorder used for the graphic display of analog (continuous) data. After amplification the electronic signal runs a pen motor that causes the recording pen to be deflected and make a vertical mark on the recording paper. Thus the vertical deflections represent the magnitude of the transducer input, while the horizontal distance represents time as determined by the speed of the paper movement through the recorder.

Gas analyzer (to measure energy expended)
  Beckman O_2 analyzer
  Capnograph (CO_2 analyzer)
  Oxygen-consumption computer
Recorder (in individual instruments, Norco Physiograph, or digital computer)

The electronic transducers involved in the analysis systems listed above can be connected either directly to their signal conditioners and recorders or indirectly through telemetry to these conditioners and recorders. A typical telemetry system, which involves the transmission of data via radio signals, is illustrated and described in Fig. 3-14.

Biological and behavioral types of electronic data-acquisition systems. The complete analysis of motor performance variables, as mentioned in the last two lessons, also involves the analytical devices and techniques associated with the subject matters of physiological kinesiology and psychological kinesiology. These devices and analysis systems are identified in the UNC kinesiology laboratory as biological and behavioral, respectively. Many of the devices identified here, however, are used by scholars working in all three of the major areas of kinesiology. Electromyography, for example, is used by the physiological kinesiologist in the study of muscle bioelectrical functions; it is also used by the psychological kinesiologist in the study of motivational (arousal) states through biofeedback techniques; and it is used by the mechanical

Text continued on p. 40.

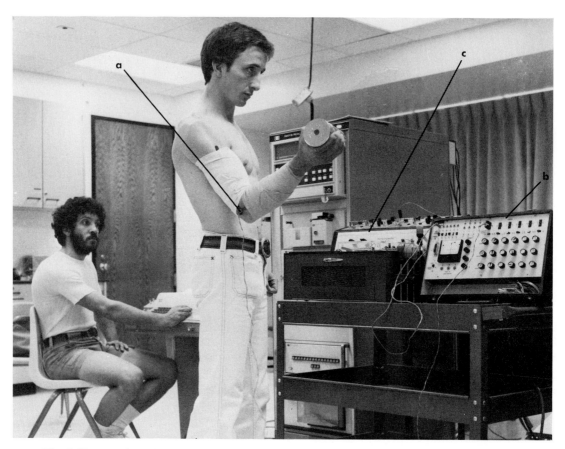

Fig. 3-10. A typical electrogoniometry system consists of *(a)* an elgon, which is an electronic transducer (potentiometer) for the measurement of angular position and movement of body segments, *(b)* an amplifying module, and *(c)* a recorder. The recorder is also interfaced with the computer for an on-line real-time analysis of angular motion.

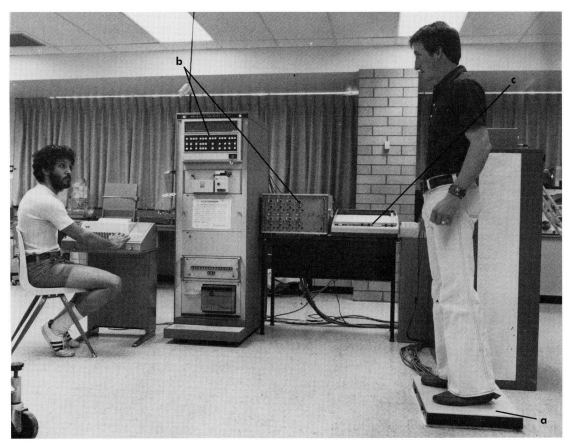

Fig. 3-11. Typical electrodynamometry system. This consists of *(a)* a transducer, which has an electrical output in direct proportion to the applied forces, *(b)* a group of charge and summing amplifiers, and *(c)* a recording system. Shown here is the Kistler piezoelectric force platform interfaced with the digital computer. A graphic recording is being made on the computer's plotter.

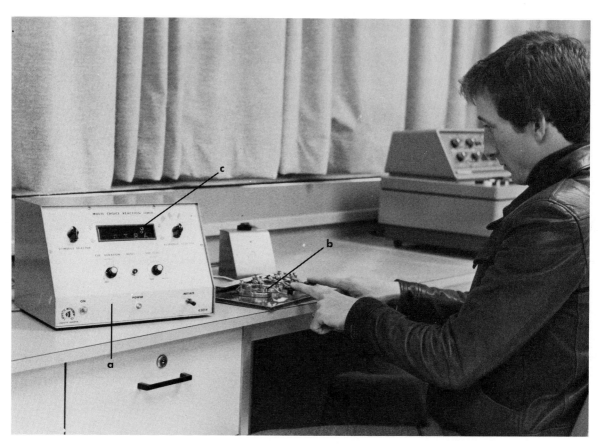

Fig. 3-12. Typical electrochronoscopy system. Shown here are *(a)* an electronic timer, *(b)* a means of stopping and starting the timer, and *(c)* a display or recording device.

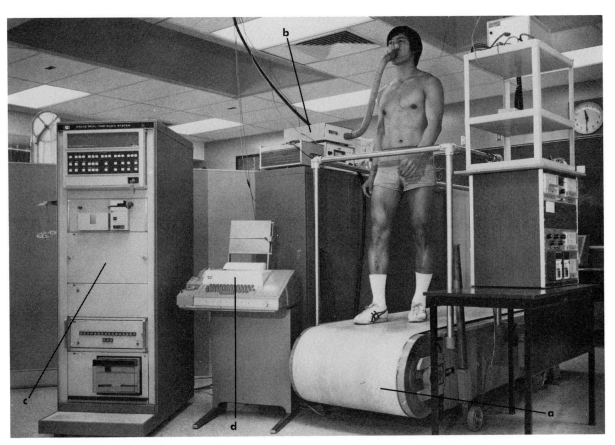

Fig. 3-13. Typical electrocalorimetry system. The components of this system consist of *(a)* a work-measuring device such as a treadmill, *(b)* a gas-measuring device, *(c)* electronic digitizing of data, and *(d)* a hard copy printout on the teleprinter.

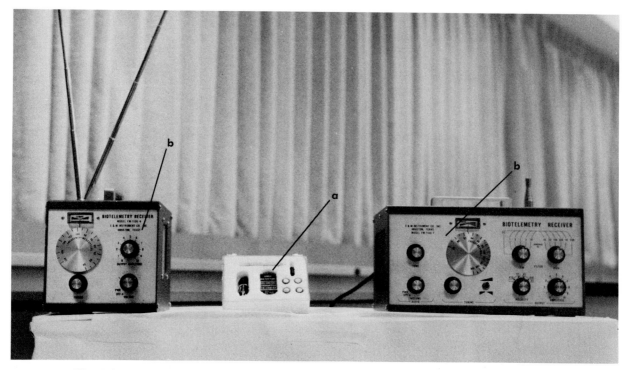

Fig. 3-14. A typical telemetry system consists of an electronic analysis system of transducers, amplifiers, and recorders plus *(a)* a radio transmitter and *(b)* radio receivers that can be interfaced with the system amplifiers. Shown here are the Norco transmitter and receivers.

kinesiologist in the study of muscle forces and their effects.

Data reduction

The electronic data-reduction systems are basically the same as those for the photographic systems; that is, data can be digitized and mathematically analyzed by hand or through the use of electronic calculators and/or digital computers.

The digitizing system developed for the UNC kinesiology laboratory involves the direct interfacing of transducers and signal conditioners to the laboratory's digital computer or the interfacing of the Physiograph recorder to the computer. In either case the analog-to-digital converter card of the computer performs the function of converting the electrical signals into digital form. Once the data are digitized, they are mathematically analyzed by the computer and the results are output in one or more of the following ways: (1) printed out in hard copy form on the teletypewriter, (2) graphically recorded in hard copy form on either the computer's plotter or the Physiograph recorder, (3) graphically displayed on an oscilloscope.

INTEGRATED ANALYSIS APPROACH OF KINESIOLOGY

The complete analysis of human movement behavior, it must be emphasized, requires the integration of knowledge acquired through the three major divisions of kinesiology. In other words, since human motor behavior is not solely a physiological, psychological, or mechanical phenomenon, its comprehension requires an integrated approach. This approach is academically unique and qualifies kinesiology as a distinct academic discipline.

The kinesiology laboratory at UNC was planned according to the integration principle. Thus this laboratory is not a physiology laboratory, or a psychology laboratory, or a mechanics laboratory; it is a kinesiology laboratory that is designed for the integrated study of human motor performance.

STUDY GUIDELINES
Introduction
Specific student objectives

1. Differentiate between motor performance effectiveness and motor performance efficiency.
2. Describe motor skill.
3. Describe the analyses of live and recorded performances.
4. Identify the two processes involved in a mechanical analysis system.
5. Identify examples of software analysis systems.
6. Identify the three types of hardware analysis systems.

Summary

The parameters of motor performance are effectiveness (accomplishing the goal), efficiency (the ratio of work done to energy expanded), and motor skill (the process of a motor performance).

A movement performance is analyzed by studying live or recorded performances. Software analysis instruments are exemplified by charts, rating scales, and tests. Hardware instruments are classified as mechanical, electronic, and optical or cinematographic.

A mechanical analysis system is an assembly of units combined to work as an integrated unit. The two processes involved in this system are data acquisition (gathering, measuring, and recording of information) and data reduction (transforming of raw data into a more condensed, organized, or useful form).

Performance tasks

1. Hitting the bull's-eye in archery would demonstrate motor performance (effectiveness/efficiency) _____.
2. The process which leads to performance products is called motor
   a. effectiveness
   b. efficiency
   c. skill
   d. none of the above
3. A cinematographic analysis would focus on motor
   a. effectiveness   c. skill
   b. efficiency   d. all of the above
4. The ratio of mechanical work done to the total energy expanded is called motor (effectiveness/efficiency) _____.
5. Identify the three motor performance parameters.
6. Develop a written list of descriptions for
   a. motor effectiveness   c. motor skill
   b. motor efficiency
7. Develop a mechanical analysis system.
8. An example (or examples) of hardware instrumentation used in the recording of a motor performance would be
   a. a camera
   b. rating scales
   c. both a and b
   d. neither a nor b
9. Identify three examples each of hardware and software instruments used for recording motor performances.
10. Consult a dictionary for definitions of a parameter.

Introduction to mechanical hardware and software analysis systems
Specific student objectives

1. Identify examples of software and mechanical instruments used to record movement parameters.
2. Describe the meaning of absolute and relative types of evaluative analysis.
3. Describe the analysis of motor effectiveness through the use of motor performance tests.
4. Describe the analysis of motor effectiveness through the use of performance charts.
5. Describe the analysis of motor effectiveness through the use of rating scales.

Summary

Software analysis systems include test administration directions as well as the actual tests, performance charts, and rating scales. Examples of mechanical devices used to record movement parameters include chronographs, measuring wheels, other distance-recording instruments, force dynamometers, and goniometers.

Measures of performance results can be accomplished by objective means (performance charts and tests) and by subjective means (rating scales).

Evaluative analysis of motor effectiveness can

be done on an absolute basis (with universal norms) or on a relative basis (with personal norms or standards).

The two basic purposes of performance are (1) prevention of motion and (2) production of motion. Dynamic performance quantities which can be measured are listed as follows:

1. Speed of performance
2. Distance covered by a performance
3. Force employed in a performance
4. Accuracy of a performance
5. Difficulty of a performance
6. Competitive standings

A performance chart is a form for recording specified incidents or occurrences. Recordings can be made of such items as the number of times a specified action occurs, the location where it occurs, and the success of the specified action.

Rating scales are subjective evaluations of motor performance in which a judgment is made by an observer comparing the subject with other subjects or with a theoretical model or some other standard.

Performance tasks

1. Make a list of five mechanical devices used to record movement parameters.
2. Comparing performance goals to performance results is known as the analysis of motor
   a. effectiveness
   b. efficiency
   c. skill
   d. none of the above
3. Measuring performance results can be done objectively and subjectively. Categorize performance charts, performance tests, and rating scales in one of these categories.
4. Identify the following examples as resulting from the administration of a test, chart, or rating scale:
   a. Speed of a throw
   b. Height of a jump
   c. Gymnastics score on the vault
   d. Score for a round in golf
   e. Number of rebounds in a basketball game
5. Write out descriptions for each of the following:
   a. Rating scales
   b. Performance charts
   c. Performance tests

d. Motor effectiveness
e. Software routines
f. Dynamic performance quantities

Introduction to photographic analysis systems
Specific student objectives

1. Identify the categories of photographic devices involved in data acquisition.
2. Describe photographic data-acquisition and data-reduction systems.
3. Identify examples of data-acquisition and data-reduction systems.

Summary

Photography can be used to analyze performance parameters and is usually divided into devices concerned with data acquisition and data reduction.

A photographic data-acquisition system includes the software routines and hardware devices needed to record, process, edit, and project filmed data. The three major systems are (1) still photography, (2) motion picture photography, and (3) television or videotape recording.

Data reduction includes the devices and routines involved in digitizing and mathematically analyzing filmed data. Digitizing systems convert the collected data into digital form for interpretation. These systems can be classified as employing (1) paper and pencil procedures, (2) mechanical devices, and (3) electronic methods. Body contour and point-and-line techniques are the two paper and pencil procedures commonly used. There are a variety of mechanical and electronic digitizers on the commercial market today.

Further analysis of the digitized data is carried out through mathematical systems such as calculators and computers.

Performance tasks

1. Describe data-acquisition and data-reduction systems for cinematography.
2. Identify and list an example of each of the three types of data-acquisition systems for cinematography.
3. The two types of photographic data-reduction systems are the devices and routines involved in
   a. digitizing and mathematical analysis

b. photography and television
c. both a and b
d. neither a nor b

4. Make a written list of devices used for the categories of
   a. data acquisition
   b. data reduction

5. Classify the following devices in photographic data-acquisition or data-reduction categories:
   a. Calculator
   b. Still camera
   c. Digitizer
   d. Computer
   e. TV camera
   f. Film developer
   g. Film projector
   h. Slide rule

6. Inspect the photographic equipment in a kinesiology laboratory.

Introduction to electronic analysis systems
Specific student objective

1. Describe electronic data-acquisition and data-reduction systems.

Summary

The electronic devices used in the analysis of performance parameters are concerned with data acquisition and data reduction. An electronic data-acquisition system consists of three major parts: (1) transducers which convert physical phenomena into electrical signals, (2) a signal conditioner which acts to modify the signal, and (3) a recorder which displays the electronic data.

Data reduction is carried out on digitizers and through mathematical analysis by electronic calculators or digital computers.

Performance tasks

1. Write out descriptions for electronic data acquisition and data reduction.
2. Develop an outline of electronic data-acquisition and data-reduction systems.
3. Make a list of electronic data-acquisition and data-reduction devices.
4. Inspect a kinesiology laboratory equipped with electronic data-acquisition and data-reduction capabilities to examine the systems.

Integrated analysis approach of kinesiology
Specific student objective

1. Describe the integrated analysis approach of kinesiology.

Summary

Since human motor behavior is not solely a physiological, psychological, or mechanical phenomenon, its comprehension requires an integrated approach. The integrated approach is what is academically unique and what qualifies kinesiology as a distinct academic discipline.

Performance tasks

1. Develop a rationale for the justification of kinesiology as a distinct academic discipline.
2. Kinesiology (is/is not) _____ primarily concerned with the mechanical aspects of human movement.
3. Kinesiology (is/is not) _____ just as much concerned with the physiological and psychological components of human movement as with the mechanical components.

SELF-EVALUATION

Students should use no reference materials for this progress test, and they can check their answers by referring to Appendix A.

1. Comparing motor performance goals to results is called motor
   a. behavior analysis
   b. effectiveness analysis
   c. efficiency analysis
   d. skill analysis
2. The process of a motor performance is known as motor (efficiency/skill) _____.
3. Identify the software instruments used to record motor performance.
4. The two processes involved in a kinematic analysis system are
   a. continuous and discrete
   b. qualitative and quantitative
   c. both a and b
   d. neither a nor b
5. Describe data acquisition and data reduction.
6. Develop a summary of the measurement of dynamic performance quantities through the use of performance charts, tests, and rating scales.
7. A form for recording specified motor performance incidents or occurrences is a
   a. performance chart
   b. performance test
   c. rating scale
   d. none of the above

8. The software routines and hardware devices needed to record, process, edit, and project filmed data are what make up the (photographic/electronic)_____ data-acquisition system.

9. Identify the three major systems for acquiring cinematographic motion data.

10. The two types of photographic data-reduction systems are the devices and routines involved in

    a. digitizing and math analysis
    b. photography and television
    c. both a and b
    d. neither a nor b

11. Describe electronic data-acquisition and data-reduction systems.

12. Kinesiology (is/is not)_____ concerned with the total study of human movement.

PART II
KINEMATIC ANALYSIS OF HUMAN MOTION

SECTION ONE

Kinematic analysis of fundamental qualities

Nominal analysis of the skeletal system

INTRODUCTION

The nominal analysis of human movement is mostly concerned with identifying and naming the fundamental segmental movements of the body. These segmental movements are defined as the fundamental qualities of human motion.

Eight major segments of the body

The eight major or fundamental segments of the body are illustrated in Fig. 4-1. These segments are classified according to location as follows:

    Lower extremity: foot, leg, and thigh
    Axial skeleton: trunk and head-neck
    Upper extremity: arm, forearm, and hand

The three anatomical components involved in the movement of these segments are (1) the skeletal muscles which provide the propulsive forces, (2) the bones of the body to which the muscle forces are applied, and (3) the joints of the body around which the bones of the segment rotate. The kinematic analysis of fundamental qualities is not concerned with the analysis of muscle actions but with the description of bone and joint movements. In this lesson we will be concerned with naming the bones involved in the movement of body segments. Classification of joints and of segmental movements will be the concern of the next lesson. The naming of bones is presented first because these names are used in defining the specific segmental movements that occur at each of the joints of the body.

Bones of the body

The bones of the body are illustrated in Fig. 4-2 and are divided as follows:

Lower extremity
  Foot
    Phalanges: bones of the toes
    Metatarsals: bones of the sole and lower instep
    Tarsals: bones of the ankle
  Leg
    Tibia: large bone of the leg
    Fibula: small bone of the leg
  Thigh
    Patella: knee cap
    Femur: thigh bone
Axial skeleton
  Trunk
    Pelvis
      Hip bones
      Sacrum and coccyx

LONGITUDINAL AXIS

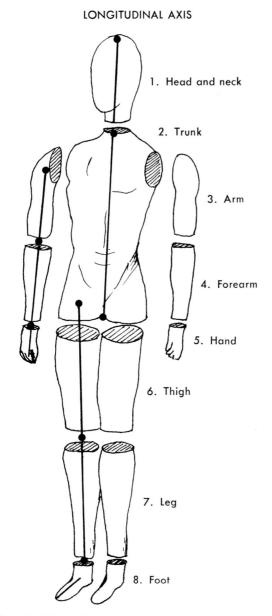

1. Head and neck

2. Trunk

3. Arm

4. Forearm

5. Hand

6. Thigh

7. Leg

8. Foot

Fig. 4-1. The eight major segments of the human body.

Lower back
  Lumbar vertebrae
Thorax: chest
  Thoracic vertebrae
  Ribs
  Sternum: breastbone
Head-neck
  Cervical vertebrae: bones of the neck
  Skull: bones of the head

Upper extremity
  Arm region
    Shoulder girdle
      Scapula: shoulder blade
      Clavicle: collarbone
    Humerus: arm bone
  Forearm
    Radius: bone which is free to rotate
    Ulna: bone which is not free to rotate
  Hand
    Carpals: wrist bones
    Metacarpals: bones of the palm
    Phalanges: bones of the fingers

Reference terminology

In order to properly study the skeletal system, it is important to use certain descriptive terms. The terms medial and lateral, for example, are often used in the study of bones. The body parts nearest the midline of the body are *medial;* the parts farther away are *lateral.* Using these terms, we can now identify the tibia as the medial bone of the leg and the fibula as the lateral bone of the leg.

The terms *anterior* and *posterior* refer to the front and rear sides of the body respectively. A location near the trunk or the beginning of a body part is *proximal;* farther away is *distal.* A body part above another is said to be *superior;* a part below another is said to be *inferior.* Therefore, using these terms we can correctly describe the location of the clavicle as being on the anterior side of the thorax, lateral to the sternum, inferior to the neck, and superior to the proximal end of the humerus.

BONES OF THE LOWER EXTREMITY

The bones of the lower extremity are those found in the foot, leg, and thigh.

Bones of the foot

The bones of the foot are illustrated in Fig. 4-3, which should be used as a reference in this section. The five toes of the foot are called digits. The word digit is a general term that applies also to the fingers.

The digits are numbered from the medial to the lateral side, with the big toe being numbered one. The big toe is also called the hallux, and the little toe is also called the digitus minimus (littlest toe) or digitus quintus (fifth toe). The bones of the toes are called *phalanges* (a single bone is a

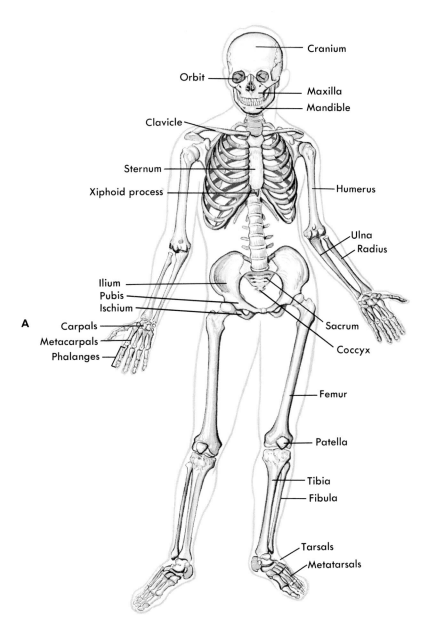

Fig. 4-2. Major bones and bone groups in the human skeleton. **A,** Anterior view. **B,** Posterior view. (Modified from Anthony, C. P., and Kolthoff, N. J.: Textbook of anatomy and physiology, ed. 9, St. Louis, 1975, The C. V. Mosby Co.)

Continued.

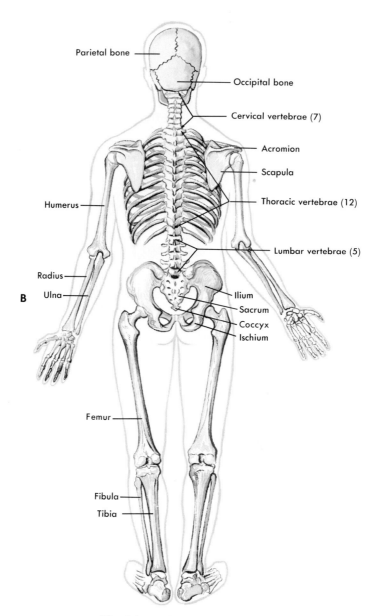

Parietal bone

Occipital bone

Cervical vertebrae (7)

Acromion

Scapula

Thoracic vertebrae (12)

Humerus

Lumbar vertebrae (5)

Radius

Ulna

Ilium

Sacrum

Coccyx

Ischium

B

Femur

Fibula

Tibia

Fig. 4-2, cont'd. For legend see p. 49.

phalanx). There are two phalanges in the big toe and three in each of the other toes—the first or proximal, the second or middle, and the third or distal.

The metatarsus is the sole and lower instep of the foot and is composed of five bones called the *metatarsal bones*. These are numbered, like the digits, from the medial to the lateral side. The first metatarsal bone is the most massive because it serves as the main support of the body weight when the foot is used in locomotion activities.

The tarsus or ankle is composed of seven bones known as the *tarsal bones*. These bones, important in body support and locomotion, are the calcaneus, the talus, the navicular, the cuboid, and the three cuneiform bones.

1. The largest of the bones is the *calcaneus*. The posterior extremity of the calcaneus projects backward behind the ankle and forms the heel of the foot.

2. The *talus* is the most superior of the tarsal bones. It rests upon the upper surface of the calcaneus.

3. The *navicular* is located anterior to the talus on the medial side of the foot. As its name implies, it is somewhat boat shaped (Latin *navicula*, little boat).

4. The *cuboid* is lateral to the navicular. It articulates posteriorly with the calcaneus. As its name implies, it is roughly cube shaped. It has a proximal articular surface for the calcaneus, a distal surface for the two lateral metatarsals, and a small medial surface for the lateral cuneiform.

5. The *three cuneiforms* are located between the navicular and the bases of the first three metatarsals on the medial to the lateral side of the foot. As their name implies (Latin *cuneus,* wedge)—these bones take on a wedge-shaped configuration, a rather broad superior surface and a narrower inferior surface. The first or medial cuneiform, the largest of the three bones, articulates anteriorly with the base of the first metatarsal. The second or middle cuneiform, the smallest of the three, articulates anteriorly with the base of the second metatarsal. The third or lateral cuneiform articulates anteriorly with the base of the third metatarsal.

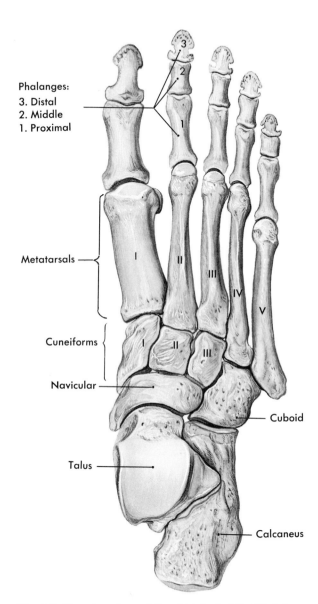

Fig. 4-3. Dorsal surface of the right foot. (Modified from Anthony, C. P., and Kolthoff, N. J.: Textbook of anatomy and physiology, ed. 9, St. Louis, 1975, The C. V. Mosby Co.)

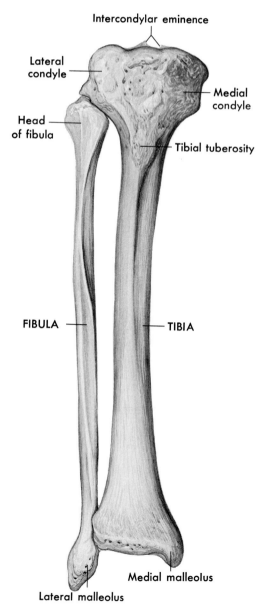

Intercondylar eminence

Lateral
condyle

Head
of fibula

Medial
condyle

Tibial tuberosity

FIBULA

TIBIA

Medial malleolus

Lateral malleolus

Fig. 4-4. Anterior surface of the right fibula and tibia. (Modified from Anthony, C. P., and Kolthoff, N. J.: Textbook of anatomy and physiology, ed. 9, St. Louis, 1975, The C. V. Mosby Co.)

Bones of the leg

Fig. 4-4 depicts the two bones of the leg: the tibia and the fibula. They are connected by an interosseous membrane, and they articulate with each other at their superior and inferior ends. Distally the tibia and fibula form an arch or

mortise to accommodate the talus, the highest of the tarsal bones, and the strong hinge so formed is the ankle joint.

The *tibia* is the medial of the two bones of the leg, and it is commonly referred to as the shinbone. This bone alone articulates with the femur to form the knee joint.

The *fibula* is the lateral of the two bones of the leg. It does not sustain any of the weight of the body because it does not articulate with the femur at the knee joint, but it does enter into the formation of the ankle joint and it serves to secure the ankle laterally. The fibula is primarily important, however, as the site of muscle attachments.

Bones of the thigh region

The two bones of the thigh region are the patella and femur, which are shown in Fig. 4-5. The *patella,* also called the kneecap, is a sesamoid bone located anterior to the knee joint. It develops within the tendon of the quadriceps femoris muscle. Its posterior surface articulates with the femur. The *femur,* located in the thigh, is the longest and largest bone in the body.

BONES OF THE AXIAL SKELETON

The bones of the axial skeleton are those found in the head, vertebral column, thorax, and pelvis.

Bones of the head

The bones of the head or skull are illustrated in Fig. 4-6 and are divided as follows:

Cranium: brain case
  Occipital: base of the skull
  Parietal: crown
  Frontal: forehead
  Temporal: ear region
  Ethmoid: between the cranial and nasal cavities
  Sphenoid: base of the brain and back of the eyes
Face: anterior skull
  Nasal
  Vomer
  Inferior nasal concha or inferior turbinate
  Lacrimal (not shown in Fig. 4-6)
  Zygomatic or malar
  Palatine: palate (roof of the mouth) (not shown in Fig. 4-6)
  Maxilla: upper jaw
  Mandible: lower jaw

These bones are not studied in great detail in mechanical kinesiology because most are not significantly involved in gross body movement.

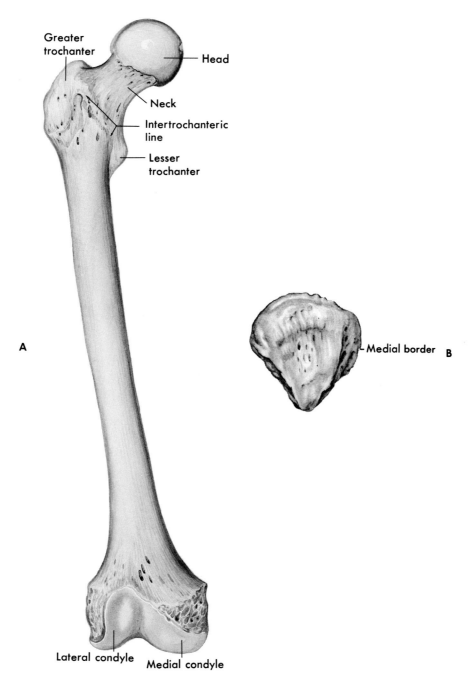

Fig. 4-5. Anterior surfaces of the right femur and patella. **A,** Right femur. **B,** Right patella. (**A** modified from Anthony, C. P., and Kolthoff, N. J.: Textbook of anatomy and physiology, ed. 9, St. Louis, 1975, The C. V. Mosby Co.; **B** modified from Francis, C. C., and Martin, A. H.: Introduction to human anatomy, ed. 7, St. Louis, 1975, The C. V. Mosby Co.)

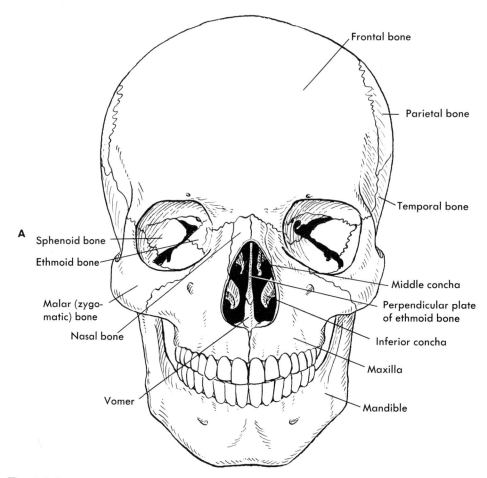

Fig. 4-6. Bones of the skull. **A,** Front view. **B,** Side view. (Modified from Anthony, C. P., and Kolthoff, N. J.: Textbook of anatomy and physiology, ed. 9, St. Louis, 1975, The C. V. Mosby Co.)

Bones of the vertebral column

The vertebral column is formed of a series of bones called vertebrae. The skeletal framework of the neck consists of seven *cervical* vertebrae; the next twelve vertebrae are known as *thoracic* because they help to form the thorax and are attached to the ribs; the next five, which support the small of the back, are called *lumbar* vertebrae; below the lumbar vertebrae are the *sacrum* and the *coccyx* (Fig. 4-7).

The first and second cervical vertebrae differ considerably from the rest (Fig. 4-8). The first cervical vertebra, or *atlas,* so named because it supports the globe of the head, is a bony ring. The second cervical vertebra, or *axis,* is characterized by the presence of a toothlike process called the dens or odontoid process. This process forms a pivot around which the atlas rotates when the head is turned from side to side.

The *sacrum* is formed by the union of five sacral vertebrae. It is a large triangular bone situated like a wedge between the hip bones to form the posterior wall of the pelvis. The *coccyx* is usually formed of four small segments of bone (but there may be three or five segments) and is the most rudimentary part of the vertebral column. It represents the atrophied human tail.

Bones of the thorax

The thorax is a bony cage formed by the sternum and costal cartilages, the ribs, and the bodies of the thoracic vertebrae as shown in Fig.

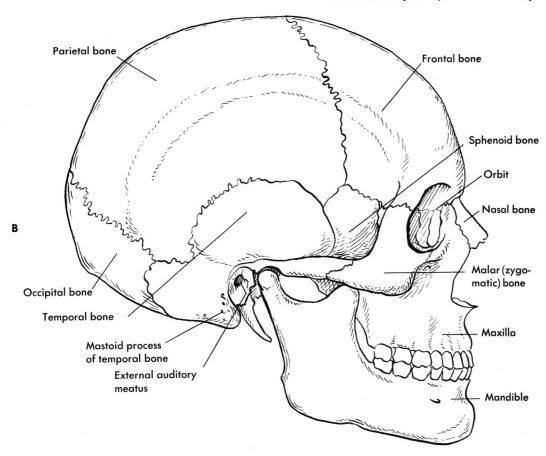

B

Parietal bone

Frontal bone

Sphenoid bone

Orbit

Nasal bone

Malar (zygomatic) bone

Maxilla

Mandible

Occipital bone

Temporal bone

Mastoid process
of temporal bone

External auditory
meatus

Fig. 4-6, cont'd. For legend see opposite page.

4-9. The unique bones of the thorax, however, are the sternum and the ribs.

The *sternum,* or breastbone, is a flat narrow bone about 6 inches long situated in the median front of the chest. It is somewhat dagger shaped and consists of three parts: the upper handle part (manubrium), the middle blade (body), and the blunt cartilaginous lower tip (xiphoid process). The xiphoid ossifies during adult life.

The greater portion of the thoracic wall consists of the twelve *ribs* and the costal cartilages (Latin *costa,* rib) that connect them to the sternum. The ribs articulate posteriorly with the thoracic vertebrae—the first, eleventh, and twelfth with the corresponding vertebra and the others with the corresponding vertebra as well as with the vertebra above. The first seven pairs are called *true* or *sternal* ribs because their anterior extremities articulate by means of costal cartilages with the sternum. The remaining five pairs of ribs are called *false* or *asternal* ribs because they do not articulate with the sternum. The costal cartilages of the eighth, ninth, and tenth ribs are joined together—that of the eighth with that of the seventh, etc. The anterior ends of the eleventh and twelfth ribs are free, and these are therefore called *floating ribs.*

Bones of the pelvis

As shown in Fig. 4-10, the hips are called the pelvis because they resemble a basin (Latin *pelvis,* basin).

The pelvis is composed of four bones—the two hipbones forming the sides and front, and the sacrum and coccyx completing it behind. It is divided by a narrow bony ring, called the *brim of the pelvis,* into an upper part, the *greater* or *false pelvis,* and a lower part, the *lesser* or *true pelvis.*

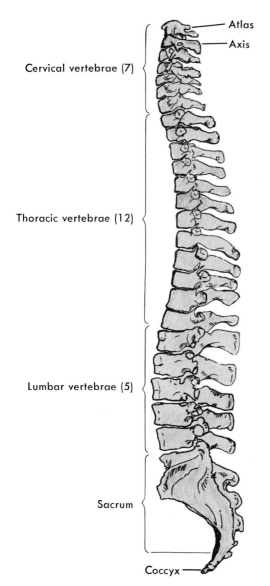

Fig. 4-7. Bones of the vertebral column.

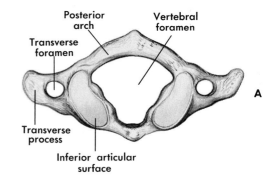

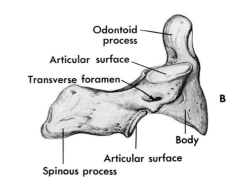

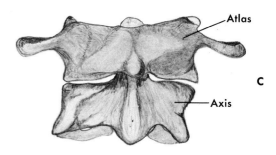

Fig. 4-8. First and second cervical vertebrae. **A,** Superior view of atlas. **B,** Lateral view of axis. **C,** Articulation of atlas and axis. (**A** and **B** modified from Anthony, C. P., and Kolthoff, N. J.: Textbook of anatomy and physiology, ed. 9, St. Louis, 1975, The C. V. Mosby Co.)

The *os coxa* or *hip bone* is a large irregularly shaped bone that, together with the corresponding bone of the opposite side, forms the side and front walls of the pelvic cavity. In youth it consists of three separate parts separated by a Y-shaped cartilage. In adulthood these parts are united by ossification of the cartilage, but the bone is still described as being divisible into three portions: (1) the *ilium* or upper expanded portion forming the prominence of the hip, (2) the *ischium* or lower strong portion, and (3) the *pubis* or portion helping to form the front of the pelvis.

BONES OF THE UPPER EXTREMITY

The bones of the upper extremity are those found in the arm region, forearm, and hand.

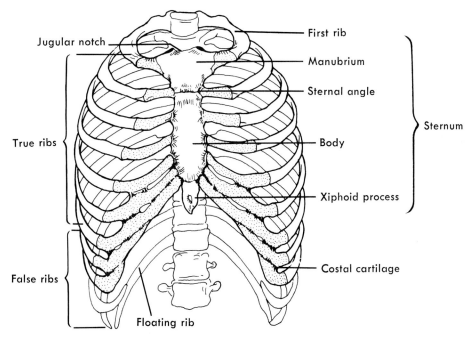

Jugular notch

First rib

Manubrium

Sternal angle

True ribs

Sternum

Body

Xiphoid process

Costal cartilage

False ribs

Floating rib

Fig. 4-9. Bones of the thorax. (Modified from Francis, C. C., and Martin, A. H.: Introduction to human anatomy, ed. 7, St. Louis, 1975, The C. V. Mosby Co.)

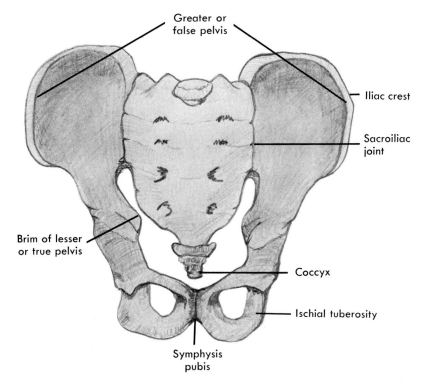

Greater or false pelvis

Iliac crest

Sacroiliac joint

Brim of lesser or true pelvis

Coccyx

Ischial tuberosity

Symphysis pubis

Fig. 4-10. Bones of the pelvis.

Bones of the arm region

The bones of the arm region consist of the scapula and clavicle, which form the shoulder girdle (Fig. 4-11), and the humerus or arm bone (Fig. 4-12).

The *scapula* or shoulder blade is located on the posterior aspect of the thorax and is attached to the thorax only by muscles. The *clavicle* or collarbone is an f-shaped bone located horizontally at the upper and anterior part of the thorax, just above the first rib. Its medial extremity ar-

ticulates with the sternum, while its lateral extremity articulates with the scapula.

The *humerus* or arm bone is the longest and largest bone of the upper extremity.

Bones of the forearm

The two bones of the forearm are the ulna and the radius (Fig. 4-13).

The *ulna,* or elbow bone, is the medial of the two bones and is slightly longer than the radius. The *radius* is located on the lateral side of the ulna

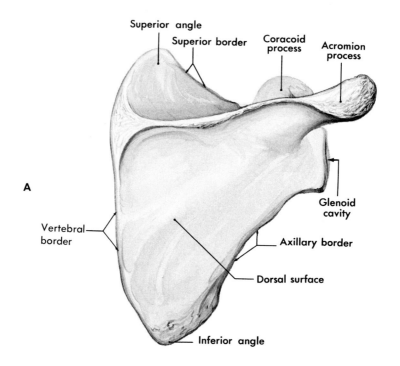

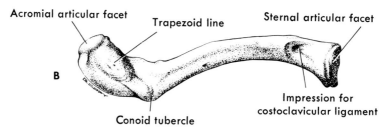

Fig. 4-11. Bones of the right shoulder girdle. **A,** Scapula. **B,** Clavicle. (**A** modified from Anthony, C. P., and Kolthoff, N. J.: Textbook of anatomy and physiology, ed. 9, St. Louis, 1975, The C. V. Mosby Co.; **B** modified from Francis, C. C., and Martin, A. H.: Introduction to human anatomy, ed. 7, St. Louis, 1975, The C. V. Mosby Co.)

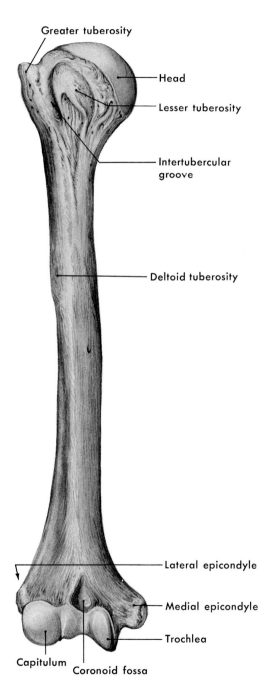

Greater tuberosity

Head

Lesser tuberosity

Intertubercular groove

Deltoid tuberosity

Lateral epicondyle

Medial epicondyle

Trochlea

Capitulum

Coronoid fossa

Fig. 4-12. Anterior surface of the right humerus. (Modified from Anthony, C. P., and Kolthoff, N. J.: Textbook of anatomy and physiology, ed. 9, St. Louis, 1975, The C. V. Mosby Co.)

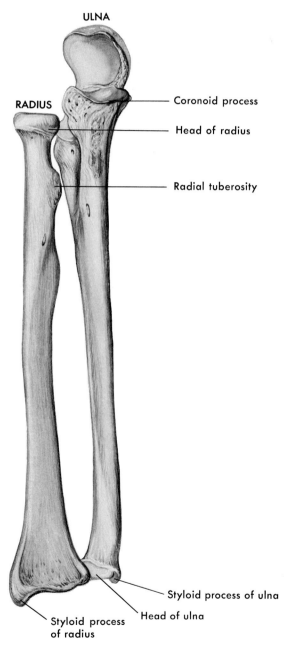

ULNA

RADIUS

Coronoid process

Head of radius

Radial tuberosity

Styloid process of ulna

Head of ulna

Styloid process of radius

Fig. 4-13. Anterior surfaces of the right radius and ulna. (Modified from Anthony, C. P., and Kolthoff, N. J.: Textbook of anatomy and physiology, ed. 9, St. Louis, 1975, The C. V. Mosby Co.)

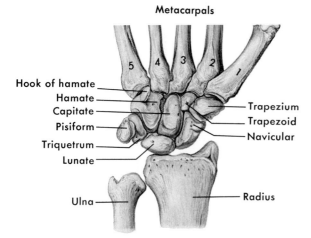

Metacarpals

Hook of hamate
Hamate
Capitate
Pisiform
Triquetrum
Lunate
Ulna

Trapezium
Trapezoid
Navicular
Radius

Fig. 4-14. Bones of the right wrist (anterior view). (Modified from Anthony, C. P., and Kolthoff, N. J.: Textbook of anatomy and physiology, ed. 9, St. Louis, 1975, The C. V. Mosby Co.)

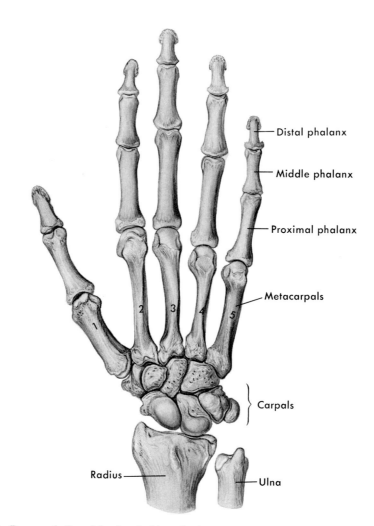

Distal phalanx
Middle phalanx
Proximal phalanx
Metacarpals
Carpals
Radius
Ulna

Fig. 4-15. Bones of the right hand (dorsal view). (Modified from Anthony, C. P., and Kolthoff, N. J.: Textbook of anatomy and physiology, ed. 9, St. Louis, 1975, The C. V. Mosby Co.)

and is shorter and smaller than the ulna. The radius is free to rotate during supination and pronation of the forearm, while the ulna is not. Both bones articulate with the humerus to form the elbow and with the carpal bones to form the wrist joint.

Bones of the hand

The carpus or wrist is composed of eight short *carpal bones* arranged in proximal and distal rows with four bones in each (Fig. 4-14).

Passing from the radial to the ulnar side, the bones of the *proximal* row are the navicular, the lunate, the triquetral, and the pisiform. Because it actually sits upon the triquetral bone, the pisiform can be observed only on the anterior surface of the wrist. From the radial to the ulnar side, the bones of the *distal* row are the trapezium, the trapezoid, the capitate, and the hamate.

The metacarpus or body of the hand consists of five *metacarpal bones* (Fig. 4-15) numbered from the radial or thumb side to the ulnar or little finger side of the hand. The *phalanges* are the bones of the digits or fingers. Each digit has three phalanges with the exception of the thumb, which has only two. The phalanges are called the first or proximal, the second or middle, and the third or distal. Pollicis is another name for the thumb or first digit.

STUDY GUIDELINES
Introduction
Specific student objectives

1. Define the fundamental qualities of human motion.
2. Classify the eight major segments of the body by location.
3. Identify the three anatomical components involved in the movement of body segments.
4. List the bones and bone groups of the body that comprise the eight major segments.
5. Describe the terms medial, lateral, anterior, posterior, proximal, distal, superior, and inferior.

Summary

A basic descriptive analysis task involved in the study of human movement is to identify and name the fundamental segmental movements of the body. These movements are termed the fundamental qualities of human motion.

The eight major segments of the body are (1) the lower extremity segments: foot, leg, and thigh, (2) the upper extremity segments: arm, forearm, and hand, and (3) the segments comprising the axial skeleton: trunk and head-neck.

Muscles, bones, and joints are the anatomical components which provide the respective propulsive forces, levers, and rotation points involved in human motion. Kinematic analysis of fundamental qualities is composed of descriptions of bone and joint movements.

The bones of the body are summarized by location groupings in Fig. 4-2.

Reference terms used in describing the upright body and its parts follow:

1. Anterior (front); example: the phalanges of the foot are anterior to the metatarsals.
2. Posterior (rear); example: the calcaneus is on the posterior region of the foot.
3. Inferior (lower, i.e., away from the head); example: the hip is inferior to the chest.
4. Superior (higher, i.e., toward the head); example: the femur is superior to the tibia.
5. Lateral (away from the midline of the body); example: the fibula is located on the lateral side of the leg.
6. Medial (toward the midline of the body); example: the big toe is on the medial side of the foot.
7. Proximal (near or toward a point of reference); example: the proximal end of the tibia is at the knee joint.
8. Distal (away from the point of reference, usually the trunk or body midline or source of a part); example: the distal end of the humerus is located at the elbow joint.

Performance tasks

1. The segmental movements of the body are called the fundamental (qualities/quantities) _____ of motion.
2. Develop a written list of lesson definitions.
3. Identify the major body segments in (a) the lower extremity, (b) the axial skeleton, and (c) the upper extremity.
4. In order to identify and name the segmental movements, it is first necessary to identify and name the involved body structures. List the

two areas of primary concern that must be identified prior to carrying out a nominal analysis of segmental movements.

5. Identify the bones of the hand involved in the motor task of writing.
6. Identify the bones of the forearm and upper arm used in the task of bowling.
7. Human beings are many times incapacitated with problems of the "lower back" area. What bones are involved?
8. A broken leg might involve what bones?
9. If a person states that he or she has suffered a broken ankle, what bones might be fractured?
10. Identify by palpation on your own body as many of the bones shown in Fig. 4-2 as possible.
11. Identify the *true* statements below:
    a. The calcaneus is in the posterior area of the foot.
    b. The sternum is superior to the pelvis.
    c. The sternum is posterior to the vertebrae.
    d. The thumb is lateral to the other digits when the arms are at the side and the palms are facing forward.
    e. A metacarpal is proximal to a carpal.
    f. The hand is located distally on the upper limb.
    g. The collarbone is inferior to the ribs.
    h. The fibula is lateral to the tibia.
    i. Lumbar vertebrae are superior to the coccyx.
    j. The radius articulates with the proximal end of the humerus.
12. Using a skeletal model (or its constituent parts), identify the bones of the three major body segments as outlined in the summary.
13. Using skeletal models, identify and draw diagrams of the bones in each segment below:
    a. Upper extremity (arm, forearm, and hand)
    b. Lower extremity (thigh, leg, and foot)
    c. Axial skeleton (trunk and head-neck)

Bones of the lower extremity
Specific student objectives

1. Given a pictorial diagram or the model of a skeleton, identify the bones of the lower extremity.
2. Identify the bones of the foot, leg, and thigh.
3. Given the names of the lower extremity bones,

be able to palpate on your own body as many as possible.

Summary

The bones of the lower extremity include the following:
1. Foot
   a. Phalanges—five toes or digits
   b. Metatarsals—five bones of the sole and lower instep
   c. Tarsals—seven ankle bones (calcaneus, talus, navicular, three cuneiforms, and cuboid)
2. Leg
   a. Tibia—larger medial shinbone
   b. Fibula—smaller lateral non–weight-bearing bone
3. Thigh
   a. Patella—kneecap
   b. Femur—longest and largest bone in the body

Performance tasks

1. Using a mounted skeleton, identify and palpate the bones of the thigh, leg, and foot.
2. Write out a listing of the bones in the thigh, leg, and foot.
3. Describe the location of each bone in the thigh, leg, and foot using reference terminology. An example would be the location of hallux (distal portion of the foot, or the most medial of the digits).
4. Select at least three movement tasks such as walking and identify the primary lower extremity bones involved in the motion.

Bones of the axial skeleton
Specific student objectives

1. Given a pictorial diagram or the model of a skeleton, identify the bones of the axial skeleton.
2. Identify the bones of the head (two groups), vertebral column, thorax, and pelvis.
3. On your own body, palpate as many of the axial skeleton bones as possible.

Summary

The bones of the axial skeleton include the following:
1. Head

a. Cranium—brain case
b. Face—anterior skull

2. Vertebral column
   a. Cervical vertebrae—seven bones of the neck
   b. Thoracic vertebrae—twelve bones of upper back
   c. Lumbar vertebrae—five bones in the small of the back
   d. Sacrum and coccyx—usually nine small bones of the posterior pelvis

3. Thorax
   a. Sternum—breastbone
   b. Ribs—twelve bones of anterior chest area

4. Pelvis
   a. Hip bone (left and right)—side and front of the pelvic cavity
   b. Sacrum and coccyx—lowest bones of the vertebral column

Performance tasks

1. Using a mounted skeleton, identify and palpate the bones of the head, vertebral column, thorax, and pelvis.
2. Write out a listing of the major bones in the head (two groups), vertebral column (four groups), thorax (two groups) and pelvis (two parts).
3. Describe the location of each group of major bones in the axial skeleton. Use reference terminology.
4. Identify a movement of the body that involves motion of each group of bones in the axial skeleton.

Bones of the upper extremity
Specific student objectives

1. Given a pictorial diagram or the model of a skeleton, identify the bones of the upper extremities.
2. Identify the bones of the arm, forearm, and hand.
3. On your own body, palpate as many of the upper extremity bones as possible.

Summary

The bones of the upper extremity include the following:
1. Arm region

a. Scapula—shoulder blade
b. Clavicle—collarbone
c. Humerus—upper arm

2. Forearm
   a. Radius—bone free to rotate located on the thumb side
   b. Ulna—elbow bone, medial to the radius and longer

3. Hand
   a. Carpals—eight wrist bones
   b. Metacarpals—five bones in the palm of the hand
   c. Phalanges—fourteen bones of the fingers

Performance tasks

1. Using a mounted skeleton, identify and palpate the bones of the arm, forearm, and hand.
2. Write out a listing of the major bones of the arm (one bone), forearm (two bones), and hand (twenty-seven bones).
3. Describe the location of each group of bones in the upper extremity. Use reference terminology.
4. Identify a movement of the body that involves motion of each group of bones in the upper extremity.

SELF-EVALUATION

Students should use no reference materials for this progress test, and they can check their answers by referring to Appendix A.

1. The fundamental qualities of human motion are the (segmental/whole)_____ movements of the body.
2. Define the fundamental *qualities* of human motion.
3. Match the following body segments with their appropriate areas by placing the correct letter in the blanks at the right:
   a. Axial skeleton   _____Trunk
   b. Upper extremity   _____Hand
   c. Lower extremity   _____Leg
                          _____Forearm
                          _____Head-neck
                          _____Foot
4. The three anatomical components involved in the movement of body segments are
   a. axial skeleton, lower extremity, and upper extremity
   b. arms, legs, and trunk

c. bones, joints, and muscles
d. head, trunk, and extremities

5. Complete the outline by naming the bones of the eight major segments below:
   a. Lower extremity
      Foot
      _____ Bones of the toes
      _____ Bones of the sole and lower instep
      _____ Bones of the ankle
      Leg
      _____ Large weight-bearing bone
      _____ Small bone of the leg
      Thigh
      _____ Kneecap
      _____ Longest bone in the body
   b. Axial skeleton
      Trunk
      _____ Hip bones, sacrum, and coccyx
      _____ Lower back
      _____ Chest region
      _____ Cage structure
      _____ Breastbone
      Head-neck
      _____ Neck bones
      _____ Bones of the head
      _____ Brain case
      _____ Anterior skull
   c. Upper extremity
      Arm region
      Shoulder girdle
      _____ Shoulder blade
      _____ Collarbone
      Arm bone _____
      Forearm
      _____ Thumb side (free to rotate)
      _____ Little finger side (elbow bone)
      Hand
      _____ Wrist bones
      _____ Bones of the palm
      _____ Finger bones

6. List the number of bones and the names of each bone in the following groups:
   a. Toes
   b. Metatarsals
   c. Tarsals
   d. Hip bones

e. Vertebral column
f. Shoulder girdle
g. Skull
h. Carpals
i. Metacarpals
j. Phalanges of the fingers

7. Match the terms on the left with the correct meaning on the right.
   a. Medial _____Front
   b. Lateral _____Away from midline
   c. Anterior _____Above
   d. Posterior _____Below
   e. Proximal _____Toward midline
   f. Distal _____Near the source
   g. Superior _____Rear
   h. Inferior _____Away from the source

8. Using the illustration of the skeleton (Fig. 4-16), match the technical names of the bones with their corresponding numbers.

Phalanges_____   Femur _____
Tarsals _____   Metacarpals _____
Carpals _____   Clavicle _____
Humerus _____   Patella _____
Ribs _____   Ulna _____
Sternum _____   Radius _____
Skull _____   Vertebrae _____
Fibula _____   Hips _____
Tibia _____

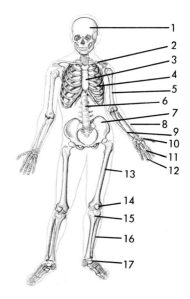

Fig. 4-16. Anterior view of the human skeleton.

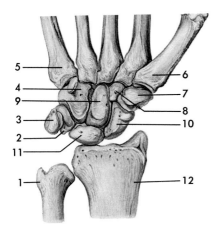

Fig. 4-17. The wrist. (Modified from Anthony, C. P., and Kolthoff, N. J.: Textbook of anatomy and physiology, ed. 9, St. Louis, 1975, The C. V. Mosby Co.)

9. Using the illustration of the wrist (Fig. 4-17), match the technical names of the bones with their corresponding numbers.
    _____Radius
    _____Capitate
    _____Navicular
    _____Pisiform
    _____Ulna
    _____First metacarpal
    _____Trapezium
    _____Trapezoid
    _____Lunate
    _____Hamate
    _____Triquetrum
    _____Fifth metacarpal
10. Using the illustration of the foot (Fig. 4-18), match the technical names of the bones with their corresponding numbers.
    _____Phalanges of the hallux
    _____ Talus
    _____ Cuboid
    _____ Middle phalanx
    _____ Medial cuneiform
    _____ Third metatarsal
    _____ Navicular
    _____ Middle cuneiform
    _____ Calcaneus
    _____ Lateral cuneiform
    _____ First metatarsal

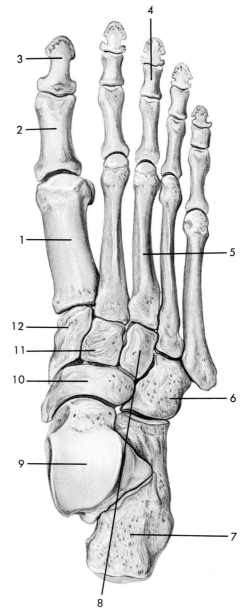

Fig. 4-18. Dorsal surface of the right foot. (Modified from Anthony, C. P., and Kolthoff, N. J.: Textbook of anatomy and physiology, ed. 9, St. Louis, 1975, The C. V. Mosby Co.)

11. Using the illustration of the skeleton (Fig. 4-19, *A*), match the technical names of the bones with their corresponding numbers.

   _____Tarsals
   _____Humerus
   _____Sternum
   _____Fibula
   _____Tibia
   _____Metacarpals
   _____Patella
   _____Radius
   _____Hips
   _____Phalanges
   _____Carpals
   _____Ribs
   _____Skull
   _____Femur
   _____Clavicle
   _____Ulna
   _____Vertebrae

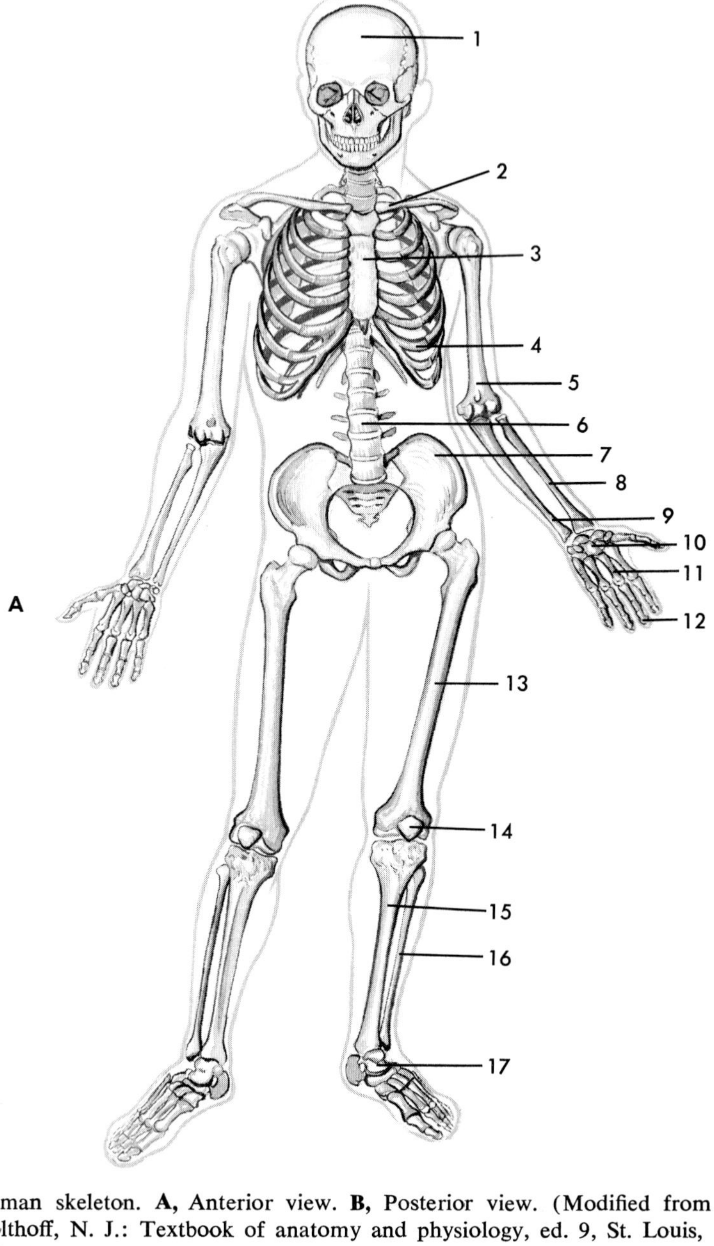

Fig. 4-19. Human skeleton. **A,** Anterior view. **B,** Posterior view. (Modified from Anthony, C. P., and Kolthoff, N. J.: Textbook of anatomy and physiology, ed. 9, St. Louis, 1975, The C. V. Mosby Co.)

12. Using the illustration of the skeleton (Fig. 4-19, *B*), match the technical names of the bones with their corresponding numbers.

_____ Tarsals
_____ Humerus
_____ Scapula
_____ Fibula
_____ Tibia
_____ Metacarpals
_____ Calcaneus
_____ Radius
_____ Hips
_____ Phalanges
_____ Carpals
_____ Ribs
_____ Skull
_____ Femur
_____ Clavicle
_____ Ulna
_____ Vertebrae

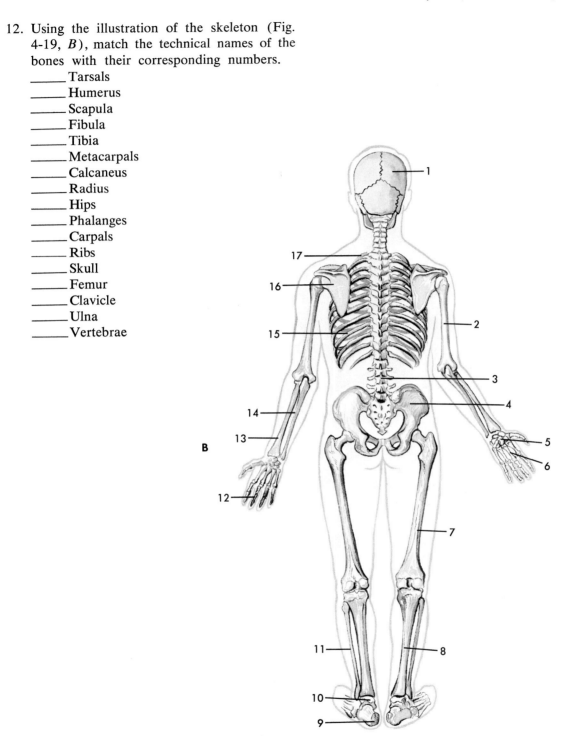

Fig. 4-19, cont'd. For legend see opposite page.

LESSON 5

Nominal analysis of joint and segmental movement classifications

Body of lesson
Study guidelines
Self-evaluation

PLANES AND AXES OF DIARTHRODIAL JOINT MOVEMENTS

The junction between two bones or between a bone and cartilage is called an articulation or joint. The three types of joints found in the human body are identified as synarthrodial or immovable, amphiarthrodial or slightly movable, and diarthrodial or freely movable. The type of joint with which we are primarily concerned in the analysis of human motion is in the diarthrodial or freely movable category.

Diarthrodial joints are of importance in the study of human motion because they form the centers of rotation around which the bones and body segments turn, i.e., the axes of motion. The cardinal planes and the three cardinal axes of human motion are illustrated in Fig. 5-1. In this illustration it can be seen that the midline of the body can be expanded into a flat surface called the sagittal plane (Latin *sagitta,* arrow). This plane divides the body into right and left sides. The body is divided into front and back sides by the frontal plane, and into upper and lower parts by the transverse plane.

The three axes of human motion, called the x, y, and z axes, are located at the intersection of the three planes of the body as shown in Fig. 5-1. Each of these axes is perpendicular to the plane in which the angular motion occurs. As can be seen in Fig. 5-1, the x axis (also called the *frontal axis*) passes horizontally from side to side perpendicular to the sagittal plane. Likewise, the y axis (also called the *vertical* or *longitudinal axis*) passes perpendicular to the ground and to the transverse plane. The z axis (also called the *sagittal axis*) passes horizontally from front to back and is perpendicular to the frontal plane.

CLASSIFICATION OF DIARTHRODIAL JOINTS

The diarthrodial type of joint can be further divided into six categories according to their articulating surfaces and the number of axes around which they allow movement. A joint that allows movement in only one plane, for instance, is called *uniaxial*. Thus a *biaxial* joint allows movement in two planes, and a *triaxial* joint allows movement in three different planes.

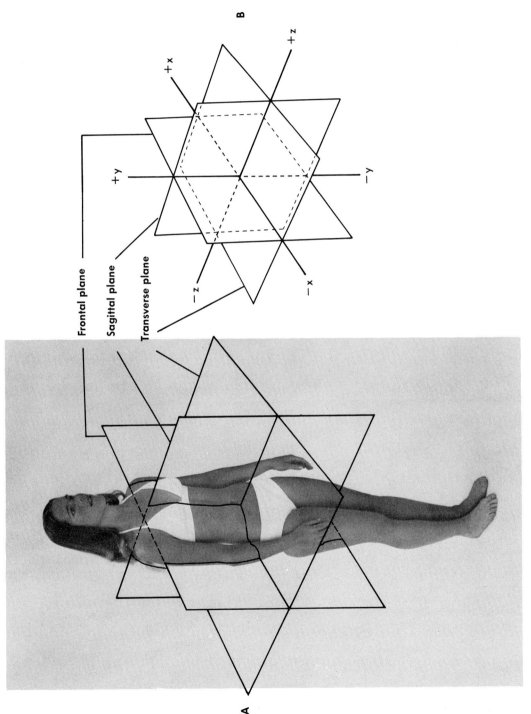

Frontal plane

Sagittal plane

Transverse plane

Fig. 5-1. A, The cardinal planes. **B,** Cardinal planes and axes of motion.

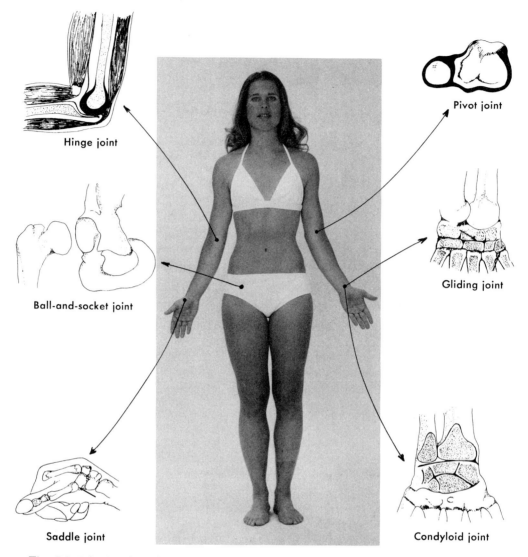

Fig. 5-2. Diarthrodial joint classification. A *hinge joint* has a spool-shaped surface that fits into a concave surface, e.g., elbow joint. A *ball-and-socket joint* has a ball-shaped head that fits into a cuplike socket, e.g., hip joint. A *saddle joint* has a saddle-shaped bone that fits into a saddle-shaped socket, e.g., carpometacarpal joint of the thumb. A *condyloid joint* has an oval surface that fits into an elliptical cavity, e.g., wrist joint. A *gliding joint* has an articulation between flat or slightly concave surfaces, e.g., intercarpal joints. A *pivot joint* has an arch-shaped surface that rotates about a rounded pivot, e.g., radioulnar joint. These joints are shown in the figure.

The six types of diarthrodial joints are illustrated and defined in Fig. 5-2. The only type of diarthrodial joint that does not allow rotation is the gliding type. *Gliding joints* are located between the carpal bones (of the wrist) and between the tarsal bones (of the foot). Since, as their name implies, these joints allow only gliding types of movement with no rotation, they are classified as nonaxial.

The elbow, knee, and ankle are examples of simple *hinge joints,* which permit movement in only one plane. These joints are therefore uni-

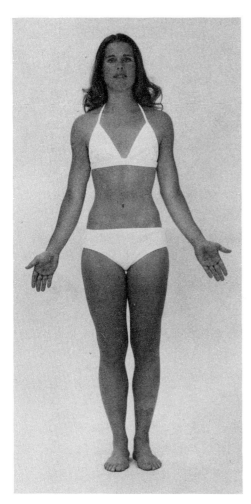

Fig. 5-3. The anatomical position has the individual standing upright, arms at the sides, and the palms of the hands facing forward.

The *ball-and-socket joint* is a triaxial joint because it allows movement in all three planes. The two ball-and-socket joints in the body are the shoulder joint and the hip joint.

Anatomical position

The definitions given above for the axes of motion, as well as the definitions given for the terms used in the classification of segmental movements discussed in the next section, are made with the assumption that the individual is in the anatomical position. This position, illustrated in Fig. 5-3, has the individual upright, feet together, arms at the sides, and the palms of the hands facing forward.

CLASSIFICATION OF SEGMENTAL MOVEMENTS

The two basic types of segmental movements, as defined before, are linear and angular.

Linear or *gliding movement* is the simplest kind of motion that can take place in a joint and is the characteristic movement of the gliding joints. In this type of movement one surface glides over another without any rotation occurring. Examples of this type of movement would be the motion occurring between the different tarsal bones and between the different carpal bones.

Angular movement is the type of motion that occurs in the spinal column and between long bones, and by it the angle between two bones is either increased or decreased. When the anatomical position is used as a frame of reference, we can see that these movements occur in a definite plane and around a definite axis. For example, movement of the arm forward or backward (which is called *flexion* or *extension* respectively) occurs in the sagittal plane about a frontal axis (Fig. 5-4). Likewise, the sideward raising or lowering of the arm (which is called *abduction* or *adduction* respectively) occurs in the frontal plane about a sagittal axis. Horizontal flexion and extension (Fig. 5-5) as well as inward and outward rotation of the arm occur in the transverse plane about a longitudinal axis. Since the swinging of an arm in a circle (called *circumduction*) involves flexion, abduction, extension, and adduction performed in sequence, it occurs in both the sagittal and the frontal planes.

axial. The only other type of uniaxial joint is the pivot type. *Pivot joints* allow rotation of one bone about the projection of another. A pivot joint, such as the radioulnar joint in the forearm, is uniaxial because the rotation occurs in only one plane.

The two types of biaxial joints are the condyloid and saddle types. In *condyloid joints* we find a convex surface fitting into a concave surface. The condyloid type, as evidenced by the wrist, permits motion backward and forward and from side to side; hence it is a biaxial joint. The only *saddle joint,* which receives its name from its shape, is located at the base of the thumb. It is a biaxial joint.

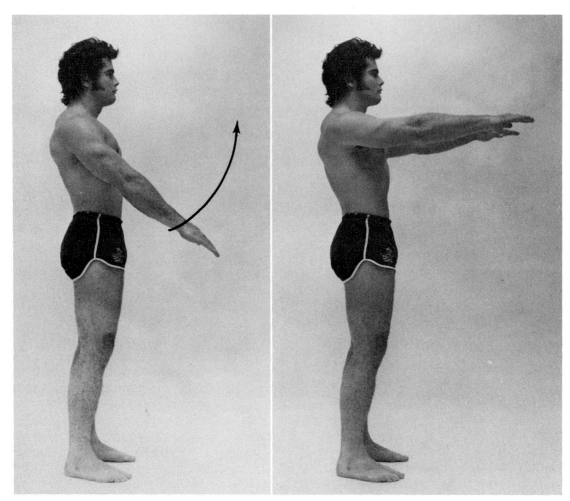

Starting position and
direction of movement

Final position

Fig. 5-4. Angular movement in the sagittal plane around a frontal axis.

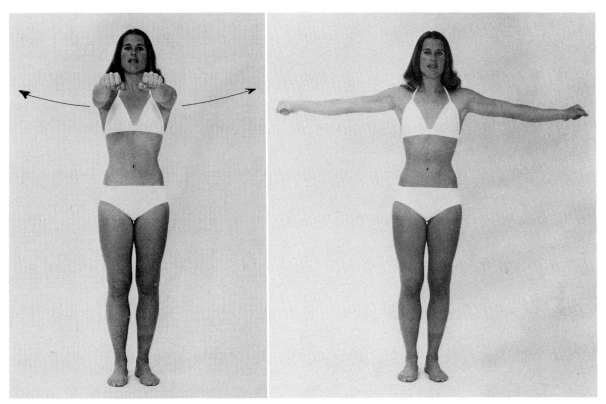

Starting position and
direction of movement

Final position

Fig. 5-5. Angular movement in the transverse plane around a vertical axis.

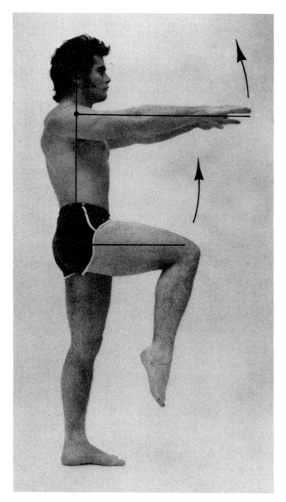

Fig. 5-6. Flexion occurs between adjacent longitudinal axes.

The paired movements of flexion-extension and abduction-adduction occur *between* the longitudinal axes of two or more adjacent body segments (Fig. 5-6). Inward and outward rotation, however, occurs *around* the longitudinal axes of one or more adjacent body segments (Fig. 5-7). The longitudinal axis of a bone or body segment can be specifically defined as a straight line which connects the midpoint of the joint at one end of the bone or segment with the midpoint of the joint at the other end, or in the case of a terminal segment with its distal end.

Flexion and extension

Flexion and extenison are forward and backward movements of body segments which occur in the sagittal plane around a frontal (x) axis.

Flexion is a movement where the anterior or posterior surface of one segment approaches the anterior or posterior surface of another segment and thereby produces a *decrease in the angle* between the longitudinal axes of the two segments. An example of this type of movement is hip flexion, which involves a movement of the anterior surface of the thigh toward the anterior surface of the trunk or abdomen. In this movement the angle between the longitudinal axis of the thigh and the longitudinal axis of the trunk is decreased (Fig. 5-8).

The term *extension* means to increase or to lengthen. Thus the movement of extension involves an *increase in the angle* between the longitudinal axes of two adjacent segments and is, therefore, the opposite of flexion. The movement is actually a return to the anatomical position from flexion. Movement beyond the anatomical position in which the angle between the longitudinal axes of body segments is increased beyond that found in the anatomical position is called *hyperextension.*

Flexion and extension (and hyperextension) of the ankle joint are often referred to as *dorsal flexion* (dorsiflexion) and *plantar flexion,* respectively. These movements begin from the anatomical position, in which the longitudinal axis of the foot is at a right angle to the longitudinal axis of the leg. When the foot is raised and leads with its dorsal (upper) surface, the movement is called dorsal flexion (Fig. 5-9). When it is lowered and leads with its plantar (lower) surface, the movement is called plantar flexion.

Abduction and adduction

Abduction and adduction (Latin *ab,* from; *ad,* to or toward; *ductus,* having been led) are sideward movements of body segments which occur in the frontal plane around a sagittal (z) axis. *Abduction* (Fig. 5-10, *A*) is the movement of a body segment *away from the midline* of the body or body part to which it is attached. *Adduction* is the movement of a body segment *toward the midline* of the body or body part to which it is attached (Fig. 5-10, *B*). An example of adduction would be the lowering of the arm back to the side again from the abducted postion.

The movement of abduction, once started, is

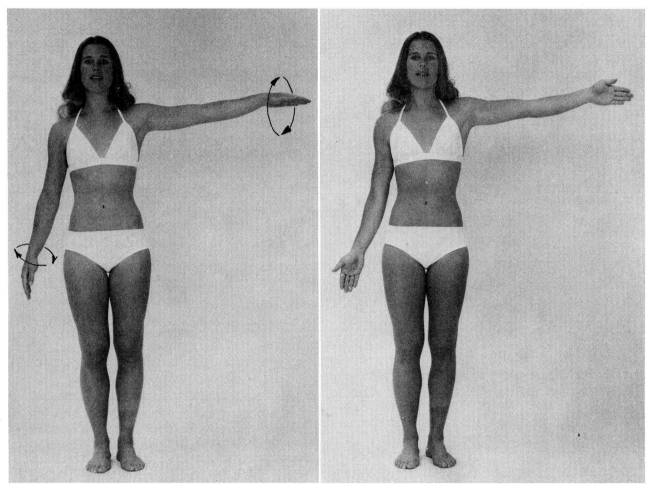

<table>
<tr><td align="center">Starting positions and
directions of movement</td><td align="center">Final positions</td></tr>
</table>

Fig. 5-7. Outward rotation occurs around the longitudinal axis of the segment(s).

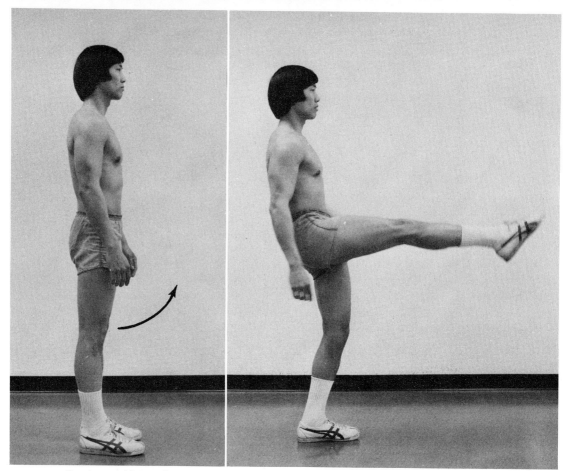

Starting position and
direction of movement

Final position

Fig. 5-8. Flexion of the thigh at the hip joint.

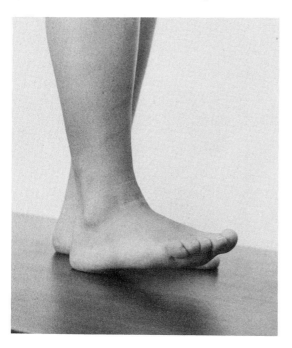

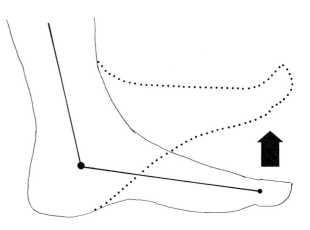

Fig. 5-9. Dorsal flexion of the foot.

called abduction throughout its entire range. This is true even though, as in the case of the abducting arm at the shoulder joint, the segment actually comes back toward the body after passing above the horizontal or during the second 90° of the movement.

At the wrist, abduction is also called *radial flexion*; and adduction is also called *ulnar flexion.* The radius and ulna are, of course, the two bones located in the forearm. The radius is located on the thumb side of the forearm, while the ulna is located on the little finger side. Thus movement of the hand toward the thumb side is called

radial flexion, while movement of the hand toward the little finger side is called ulnar flexion.

In the hand, abduction and adduction of the second, fourth, and fifth fingers relate to an imaginary line (longitudinal axis) drawn through the middle finger. Thus abduction of, for instance, the little finger is movement of this finger away from the middle finger. Correspondingly adduction would be a movement toward the middle finger. Likewise in the foot, abduction and adduction of the first, second, fourth, and fifth toes relate to an imaginary line drawn through the middle toe. In all other instances (except the

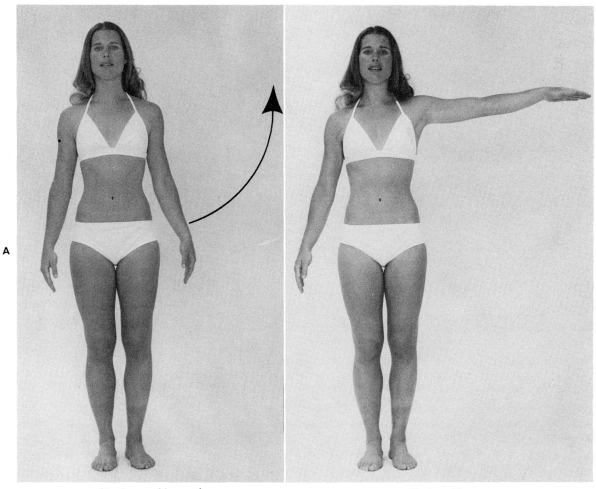

A

Starting position and
direction of movement

Final position

Fig. 5-10. A, Shoulder joint abduction.

movements of the thumb, which do not follow any of these rules) the midline of the body as a whole, when in the anatomical position, is used as a frame of reference in identifying the abduction and adduction of body parts. This is the frame of reference used, for example, when describing the abductions and adductions of the middle finger and of the middle toe. Movements of the middle finger can also be identified as being either radial flexion or ulnar flexion according to the rules already given regarding the movements of the wrist.

Sideward movements of the trunk are often called *lateral flexions*. In other words, bending the trunk laterally to the right can be labeled as either abduction of the trunk to the right or lateral flexion of the trunk to the right.

Rotation

Rotation is a movement in which a bone or body segment moves *around* its longitudinal axis without displacing the axis. In the anatomical position this movement occurs in the transverse plane around a longitudinal axis.

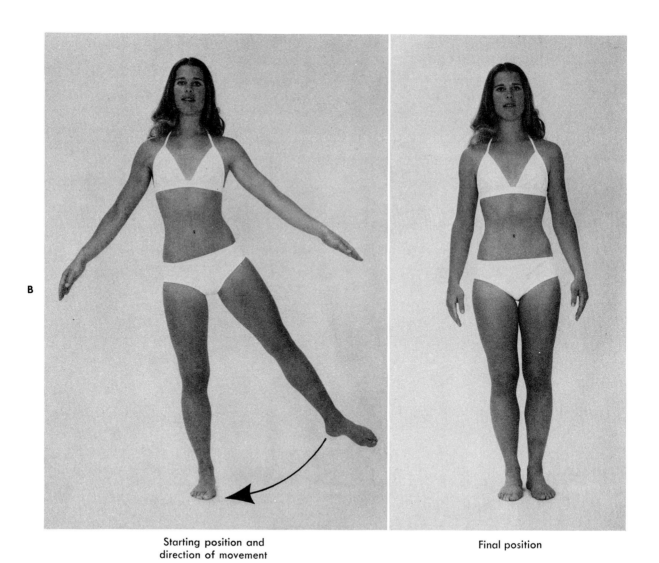

B

Starting position and
direction of movement

Final position

Fig. 5-10, cont'd. B, Hip joint adduction.

Inward or *medial rotation* (Fig. 5-11) occurs when the anterior surface of a bone or body segment turns inward. Likewise *outward* or *lateral rotation* occurs when the anterior surface of a bone or body segment turns outward.

The rotations of the forearm and of the foot are often identified through special names. Inward rotation of the forearm is called *pronation,* while outward rotation is called *supination.* Inward rotation of the foot, which involves lifting the lateral border of the foot, is called *eversion;*

outward rotation, which involves lifting the medial border of the foot, is *inversion.*

Combined movements

The two major combined movements are circumduction and the horizontal movements of the arm at the shoulder joint.

Circumduction is a movement in which a body part describes a cone, with the apex of the cone located at the joint and the base of the cone located at the distal end of the part. Since this

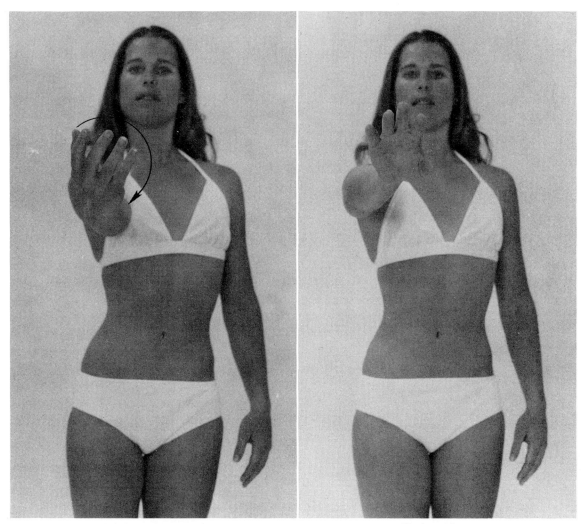

Starting position and
direction of movement

Final position

Fig. 5-11. Inward rotation of the right forearm.

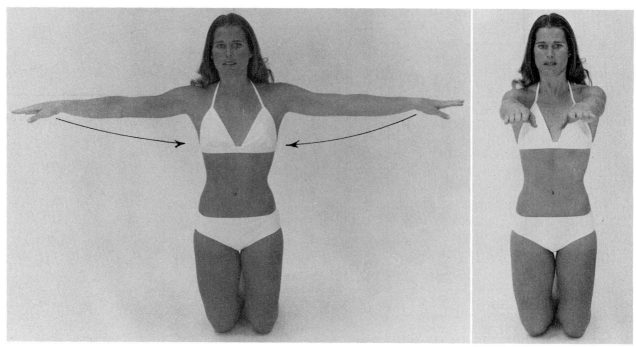

Starting position and
direction of movement

Final position

Fig. 5-12. Horizontal flexion of the shoulder joint.

movement does not involve rotation, it may occur in biaxial joints by a combination of flexion, abduction, extension, and adduction performed sequentially. Thus, when performed in the anatomical position this movement occurs in both the sagittal and the frontal planes. An example of circumduction would be the swinging of an arm in a circle.

The forward movement of the abducted arm in the transverse plane (Fig. 5-12), often called simply *horizontal* or *transverse flexion,* involves the combined movements of shoulder joint flexion and adduction. The opposite movement, called *horizontal* or *transverse extension,* is a backward movement of the flexed arm in the transverse plane which involves the combined movements of shoulder joint extension and abduction.

Diagonal movements

Very seldom do segmental movements occur parallel with the orientation planes and perpen-

dicular to the axes of the body as described thus far. In fact, most movements occur diagonally to these planes and from many body positions which do not simulate the anatomical position. The anatomically referenced names given to the movements, however, continue to be used as before. Thus elbow flexion is called elbow flexion regardless of the position of the individual or the position of his limbs.

STUDY GUIDELINES
Planes and axes of diarthrodial joint movements
Specific student objectives

1. Describe an articulation or joint.
2. Identify and define the three types of joints found in the human body.
3. State why diarthrodial joints are of importance in the study of human motion.
4. Identify and describe the three cardinal planes of the human body.

5. Identify and describe the three cardinal axes of human motion.

Summary

The junction between two bones or between a bone and cartilage is called a joint. The three types of joints found in the human body are identified as synarthrodial or immovable, amphiarthrodial or slightly movable, and diarthrodial or freely movable. Diarthrodial joints are of primary importance in studying human motion because they act as the centers of rotation (axes) about which the bones and body segments turn.

The three cardinal planes of motion which pass through the center of gravity of the body are the sagittal plane (dividing the body into left and right sides), the frontal plane (dividing the body into front and back areas), and the transverse plane (dividing the body into upper and lower parts). The three axes of human motion are each perpendicular to the plane in which angular motion occurs. The frontal or x axis passes horizontally from side to side perpendicular to the sagittal plane. The longitudinal or y axis passes vertically perpendicular to the transverse plane. The sagittal or z axis passes horizontally from front to back perpendicular to the frontal plane.

Performance tasks

1. State a synonym for "joint."
2. Match the terms on the left with the correct meaning on the right.
   a. Synarthrodial _____ Slightly movable
   b. Amphiarthrodial _____ Freely movable
   c. Diarthrodial _____ Immovable
3. Use a mounted skeleton to identify at least two joints of each type listed in the preceding question in the human body.
4. The most important sites of human motion occur at (amphiarthrodial/diarthrodial) _____ joints.
5. Diarthrodial joints are the centers of rotation about which the bones turn, i.e., the _____ _____ of motion.
6. Horizontal motion occurs in the _____ cardinal plane.
7. The two vertical cardinal motion planes are the _____ and _____ planes.

8. Match the planes on the left with the correct meaning on the right.
   a. Frontal _____ Divides upper and lower
   b. Sagittal _____ Divides front and back
   c. Transverse _____ Divides left and right
9. The cardinal axes of motion are (parallel/perpendicular) _____ to the cardinal planes of motion.
10. Match the axes of motion on the left with the correct meaning on the right.
    a. x axis _____ Horizontal—front to back
    b. y axis _____ Horizontal—side to side
    c. z axis _____ Vertical

Classification of diarthrodial joints
Specific student objectives
1. Describe the terms uniaxial, biaxial, and triaxial.
2. Describe the six types of diarthrodial joints and give an example of each.
3. Describe the anatomical position.

Summary

Diarthrodial joints are further divided into six categories according to their articulating surfaces and the number of axes around which they allow movement:
1. Gliding—allows no rotation (nonaxial)
2. Hinge—works like a door hinge and permits motion in one plane (uniaxial)
3. Pivot—allows rotation of one bone about the projection of another in a single plane (uniaxial)
4. Condyloid—a convex surface fits into a concave surface; allows movement in two planes (biaxial)
5. Saddle—shaped like a saddle; allows movement in two planes (biaxial)
6. Ball-and-socket—a ball-like structure fits into a socket-like structure; allows movement in three planes (triaxial)

The anatomical position has the individual upright, feet together, arms at the side with the palms of the hands supinated and facing forward.

Performance tasks

1. Match the terms on the left with the correct meaning on the right:
   a. Uniaxial _____ Motion in three planes
   b. Biaxial _____ Motion in two planes
   c. Triaxial _____ Motion in one plane

2. Using a mounted skeleton or skeletal parts, identify examples of each type of diarthrodial joint. Determine the number of axes of motion allowed by manipulation of the skeletal parts.
3. Identify the two ball-and-socket diarthrodial joints in the human body.
4. Identify the only saddle joint found in the body.
5. Using a mounted skeleton, classify the following diarthrodial joints and determine the number of axes of motion allowed.
   a. Interphalangeal joints of the foot
   b. Metatarsophalangeal joints
   c. Intertarsal joints
   d. Ankle joint
   e. Knee joint
   f. Hip joint
   g. Sacroiliac articulation
   h. Atlantodental joint
   i. Sternoclavicular joint
   j. Elbow joint
   k. Superior and inferior radioulnar joints
   l. Radiocarpal joint
   m. Intermetacarpal joints
   n. Interphalangeal joints of the fingers
   o. Carpometacarpal joint of the thumb

Classification of segmental movements
Specific student objectives

1. Describe the linear or gliding type of joint motion.
2. Describe the angular types of joint motion.
3. Define the longitudinal axis of a body segment.
4. List the types of motion that occur between the longitudinal axes of adjacent segments.
5. List the type of motion that occurs around the longitudinal axis of a segment.
6. Describe flexion and extension.
7. Describe abduction and adduction.
8. Describe rotation.
9. Describe circumduction, horizontal flexion, and horizontal extension.
10. Describe diagonal types of motion.
11. Illustrate the segmental movements as defined above with specific examples in the human body.

Summary

The two basic types of segmental movements are linear or gliding, in which one surface glides over another (without rotation), and angular, in which the angle between two bones is increased or decreased.

A straight line connecting the midpoint of the joints at both ends of a bone is called the longitudinal axis of that bone or body segment. Flexion-extension and abduction-adduction occur between the longitudinal axes of two or more adjacent body segments. Rotation occurs around the longitudinal (y) axis of one or more adjacent body segments.

Flexion and extension are forward and backward angular movements of body segments that occur in the sagittal plane around a frontal (x) axis. Flexion is a movement in which the respective anterior or posterior surface of one segment approaches the anterior or posterior surface of another segment and produces a decrease in the angle between the longitudinal axes of the two segments. Extension is the opposite of flexion and involves an increase in the angle between the longitudinal axes of two adjacent segments. Movement beyond the return to anatomical position in extension is termed hyperextension. In the ankle joint, flexion and extension are designated by the respective terms dorsal flexion and plantar flexion.

Abduction is the movement of a body segment away from the body midline or body part to which it is attached. Adduction is the opposite return movement toward the body or body part. Both occur in the frontal plane around a sagittal (z) axis. At the wrist, abduction is called radial flexion and adduction is called ulnar flexion. Sideward trunk movements are termed lateral flexion.

Inward or medial rotation occurs when the anterior surface of a bone or body segment turns inward. Outward or lateral rotation is the opposite motion that occurs when the anterior surface turns outward. Inward rotation of the forearm is identified by the special name pronation, while outward rotation of the forearm is called supination. At the foot, inward rotation is termed eversion and outward rotation is called inversion.

Circumduction occurs in both the sagittal and

the frontal plane and is a movement in which a body part describes a cone. Horizontal flexion is the forward angular movement of the abducted arm in the transverse plane. Horizontal extension is the return movement. Diagonal movements occur at other than right angles to one of the cardinal planes.

Performance tasks

1. Develop a written list of section definitions.
2. In your own words, describe a joint.
3. Match the type of diarthrodial joint with its appropriate description.
   a. Ball-and-socket
   b. Condyloid
   c. Gliding
   d. Hinge
   e. Pivot
   f. Saddle
   _____No rotation
   _____Opens like a door and is uniaxial
   _____ One bone rotates about another's projection (uniaxial)
   _____Convex surface in a concave surface
   _____Biaxial
   _____Motion in three planes
4. Describe the three cardinal planes and the three cardinal axes of motion.
5. Illustrate linear or gliding movement and angular movement with a motion example.
6. Describe the position of the trunk and extremities when the body is in the anatomical position.
7. Describe the relationship of the movements of flexion-extension, abduction-adduction, and rotation to the longitudinal axes of adjacent body segments.
8. Identify the cardinal plane and cardinal axis of motion for the following movements:

	Plane	Axis
Flexion-extension	_____	_____
Abduction-adduction	_____	_____
Inward and outward rotation	_____	_____

9. Identify the movements, planes, and axis for the motion description:
   a. Anterior surface of the forearm moving toward the anterior surface of the humerus
   b. Arm moving toward the body midline during angular motion about the shoulder joint
   c. Palm of the hand turning inward
   d. Sole of the foot lowered with the body in anatomical position
   e. From the anatomical position, arching the back in a backward manner
   f. Movement at the shoulder joint of a discus thrower during the final throwing action
   g. Movement of the thigh forward during a kicking motion
   h. Return of the fingers to the anatomical position from a fully spread position
   i. Swinging the arm about the shoulder joint in a conical path
   j. Tilting the trunk to the side from an anatomical starting position

SELF-EVALUATION

Students should use no reference materials for this progress test, and they can check their answers by referring to Appendix A.

1. A synonym for a skeletal joint is an (articulation/segment)_____ .
2. Describe a joint._____
   _____
   _____
   _____
3. The three types of joints in the human body are
   a. amphiarthrodial, diarthrodial, and synarthrodial
   b. gliding, hinge, and pivot
   c. both a and b
   d. neither a nor b
4. Define the three types of body joints.
5. The study of human motion centers upon joints called
   a. amphiarthrodial   c. both a and b
   b. diarthrodial   d. neither a nor b
6. The three cardinal planes of the body are
   a. x, y, and z
   b. medial, lateral, and horizontal
   c. both a and b
   d. neither a nor b
7. Describe the three cardinal planes of motion.
   _____
   _____
   _____
8. Describe the three axes of motion._____
   _____
   _____
   _____

9. Match the terms on the left with the correct meaning on the right:
   a. Uniaxial _____ Motion in three planes
   b. Biaxial _____ Motion in one plane
   c. Triaxial _____ Motion in two planes
10. Match the joints listed below with their correct descriptions.
    a. Gliding
    b. Hinge
    c. Pivot
    d. Condyloid
    e. Ball-and-socket
    f. Saddle
    _____ Rotates one bone on another's projection
    _____ Convex surface fits into a concave surface
    _____ Motion in three planes
    _____ Biaxial joint
    _____ Uniaxial
    _____ Allows no rotation
11. Give an example of each type of diarthrodial joint.
12. Describe the anatomical position. _____
    _____
    _____
    _____
13. Gliding motion produces (angular/linear) _____ movement.
14. Angular motion occurs when the angle between two bones is

a. decreased
b. increased
c. both a and b
d. neither a nor b
15. A straight line connecting the midpoint of the joints at both ends of a bone is an axis termed
    a. x
    b. y
    c. both a and b
    d. neither a nor b
16. The motions of body segments that occur between their longitudinal axes are
    a. abduction-adduction
    b. flexion-extension
    c. both a and b
    d. neither a nor b
17. Describe the following terms:
    a. Flexion
    b. Extension
    c. Hyperextension
    d. Abduction
    e. Adduction
    f. Rotation
    g. Horizontal flexion and extension
    h. Circumduction
    i. Lateral flexion
    j. Diagonal motion
18. Illustrate the segmental movements listed in the preceding question with specific joint movement examples.

Nominal analysis of segmental movements

MOVEMENTS OF THE LOWER EXTREMITY

The three major segments of the lower extremity are the foot, leg, and thigh.

Joints of the lower extremity

1. As their name indicates, the *interphalangeal joints* are the articulations between the phalanges of the toes. These joints are found between the proximal and middle phalanges and between the middle and distal phalanges. This means that there are nine interphalangeal joints in each foot. Since these joints are of the hinge type, they are uniaxial and allow only the movements of flexion and extension. Flexion consists of curling the toes, while extension consists of straightening the toes.

2. The *metatarsophalangeal joints* involve the distal ends of the metatarsals and the proximal phalanges. These condyloid-type joints allow motion in two planes. In the sagittal plane they allow flexion and extension. In the transverse plane they allow a slight spreading of the toes, which is called abduction; bringing the toes back together again is adduction.

3. The name *tarsometatarsal* tells us that these joints involve the tarsal and metatarsal bones. Tarsometatarsal joints are at the proximal ends of the metatarsal bones. Since these joints are of the gliding type, we know that they allow only linear motion.

4. As their name indicates, the *intertarsal joints* are the articulations among the seven tarsal bones. The movements of eversion and inversion occur at these joints. Inversion consists of outward rotating or lifting of the medial border of the foot, while eversion consists of inward rotating or lifting the lateral border of the foot.

5. The *ankle joint* is the hinge joint formed by the tibia and fibula fitting over the talus (as shown in Fig. 6-1, *6*). The only true movements of the ankle joint are flexion, also called dorsal flexion, and extension, also called plantar flexion.

6. The *tibiofibular joints* (Fig. 6-1, *4*) include three separate articulations: the *inferior tibiofibular joint,* the *tibiofibular union,* and the *superior tibiofibular joint.* The tibiofibular union is the joining of the shafts of the tibia and fibula by an interosseous membrane. The superior and inferior tibiofibular joints are the articulations between the superior and inferior

85

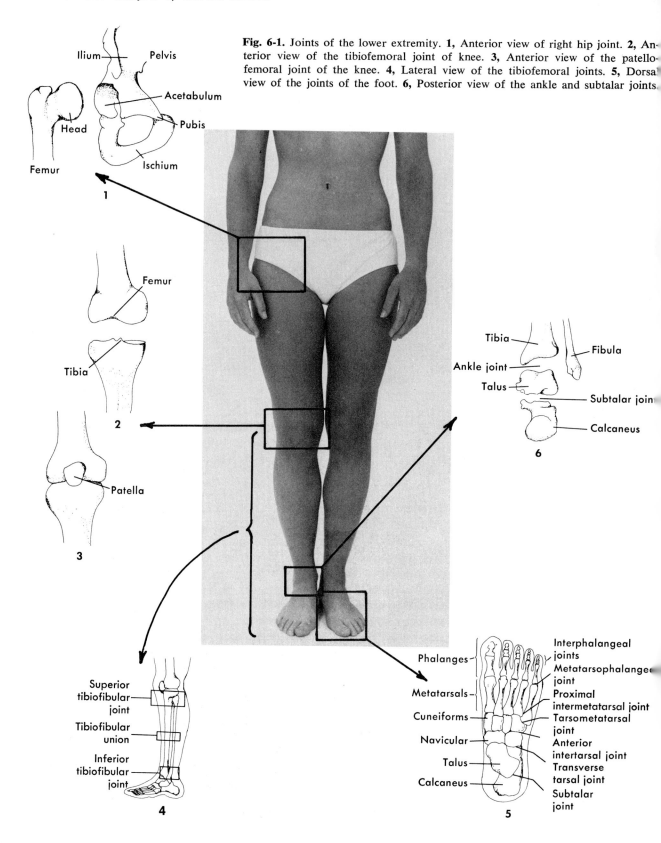

Fig. 6-1. Joints of the lower extremity. **1,** Anterior view of right hip joint. **2,** Anterior view of the tibiofemoral joint of knee. **3,** Anterior view of the patellofemoral joint of the knee. **4,** Lateral view of the tibiofemoral joints. **5,** Dorsal view of the joints of the foot. **6,** Posterior view of the ankle and subtalar joints.

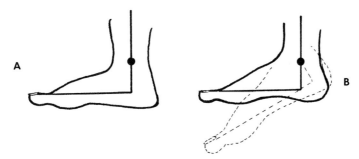

Fig. 6-2. Illustration of the longitudinal axis of the foot **(A)** and plantar flexion of the ankle joint **(B).**

extremities of the tibia and fibula, and the joints formed are of the gliding category. Thus, these joints allow only sliding movements, which are passive in character because they complement the movements of the ankle joint.

7. The *knee joint* (Fig. 6-1, *2*) is a hinge joint that allows motion in only one plane. The movements allowed at this uniaxial joint are flexion and extension. Although the fibula and patella are part of the knee area, as shown in Fig. 6-1, *3,* the knee joint is formed only between the condyles of the tibia and femur. Inward and outward rotation at the knee is possible only when the knee is flexed and when not bearing weight.

8. The *hip joint* (Fig. 6-1, *1*) is formed between the femur and the hip bone. This joint is of the ball-and-socket type. The movements possible at the hip joint in the sagittal plane are flexion and extension. The movements possible in the frontal plane are abduction and adduction. The combined movements of flexion, extension, abduction, and adduction are called circumduction (Latin *circum,* around). This movement occurs at the hip joint when the thigh is swung in a circle. The thigh then describes a cone with the apex at the hip joint and the base at the distal end of the thigh. Turning the thigh around its longitudinal axis is called either inward or outward rotation at the hip joint.

Segmental movements of the foot

The longitudinal axis of the foot illustrated in Fig. 6-2 is the only longitudinal axis which is not vertically aligned in the anatomical position.

Therefore, in the anatomical position, it is defined as a horizontal line that divides the foot and that is perpendicular to the extended longitudinal axis of the leg.

The movements of the foot are dorsal and plantar flexion and inversion-eversion.

In *dorsal flexion* the foot and leg segments change their positions relative to each other so the angle formed between them is made to decrease. One or both of the segments may move to accomplish this change. In *plantar flexion* (Fig. 6-2) the angle between the longitudinal axes of the foot and leg segments increases.

The movements of *inversion* and *eversion* are rotations of the foot around its longitudinal axis. Lifting the medial border of the foot is inversion, and lifting the lateral border of the foot is eversion.

Segmental movements of the leg

The movements of the leg can occur at either the ankle joint or the knee joint. At the ankle joint the movements are *plantar* and *dorsal flexion* as defined above for the foot. These movements can occur when the leg or the foot or both are moved, and it is customary to identify which of the segments moves predominantly on the other. The movements of the leg at the knee joint, *flexion* and *extension,* are shown in Fig. 6-3. Flexion results in a decrease in the angle between the longitudinal axes of the leg and thigh, while extension results in an increase in this angle. *Inward* and *outward rotation* of the leg around its longitudinal axis is possible only when the knee is flexed and when not bearing weight.

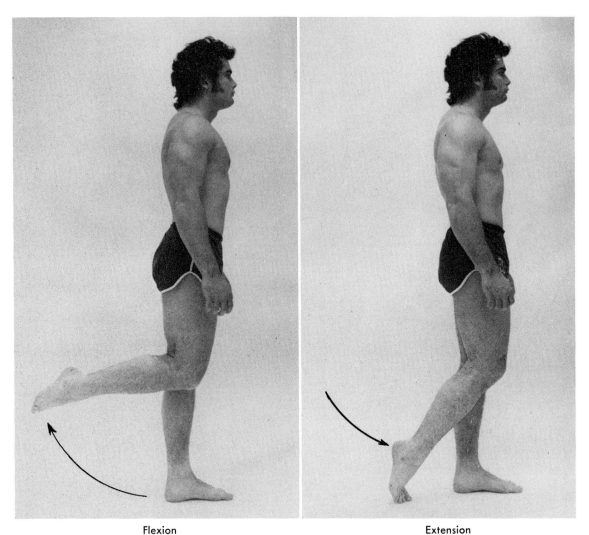

Flexion   Extension

Fig. 6-3. Flexion and extension at the knee joint.

Segmental movements of the thigh

The movements of the leg can occur at either the knee joint or the hip joint.

At the knee joint the movements of *flexion* and *extension* as defined for the leg can occur by moving the thigh on the leg or by a combination of the movements of both segments toward each other.

At the triaxial hip joint the movements of the thigh are *flexion* (swinging the thigh forward), *extension* (swinging the thigh backward to the anatomical position), *hyperextension* (swinging the thigh backward beyond the anatomical position), *abduction* (swinging the thigh laterally), *adduction* (swinging the thigh medially), and *rotation* (turning the thigh around its longitudinal axis) *inward* and *outward*.

Several of these hip joint movements are illustrated in Fig. 6-4.

MOVEMENTS OF THE AXIAL SKELETON

The two major segments of the axial skeleton are the trunk and the head-neck.

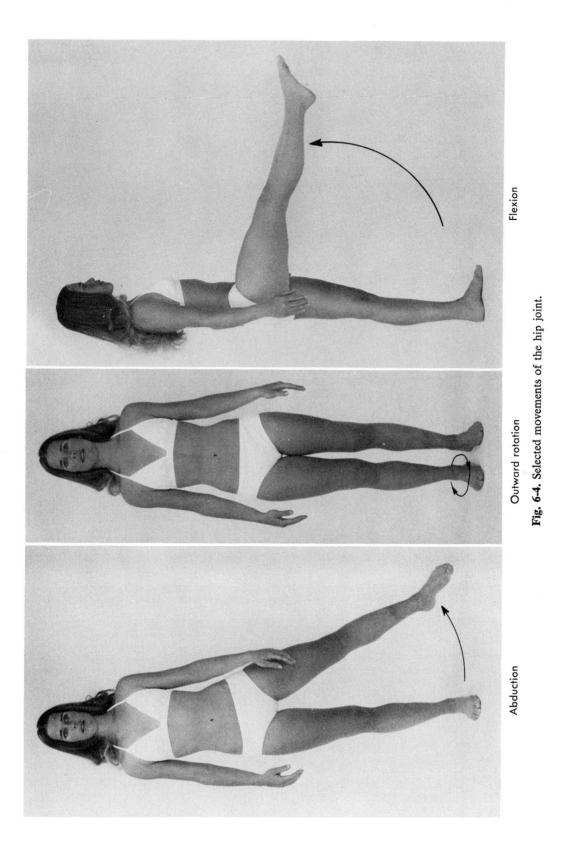

Flexion

Outward rotation

Abduction

Fig. 6-4. Selected movements of the hip joint.

Fig. 6-5. Joints of the axial skeleton. **1,** Joints of the anterior thorax. **2,** Costovertebral joints. **3,** Intervertebral joints. **4,** Atlantoaxial joints. **5,** Atlanto-occipital joint.

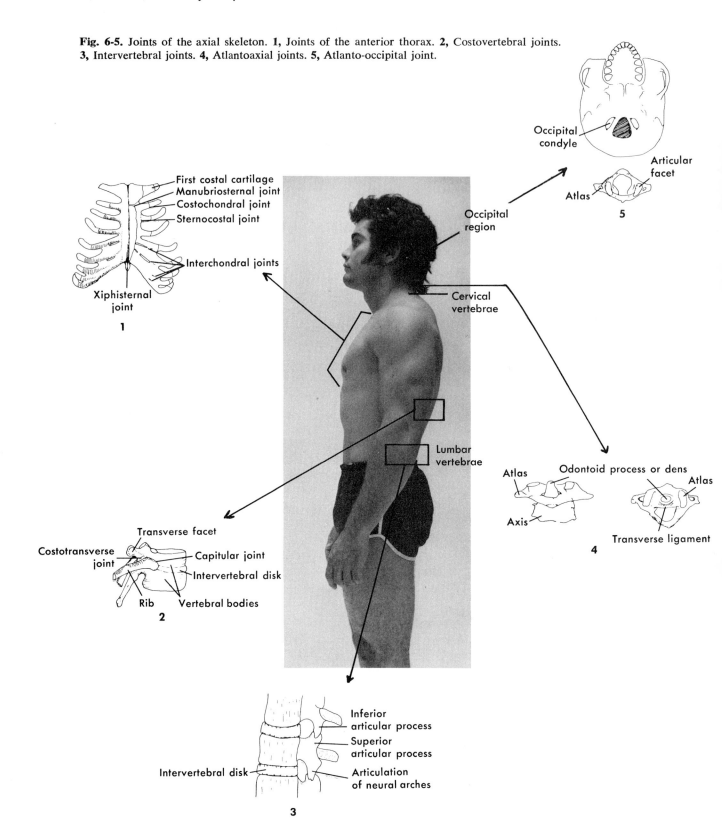

Joints of the axial skeleton

The joints of the axial skeleton include those of the skull, thorax, vertebral column, and pelvis. Since we will not be concerned in this text with the movements of the facial joints or any other specific joints of the skull, or with the thoracic joints which are involved in the act of breathing, our present consideration will be limited to the joints of the vertebral column and of the pelvis.

The joints of the vertebral column, illustrated in Fig. 6-5, can be listed as follows:

Atlanto-occipital articulation (Fig. 6-5, *5*)
Intervertebral articulations
  Atlantoaxial articulations (Fig. 6-5, *4*)
  Articulations of the bodies of the vertebrae (Fig. 6-5, *3*)
  Articulations of the neural arches (Fig. 6-5, *3*)

1. The *atlanto-occipital articulation* (Fig. 6-5, *5*) is a condyloid joint formed by the condyles of the occipital bone and the superior articular processes of the atlas. The movements of the skull permitted at the atlanto-occipital joint are *flexion* (bending forward), *extension* (returning to the anatomical position), *hyperextension* (bending backward from the anatomical position), *abduction* or *lateral flexion* (bending sideward), *adduction* (returning to the anatomical position), and *circumduction*.
2. The *intervertebral articulations* (with the exception of the atlantodental joint) are gliding-type joints. They exist between the vertebral bodies and between the neural arches.
   a. The *atlantoaxial articulations* (Fig. 6-5, *4*) consist of two *lateral atlantoaxial joints* and the atlantodental joint. The lateral atlantoaxial joints are gliding joints formed between the superior articular processes of the axis and the inferior articular processes of the atlas. The *atlantodental joint* is a pivot joint formed by the odontoid process of the axis and the articular facet of the atlas.

   The chief movement between the atlas and the axis is horizontal *rotation*. Owing to the ligamentous attachments and the nature of the articular surface, the head and atlas rotate as one around the dens of the axis. At the same time the inferior articular processes of the atlas glide, one forward and the other backward, on the superior articular processes of the axis.

b. The *articulations of the bodies of the vertebrae* (Fig. 6-5, *3*) are formed by the intervertebral fibrocartilaginous disks located between the bodies of adjacent vertebrae. Separately these synarthrodial joints allow slight movements through compressions of the disks. Collectively they allow the biaxial movements of flexion, extension, and lateral flexion.
   c. The *articulations of the neural arches* (Fig. 6-5, *3*) are gliding joints formed between the articular processes of adjacent vertebrae.

   The movements of the vertebral column as a whole are *flexion, extension, hyperextension, abduction* (or lateral flexion), *adduction* (or return to the anatomical position), and *rotation* (twisting) ı the cervical and thoracic regions.

The two joints of the pelvis are the *symphysis pubis* and the *sacroiliac joint*.
1. The *symphysis pubis* is formed by the pubic bones where these come together and complete the pelvis anteriorly. Very little movement occurs at this joint.
2. The *sacroiliac* is a gliding joint formed by the articulation of the sacrum and the ilium. The only movement allowed at this joint is a small amount of upward and downward, as well as forward and backward, sliding plus some rotation around a transverse axis so the pelvis can be tilted slightly.

   The movements of the pelvis are forward tilt, backward tilt, lateral tilt, and rotation. None of the muscles that perform these movements are, strictly speaking, muscles of the pelvis; rather they are muscles of the hip joint or the trunk. This is true because the movements of the pelvis take place at the lumbar spine, the lumbosacral junction, and the hip joints. *Forward tilt* of the pelvis is associated with hyperextension of the lumbar spine and flexion of the hip joints. *Backward tilt* is associated with flexion of the lumbar spine and extension of the hip joints. *Lateral tilt* is associated with lateral tilt of the lumbar spine. *Rotation* of the pelvis and trunk as a whole occurs around the femoral heads. Thus rotation of the pelvis to the right involves a rotation of the right femur to the left.

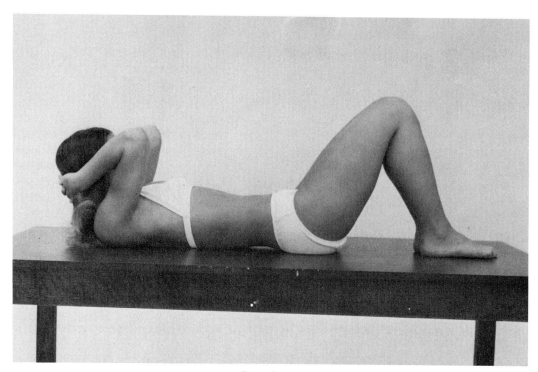

Extension

Flexion

Fig. 6-6. Selected movements of the trunk.

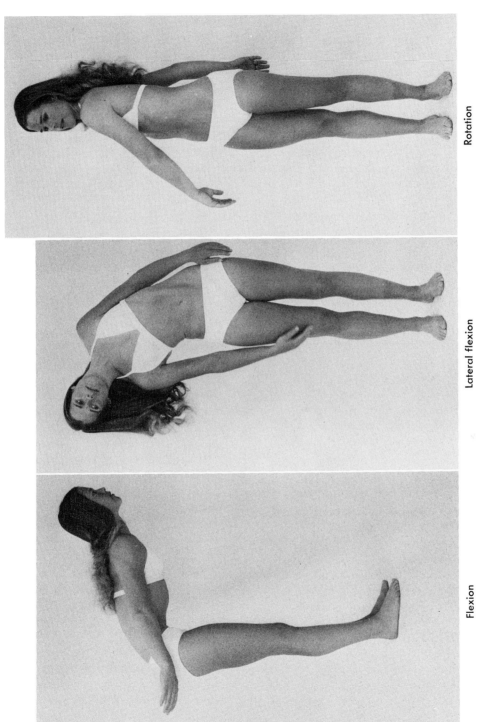

Rotation

Lateral flexion

Flexion

Fig. 6-6, cont'd. Selected movements of the trunk.

Segmental movements of the trunk

The movements of the trunk (Fig. 6-6) can occur at the hip joints and at the trunk's own intervertebral joints.

At the hip joints the pelvis can be *flexed* on the thigh, thereby decreasing the angle between the longitudinal axes of the trunk and thigh as in doing a sit-up exercise. The pelvis can also, of course, be *extended* on the thigh (increasing the trunk-thigh angle) as well as *abducted* (called lateral tilt) and *adducted* (called reduction of lateral tilt). *Rotation* of the pelvis right and left occurs at both the hip joints and the lumbar intervertebral joints (as do all movements of the pelvis).

At the intervertebral joints the movements of the trunk are *flexion* (bending forward), *extension-hyperextension* (straightening up and bending backward), *lateral flexion* (bending sideward), and *rotation* (turning the trunk around its longitudinal axis) *right* and *left*. In the thoracic region all these movements are limited by the presence of the ribs, with rotation being the most free. In the lumbar region there is more freedom of movement than in the thoracic region (with the exception of rotation, which is practically negligible) but less than in the neck or cervical region.

Thus rotation of the trunk is mostly associated with the rotation of the pelvis which occurs at the hip joints.

Segmental movements of the head-neck

The movements of the head-neck segment are the same as those of the trunk occurring at the intervertebral joints; and the definitions given for *flexion, extension-hyperextension, lateral flexion,* and *rotation* of these joints in the trunk also apply to the head-neck. The movements are much freer in the neck, however, than they are in the trunk.

MOVEMENTS OF THE UPPER EXTREMITY

The three major segments of the upper extremity are the arm, forearm, and hand.

Joints of the upper extremity

1. The three joints of the *shoulder girdle* (Fig. 6-7, *1*) are the *sternoclavicular joint,* the

acromioclavicular joint, and the *coracoclavicular union.* The movements of the shoulder girdle occur at the sternoclavicular and acromioclavicular joints. These movements are *elevation* (hunching the shoulder), *depression* (returning from the position of elevation), *abduction* (moving the scapula laterally away from the spinal column), *adduction* (moving the scapula medially toward the spinal column), *upward rotation* (moving the lower end of the scapula laterally), *downward rotation* (moving the lower end of the scapula medially), and *circumduction.* Selected shoulder girdle movements are shown in Fig. 6-8.

2. The *shoulder joint* (Fig. 6-7, *1*) is a ball-and-socket joint formed by the head of the humerus articulating with the scapula. Owing to the looseness of the articulation and mobility of the shoulder girdle, the range of movement in the shoulder joint is greater than in any other joint. The movements possible at the shoulder joint (Fig. 6-9) are *flexion* (swinging forward), *hyperflexion* (continuing forward beyond the vertical), *extension* (swinging backward), *hyperextension* (continuing backward beyond the anatomical position), *abduction* (swinging sideward), *adduction* (returning from abduction), *outward* and *inward rotation* (rotating around the longitudinal axis of the humerus), *horizontal flexion-adduction* (moving the abducted arm forward and sideward in the transverse plane), *horizontal extension-abduction* (moving the flexed humerus backward and sideward in the transverse plane). Swinging the arms in a circle is called *circumduction.*

3. The *elbow joint* (Fig. 6-7, *2*) is a hinge joint formed between the humerus and the radius and ulna. The movements possible at this joint are only flexion and extension. *Flexion* is the movement of the anterior forearm toward the anterior surface of the arm. *Extension* is the reverse of flexion.

4. The *radioulnar joints* (Fig. 6-7, *3*) include the superior and inferior articulations of the radius and ulna. The pivot joints formed at these locations allow only the movements of pronation and supination. *Pronation* involves inwardly rotating the radius over the ulna.

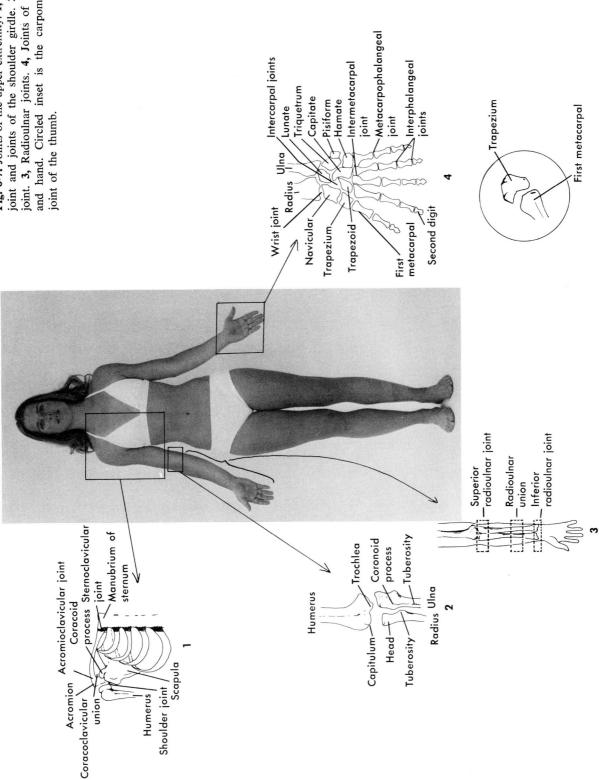

Fig. 6-7. Joints of the upper extremity. **1,** Shoulder joint and joints of the shoulder girdle. **2,** Elbow joint. **3,** Radioulnar joints. **4,** Joints of the wrist and hand. Circled inset is the carpometacarpal joint of the thumb.

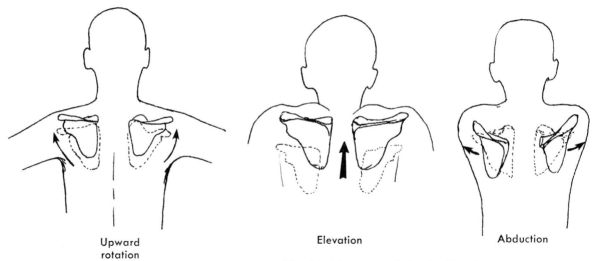

Upward
rotation

Elevation

Abduction

Fig. 6-8. Movement of the shoulder girdle.

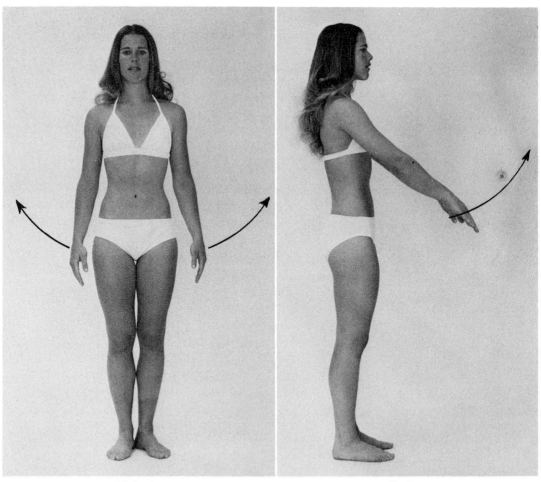

Abduction

Flexion

Fig. 6-9. Selected movements at the shoulder joint.

Supination involves outwardly rotating the radius.

5. The *radiocarpal* or *wrist joint* (Fig. 6-7, *4*) is a condyloid joint formed by the distal end of the radius and articular disc superiorly and the navicular, lunate, and triquetral bones inferiorly. The movements of the wrist joint are *flexion* (decreasing the hand-forearm angle), *extension* (reverse of flexion), *hyperextension, abduction* or *radial flexion* (bending the hand laterally), *adduction* or *ulnar flexion* (bending the hand medially), and *circumduction.*

6. The *intercarpal joints* (Fig. 6-7, *4*) are gliding joints located between the individual carpal bones. The gliding movements possible at these joints complement those of the wrist joint.

7. The *carpometacarpal joints* (Fig. 6-7, *4*) are the articulations between the bases of the metacarpals and the distal row of carpal bones. The medial four articulations are gliding joints, whereas the articulation of the thumb (circled inset in Fig. 6-7, *4*) is a saddle joint. The medial four articulations permit only slight flexion and extension, which supplement the movements of the wrist. The saddle joint at the thumb (Fig. 6-10) allows *abduction* (moving forward), *adduction* (movement backward), *extension* (moving laterally), *flexion* (moving medially), and *opposition* (touching the thumb tip to a finger tip).

8. The *intermetacarpal joints* (Fig. 6-7, *4*) are gliding joints located between the adjacent sides of the proximal ends of the medial four metacarpal bones. These gliding joints permit only slight amounts of flexion and extension, which supplement the movements of the wrist.

9. The *metacarpophalangeal joints* (Fig. 6-7, *4*)

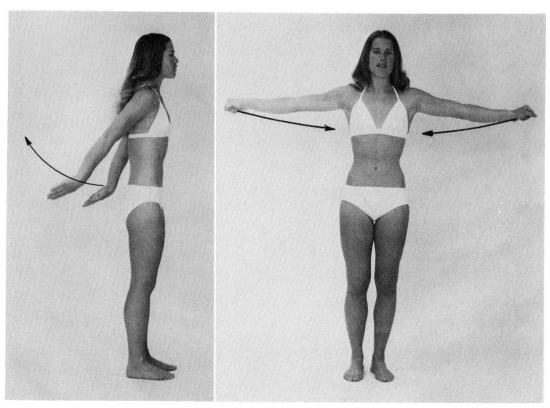

Hyperextension

Horizontal
flexion-adduction

Fig. 6-9, cont'd. Selected movements at the shoulder joint.   *Continued.*

Outward rotation

Fig. 6-9, cont'd. Selected movements at the shoulder joint.

are formed between the metacarpal bones and the proximal phalanges. The medial four articulations are condyloid joints, whereas the articulation of the thumb is a hinge joint.

10. The *interphalangeal joints* (Fig. 6-7, *4*) are the hinge joints formed between the proximal and middle phalanges and between the middle and distal phalanges. *Flexion* (curling the fingers) and *extension* (straightening the fingers) occur at both the metacarpophalangeal and interphalangeal joints. *Abduction* and *adduction* of the fingers occur only at the metacarpophalangeal joints.

Segmental movements of the arm

The movements of the arm can occur at either the shoulder joint or the elbow joint.

At the triaxial shoulder joint the movements of the arm are *flexion, extension-hyperextension, abduction, adduction, horizontal flexion-adduction, horizontal extension-abduction, rotation inward* and *outward,* and *circumduction* as already defined.

At the hinged elbow joint the movements of the arm are *flexion* (decreasing the angle between the longitudinal axis of the arm and forearm) and *extension* (increasing the arm-forearm angle).

The movements of the shoulder joint are ordinarily accompanied by movements of the shoulder girdle. The movements of the shoulder girdle increase not only the force of arm movements but also, by tilting the scapula in the desired direction, the range of arm movements. As the arm is abducted, for instance, there is an accompanying *upward rotation* of the scapula (more marked as the arm reaches and passes the horizontal plane). Likewise bringing the arm back to the side (adduction) is accompanied by *downward rotation* of the scapula. When the arms are used in front of the body, the scapula is *abducted;* when the arms are moved backward, the scapula is *adducted.*

Segmental movements of the forearm

The movements of the forearm occur at the elbow, radioulnar, and wrist joints. At the elbow joint the angle between the longitudinal axes of the forearm and arm can be either increased *(extension)* or decreased *(flexion).* At the radioulnar joints the forearm can be rotated around its longitudinal axis either inward *(pronation)* or outward *(supination).* At the wrist joint the forearm can be flexed, extended, abducted, and adducted on the hand, as in trying to maintain a handstand. These movements are defined the same as are the movements of the hand occurring at the wrist.

Segmental movements of the hand

The movements of the hand occur mainly at the wrist joint. The movements of the hand occurring at the wrist joint are *flexion* (decreasing the angle between the longitudinal axes of the hand and forearm), *extension* (increasing the hand-forearm angle), *hyperextension* (moving back beyond the anatomical position), *abduction* or *radial flexion* (moving laterally), and *adduction* or *ulnar flexion* (moving medially). The com-

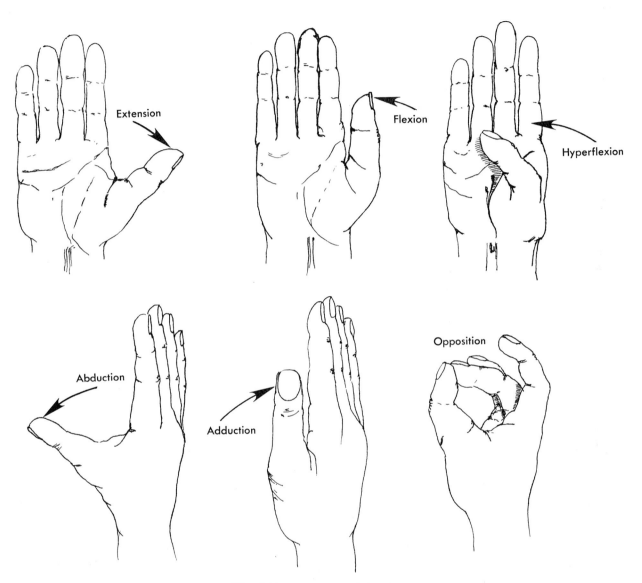

Fig. 6-10. Movements of the thumb.

bined movement of flexion, extension, abduction, and adduction is called *circumduction*.

STUDY GUIDELINES
Movements of the lower extremity
Specific student objectives

1. Identify and describe the joints of the lower extremity.
2. Describe the movements that are possible at each joint of the lower extremity.
3. Describe the segmental movements of the foot.
4. Describe the segmental movements of the leg.
5. Describe the segmental movements of the thigh.

Summary

The joints and segmental movements of the lower extremity (foot, leg, and thigh) are outlined as follows:

1. Interphalangeal—between the phalanges of the toes; hinge joints; uniaxial
   a. Flexion—curling the toes
   b. Extension—straightening the toes
2. Metatarsophalangeal—between the distal ends of the metatarsals and proximal phalanges; condyloid joints; biaxial
   a. Flexion and extension
   b. Abduction—spreading the toes
   c. Adduction—bringing the toes back together
3. Tarsometatarsal—between the tarsal and metatarsal bones; gliding joints; nonaxial
   a. Linear motion only
4. Intertarsal—between the seven tarsals; gliding joints; nonaxial
   a. Eversion—lifting the lateral border of the foot
   b. Inversion—lifting the medial border of the foot
5. Ankle—tibia and fibula fitting over the talus; hinge joint; uniaxial
   a. Flexion—dorsal flexion, lifting the foot up
   b. Extension—plantar flexion, putting the foot down
6. Tibiofibular—between the tibia and fibula at three locations; gliding joints; nonaxial
   a. Inferior and superior joints
   b. Tibiofibular union—joining of the bone shafts by an interosseous membrane
7. Knee—between the tibia and femur; hinge joint; uniaxial
   a. Flexion—bending the knee
   b. Inward-outward rotation (possible only in the flexed and non–weight-bearing knee)
8. Hip—between the femur and hip bone; ball-and-socket joint; triaxial
   a. Flexion-extension
   b. Abduction-adduction
   c. Circumduction
   d. Inward-outward rotation

Performance tasks

1. The joints of the lower extremity are motion sites for segmental movements of the
   a. arm, forearm, and hand
   b. foot, leg, and thigh
   c. trunk and head-neck
2. Using a mounted skeleton or pictorial diagram, identify the joints of the lower extremity.
3. On a mounted skeleton or your own body, carry out the following segmental movements:
   a. The foot
      (1) Flexion and extension of the interphalangeal joints
      (2) Abduction and adduction of the metatarsophalangeal joints
      (3) Inversion and eversion of the foot
   b. The ankle joint
      (1) Plantar and dorsal flexion of the ankle joint
   c. The knee joint
      (1) Flexion and extension
      (2) Inward and outward rotation of the flexed knee
   d. The hip joint
      (1) Flexion, extension, and hyperextension
      (2) Abduction and adduction
      (3) Inward and outward rotation
      (4) Circumduction

Movements of the axial skeleton
Specific student objectives

1. Describe the joints of the axial skeleton.
2. Describe the segmental movements of the trunk.
3. Describe the segmental movements of the head-neck.

Summary

The joints and segmental movements of the axial skeleton are outlined as follows:

1. Skull
2. Thorax—involved in breathing
3. Vertebral column
   a. Atlanto-occipital—between the occipital bone of the skull and the atlas; condyloid joint; biaxial
      (1) Flexion—bending forward
      (2) Extension—returning to the anatomical position
      (3) Hyperextension—bending backward
      (4) Lateral flexion—bending sideward
      (5) Adduction—returning to the anatomical position
      (6) Circumduction
   b. Atlantoaxial—between the atlas and the axis; gliding and pivoting joints; nonaxial and uniaxial
      (1) Gliding (two lateral atlantoaxial)
      (2) Rotating (atlantodental)
   c. Intervertebral—between the bodies of the vertebrae and between the neural arches
      (1) Flexion, extension, hyperextension, and rotation
      (2) Gliding (sacroiliac joint)
4. Pelvis
   a. Forward and backward tilt
   b. Lateral tilt
   c. Rotation

Performance tasks

1. The joints of the axial skeleton are motion sites for segmental movements of the
   a. Arm, forearm, and hand
   b. Foot, leg, and thigh
   c. Trunk and head-neck
2. Using a mounted skeleton or pictorial diagram, diagram, identify the joints of the axial skeleton.
3. On a mounted skeleton or your own body, carry out the following segmental movements:
   a. Vertebral column
      (1) Flexion, extension, and hyperextension
      (2) Lateral flexion and adduction
      (3) Rotation and circumduction
   b. Pelvis
      (1) Flexion—forward tilt
      (2) Extension—backward tilt
      (3) Lateral tilt
      (4) Rotation—lateral twist
      (5) Circumduction

Movements of the upper extremity
Specific student objectives

1. Identify and describe the joints of the upper extremity.
2. Describe the movements that are possible at each joint of the upper extremity.
3. Describe the segmental movements of the arm.
4. Describe the segmental movements of the forearm.
5. Describe the segmental movement of the hand.

Summary

The joints and segmental movements of the upper extremity (arm, forearm, and hand) are outlined as follows:

1. Shoulder girdle
   a. Sternoclavicular joint
   b. Acromioclavicular joint
   c. Coracoclavicular union
   d. Movements of the shoulder girdle—elevation (hunching the shoulders), depression (returning from elevation), abduction (moving the scapula laterally from the spinal column), adduction (moving the scapula medially), upward rotation (moving the inferior end of the scapula laterally), downward rotation (moving the inferior end of the scapula medially), and circumduction.
2. Shoulder joint—between the humerus and the scapula; ball-and-socket type; triaxial
   a. Flexion-hyperflexion and extension-hyperextension
   b. Abduction-adduction
   c. Inward and outward rotation
   d. Horizontal flexion and extension
   e. Circumduction
3. Elbow joint—between the humerus and the radius and ulna; hinge type; uniaxial
   a. Flexion-extension
4. Radioulnar joints—formed by the superior and inferior articulations of the radius and ulna; pivot type; uniaxial
   a. Pronation—inward rotation of radius over ulna

b. Supination—outward rotation
5. Radiocarpal (wrist) joint; condyloid type; biaxial
   a. Flexion, extension, and hyperextension
   b. Radial flexion (abduction) and ulnar flexion (adduction)
   c. Circumduction
6. Intercarpal joints—between the carpal bones; gliding type; nonaxial
   a. Complement all movements of the wrist joint
7. Carpometacarpal joints—between the bases of the metacarpals and the distal row of the carpals
   a. Medial four (gliding joints that allow only slight flexion and extension)
   b. Thumb (saddle joint; biaxial)
      (1) Abduction (forward) and adduction (backward)
      (2) Flexion (medial) and extension (lateral)
      (3) Opposition—touching the thumb to the fingertip
8. Intermetacarpal joints—between the adjacent sides of the proximal ends of the medial four metacarpals; gliding type; nonaxial
   a. Slight flexion-extension
9. Metacarpophalangeal—between the metacarpals and the proximal phalanges
   a. Medial four (condyloid joints)
   b. Thumb (hinge joint)
   c. Flexion and extension of all joints; abduction and adduction of the medial four
10. Interphalangeal joints—hinge type; uniaxial
    a. Flexion-extension

Performance tasks

1. The joints of the upper extremity are motion sites for segmental movements of the
   a. arm, forearm, and hand
   b. foot, leg, and thigh
   c. trunk and head-neck
2. Using a mounted skeleton or pictorial diagram, identify the joints of the upper extremity.
3. The summaries of the three regions of the body can be utilized as an outline of the joints and segmental movements of the lower extremity, axial skeleton, and upper extremity. With the aid of these summaries, locate each joint on a mounted skeleton and do the following:
   a. Name and classify the joint
      (1) Name (e.g., the shoulder joint)
      (2) Classify (gliding, hinge, pivot, condyloid, saddle, or ball-and-socket)
   b. Determine the axes of motion
      (1) Nonaxial, uniaxial, biaxial, or triaxial
   c. Determine the type of movement allowed
      (1) Linear or angular
   d. Identify the specific actions allowed at the joint
      (1) Flexion, extension, hyperextension
      (2) Abduction, adduction
      (3) Rotation
      (4) Circumduction
   e. Identify the specific plane and axes of motion for each joint action
      (1) Sagittal, frontal, transverse, or diagonal plane
      (2) Frontal, longitudinal, sagittal, or diagonal axis
4. Perform specific joint actions yourself and
   a. name and classify the joint
   b. determine the type of motion (linear or angular)
   c. identify the joint action
   d. identify the plane and axis of motion
5. Select a specific motor task such as walking, throwing, jumping, catching, and pushups. Carry out the nominal analysis as listed in the preceding task.

SELF-EVALUATION

Students should use no reference materials for this progress test, and they can check their answers by referring to Appendix A.
1. The bones of the lower extremity include the
   a. trunk and head-neck
   b. foot, leg, and thigh
   c. both a and b
   d. neither a nor b
2. Complete the descriptive outline of segmental movements of the lower extremity:
   a. Interphalangeal joints (nine)
      (1)_____—curling the toes
      (2)_____—straightening the toes
   b. Metatarsophalangeal joints
      (1) Flexion and extension
      (2)_____—bringing the toes together

(3)_____—spreading the toes

c._____joints—between the tarsal and the metatarsal bones

  (1) Linear motion only

d. Intertarsal joints

  (1)_____—lifting the lateral border of the foot

  (2)_____—lifting the medial border of the foot

e._____joint—tibia and fibula articulating with the talus

  (1)_____flexion—lifting the foot up

  (2)_____flexion—putting the foot down

f._____joint—between the tibia and the femur

  (1)_____—bending the joint

  (2)_____—straightening the joint

  (3)_____and _____rotation

g. Hip joint—between the femur and the hip bone

  (1) Flexion-extension

  (2)_____—past extension

  (3) Abduction-adduction

  (4) Circumduction

  (5) Inward and outward _____

3. List the segmental movements of the following:

  a. Foot _____

  b. Leg _____

  c. Thigh _____

4. Complete the descriptive outline of segmental movements of the axial skeleton:

  a. Skull

  b. Thorax—involved in _____

  c. Vertebral column

    (1) Atlanto-occipital joint—between the occipital bone of the skull and the atlas

      (a) _____—bending forward

      (b) Extension— _____

      (c) _____—bending backward

      (d) _____—bending sideward

      (e) _____—returning from sideward

      (f) Circumduction _____

    (2) Atlantoaxial articulations

      (a) _____—motion around the longitudinal axis

    (3) Intervertebral—between the vertebral bodies and between the neural arches

      (a) _____ - _____

      (b) Hyperextension

      (c) _____ - _____

      (d) Rotation

  d. Pelvis

    (1) Flexion—_____

    (2) Extension—_____

    (3) Lateral tilt—_____

    (4) Rotation—twist left or right

5. The bones of the upper extremity are in the areas of the _____, _____, and _____.

6. Complete the descriptive outline of segmental movements of the upper extremity:

  a. Shoulder girdle

    (1)_____—hunching the shoulders

    (2)_____—returning from the hunched position

    (3)_____—moving the scapula laterally away from the spinal column

    (4)_____ —moving the scapula medially toward the spinal column

    (5)_____—moving the inferior end of the scapula laterally

    (6)_____—moving the inferior end of the scapula medially

    (7) Circumduction

  b. Shoulder joint—between the _____and the _____

    (1) Flexion-extension

    (2) Hyperextension

    (3) Inward and outward rotation

    (4) Horizontal _____and _____

    (5) _____ —combination of the above (excluding rotation)

  c. _____joint—between the humerus and the radius and ulna

    (1) _____ - _____

  d. Radioulnar joints—formed by the superior and inferior articulations of the radius and ulna

    (1) _____—inward rotation of the radius over the ulna

    (2) _____—outward rotation

  e. Radiocarpal or _____ joint

    (1) Flexion-extension

    (2) Hyperextension

    (3) Radial flexion or _____

    (4) Ulnar flexion or _____

f. Intercarpal joints—gliding
g. Carpometacarpal—between the bases of the _____ and the distal row of the _____
  (1) Medial four—slight flexion and extension
  (2) Thumb
    (a) Abduction-adduction
    (b) Flexion-extension
    (c) _____—touching the thumb to the fingertip

h. Intermetacarpal—Slight flexion and extension
i. _____—between the metacarpals and the proximal phalanges
  (1) _____-_____
  (2) Hyperextension
  (3) Circumduction
  (4) Abduction- _____
  (5) Thumb only—allows _____ and _____
j. Interphalangeal joints
  (1) _____ and _____

Kinematic analysis of fundamental quantities

LESSON 7

Measurement of motion quantities

FUNDAMENTAL QUANTITIES OF MOTION

The three measurable mechanical quantities in human motion are the body in motion (matter), the distance or angle through which the body travels (space), and the duration of the motion (time). The definitions of these three quantities are as follows:

time Period during which an action or process continues, i.e., a measurable duration

space Area or volume occupied by a body and the distance· or angle through which it travels during movement; *area* of the body is determined by multiplying its length by its width; *volume* determined by multiplying its area by its height

matter Anything that has mass and that occupies space; i.e., the substance of the physical universe, which with energy forms the basis of all objective phenomena (the amount of matter in a body is quantified by either its weight or its mass)

DESCRIPTION OF MOTION QUANTITIES

The process of quantification is defined simply as the assignment of numbers and units to objects according to rules. When we count or measure a quantity, we single out some property such as length or weight for comparison with a standard. Every quantification requires two things: a number and a unit. In expressing the amount of something, the unit is just as essential as the number. For example, the number three is meaningless without the specification of a unit, e.g., 3 feet, 3 inches, 3 seconds, 3 gallons, or 3 dogs.

Although there are a variety of different units used in the quantification of continuous variables, each can be expressed in terms of not more than three special units. These units, called *fundamental* units, are the units of weight (matter), space, and time. Examples of these fundamental units would be pounds, feet, and seconds, respectively. All other units are called *derived* units since, as we shall see later, they can always be written as some combination of the three fundamental units. Examples of derived units are miles per hour, cubic inches, and feet per second per second (ft/sec/sec). Within the context of this discussion, we can now define measurement as the process of quantification requiring the use of fundamental units. The alternative form of quantification is, of course, the process of counting.

Since ancient times, people have developed methods of describing the three fundamental units of time, space, and weight. Today, however, there are only two basic systems still in existence: the

metric, which is the system used throughout most of the world, and the *British engineering system,* which is the system currently used in the United States. The fundamental units of these systems are presented in Table 7-1.

Description of space

The *linear* distance a moving body travels can be determined through the use of such instruments as a ruler, caliper, tape measure, or measuring wheel that is calibrated in either metric or British units. Motion pictures, as will be discussed later, can also be used as a medium for the measurement of linear displacement.

Table 7-1. Metric and British engineering system units for the fundamental quantities of length, weight, and time

Length			
Metric		**British**	
Unit	**Part of standard**	**Unit**	**Part of standard**
Myriameter	10,000	Mile	5,280
Kilometer (km)	1,000	Yard (yd)	3
Hectometer	100	Foot (ft)	1
Decometer	10	Inch (in)	$\frac{1}{12}$
Meter (m)	1		
Decimeter	1/10		
Centimeter (cm)	1/100		
Millimeter (mm)	1/1,000		
Micron	1/1,000,000		

Weight			
Metric		**British**	
Unit	**Part of a gram**	**Unit**	**Part of standard**
Ton (T)	1,000,000	Ton (T)	2,000
Quintal	100,000	Hundredweight	100
Myriogram	10,000	Quarter	25
Kilogram (kg)	1,000	Pound (lb)	1
Hectogram	100	Ounce (oz)	1/16
Decagram	10	Dram (dr)	1/256
Gram (g)	1	Grain	1/7,000
Decigram	1/10		
Centigram (cg)	1/100		
Milligram (mg)	1/1,000		
Microgram	1/1,000,000		

Time	
Metric and British	
Unit	**Part of standard**
Second	1
Minute	60
Hour	3,600
Day	86,400

The metric system uses the standard meter as the unit of distance or length. The meter was originally defined as one ten-millionth of the distance from the equator to the North Pole. In the United States the "foot" is used as the basic standard of length and is defined as 1200/3937 of a standard meter. Since the foot is divided into 12 inches, 1 meter equals 39.37 inches.

The meter is divided into 100 equal parts, each of which is called a centimeter. The centimeter is further divided into ten equal parts, each of which is called a millimeter. On a larger scale, 1000 meters equals 1 kilometer and 1 mile equals 5280 feet.

Angular positions and displacements are measured through the use of such devices as magnetic compasses, protractors, goniometers, and flexometers and by cinematography techniques. In both the metric and the British systems angular space is measured in terms of the *degrees* of a circle or, as we shall see later in this lesson, in *radians.* Angular motion is measured by the angle turned through.

Description of matter

The amount of matter in a body at a given location on the earth is given by its weight. Since the weight of a body is actually a quantification of the gravitational forces acting upon it and since the magnitude of gravitational forces varies with location in the universe, matter is often quantified in terms of mass—defined as weight per unit of gravitational acceleration (m = w/g). Thus weight is a fundamental quantity while mass is a derived quantity. The equation m = w/g shows that mass and weight are directly proportional to each other. In other words, when we double the weight of a body we also double its mass. Thus, if we standardize the location of a body on the surface of the earth and thereby standardize the magnitude of the gravitational acceleration (g), we can use weight as the quantification of matter. The weight of an object can, of course, be determined through the use of a variety of measuring devices, the most usual being the common bathroom scale.

The standard unit of weight in the metric system is the *kilogram,* which can be divided into 1000 equal parts called grams; i.e., 1 kilogram equals 1000 grams. The standard *pound,* the unit

of weight in the British system, is defined in terms of the standard kilogram by the relation 1 pound equals 0.4536 kilogram. We also know that 1 oz = 28.35 gm, 1 lb = 16 oz, and 1 ton = 2000 lb.

All forces in the British system are measured in terms of gravitational units; that is, the units in which forces are described are the units of weight which represent the force of gravity—the force of attraction exerted by the earth on every object in the earth's atmosphere.

The force with which the earth attracts a body, at some specified point on the earth's surface where a freely falling body is given an acceleration of 32.1740 ft/sec/sec, is a perfectly definite reproducible force and is called a force of *one pound* in the British system.

In order that an unknown force can be compared with the force unit and thereby be measured, some measurable effect produced by the force must be used. One such effect is to alter the dimensions of the body on which the force is exerted, while another is to alter the state of motion of the body. Both these effects are used to measure the forces affecting human motion. For example, the dynamometer measure of isometric muscle strength (Fig. 7-1, *A*) deals with the deformation produced in an instrument by the muscular contractions of an individual. A measure of isotonic muscle strength, weight lifting (Fig. 7-1, *B*), deals with the ability of the individual to impart motion to an object. In an isometric muscle contraction the propulsive force of the contraction is not sufficient to overcome the resistive forces. In an isotonic muscle contraction, however, the propulsive force is sufficient to overcome the resistance and thus impart motion (to a body). Both isometric and isotonic measures of strength, however, are expressed in terms of the force unit, i.e., in pounds or kilograms.

Description of time

The fundamental unit of time in both the metric and the British systems is the *second,* which is defined as one 86,400th of a mean solar day. Time is expressed in seconds, minutes, and hours. The stopwatch is, of course, the standard timing device used in kinesiology. Motion pictures, as will be discussed later, and various modern elec-

Fig. 7-1. Examples of isometric **(A)** and isotonic **(B)** strength tests.

tronic devices (e.g., the digital stopwatch shown in Fig. 7-2) can also be used to measure time.

The two metric systems

The two metric systems of units are identified as the *meter-kilogram-second* (mks) system and the *centimeter-gram-second* (cgs) system. Each of these systems uses different names for derived units, as will be demonstrated in later lessons of this text. In Lesson 18, for instance, we will see that the two systems even use different names for the force unit: the force unit is called a *newton* in the mks system, and a *dyne* (pronounced "dine") in the cgs system.

SCALAR AND VECTOR QUANTITIES

In the study of motion quantities it is important to distinguish between scalar and vector quantities because, as we shall see in the next lesson,

they obey different rules of arithmetic. A scalar quantity can be specified by stating only its *magnitude*. For example, time is a scalar quantity. A quantity that has a *direction* as well as a *magnitude* is called a vector quantity. Force and displacement are the two fundamental quantities usually expressed as vectors.

Graphical representation of vectors

The graphical description of vectors can be illustrated by using force as an example. The three qualities of a force which must be accurately described in the study of human motion are its magnitude, its direction, and the point at which it is applied in a body. All three of these qualities can be graphically represented.

The graphical representation of a force is called a vector. A vector consists of an arrow drawn from a dot on a sheet of paper. The dot

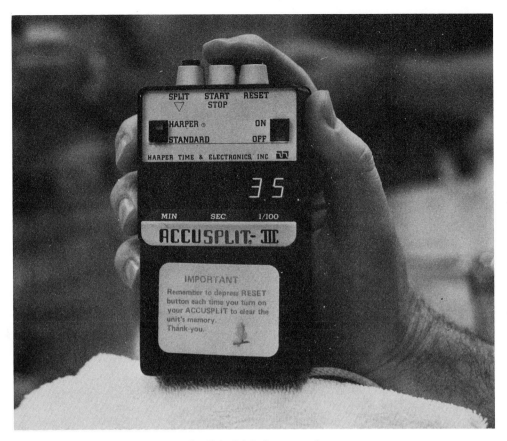

Fig. 7-2. Digital stopwatch.

indicates the point of application of the force. The length of the arrow, drawn to some specified scale, indicates the magnitude of the force. The head of the arrow is used to indicate the direction of the force. As an example of the methods involved in the graphical description of forces and other vectors, we will consider body weight as both a force and a vector.

The instrument most commonly used to measure body weight is the balance scale (Fig. 7-3, *A*). Most people are aware from an early age that when they weigh themselves they are actually measuring the downward force they exert on the footboard of the scale and that this force causes some mechanism within the scale to indicate their weight. We are not concerned here with how the scale works but rather with the downward force we call our *weight*.

As illustrated in Fig. 7-3, *B,* gravitational forces always act in the direction of a line joining the body with the center of the earth. The forces thus act, for all practical purposes, perpendicular to the earth's surface. The direction in which a force acts is called its *line of action* or, with special reference to gravity, its *line of gravity* or *gravitational line.* The line of gravity is actually an imaginary vertical line connecting the center of gravity of a body with the center of the earth.

Every object on the surface of the earth has a *center of gravity,* defined as an imaginary point representing the weight center of the body. The center of gravity can be located within or outside the body, depending on the body's configuration and position. At this point all the parts of the body exactly balance each other. Since the center of gravity can be considered the point in the body at which its weight is concentrated, the line of action of gravity can be represented by an

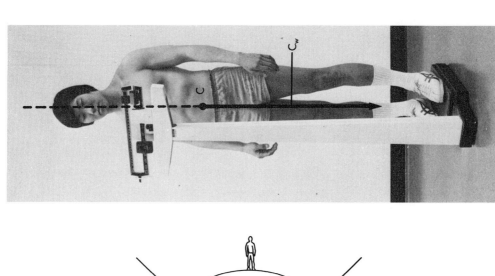

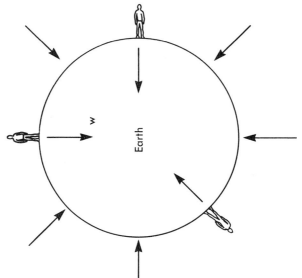

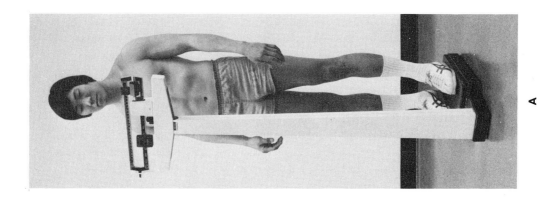

Fig. 7-3. Graphical representation of weight as a vector quantity. The magnitude of the weight is determined through measurement **(A)**. The direction of the weight is toward the center of the earth **(B)**. The two dimensions of weight, magnitude and direction, can be represented by an arrow drawn downward from the center of gravity **(C)**.

imaginary line passing through this point and is perpendicular to the surface of the earth.

Fig. 7-3, *C,* can be used to illustrate the above concepts. The point *C* represents the center of gravity of the man and the line *Cw* represents the line of gravity drawn as a vector to represent the weight of the man. We can assume that the weight of the man has been determined to be 150 lb. Therefore if we choose a scale whereby 1 inch equals 75 lb, the length of line *Cw* will be 2 inches and the line will originate from the center of gravity as shown in the illustration.

Also, as shown in the illustration, the line of gravity can be extended indefinitely in both directions. Thus we see the difference between the force vector and the gravitational line. It should now be evident that all the necessary information about a force or vector can be conveyed by conveniently using an arrow. The length of the arrow, drawn to some chosen scale, indicates the size or magnitude of the vector, and the direction in which the arrow points indicates the direction in which the vector is acting.

INTRODUCTION TO THE SPATIAL ANALYSIS OF HUMAN MOTION

The quantitative analysis of human motion usually begins with a consideration of its spatial and temporal dimensions. Methods of determining the spatial dimensions of a motion are the topic of this section, while the methods of determining its temporal dimensions will be taken up in the following section. The key concept in the spatial analysis of human motion is *displacement,* which is the change in position of a body.

Spatial components of linear displacement

The two measurable components of linear displacement are distance and direction. In other words, the linear displacement of a body occurs not only over a certain distance but also in a certain direction. Thus linear displacement is a vector quantity (i.e., it has both magnitude and direction).

The distance any object moves can be measured, as already mentioned, by yardsticks, tape measures, measuring wheels, and so forth. The direction of the movement can be determined through the use of a magnetic compass, a protractor, or some other angle-measuring device. Thus direction is described in terms of the degrees of a circle. The direction in which a vector is acting is called its *line of action.*

Perhaps the difference between displacement, a vector quantity, and the total distance traveled, a scalar quantity, can be clarified through the use of the following hypothetical experiment: Suppose we have a sheet of graph paper on the table

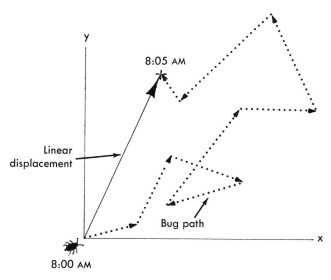

Fig. 7-4. Example of linear displacement.

in front of us and we notice a bug randomly crawling upon it. Let us further suppose that for some strange reason we decide to trace the path which the bug follows for 5 minutes starting at 8:00 AM. The path we trace might be similar to the path shown in Fig. 7-4. In this illustration the point of beginning is indicated by the starting time of 8:00 AM and the point of termination by the time 8:05 AM. The linear displacement of the bug during the 5 minutes is given by a straight line drawn to connect the point of beginning with the point of termination. The displacement is measured and found to be 2.5 inches at an angle of 25° from the reference line. The reference line, as can be seen in the illustration, is the y axis, which passes through the point of beginning. The total distance the bug travels, however, is given by the length of the random path.

Measurement of linear displacement

The measurement of linear displacement through the use of mechanical and electronic de-vices is a fairly straightforward procedure. The chief cause of error in these situations is usually carelessness. Thus if proper care is taken, errors of measurement can be minimized.

The measurement of linear displacement through the use of cinematography, however, does require some special considerations to ensure accuracy. This is due to the fact that the projected images on the screen are usually not true life size. Since projected images are generally smaller than in real life, all errors of measurement are magni-fied. A 1 mm error might equal a 0.1% error in real distance but be as much as a 10% error when obtained from filmed images. If an object of known length is photographed in the plane of the motion, then this length can be measured and used to convert the linear dimensions obtained from the film to real-life size. For example, if a 3-foot rod measures 1 foot when projected, then the *multiplier* that must be used to obtain life-size dimensions is 3. Each foot on the screen is therefore equivalent to 3 feet of real distance.

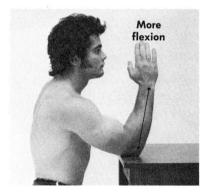

A   Final position

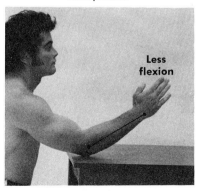

Starting position

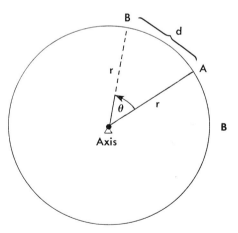

Fig. 7-5. Angular displacement of the forearm at the elbow joint.

The multiplier used in this situation is the ratio of the true length to the projected length:

$$\frac{\text{True length}}{\text{Projected length}} = \frac{3 \text{ ft}}{1 \text{ ft}} = 3 = \text{Multiplier}$$

$$\text{(Equation 7-A)}$$

Spatial components of angular motion

The angular displacement of the forearm lever at the elbow joint is shown in Fig. 7-5, *A*. In this illustration the lever is moved by muscles from position *A* to position *B*. As shown in the analog drawing (Fig. 7-5, *B*), the length of the lever can be considered the radius of a circle. This illustration also shows that the difference between positions *A* and *B*, which is the angular displacement of the lever, can be measured by the angle turned through (symbolized by the Greek letter theta, θ). The actual distance (*d*) the end of the lever travels is given by the arc of the circle cut by the angle.

The four spatial components of angular motion therefore are direction, radius, angle turned through, and arc of the circle cut by the angle. These four components are illustrated in Fig. 7-5, *B*.

The direction of angular displacement is given by specifying the axes of the motion and by indicating whether the motion is clockwise or counterclockwise, as we will discuss in Lesson 9. The motion shown in Fig. 7-5, *A,* is elbow flexion, which is (in this illustration) a counterclockwise rotation around the x axis.

In the formulation of mechanical laws, it has been found necessary to express the angle turned through (θ) in radians—not in degrees or revolutions, in which angles are usually measured. The radian is a unit of angular measure just as the inch is a unit of linear measure. Actually the radian is an arbitrary unit, as will be explained in the following discussion.

Definition of a radian

A radian is defined as the ratio of the arc length (*d*) to the length of the radius (*r*) for any angle cutting a circle. Therefore the angle turned through (θ) in radians can be obtained by the following formula:

$$\text{Angular displacement} = \theta = d/r$$

$$\text{(Equation 7-B)}$$

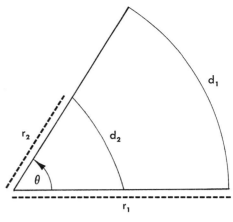

Fig. 7-6. An angle of 1 radian has an arc that is the same length as its radius. The lengths d_1 and d_2 equal r_1 and r_2, and the angle θ is 1 rad or 57.3°.

Thus we can see that an angle of 1 radian (abbreviated 1 rad) has an arc the same length as its radius (Fig. 7-6). For example, if the arc length of an angle is 10 inches and the length of the radius is also 10 inches, then we know the angle is 1 rad in size:

$$\theta = d/r = \frac{10 \text{ in}}{10 \text{ in}} = 1$$

We should note here that the units of length are removed from the quotient. Therefore the radian is an arbitrary unit because it does not appear in the definition. We are calling it arbitrary because it can be omitted as a unit when convenient.

For example, since $\theta = d/r$, we know that this formula can be rearranged to give $d = r\theta$. Thus the arc length (d) for an angle with a radius of 2 feet and an angle of 2 radians is 4 feet.

Finding the conversion factor between degrees and radians

It is relatively easy to find the conversion factor between degrees and radians by remembering the following relationships:
1. There are 360° in a complete circle.
2. The circumference (d) of a circle with a given radius (r) is $2\pi r$.

If we substitute the circumference of the circle ($2\pi r$) for d in the angular displacement equation $\theta = d/r$, we obtain $\theta = 2\pi r/r$, which simplifies

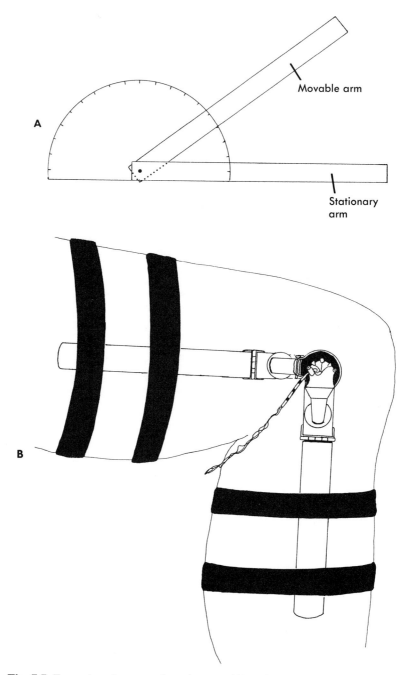

Fig. 7-7. Examples of a manual goniometer **(A)** and an electrogoniometer **(B).**

to 2π. Remembering that π equals 3.1416, we can see that 2π equals 6.283. Thus there are about 6.283 rad in one complete circle. From this relationship it can be seen that

$$360° = 6.283 \text{ rad}$$

(Equation 7-C)

By rearranging this relationship, we obtain

$$1 \text{ rad} = \frac{360°}{6.283} = 57.3°$$

Similarly

$$1° = \frac{6.283}{360} = 0.01745 \text{ rad}$$

Therefore

$$1 \text{ rad} = 57.3° \text{ and } 1° = 0.01745 \text{ rad}$$

Often it is useful to express angles in radians measured in terms of π alone. For example, if one revolution equals 2π rad, then two revolutions equal $2 \times 2\pi$ rad. Also an angle of $90°$ is one quarter of a complete circle. Thus one quarter of a revolution equals $\frac{1}{4} \times 2\pi$ rad, or $\pi/2$ rad and, of course, this has approximately the same numerical value as $90 \times 0.01745 = 1.571$ rad.

Mechanical and electronic measurements of angular displacements

The angular displacement of body segments can be measured either in two dimensions or in three dimensions. Two-dimensional measures of angular displacements in a single plane will be discussed in this lesson, while the three-dimensional measure will be discussed in Lesson 9.

Two-dimensional measures of angular displacement in a single plane involve the use of an instrument called a goniometer (Greek *gōnia,* angle; *metrein,* to measure). The displacement is in terms of degrees of a circle. A mechanical goniometer is illustrated in Fig. 7-7, *A.* This instrument consists of a $180°$ protractor with extended arms made from Plexiglas. One arm is movable, and the other is fixed at the 0 line of the protractor.

The application of this goniometer is simple. Let us assume we wish to measure the range of motion possible at the knee joint. The arms of the goniometer are placed parallel with the thigh and the leg. The center of the device is at the knee joint. Readings are taken with the leg at different positions. The differences between these readings represent the angular displacements.

There are, of course, many types of goniometers currently in use. An interesting innovation is the electrogoniometer developed by Karpovich at Springfield College. This device (Fig. 7-7, *B*) can be used to continuously measure changes in joint angles during the performance of various body movements and, needless to say, it has proved to be a most valuable aid in the study of human movement.

Use of cinematography in the measurement of angular displacement

Angular measurement through the use of cinematography has the advantage of recording actual performances. The angles can be scaled directly from the projected image with a protractor, and no multiplier or correction factor is required to obtain true size dimensions as long as there are no perspective errors. The correction for perspective errors will be discussed in Lesson 9.

The cinematographic recording of angles can effectively measure a wide variety of body angles —such as the angular relations of the limbs and trunk, the body lean in running, the angle of take-off in a jump or dive, and the angle of release in throwing. In Fig. 7-8 a shot-putter is projecting the shot at an angle of release of $42°$. This angle can be obtained directly from a photograph through the use of a protractor.

Describing the spatial components of segmental movements

The actual lever which comprises a body segment can be schematically represented by the longitudinal axis of the segment. As defined in Lesson 5, the longitudinal axis of a body segment is an imaginary straight line connecting the midpoint of the joint at one end of the segment with the midpoint of the joint at the other end or, in the case of a terminal segment, with its distal end.

Fig. 7-9, *A,* is a drawing made from a movie frame which shows a man engaged in the task of lifting a weight through the action of elbow flexion. Dots have been drawn on his skin with a grease pencil to approximate the centers of his shoulder joint, elbow, and wrist. The two straight lines connecting these dots represent the longitudinal axes and therefore the levers of the arm and forearm, respectively.

In the top illustration there is more flexion, and in the bottom there is less (corresponding to a move from position *B* to position *A*), with the resulting angular displacement *(θ)* and linear displacement *(d = rθ)* as diagramed in Fig. 7-9, *B*. The radius of rotation is the length of the longitudinal axis of the forearm. The resistance overcome, of course, is the weight of the forearm and hand plus the weight carried in the hand.

INTRODUCTION TO THE TEMPORAL ANALYSIS OF HUMAN MOTION

As noted before, the mechanical analysis of a motor performance usually begins with the quantification of its temporal dimensions. This quantification can occur through the use of (1) mechanical and electronic devices and/or (2) optical devices which yield cinematographic data.

MECHANICAL AND ELECTRONIC MEASUREMENT OF TIME

The use of mechanical and electronic devices for the measurement of time has the advantage over filmed cinematography of being more readily available. As a result a variety of timing devices has been developed for the measurement of selected phases of human movement. These devices can be classified as either chronoscopes or chronographs. A chronoscope is a mechanical or electronic timer that measures time directly, while a chronograph measures time through the use of a recording called a chronogram.

In sports events that are started with a signal (usually a gun) and ended when the performer completes a required distance, the timer's stopwatch serves as a response time chronoscope. A

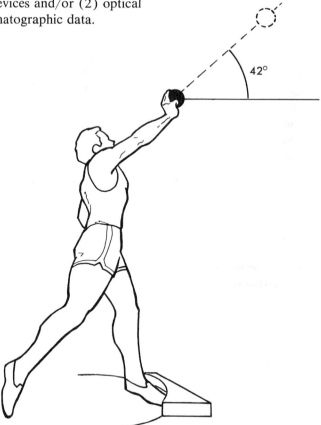

Fig. 7-8. Angle of release of a shotput.

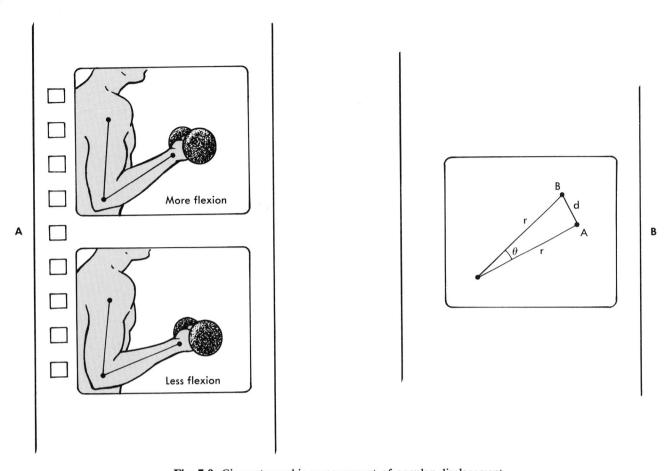

Fig. 7-9. Cinematographic measurement of angular displacement.

more sophisticated arrangement for the measurement of response time characteristics is illustrated in Fig. 7-10, *C*. The instruments shown in this illustration are

Two chronoscopes (timers) to record time in hundredths of a second

A stimulus source (buzz or light) to start a timer

A switch that, when released (or depressed, depending upon the model), stops one timer and starts the other timer

Photoelectric cells to stop the second timer when the light beam is broken

In an experiment with this equipment, a stimulus activates *Chronoscope RT*. It is stopped when the subject initiates movement (indicating reaction time). *Chronoscope MT* is started with the initiation of movement and is stopped when a light beam of the photoelectric cell is broken by the subject's arm, hand, or leg (indicating movement time). The sum of the readings on the two chronoscopes provides the response time. A third chronoscope may be used to start with the stimulus and terminate with the movement through the light beam of the photoelectric cell. This would record the response time on one chronoscope.

A very versatile chronoscope unit that meets a wide range of practical timing requirements is the Automatic Performance Analyzer (APA) manufactured by Dekan Timing Devices. This portable device can be started and stopped by a number of attachments such as switch mats, push buttons, control cords, and external switches. Time is measured in hundredths of a second. Fig. 7-11 shows the Dekan automatic performance analyzer being used to measure a subject's

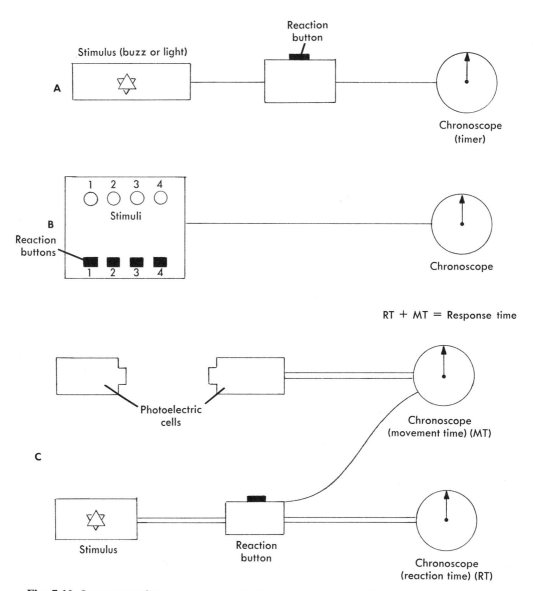

Fig. 7-10. Instruments for measuring speed of response characteristics. **A,** Simple reaction. **B,** Choice reaction. **C,** Reaction time–movement time.

shoulder joint flexion time. The subject starts the timer by moving his hand from a switch mat on the table and stops it by moving his hand through the string located at shoulder height which pulls an attached plug from the stop switch.

Chronographs are used in connection with electronic analysis systems. In Fig. 7-12 a recording of the angular displacements of a joint is obtained through the use of an electrogoniometer. As can be seen in the illustration, this recording has two dimensions: a vertical dimension and a horizontal dimension. The vertical dimension represents the angular displacements of the joint as measured by the potentiometer of the electrogoniometer. The horizontal dimension represents time as determined by the paper speed set on the recorder. The paper speed is used as the divisor, which in turn is used to convert horizontal linear

Fig. 7-11. The Dekan performance analyzer can be used for the measurement of movement time.

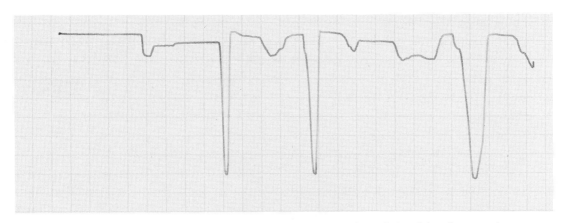

Fig. 7-12. Electrogoniogram obtained for flexion and extension of the right elbow as shown in Fig. 3-10. The downward deflections are produced by flexion, and the upward deflections are produced by extension. The horizontal dimension represents time.

dimensions of the recording to be representative of the true temporal dimensions. If the paper speed is set, for example, at 0.25 cm/sec, then each centimeter of horizontal length in the recording is representive of 4 seconds. The time of 4 seconds was obtained as follows:

$$\text{Time} = \frac{\text{Horizontal length}}{\text{Paper speed}} \quad \text{(Equation 7-D)}$$

$$4 \text{ sec} = \frac{1 \text{ cm}}{0.25 \text{ cm/sec}}$$

Measurement of time through the use of cinematographic techniques

A temporal analysis can be performed through cinematography by (1) placing a chronoscope in the field of view of the camera during the filming of the performance, (2) using timing lights in the camera to record time directly on the film, or (3) using knowledge of camera speed to calculate the time. The most typical is the third method. Time is calculated from the frame rate at which the camera is operating. If the camera is operating at 100 frames per second, the time per frame is 1/100 or 0.01 second; if it is operating at fifty frames per second, the time per frame is 0.02 second. Thus if a filmed sequence contains thirty frames and the filming rate was 100 frames per second, we know that the time during which the sequence occurred was 0.3 second. This time is determined as follows:

$$\text{Time} = \text{Number of frames} \times \text{Time per frame}$$
$$0.3 = 30 \times 0.01 \quad \text{(Equation 7-E)}$$

STUDY GUIDELINES
Fundamental quantities of motion
Specific student objective

1. Identify and describe the three fundamental quantities of motion.

Summary

The fundamental quantities of motion are the three measurable mechanical quantities in human motion: (1) the body or object in motion (matter is quantified by either its weight or its mass), (2) the distance or angle through which the body travels (space), and (3) the duration of motion (time).

Matter is defined as anything that has mass and occupies space, while space is the area or volume occupied by a body and the distance or angle through which it travels during movement. Time is the period during which an action or process continues, i.e., a measured or measurable duration.

Performance tasks

1. The three fundamental quantities of motion are _____ , _____ , and _____ (in any order).
2. Consult a dictionary for definitions of the three quantities of motion. Compare them to the descriptions given in this text.
3. Write out the three definitions for the fundamental quantities of motion.
4. Illustrate the three fundamental quantities of motion with a specific movement example.

Description of motion quantities
Specific student objectives

1. Define quantification.
2. Identify the two forms of quantification.
3. Differentiate between fundamental and derived units of measurement.
4. Identify the two basic methods of describing space.
5. Identify the two basic systems of measurement currently in existence.
6. Identify the methods of describing matter.
7. Identify the two effects of force applications used to measure the magnitudes of forces.
8. Identify the methods of describing time.
9. Identify the two metric systems of units.
10. List the standard units of length, weight, and time in both the metric and the British systems.

Summary

Quantification is defined as the assignment of numbers and units to objects according to rules. When a quantity is counted or measured, a property is selected and compared to some standard. For example, if distance is selected it can be compared to a standard length of 1 foot. Quantification of continuous variables is carried out and expressed in terms of the fundamental units of weight, space, and time. All other such measurement units are called derived units and are developed as combinations of the fundamental units. Counting is the alternate form of quantification.

The two basic measurement systems in use today are (1) the metric system and (2) the British system.

Space descriptions usually begin with the fundamental unit of length or linear distance. This can be determined by such instruments as rulers, tape measures, and measuring wheels. Cinematographic techniques can also be used for such descriptions. The metric system uses the meter as the standard unit of distance. A meter was originally defined as one ten-millionth of the distance from the equator to the North Pole. In the British system the foot is the basic standard of length and is defined as 1200/3937 of a standard meter. Angular space dimensions are described by such angle-measuring instruments as magnetic compasses, protractors, and goniometers and through cinematographic techniques. In both metric and British systems angular measures are made in terms of the angle turned through in degrees or radians.

The amount of matter in an object at a given location on the earth is given by the weight of the object. Weight is a quantification of the gravitational forces acting on an object and varies with location in the universe. Therefore matter is often quantified in terms of either its weight or its mass. Mass is defined as weight per unit of gravitational acceleration, i.e., $m = w/g$. This implies that mass and weight are in direct proportion. Thus, if the location of a body on the earth's surface is standardized (which would standardize "g"), then weight can be used to quantify matter. Weight can be measured by such instruments as tension or compression scales. The standard unit of weight in the metric system is the kilogram, and in the British system the pound. One pound is defined as the force that attracts an object so as to cause an acceleration of 32.1740 ft/sec/sec. In general, forces are measured directly like weight by tension or compression scales.

Time is described in both the metric and the British systems by the fundamental unit of seconds, defined as one 86,400th of a mean solar day. Stopwatches, electronic timing devices, and motion pictures are used to measure time.

In actual practice the metric system is used in either the meter-kilogram-second (mks) form or the centimeter-gram-second (cgs) form. The derived units in these two forms differ; the force units are the newton and the dyne, respectively.

Performance tasks

1. The process of assigning numbers and units to objects according to rules is called
   a. qualification
   b. quantification
   c. both a and b
   d. neither a nor b
2. The two forms of quantification are
   a. count and measure
   b. nominal and evaluate
   c. both a and b
   d. neither a nor b
3. List three fundamental and three derived units used in describing weight, space, and time.
4. Using Table 7-1, list the fundamental units of measure in both the metric and the British systems for weight, space, and time.
5. A comparison of selected units of measure in the two systems and their equivalancy relations is given below. Complete the missing items. Carry out the conversions mathematically or consult a reference table.

Equivalence

Metric	Length	British
Kilometer	1 km = 0.62 mi; 1 mi = _____ km	Mile
Meter	1 m = _____ yd; 1 yd = 0.91 m	Yard
Meter	1 m = _____ ft; 1 ft = 0.30 m	Foot
Centimeter	1 cm = 0.39 in; 1 in = 2.54 cm	Inch

Metric	Weight	British
Kilogram	1 kg = _____ T; 1 T = 907.18 kg	Ton
Gram	1 g = _____ lb; 1 lb = 453.59 g	Pound
Gram	1 g = _____ oz; 1 oz = 28.35 g	Ounce

6. Miscellaneous units of measure are listed below. Can you identify the basic units and the derived units of measure?

```
1 hand = 4 in
1 fathom = 6 ft
1 knot (nautical mile) = 6080 ft
3 barleycorns = 1 in
1 pace = 3 ft
8 furlongs = 1 mile (statute)
1 acre = 160 square rods
1 cubic ft = 1728 in
1 cord = 128 cu ft
1 gallon (liquid) = 231 cu in
1 span = 9 in
1 cubit (distance from elbow to end of middle
    finger) = 18 in
1 square ft = 144 square in
1 square mi = 640 acres
1 township = 36 square mi
1 stone = 14 lb
1 league = 3 m
1 liter (liquid) = 1.06 qt
1 pint (liquid) = 4 gills
```

NOTE: Complete tables of measures are given in nearly every standard dictionary. For equivalence conversion units, consult such a table.)

7. Space descriptions begin with the fundamental units for linear motion of _____ and for circular motion of _____ .

8. Matter is usually described by the quantity called
   a. distance
   b. length
   c. time
   d. weight (or mass)

9. List two methods of measuring weight, space, and time.

10. What do the abbreviations *mks* and *cgs* represent?

Scalar and vector quantities

Specific student objectives

1. Describe the concept of force.
2. Differentiate between scalar and vector quantities.
3. Describe the graphical representation of vectors.
4. Given a vector quantity, diagram its graphical representation.
5. Define the line of action of a vector.
6. Define center of gravity and line of gravity.

Summary

Scalar quantities are measurable quantities that can be specified by stating only their magnitude. Quantities that have both magnitude and direction are called vector quantities. Force and displacement are the two fundamental quantities usually expressed as vectors.

The graphical representation of a vector quantity (magnitude *and* direction) consists of an arrow drawn from a dot which indicates the origin. The length of the arrow, drawn to some specified scale, indicates the magnitude of the vector while the head of the arrow indicates vector direction. Thus it can be seen that an arrow graphically represents the three essential elements of a vector: point of origin, magnitude, and direction.

The direction in which a vector quantity such as force acts is called the line of action of that vector. With special reference to a gravitational force, the line of action of a weight is termed the line of gravity or the gravitational line. The center of gravity of any material object is defined as an imaginary point representing the weight center of the body.

Performance tasks

1. Consult a dictionary for descriptions of force and weight.

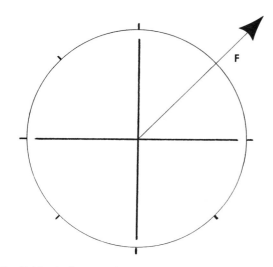

Fig. 7-13. A force acting at a direction of 45° above the horizontal.

2. The two types of measurable motion quantities are
   a. British and metric
   b. space and time
   c. both a and b
   d. neither a nor b
3. A vector has two measurable quantities: magnitude and _____.
4. Identify three examples of measurable scalar and vector quantities.
5. Demonstrate how a vector quantity can be represented by an arrow by performing the following exercise in the graphical representation of a force. (NOTE: A ruler and an angle-measuring device [protractor] are needed for this exercise.)
   Example: A force *(F)* of 45 lb is acting at a direction of 45° above horizontal (Fig. 7-13):
   a. Select a starting point to represent the point of application.
   b. Select an arbitrary scale for the measuring unit, e.g., 1 in = 20 lb.
   c. From the starting point, determine the direction of the vector with a protractor.
   d. Using a distance-measuring device and the selected scale, determine the length of the vector and draw the head of the arrow; e.g., since 1 in = 20 lb and F = 45 lb, the arrow will be 45/20 = 2¼ inches long.
6. Given the following vector quantities, depict their graphical representation with an arrow:
   a. F_1 = 100 lb; direction, horizontal and to the right
   b. F_2 = 320 lb; direction, horizontal and to the right
   c. F_3 = 84 lb; direction, south
   d. F_4 = 4.5 lb; direction, vertical and up
   e. V (velocity) = 59 mi/hr; direction, west

Introduction to the spatial analysis of human motion

Specific student objectives

1. Identify the two types of displacement.
2. Identify the two spatial components of linear displacement.
3. Differentiate between distance and displacement.
4. Identify the four spatial components of angular displacement.
5. Define a radian.
6. State the relationship between degree and radian measure.
7. Identify two methods of measuring angular displacements in two dimensions.
8. Define a goniometer.
9. Describe the measurement of linear and angular displacements through the use of cinematographic techniques.

Summary

The spatial analysis of human motion consists of describing linear and angular displacements. Linear displacement has two spatial components, distance and direction; thus it is a vector quantity. The spatial components of angular motion are direction, radius, angle turned through (θ), and arc of the circle cut by the angle.

Angular motion is circular movement of an angle (θ) turned through and takes place about an axis or center of rotation. The linear distance (d) a point on the rotating object moves is the arc of a circle cut by the angle. The linear distance of the rotating point from the axis is termed the radius (r) of its motion.

A radian, a unit of angle measure, is defined as the ratio of the arc length (d) to the length of the radius (r) for any angle, i.e., $\theta = d/r$ (in radians). One radian equals $180°/\pi$, which is approximately 57.3°, and 1 degree equals roughly 0.01745 rad.

Angular displacement can be measured in two or three dimensions, both mechanically and electronically.

Linear displacement can also be measured mechanically and electronically. Unlike angular displacement, however, a multiplier must be used to get true linear measurements with cinematographic techniques.

Performance tasks

1. Displacement is a (scalar/vector) _____ quantity.
2. Illustrate the difference between distance and displacement with a motion example.
3. The two spatial components of linear displacement are
   a. angle and radius
   b. arc and distance
   c. both a and b
   d. neither a nor b

4. Give two measurement examples of linear displacement.
5. The four spatial components of angular displacement are direction, _____ of motion, _____ turned through, and _____ of the circle cut by the angle.
6. A radian is a unit of (angle/length) _____ measure.
7. State the relationship between degree and radian measure below:
   a. 1 rad approximately equals _____ degrees
   b. 1° approximately equals _____ radian
8. List two instruments used to measure angular displacement.
9. A joint angle–measuring device is termed a _____.
10. Describe the use of an electrogoniometer for angular displacement descriptions.
11. Describe how cinematographic techniques can be used in the measurement of linear and angular displacements.
12. Define an angle of 1 radian. _____ _____
13. Convert the following angle measure into "degree" or "radian" form according to the instructions. Use 1 rad = $180°/\pi$ = 57.3° or 1° = 0.01745 rad.
    a. 90° = _____ rad
    b. $\pi/4$ rad = _____ °
    c. $\pi/2$ rad = _____ °
    d. 192° = _____ rad

Introduction to the temporal analysis of human motion

Specific student objectives

1. Describe a chronoscope.
2. Describe a chronograph.
3. Describe the cinematographic measurement of time.

Summary

Quantification of time in a mechanical analysis can be accomplished through the use of mechanical and electronic devices or through the use of cinematography.

Timing devices can be classified as being either chronoscopes (mechanical or electronic timers which measure directly) or chronographs (mechanical or electronic timers which measure time indirectly on a graph).

Cinematographic measures of time can be carried out by (1) using a chronoscope in the field of view of the camera, (2) recording time directly on the frames of the film, or (3) using the camera speed to calculate time from the frame rate.

Performance tasks

1. Consult a dictionary for definitions of chronoscope and chronograph.
2. List three methods of measuring time during a mechanical analysis of motion.
3. Examine the process of measuring time in a high-precision research situation. Use other references and sources.
4. Examine the process of time measurement in a cinematography analysis.

SELF-EVALUATION

Students should use no reference materials for this progress test, and they can check their answers by referring to Appendix A.

1. The three fundamental quantities of motion are
   a. length, width, and height
   b. time, distance, and area
   c. mass, angle, and time
   d. weight, space, and time
2. Describe the three fundamental quantities of motion.
3. Define quantification.
4. The two forms of quantification are
   a. linear and angular
   b. count and measure
   c. both a and b
   d. neither a nor b
5. The two methods of describing space are
   a. linear and angular
   b. count and measure
   c. both a and b
   d. neither a nor b
6. Matter can be described by measuring
   a. mass
   b. weight
   c. both a and b
   d. neither a nor b
7. The two basic systems of measurement are
   a. count and measure

b. add and substract

c. both a and b

d. neither a nor b

8. Example(s) of fundamental unit(s) of measure would be

a. fcct

b. pounds

c. both a and b

d. neither a nor b

9. State the difference between fundamental and derived units of measurement.

10. The two effects used to measure forces are to alter the

a. dimensions and state of motion

b. distance and angle

c. both a and b

d. neither a nor b

11. The two metric systems used in describing motion quantities are _____ and _____ (in any order).

12. List the standard units below

	Metric	British
Length	_____	_____
Weight	_____	_____
Time	_____	_____

13. Describe the concept of force.

14. A measurable quantity with magnitude only is called a (scalar/vector) _____ .

15. A vector is a scalar with the added element of _____ .

16. Describe the line of action of a vector.

17. The weight center of an object is its

a. mass

b. weight

c. both a and b

d. neither a nor b

18. Define the line of gravity of an object.

19. Draw a graphical representation of the vector quantities described below:

a. F_1 = 75 lb; direction, vertical and down

b. F_2 = 120 lb; direction, 45° above horizontal and to the left

c. A weight of 110 lb

d. A horizontal pull of 90 lb to the right

e. A velocity of 20 mi/hr in a northerly direction

20. The two types of displacements are

a. linear and angular

b. degrees and radians

c. both a and b

d. neither a nor b

21. The two spatial components of linear displacements are

a. degrees and radians

b. distance and direction

c. both a and b

d. neither a nor b

22. List the four spatial components of angular displacement.

23. Define a radian and a goniometer.

24. State the relationship between degree and radian measure:

a. 1 radian equals approximately _____ degree(s)

b. 1 degree equals approximately _____ radian(s)

25. Describe a chronoscope and a chronograph.

26. Describe how cinematographic techniques can be used in the measurement of space (linear and angular) and time.

LESSON 8

Vector analysis of motion quantities

Body of lesson
    Addition of vectors
        Composition of vectors
        Obtaining the resultant of parallel and antiparallel vectors
        Obtaining the resultant of nonparallel vectors through graphical means
    Resolution of vectors
        Pythagorean theorem
    Addition of more than two vectors
        Polygon method
        Method of components
    Subtraction of vectors
    Examples of vector components in human motion
Study guidelines
Self-evaluation

ADDITION OF VECTORS

The basic problem in the vector analysis of motion quantities is the combining of vectors. In this lesson we will be concerned with the problem of combining sets of vectors which lie in the same plane (coplanar vectors) and which have the same point of origin (concurrent vectors). The three-dimensional analysis of motion quantities will be covered in Lesson 9.

Composition of vectors

Any set of coplanar concurrent vectors can be replaced by a single vector, called their *resultant,* whose effect is the same as that of the given vectors. This process of finding the resultant of several vectors is termed the addition or composition of vectors.

Obtaining the resultant of parallel and antiparallel vectors

The addition of *parallel* vectors (as well as of antiparallel vectors) is performed in the traditional manner; that is, when two or more parallel vectors have the same origins and identical directions, their resultant is obtained through simple addition ($A + B = R$, or $4 + 2 = 6$). This effect can be seen in Fig. 8-1, which shows two men pulling on a rope in the same direction.

Antiparallel vectors are defined as two or more vectors acting at the same point but in opposite directions as exemplified by the tug-of-war between the two men in Fig. 8-2. The resultant of the forces produced by the two men is also obtained through algebraic addition. However, it must be remembered that forces acting to the left are negative and those to the right are positive. Therefore since the direction of the force of one man in Fig. 8-2 is to the left, his force is designated as negative and then added to the force of the other man ($A + (-B) = R$ or, using arbitrary figures, $4 + (-2) = 2$).

Obtaining the resultant of nonparallel vectors through graphical means

There are two generally accepted methods for the graphical addition of *nonparallel* vectors: the parallelogram method and the triangle or vector chain method.

The *parallelogram* method may be used when only two forces are applied at the same point simultaneously. In this method, as shown at the left in Fig. 8-3, the vectors are first drawn out-

128

Fig. 8-1. Addition of parallel vectors.

Fig. 8-2. Addition of antiparallel vectors.

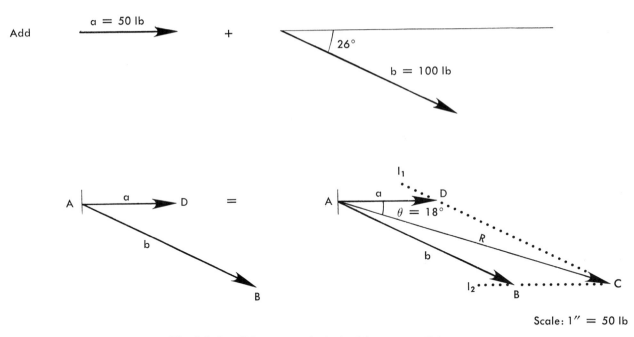

Add

a = 50 lb

+

26°

b = 100 lb

A ⟶ᵃ D = A ⟶ᵃ D l₁

θ = 18°

b

B

R

l₂ B C

Scale: 1″ = 50 lb

Fig. 8-3. Parallelogram method of adding nonparallel vectors.

ward from the same origin *(A)*. From *D* a dotted line *(l₁)* is next drawn parallel with line *b,* and from *B* a dotted line *(l₂)* is drawn parallel with line *a* as shown in the diagram on the right. From point *A* to point *C,* where the two dotted lines cross, the diagonal *AC* is drawn in and labeled with an arrowhead as the resultant *(R)*. The magnitude of *R* may be found by measuring its length on the diagram and making the appropriate conversion, and its direction with respect to *a* or *b* is as shown in the diagram. If *a* and *b* are forces acting on a body, their sum *(R)* is the net force acting on that body. The single vector *R* can replace the two vectors *a* and *b* without any change in the behavior of the body. Application of this method to the resolution of the forces acting upon an arrow in the bow of an archer is shown in Fig. 8-4.

The *triangle* method for adding vectors is illustrated in Fig. 8-5. To add *b* to *a,* we simply shift *b* parallel with itself until its tail is at the end of *a.* In its new position *b* must have its original length and direction. The sum *a + b* is a vector *(R)* drawn from the tail of *a* to the head of *b.* This procedure can be used with any number of vectors; place the tail of each vector at the head

of the previous one, keeping their lengths and original directions unchanged, and draw the resultant vector *(R)* from the tail of the first vector to the head of the last. The resultant *R* is the vector sum obtained and is designated by both magnitude and direction.

RESOLUTION OF VECTORS

We have now examined the "composition" of vectors by two graphical methods, the parallelogram and the triangle (vector chain) methods. Both methods can give us an approximation for the vector sum or resultant of several forces. Now let us work backwards from the resultant to the vectors which might compose that resultant. This is called the resolution of vectors into their components, and it can be accomplished through either a graphical method of the use of trigonometry.

Graphical method

The graphical method of resolving a vector into its two components is illustrated in Fig. 8-6. To a wagon a force is being applied through a rope at a 30° angle from the horizontal. This force has a tendency to produce two simultaneous ac-

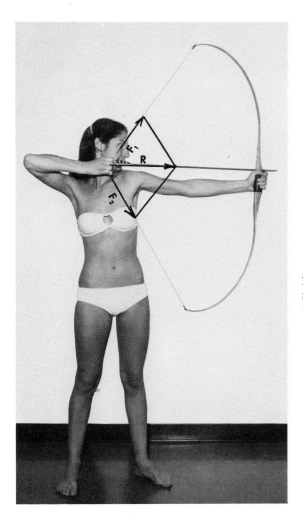

Fig. 8-4. Application of the parallelogram method to the resolution of the forces acting upon an arrow in the bow of an archer.

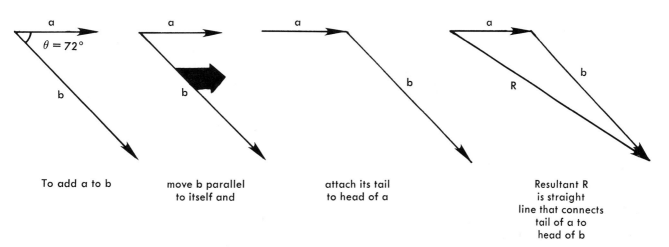

To add a to b

move b parallel to itself and

attach its tail to head of a

Resultant R is straight line that connects tail of a to head of b

Fig. 8-5. Triangle or vector chain method of adding nonparallel vectors.

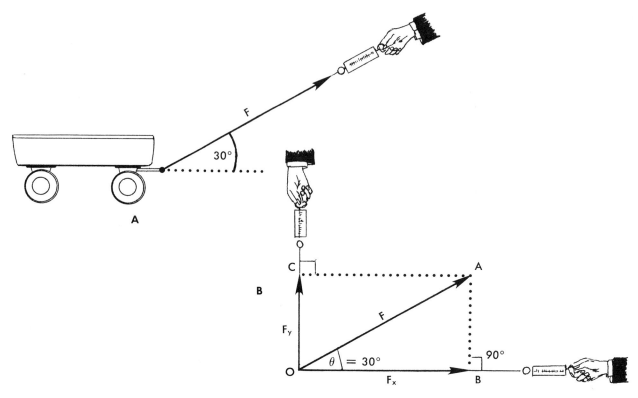

Fig. 8-6. The force *F* applied to the wagon in **A** can be resolved into its horizontal *(F$_x$)* and vertical *(F$_y$)* components as shown in **B** by drawing perpendicular lines from the tip of the force vector to the horizontal and vertical axes. The lengths of these lines represent the amount of the total force acting parallel with these axes.

tions. One action is to lift the front end of the wagon; the other is to pull the wagon forward. Therefore the force has two tendencies: one forward and the other upward. These tendencies are known as the horizontal *(F$_x$)* and the vertical *(F$_y$)* components of the force respectively.

In Fig. 8-6, *B*, the known force *F* is applied to point *O* at an angle of θ (theta) degrees with the x axis. Perpendicular lines are drawn from point *A* (the head of the vector) to the x and y axes. Note that the axes are positioned so their origins coincide with the tail of the vector. If the known force *F* has been drawn to scale, the vertical component *F$_y$* (read "F sub y") is given by the length of the line *AB,* and the horizontal component *F$_x$* by the length of line *AC*. With *F$_x$* and *F$_y$* perpendicular to each other, triangles *OAB* and *OAC* are equivalent right triangles having corresponding sides that are equal; i.e.,

both lines *AC* and *BO* are equal to *F$_x$*. Therefore the rectangular components of *F* are *F$_x$* and *F$_y$*; i.e., $\overrightarrow{F} = \overrightarrow{F_x} + \overrightarrow{F_y}$ (in a vector sense). The vector descriptions are completed by finding the magnitudes and directions of the vectors. The magnitudes can be obtained by measuring the respective lengths and applying the scale to determine the approximate sizes. The directions are obtained as horizontal and vertical by definition of the rectangular components.

Trigonometric method

The trigonometric method of resolving a force into its two components involves the solving of a right triangle and therefore the use of the three most common functions in trigonometry—sine, cosine, and tangent. This is a mathematically precise method of resolving a vector into its components.

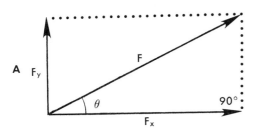 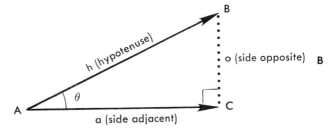

Fig. 8-7. The rectangular components of a force **(A)** form a right triangle. The three sides of a right triangle **(B)** are the hypotenuse, side opposite, and side adjacent.

A right triangle is shown in Fig. 8-7, *B,* with the letters *h, o,* and *a* representing the lengths of the three sides of the triangle and the capital letters *A, B,* and *C* representing the corresponding points from which the sides are drawn. Angle *C* equals 90°, and angle *A* is θ or the angle under consideration. Side *h* is called the hypotenuse (side opposite the right angle), side *o* is the side opposite with reference to θ, and side *a* is the side adjacent to θ. The following definitions of the relationships of a right triangle are important:

1. The sine of an angle is the ratio of the side opposite to the hypotenuse.

$$\text{Sine of } \theta = \frac{\text{Side opposite}}{\text{Hypotenuse}}, \text{ or } \sin \theta = \frac{o}{h}$$

(Equation 8-A)

2. The cosine of an angle is the ratio of the side adjacent to the hypotenuse.

$$\text{Cosine of } \theta = \frac{\text{Side adjacent}}{\text{Hypotenuse}}, \text{ or } \cos \theta = \frac{a}{h}$$

(Equation 8-B)

3. The tangent of an angle is the ratio of the side opposite to the side adjacent.

$$\text{Tangent of } \theta = \frac{\text{Side opposite}}{\text{Side adjacent}}, \text{ or } \tan \theta = \frac{o}{a}$$

(Equation 8-C)

It has been found that these ratios are constant for *all* right triangles, with any given angle. Thus these ratios can be used in many situations, and tables of the ratios or functions can be developed. If we now refer to Fig. 8-7, we see the following relationships:

Hypotenuse *(h)* = Known force *(F)*
Side adjacent *(a)* = Horizontal component *(F_x)*
Side opposite *(o)* = Vertical component *(F_y)*

Therefore we have the following equivalents:

$$\sin \theta = \frac{o}{h} = \frac{F_y}{F}, \text{ and } \cos \theta = \frac{a}{h} = \frac{F_x}{F}$$

Rearranging we obtain

$$F_x = F \cos \theta \qquad \text{(Equation 8-D)}$$
$$F_y = F \sin \theta \qquad \text{(Equation 8-E)}$$

Thus the rectangular components of a force can be found by multiplying the force by the sine and cosine functions.

Suppose we have a force of 100 lb acting on a body at an angle of 30°. Then the two components F_y and F_x can be calculated by direct substitution into Equations 8-D and 8-E.

$$F_y = F \sin \theta = 100 \text{ lb} \times \sin \theta$$
$$F_x = F \cos \theta = 100 \text{ lb} \times \cos \theta$$

A table of sines, cosines, and tangents is given in Appendix B, which shows for a 30° angle a sine of .500 and a cosine of .866. When these values are substituted into the equations, we obtain

$$F_y = 100 \times .500 = 50.0 \text{ lb}$$
$$F_x = 100 \times .866 = 86.6 \text{ lb}$$

Thus 86.6 pounds of force will be acting on the body in the horizontal direction, and 50 pounds will be acting in the vertical direction; i.e., the rectangular components of *F* are F_y (a vertical force of 50 lb) and F_x (a horizontal force of 86.6 lb).

Pythagorean theorem

Some students might wonder how a force of 100 lb can be resolved into two components with magnitudes of 50 lb and 86.6 lb. The arithmetic sum of the two components obviously does not equal the original force. The reason for this is that we are dealing with vectors, not scalars. Scalars (which involve magnitude only) can simply be added together in the traditional manner, but vector addition is entirely different because the direction of vectors as well as their magnitudes must be considered. Therefore it is possible to obtain a *vector* sum such that 100 lb = 86.6 lb + 50 lb (or $F = F_x + F_y$).

In problems dealing with the resolution of a single force where the solution of a right triangle is involved, the *Pythagorean theorem* can be used for the purpose of vector addition. This theorem states that for any right triangle the square of the hypotenuse is equal to the sum of the squares of the other two sides ($h^2 = o^2 + a^2$). Thus we find that the sum of the squares of the two components of the force discussed in the preceding section (50 and 86.6 lb) equals the square of the original force of 100 pounds:

$$100^2 = (50)^2 + (86.6)^2 = 10,000$$

This can be used to check rectangular component solutions found by the trigonometric method.

ADDITION OF MORE THAN TWO VECTORS

The methods of adding more than two vectors are identified as the polygon method and the method of components.

Polygon method

The vector chain or triangle method when used with more than two vectors is called the polygon method. Fig. 8-8 shows how four vectors whose initial magnitudes and directions differ are added together using the polygon method.

Method of components

While the polygon method is a satisfactory graphical one for finding the resultant of a number of forces, it is awkward for computation because, in general, a number of oblique triangles must be solved. Therefore the usual analytical method of finding the resultant is first to resolve all forces into rectangular components along any convenient pair of axes and then combine these into a single resultant. This makes it possible to work with right triangles only.

As an illustration of the method of components, consider the example of three forces shown graphically in Fig. 8-9, *A*: $F_1 = 120$ lb at $0°$; $F_2 = 200$ lb at $60°$; and $F_3 = 150$ lb at $225°$. The problem here is to find the resultant force equivalent to all three forces together. The operations involved in this procedure are

1. Each force is resolved into x and y components (shown in Fig. 8-9, *B*).
2. A sum of the x components (Σx) is obtained to give a resultant R_x component.
3. A sum of the y components (Σy) is obtained to give a resultant R_y component.
4. The resultant R_x and R_y components are combined at right angles to obtain their resultant R.

The first step is carried out by using the equations for the x and y components given before. Thus for F_1 the x component is 120 cos $0°$ =

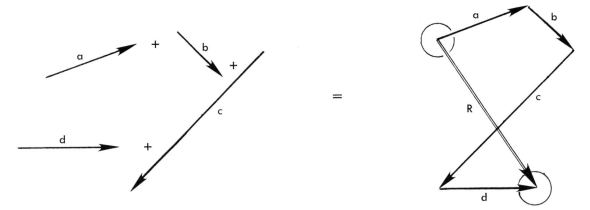

Fig. 8-8. Polygon method of adding more than two vectors.

120 lb, and the y component is 120 sin 0° = 0. Likewise, the x component for F_2 is 200 cos 60° = 100 lb and the y component is 200 sin 60° = 173 lb. For F_3 the x component is 150 cos 225° = 106 lb, and the y component is 150 sin 225° = 106 lb. Perhaps it should be noted here that when the table of trigonometric functions in Appendix B is used with angles greater than 90° the following rules should be in effect:

1. If the angle is greater than 90° but less than 180°, the value of the angle should be subtracted from 180°; or, in other words,

180° – 120° = 60°. Therefore the cos of 120° equals the cos of 60°.

2. If the angle is greater than 180° but less than 270°, 180 is subtracted from the value of the angle.

3. If the angle is greater than 270°, the value of the angle is subtracted from 360°.

It should be remembered that components to the right and up are positive in sign and components to the left and down are negative in sign. The second and third steps of adding x and y components separately are tabulated below:

Force	Angle	x component	y component
$F_1 = 120$ lb	0°	120 lb	0 lb
$F_2 = 200$ lb	60°	100 lb	173 lb
$F_3 = 150$ lb	45°	-106 lb	-106 lb
		$\Sigma \text{ x} = R_x = 114$ lb	$\Sigma \text{ y} = R_y = 67$ lb

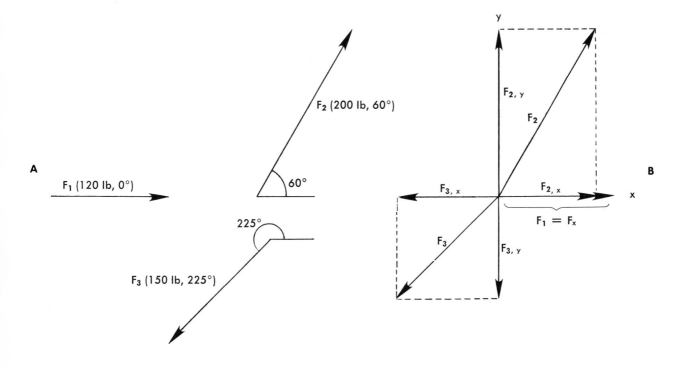

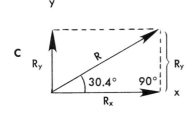

Fig. 8-9. Component method of adding more than two vectors.

Since R_x and R_y are at right angles to each other (Fig. 8-9, *C*), we are dealing again with a right triangle and the resultant *R* is the hypotenuse (given by the Pythagorean theorem as)

$$R^2 = (114)^2 + (67)^2 = 17,495$$
$$R = \sqrt{17,495}$$

By taking the square root of 17,495 (Appendix C), we obtain R = 132 lb. The angle (θ) of the

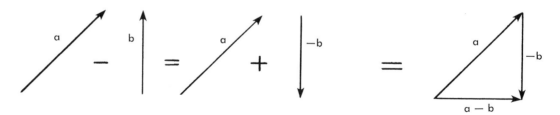

Fig. 8-10. Subtraction of vectors.

Fig. 8-11. Examples of vector components in human movements. **A,** Force components produced by a two-hand wrist pass in volleyball. **B,** The forward lean of a sprinter produces a forward component of force *(Fₓ)*. **C,** Force components acting upon a wrestler. (Modified from Armbruster, D. A., Musker, F. F., and Mood, D.: Basic skills in sports for men and women, ed. 6, St. Louis, 1975, The C. V. Mosby Co.)

resultant is obtained from the tangent relation:

$$\tan \theta = \frac{O}{A} = \frac{R_y}{R_x} = \frac{67 \text{ lb}}{114 \text{ lb}} = 0.588$$

(In Appendix B we find that the angle $\theta = 30.4°$.)

SUBTRACTION OF VECTORS

It is sometimes necessary to subtract one vector from another. If we wish to subtract A from B, for example, we first form the negative of A (denoted –A), which is a vector with the same length as A but which points in the opposite direction (Fig. 8-10). Then we add –A to B in the usual manner:

$$B - A = B + (-A)$$

Vector differences may also be found by the method of rectangular resolution. Both vectors are resolved into x and y components. The difference between the x components is the R_x component, and the difference between the y components is the R_y component.

EXAMPLES OF VECTOR COMPONENTS IN HUMAN MOTION

In Fig. 8-11, *A,* a volleyball player is executing a two-hand pass. The horizontal force component of the pass is responsible for the ball's forward movement, while the vertical force component is responsible for the ball's rise to a greater height than at release. The actual path followed by the ball will be in line with the direction of the resultant force.

In Fig. 8-11, *B,* the forward lean of the sprinter produces a forward component of force which causes his body to move down the track. The vertical component will combat the force of gravity and will make him rise into the air.

In Fig. 8-11, *C,* the horizontal displacement of the wrestler on the right is the result of the horizontal force component (F cos θ) produced by his opponent.

Other examples, of the application of the concept of force components will be given throughout the remainder of this text.

STUDY GUIDELINES
Addition of vectors
Specific student objectives

1. Define coplanar and concurrent vectors.
2. Describe the meaning of vector addition or composition.
3. Describe the addition of parallel and antiparallel vectors.
4. Given two parallel or two antiparallel vectors, find their resultant.
5. Describe the parallelogram and vector chain (triangle) methods of adding nonparallel vectors.
6. Given two or more nonparallel vectors, find their resultant by graphical methods.

Summary

One aspect of vector analysis is the combination of several vectors to find an equivalent vector that produces the same effect as the component vectors. This process is called vector addition or composition, and the single equivalent vector is termed the resultant. In Lesson 8 attention is directed only to composing vectors which lie in the same plane (coplanar vectors) and which have the same point of origin (concurrent vectors).

Parallel vectors act in the same direction and can be added in the traditional arithmetic manner. Antiparallel vectors, acting in the opposite direction, must be composed through algebraic addition, i.e., by considering the positive and negative aspects of direction.

Nonparallel vectors can be composed or added by two graphical methods: the parallelogram and the triangle methods. The parallelogram method is used to find the vector sum by forming a parallelogram with the two composite vectors. The resultant solution is the *diagonal* of the parallelogram originating at the point of application, i.e., at the intersection of the tails of the original vectors. The resultant must be completely described by both magnitude and direction. The triangle method of finding the vector sum (or resultant) is accomplished by "chaining" the composite vectors together (connecting the "head" of the first vector to the "tail" of the second). The resultant is the line drawn from the tail of the first vector (origin) to the head of the second (terminal point). Its magnitude and direction can be easily determined.

Performance tasks

1. Write out definitions of coplanar and concurrent vectors.

2. Find the vector sum (resultant) by graphical means in the following coplanar concurrent vectors:

Scale: 1″ = 40 lb

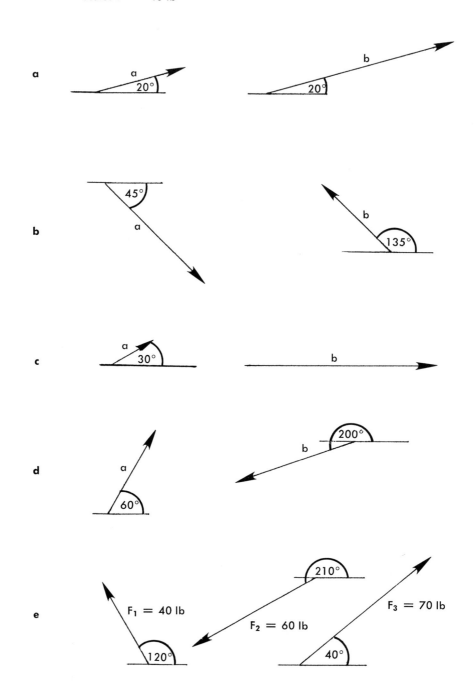

3. From the discussion of vector addition or its summary, list the step-by-step procedures for the addition of
   a. parallel and antiparallel vectors
   b. nonparallel vectors

Resolution of vectors
Specific student objectives

1. Describe the meaning of the type of analysis called the resolution of vectors.
2. Describe the graphic and trigonometric methods of resolving a vector into its two rectangular components.
3. Given a vector, resolve it into rectangular components by graphic and trigonometric methods.
4. Use the Pythagorean theorem to check a "resolution of vectors" solution.
5. State the Pythagorean theorem.

Summary

The resolution of vectors is a process of breaking down a single vector into its rectangular components, i.e., the reverse process of vector addition or composition. Any vector can be resolved into rectangular (horizontal and vertical) components by either an approximate graphical method or a precise trigonometric method.

The graphical method is utilized by drawing perpendicular lines from the head of the vector to the x and y axes after the axes are positioned so their origins coincide with the tail of the vector. The components of the original vectors are the horizontal and vertical vectors formed by the line segments from the origins to the intersections of the perpendicular lines with the axes. The directions of the vectors are determined already by definition, and the magnitudes are found by measurement and application of the scale value. In a vector sense, the two components are equivalent to the original vector.

The trigonometric method of finding the rectangular components of a vector is a more precise mathematical method of resolving a vector into its components. The respective horizontal and vertical components are found by applying the trigonometric functions as follows:

$$F_x = F \cos \theta, \text{ and } F_y = F \sin \theta$$

where F is the original vector, and θ the angle

between the vector and the x axis. Using the right triangle as shown with o as the side opposite the angle (θ) under consideration, h the hypotenuse (side opposite the right angle), and a the side adjacent, the following trigonometric definitions are made of the functions:

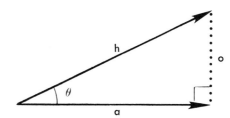

1. The sine of an angle is the ratio of the *side opposite* to the hypotenuse, or

$$\sin \theta = \frac{\text{Side opposite}}{\text{Hypotenuse}} = \frac{o}{h}$$

2. The cosine of an angle is the ratio of the *side adjacent* to the hypotenuse, or

$$\cos \theta = \frac{\text{Side adjacent}}{\text{Hypotenuse}} = \frac{a}{h}$$

3. The tangent of an angle is the ratio of the *side opposite* to the *side adjacent,* or

$$\tan \theta = \frac{\text{Side opposite}}{\text{Side adjacent}} = \frac{o}{a}$$

The Pythagorean theorem can be used for finding the resultant of vectors applied at right angles to each other or as a check on the vector resolution found by the trigonometric method. The theorem states that for any right triangle the square of the hypotenuse equals the sum of the squares of the other two sides ($h^2 = o^2 + a^2$).

Performance tasks

1. The two basic types of vector analysis for motion quantities are vector
   a. composition and resolution
   b. addition and subtraction
   c. both a and b
   d. neither a nor b
2. Develop a step-by-step procedure for the resolution of vectors by both graphical and trigonometric methods.
3. Find the rectangular components of each vector shown below by both graphical and trigonometric methods.

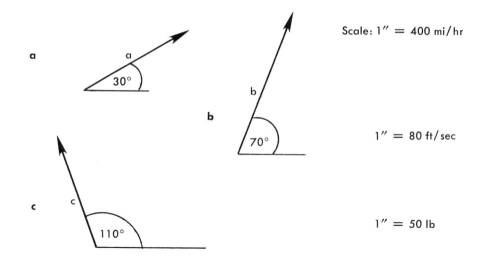

Scale: $1'' = 400$ mi/hr

$1'' = 80$ ft/sec

$1'' = 50$ lb

4. Check the solutions obtained above by using the Pythagorean theorem.

Addition of more than two vectors
Specific student objectives

1. List and describe the two methods of obtaining the resultant of more than two coplanar and concurrent vectors.
2. Given more than two coplanar and concurrent vectors, find their vector sum by the graphical polygon method and the trigonometric method of components.

Summary

Vector addition of more than two vectors can be carried out by the polygon method (approximate "chaining" method) or by the method of components (repeated applications of the trigonometric method).

Performance tasks

1. The two methods of vector composition for finding the resultant of two or more vectors are
   a. resolution and addition
   b. components and polygon
   c. both a and b
   d. neither a nor b
2. Develop a written step-by-step procedure for obtaining the resultant of more than two coplanar concurrent vectors by the polygon and component methods.

3. Using the standard reference system for angles (0°, north; 90°, east; 180°, south; 270°, west), find the resultant of the following vectors by both polygon and component methods:
   a. F_1: 50 lb, 45°; F_2: 50 lb, 90°; F_3: 150 lb, 20°
   b. F_1: 360 lb, 270°; F_2: 120 lb, 180°; F_3: 180 lb, 300°

Subtraction of vectors
Specific student objectives

1. Differentiate between vector addition and vector subtraction.
2. Given two vectors, find their difference by graphical and trigonometric methods.

Summary

Vector subtraction is carried out by forming the negative of a vector (a vector of the same length but opposite in direction); i.e., $B - A = B + (-A)$.

Vector differences may be found by graphical and trigonometric methods.

Performance tasks

1. Describe vector addition and vector subtraction.
2. Find the vector difference between two coplanar concurrent vectors by both graphical and trigonometric methods.

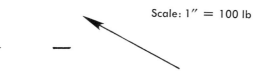

Scale: 1″ = 100 lb

Examples of vector components in human movement

Specific student objective

1. Give three human movement examples showing the effects of vertical and horizontal force components. Identify the rectangular force components in each example.

Summary

Almost all applications of force related to human movement can be analyzed by resolving the force into rectangular components. These components reveal the effects of the force in the movement situation.

Performance tasks

1. Identify the specific rectangular force components in the following human motion examples:
   a. The force applied to any projectile at the moment of release, e.g., a ball or javelin
   b. The force of a golf club striking a golf ball
   c. An offensive football player blocking an opponent
   d. The force caused by your body sliding into home plate during a softball game

SELF-EVALUATION

Students should use no reference materials for this progress test, and they can check their answers by referring to Appendix A.
1. Define coplanar and concurrent vectors.
2. The two basic problems in vector analysis are
   a. composition and resolution
   b. addition and subtraction
   c. both a and b
   d. neither a nor b
3. Vector addition is the process of finding a (component/resultant) _____.
4. Resolution of vectors involves obtaining (components/resultants)_____.
5. Describe the meaning of vector addition.
6. Describe the process of obtaining a resultant of two coplanar concurrent vectors by graphical means.
7. Describe the two methods of resolving a vector into rectangular components.
8. State the Pythagorean theorem.
9. Describe the meaning of vector subtraction.
10. Find the vector sum in the following coplanar concurrent vector situations:
    a. F_1: 100 lb, 30°; F_2: 50 lb, 60°
    b. 1: 200 ft/sec, 150°; 2: 100 ft/sec, 45°
    c. F_1: 40 lb, 0°; F_2: 80 lb, 45°; F_3: 120 lb, 120°
11. Find the rectangular components of each vector by graphical and trigonometric methods:
    a. R: 173 mi/hr, 30°
    b. R: 141 ft/sec, 45°
    c. R: 200 lb, 120°
    Check solutions by use of the Pythagorean theorem.
12. Find the vector difference (A – B) of the two coplanar concurrent vectors:
    A: 50 lb, 180°
    B: 25 lb, 45°
13. Give at least three human motion examples showing the effects of rectangular force components. Identify the rectangular components in each example.

Three-dimensional analysis of motion quantities

Body of lesson

THREE-DIMENSIONAL ANALYSIS OF LINEAR QUANTITIES

One of the first requirements for the description of linear or angular motion which might occur between two or more points in space is that these points be specified precisely. Since space has three dimensions—height, width, and depth—three separate measurements are needed to locate an object in space; that is, three numbers, and the explanation of what they mean, define any point. In certain situations, however, it may be desirable to deal with only two dimensions, in which case only two measurements are needed to locate any point in a designated plane. Therefore the description of body positions and movements can employ either a three-dimensional or a two-dimensional system.

Reference planes of the human body

The three-dimensional system for the location of body positions and the recording of body movements employs coordinates and requires that three planes perpendicular to each other be laid through the center of gravity of the body. These planes are called *cardinal* planes. When analyzing the motion of certain body segments, we normally draw similar planes through the center of gravity of the body segment, through the center of a joint, or through some other point of reference. In this event the planes are called *secondary planes*.

Following are the three cardinal planes (Fig. 9-1), which correspond to the three dimensions of space as defined in Lesson 5:

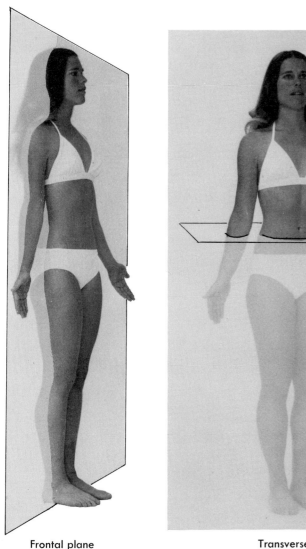

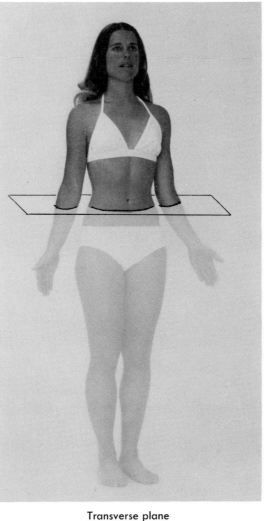

Frontal plane   Transverse plane   Sagittal plane

Fig. 9-1. The three cardinal planes.

frontal or lateral plane Passing through the body from side to side, dividing it into front and back sections
transverse plane Passing through the body, dividing it into upper and lower sections
sagittal (saj' i-tal, from Latin *sagitta,* arrow) **or anteroposterior plane** Passing through the body from front to back, dividing it into right and left sections

A two-dimensional system can also be used to describe and record certain body movements. When this is done, only one of the three planes is utilized. The two-dimensional system is typically used when movements are recorded by photography or described by line drawings. A still or motion-picture camera, for instance, can record a movement only in two dimensions. When a movement is photographed in the frontal plane (Fig. 9-2, *A*), either a front view or a back view of the body is obtained. Likewise, in the sagittal plane (Fig. 9-2, *B*) a side view is obtained and in the transverse plane (Fig. 9-2, *C*) either a top view or a bottom view. Since standard photographic procedures give only a two-dimensional

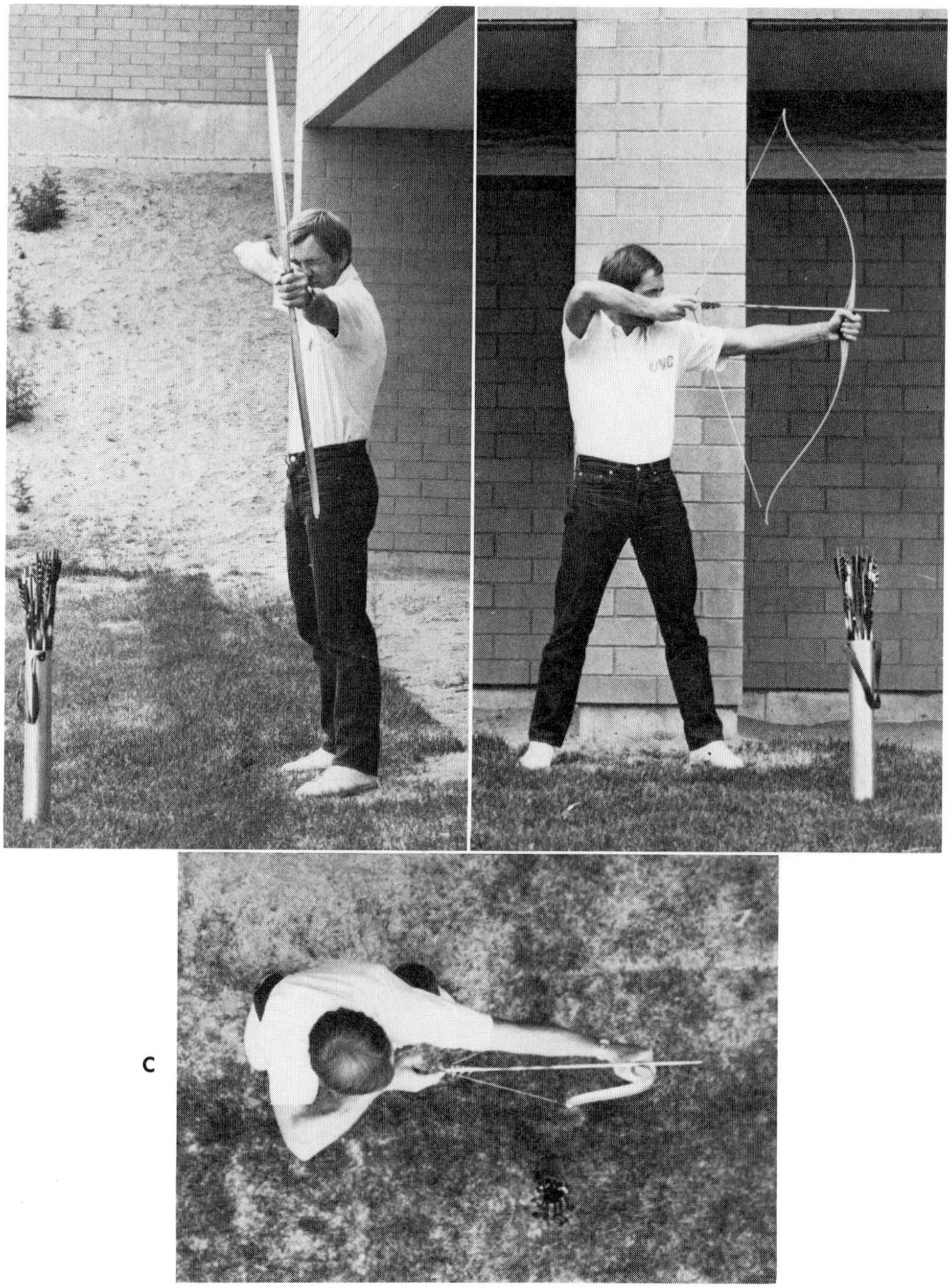

Fig. 9-2. The three cardinal views of movement. **A,** Frontal view. **B,** Sagittal view. **C,** Transverse view.

Overhead camera

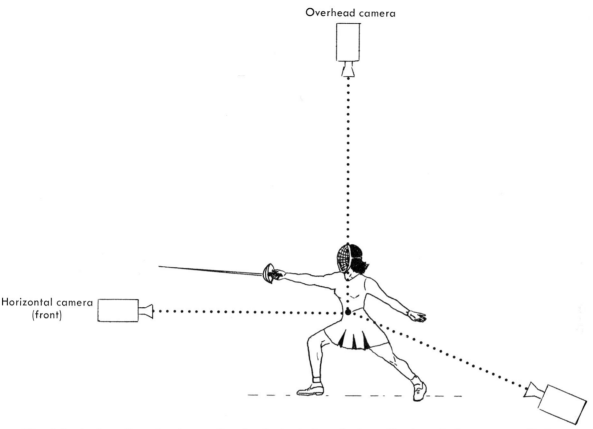

Horizontal camera
(front)

Horizontal camera
(side)

Fig. 9-3. A three-dimensional recording is obtained through the utilization of three synchronized cameras simultaneously in each of the three planes.

view, a three-dimensional recording can be acquired by using three synchronized cameras simultaneously in each of the three planes (Fig. 9-3).

The anatomical position

Perhaps it should be pointed out here that the three cardinal planes and their roles in sectioning the body are traditionally defined with reference to the anatomical position. This position, as defined in Lesson 5, has the individual upright with his feet parallel, arms at his side, and palms of his hands facing forward.

When the individual is in the anatomical position, the identification of front, back, right, left, upper, and lower sections is obvious. The important thing to remember, however, is that once these sections have been identified in the anatomical position they retain their identity regardless of what other position the body may be in. Thus, when the individual is in any position other than

upright, our definitions of the three planes with respect to dividing the body into sections are still true. For example, when the individual is lying supine as in Fig. 9-4, the transverse plane still divides the body into upper and lower sections, the sagittal plane still divides it into a right and a left section, and the frontal plane still divides it into front and back sections. Likewise, when the individual is lying on his or her side as in Fig. 9-5, the sectioning roles of the three planes are the same.

**The three cardinal directions
of linear motion**

The three cardinal directions of linear motion, called the x, y, and z directions, are defined as follows:

sideward or x direction Right-to-left or left-to-right movement perpendicular to both y and z directions
vertical or y direction Upward or downward movement

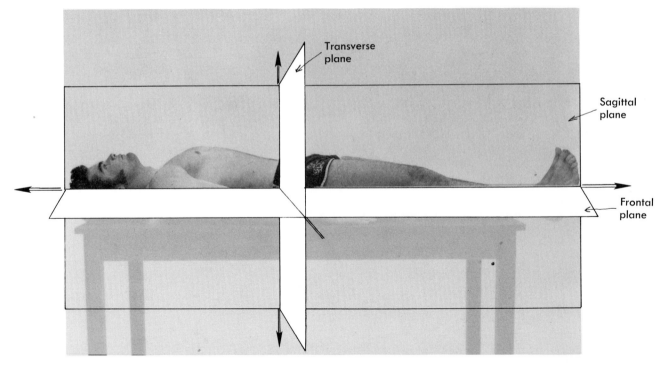

Fig. 9-4. When an individual is lying supine, the three planes continue their roles in regard to sectioning.

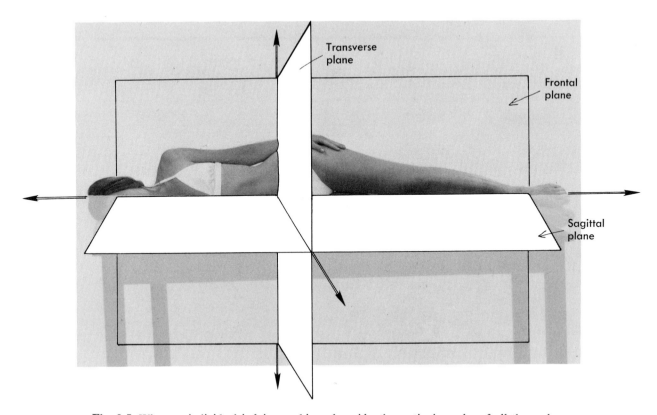

Fig. 9-5. When an individual is lying on his or her side, the sectioning roles of all three planes remain unchanged.

forward-backward or z direction Near-to-far or far-to-near movement perpendicular to both x and y directions

In the anatomical position the three cardinal directions of linear motion are located at the intersection of the three cardinal planes, as shown in Fig. 9-6. A forward or backward movement (z direction) occurs in the sagittal plane and a sideward movement (x direction) occurs in the frontal plane. A vertical movement occurs in both the frontal and the sagittal planes since the y direction is located at the intersection of these two planes.

The rectangular coordinate system

When recording the *position* of a body part, we consider the x, y, and z directions of linear motion to be coordinates in a three-coordinate

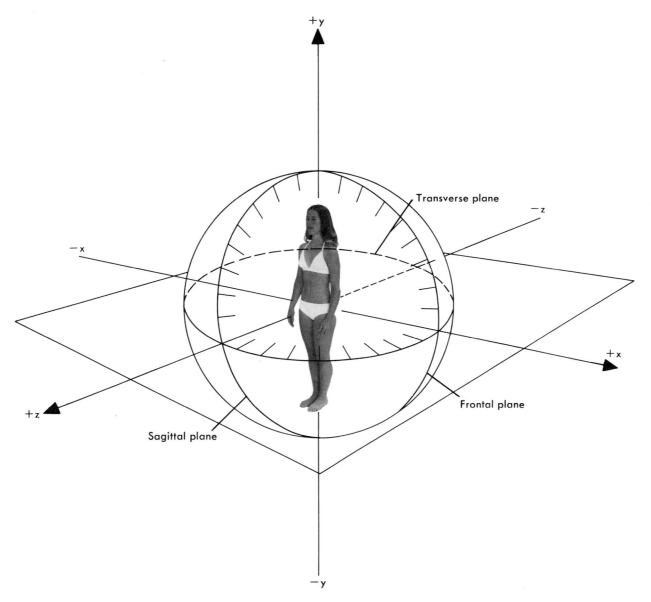

Fig. 9-6. In the anatomical position the three cardinal directions of linear motion are located at the intersection of the three cardinal planes.

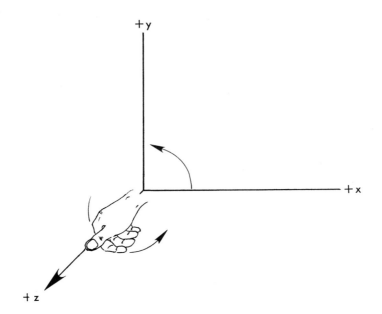

Fig. 9-7. When the fingers of the right hand are curled in the direction of a rotation of the positive *x* axis toward the positive *y* axis, the extended thumb points in the direction of the positive *z* axis.

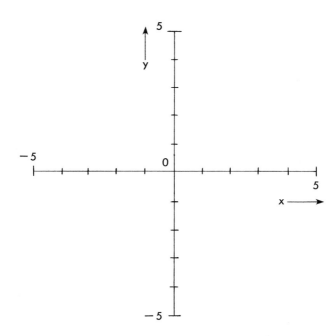

Fig. 9-8. Two-dimensional rectangular coordinate system.

system. By convention, negative and positive signs can be given to body positions in relation to the three coordinates and to the center of gravity of the body or to some other point of reference as follows:

> **width or x coordinate** To the right of some point of reference, *positive;* to the left, *negative*
> **height or vertical or y coordinate** Above some point of reference, *positive;* below, *negative*
> **depth or z coordinate** In front of some point of reference, *positive;* behind, *negative*

The x-y-z systems of this type are usually called *rectangular coordinates,* but the term *Cartesian coordinates* (after the French mathematician René Descartes) is also used. The rectangular coordinate system is a right-handed coordinate system because it conforms to the right-hand thumb rule —which states that when the fingers of the right hand are curled in the direction of a rotation of the positive x axis toward the positive y axis through the 90° angle between them the extended thumb will point in the direction of the positive z axis (Fig. 9-7).

The rectangular coordinate system can be used in either a two-dimensional or a three-dimensional analysis. A two-dimensional analysis usually uses only the x and y coordinates regardless of the

plane in which the object is viewed. The z coordinate is used only when the third dimension is being considered. In a two-dimensional graph (Fig. 9-8) it is customary to use x for the horizontal direction (or abscissa) and y for the vertical direction (or ordinate). Fig. 9-9 illustrates the common arrangement of the graphs in two and three dimensions.

The location of a point in two dimensions is specified by stating two numbers, the value of the abscissa (parallel with the x axis) and the value of the ordinate (parallel with the y axis), as illustrated by the several examples shown in Fig. 9-9, *A*. In performing a three-dimensional analysis, all three coordinates must be used to specify the location of a point. In a three-dimensional cinematographic analysis, for example, the x and y coordinates are used to analyze movements filmed in the frontal plane (Fig. 9-10). The x and z coordinates are used in the transverse plane (Fig. 9-11) while the y and z coordinates are used in the sagittal plane (Fig. 9-12).

Through the use of the three coordinates, any point in space can be determined precisely. For instance, we might specify the location of the center of a subject's right knee joint (as shown in Fig. 9-13) when in the anatomical position as

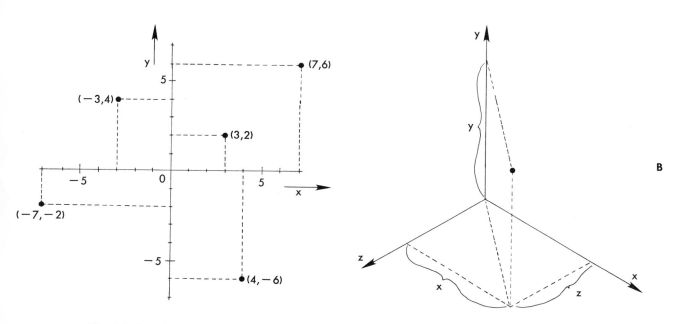

Fig. 9-9. Location of different points in two dimensions **(A)** and of a point in three dimensions **(B).**

Fig. 9-10. Frontal view of a fencing stance.

Fig. 9-11. Transverse view of a fencing stance.

Fig. 9-12. Sagittal view of a fencing stance.

follows: x = −8.5 cm; y = −52 cm; and z = 0. It should be noted that these three measures were taken relative to the body's center of gravity. This is an arbitrary starting point; the center of gravity of a body part, the center of a joint, or some other point of reference might well have been used.

Since the letters x, y, and z, which identify the three coordinates, have an alphabetical order and since measures of the coordinates are always presented in this order, the presentation of the above description can be simplified to −8.5, −52, 0. Of course, the unit of measure and the point of reference from which the measures were taken must also be specified. The unit of measure in an x-y-z description can be that representing any fundamental or derived quantity such as feet, pounds, seconds, feet per second, and foot-pounds per second.

THREE-DIMENSIONAL ANALYSIS OF ANGULAR QUANTITIES

The description of angular motion as it occurs in the human body requires not only the use of

the three planes of space already discussed but also the specification of the axes and the direction of the motion.

The axes of motion

Angular motion, by definition, is a motion that occurs around a center of rotation or axis. An axis is an imaginary straight line about which all parts turn in planes at right angles to the straight line. Flexion of the forearm at the elbow, for example, occurs in the sagittal plane and about an x axis. The sagittal plane, of course, is perpendicular to the x axis.

The three axes of motion, called the longitudinal, sagittal, and frontal axes, are located at the intersection of the three cardinal planes and coincide with the three cardinal directions, as shown in Fig. 9-14. Each of these axes is perpendicular to the plane in which the motion occurs and was defined in Lesson 5 as follows:

frontal or x axis Perpendicular to the sagittal plane
longitudinal or y axis Perpendicular to the transverse plane
sagittal or z axis Perpendicular to the frontal plane

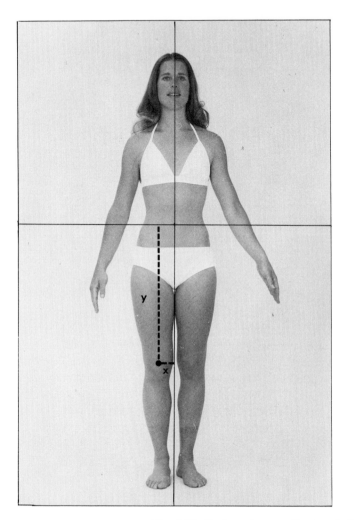

Fig. 9-13. Rectangular coordinates of the knee joint.

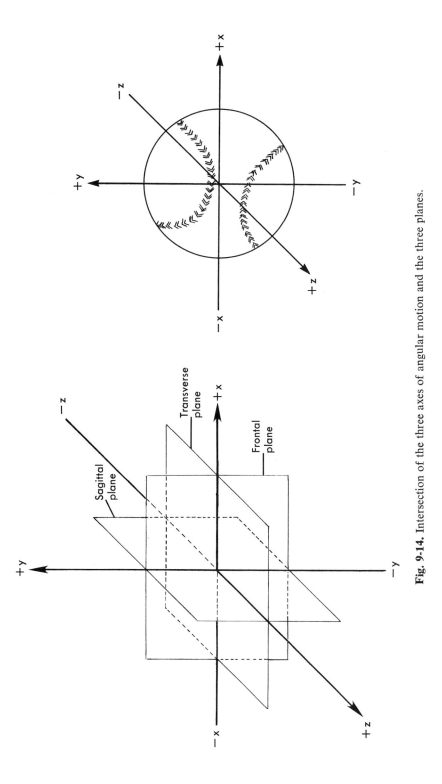

Fig. 9-14. Intersection of the three axes of angular motion and the three planes.

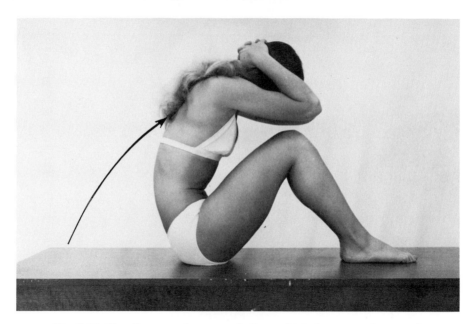

Fig. 9-15. The sit-up exercise occurs in the sagittal plane around an x axis.

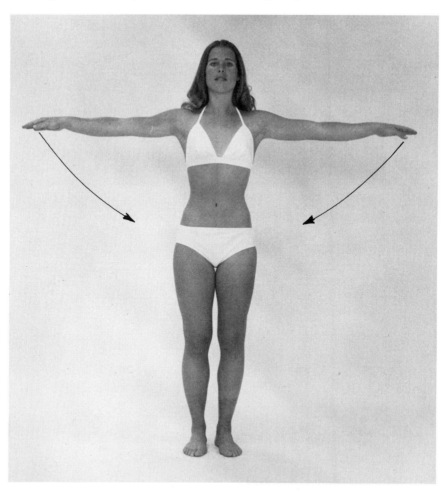

Fig. 9-16. Adduction of the arms occurs in the frontal plane around a z axis.

By convention, negative and positive signs are given to movements around the three axes—clockwise movements negative, counterclockwise movements positive.

The three cardinal directions of angular motion

The three cardinal directions of angular motion are identified according to the planes in which they occur and the axes around which they revolve:

1. Movement in the sagittal plane around a frontal (x) axis (Fig. 9-15) is a rotation either forward or backward, as in doing a forward or backward roll in tumbling.

When viewed from the right or positive side, a forward rotation is clockwise and negative while a backward rotation is counterclockwise and positive.

When describing the spin of a ball, we refer to this cardinal direction of angular motion either as forward or top spin or as back spin.

2. Movement in the frontal plane around a sagittal (z) axis (Fig. 9-16) is a rotation either to the right or to the left, as in doing a cartwheel in tumbling.

When viewed from the front or positive side, rotation to the performer's left is clockwise and negative while rotation to the per-

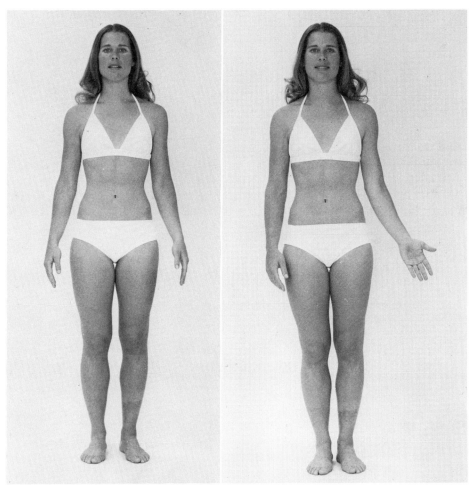

Starting position   Final position

Fig. 9-17. Supination of the forearm occurs in the transverse plane around a y axis.

Fig. 9-18. The Logan-McKinney diagonal planes include high **(A)** and low **(B)** planes for the shoulder joint and the diagonal **(C)** plane of the hip joint.

former's right is counterclockwise and positive.

3. Movement in the transverse plane around a longitudinal (y) axis (Fig. 9-17) is a rotation either to the right or to the left, as in executing a full twist in tumbling.

   When viewed from the top or positive side, rotation to the right is clockwise and negative while rotation to the left is counterclockwise and positive.

The diagonal motion planes

The three oblique motion planes identified in 1970 by Logan and McKinney to describe ballistic limb actions are illustrated in Fig. 9-18. A high and a low diagonal plane were identified for the shoulder joint and are shown in Figs. 9-18, *A,* and 9-18, *B,* respectively. The diagonal plane of motion occurring at the hip joint is shown in Fig. 9-18, *C.* Most of the ballistic actions of the limbs occur through these diagonal planes. One of the reasons presented by Logan and McKinney for this fact is that the major muscles attached to the femur and humerus provide diagonal lines of pull, which results in the limb levers' being pulled through the commonly observed diagonal planes.

The location of a diagonal motion plane is determined by measuring the angle it makes with a rectangular plane.

The polar coordinate system

The position of a point *(P)* in an x-y coordinate system can be specified, as we already know, by a pair of numbers (*x* and *y*) (Fig. 9-19, *A*). Alternatively the equivalent position information can be given by specifying two other quantities: the distance *(r)* from the origin *(O)* to the point *(P),* called the radius of the point, and the angle *(θ)* that the line *OP* (i.e., the radius *r*) makes with the positive x axis (Fig. 9-19, *B*). In other words, we can give the position of *P* in terms of *r* and *θ* instead of in terms of *x* and *y*. The numbers *r* and *θ* are called the polar coordinates of the point. The use of polar coordinates is fundamental to the study of angular motion quantities.

As can be seen in Fig. 9-20, the radial coordinate *(r)* of *P* is the hypotenuse of a right triangle in which *x* and *y* are the other two sides. Therefore *r* (which is always positive) is given in terms of x and y by the Pythagorean theorem:

$$r = \sqrt{x^2 + y^2}$$

Using the definitions of the trigonometric functions, we can express the rectangular coordinates in terms of the polar coordinates:

$$x = r \cos \theta$$
$$y = r \sin \theta$$

Likewise, since we know that tan θ = y/x, we know

$$\theta = \text{Angle whose tangent is y/x}$$

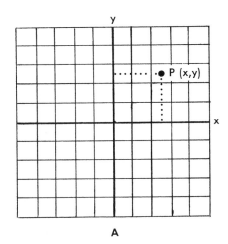

A

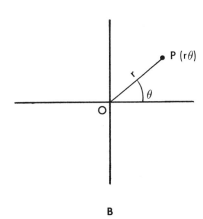

B

Fig. 9-19. The two types of coordinate systems used in mechanics are the rectangular **(A)** and the polar **(B)**.

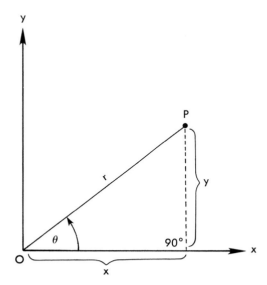

Fig. 9-20. The radial coordinate is the hypotenuse of a right triangle where *x* and *y* are the other two sides.

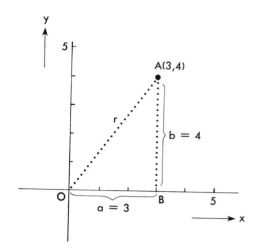

Fig. 9-21. Determining the distance of a point from the origin.

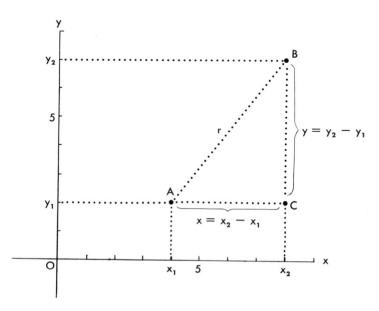

Fig. 9-22. Determining the distance between two points neither of which is the origin.

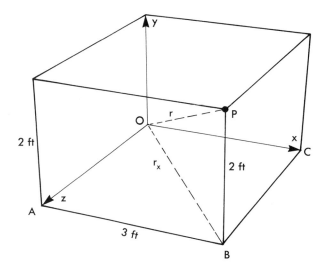

Fig. 9-23. Determining the distance between two points in three dimensions.

Determining the distance between two points in space

Because the axes of a rectangular coordinate system are at *right* angles, the Pythagorean theorem can be used to calculate the distances between any pairs of points in such a system. Consider the point *A* with coordinates *3,4* shown in Fig. 9-21. How far is *A* from the origin? The point *A* and the origin define the right triangle *OBA*. Therefore the distance *(r)* from *O* to *A* is

$$r = \sqrt{x^2 + y^2} = \sqrt{(3)^2 + (4)^2} = \sqrt{9 + 16}$$
$$= \sqrt{25} = 5 \text{ units}$$

This procedure works even if the coordinates of *A* are negative values because the squares of the coordinates are what enter into the calculations.

If we wish to find the distance between two points neither of which is the origin, we follow a similar procedure. Such a case is shown in Fig. 9-22. We require to know the distance from *A* to *B*. The two points *A* and *B* and the point *C* (which has the same x value as *B* and the same y value as *A*) define the right triangle *ACB*. Then $r = \sqrt{x^2 + y^2}$, where $x = x_2 - x_1$ and $y = y_2 - y_1$.

In order to find the distance between two points in three-dimensional space, it is necessary to use the Pythagorean theorem twice. This procedure is best illustrated by means of the example shown in

Fig. 9-23. The problem here is to determine the the distance from point *O* to point *P,* which is the diagonal of a box. The diagonal is represented in the figure by the dashed line *OP*. The procedural steps involved in solving this type of problem are as follows:

1. *Combine the horizontal vectors into a single resultant.* In this regard, consider the right triangle *OAB* and the diagonal *OB* in Fig. 9-23. The line *OA* is the z component of the total vector, the line *AB* is the x component, and the line *OB* is the hypotenuse or horizontal radial component r_x. Thus

$$OB = \sqrt{OA^2 + AB^2} \text{ or } r_x = \sqrt{z^2 + x^2}$$

2. *Combine the horizontal resultant with the vertical vector.* In this regard, consider the right triangle *OBP*. The line *OB* is the resultant or horizontal radial component (r_x) obtained in Step 1, the line *BP* is the vertical component y, and the line *OP* is the total radial component (r). Again we use the Pythagorean theorem to find the length of the side *OP* (the diagonal of the box):

$$OP = \sqrt{OB^2 + BP^2} \text{ or } r = \sqrt{r_x^2 + y^2}$$

Since $r_x^2 = x^2 + z^2$, the equation becomes

$$r = \sqrt{x^2 + z^2 + y^2}$$

Thus the distance between two points located at x_1, y_1, z_1 and x_2, y_2, z_2 is given by

$$r = \sqrt{x^2 + y^2 + z^2}$$

where $x = x_2 - x_1$, $y = y_2 - y_1$, and $z = z_2 - z_1$.

Summary

It should now be evident that the preceding discussion has been only an extension of our vector addition considerations of the last lesson. Thus in summary the steps involved in the three-dimensional addition of vectors are (1) resolve all vectors into their x, y, and z components through either graphic or trigonometric means, (2) combine the horizontal (x and z) components to obtain a horizontal resultant, and (3) combine the horizontal resultant with the vertical y component to obtain the total resultant (r).

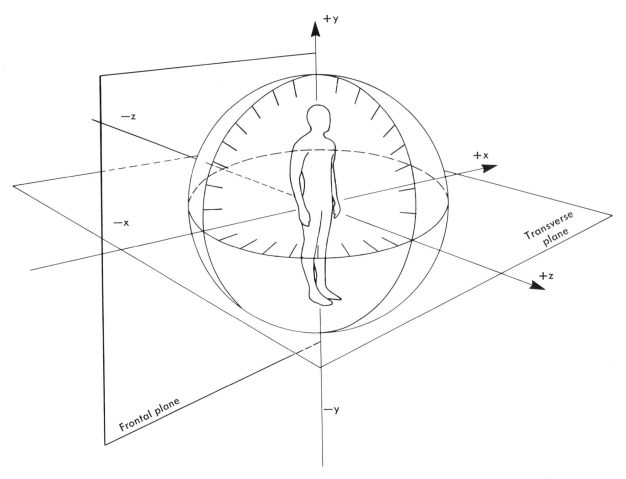

Fig. 9-24. In the three-dimensional description of segmental movements, the subject is conceived of as standing inside a globe.

Three-dimensional description of angular segmental movements

The two-dimensional methods of describing angular segmental movements discussed in Lesson 7 are useful for recording joint movements occurring in a single plane. To measure motion in a joint which has more than one axis of motion and is capable of moving in two or three planes, a three-dimensional approach is required. In this approach the individual or subject being measured is conceived of as standing inside a globe (illustrated in Fig. 9-24). The joint movements of the subject are recorded on the surface of the globe. Abduction and adduction are measured by the meridians, while flexion and extension are measured by the parallels of the globe. In this way the excursion field subscribed by a joint can be traced upon the surface of the globe as shown in Fig. 9-25.

Radial and angular coordinates in three dimensions

As can be seen in Fig. 9-25, the excursion field subscribed by a joint forms a sphere. Thus spherical coordinates are required to describe points in this field. In spherical coordinates (Fig. 9-26) locating a point *(P)* amounts to constructing a sphere whose center is the origin and whose radius is such that *P* lies on the surface of the sphere. The three spherical coordinates are

r = Radius of the sphere
θ = Angle (theta) measured from the y axis toward the x-z plane (the polar angle)
ϕ = Angle (phi) measured from the z toward the x axis (the azimuthal angle)

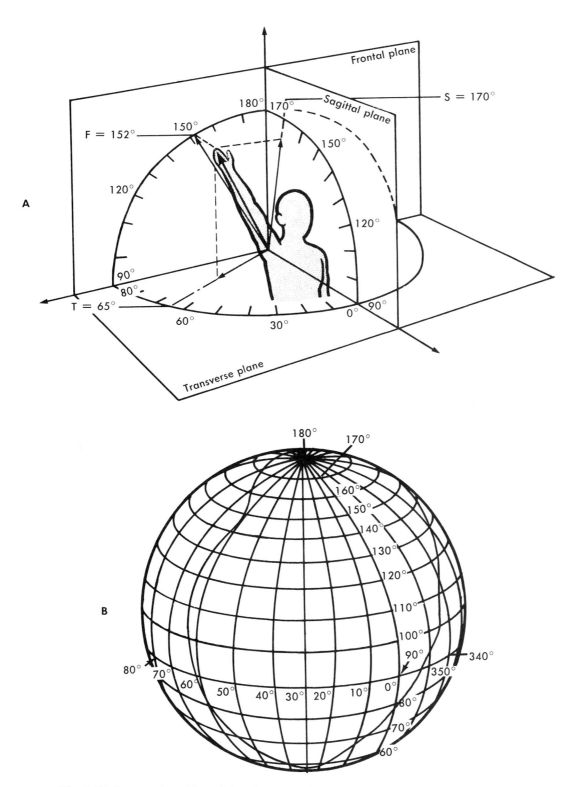

Fig. 9-25. Segmental positions **(A)** and segmental movements **(B)** of a subject are recorded on the surface of a globe. The instantaneous position of the subject **(A)** is measured by the angles the segments make with the three planes.

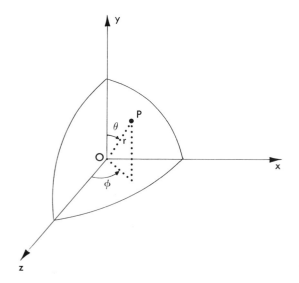

Fig. 9-26. Spherical coordinates of a point in three dimensions.

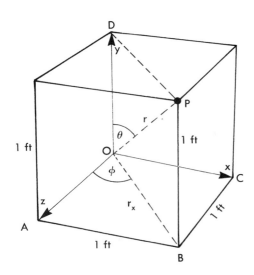

Fig. 9-27. Radial and angular coordinates of a point in three dimensions.

The relationships connecting (r, θ, ϕ) and (x, y, z) are

$$x = r \sin \theta \sin \phi$$
$$y = r \cos \theta$$
$$z = r \sin \theta \cos \phi$$

ϕ = Angle whose tangent is x/z
θ = Angle whose cosine is y/r

Perhaps the concept of radial and angular coordinates in three dimensions can be better presented by turning to Fig. 9-27, which is Fig. 9-23 redrawn as a cube to show radial and angular coordinates. The steps involved in converting spherical coordinates into rectangular coordinates are as follows:

1. *Find the horizontal component of radius r.* Fig. 9-27 shows that a line drawn from *D* to *P* will form a right triangle *ODP*. The line *DP* is equal to the horizontal component *OB* or r_x of the line *OP*, which is the radius *(r)*. Since line *DP* is the side opposite θ, it can be obtained through

$$DP = OP \sin \theta \text{ or } r_x = r \sin \theta$$

2. *Determine the values of the x, y, and z components.* Fig. 9-27 shows that

$$x = r_x \sin \phi$$
$$y = r \cos \theta$$
$$z = r_x \cos \phi$$

THREE-DIMENSIONAL CINEMATOGRAPHY

The use of a single camera, which produces two-dimensional data about three-dimensional motion, presents many problems of interpretation. The primary problem is perspective error. Everyone has seen examples of perspective error, such as pictures of people with feet larger than their bodies, trees larger than mountains, and buildings with unnaturally curved edges. These photographic distortions produce errors in the measurement of linear and angular motion quantities.

Linear errors of measurement

The cause of perspective error is the fact that some part of the object being photographed is closer to the lens than some other part in the same visual field. The part which is close appears larger than the part further away. This is illustrated by the railroad track perspective shown in Fig. 9-28. The cross ties farther away appear smaller than identical ones closer to the viewer. The difference in sizes of the images of these cross ties can be explained through use of the following equation:

$$\frac{\text{Distance of object to lens}}{\text{Size of object}} = \frac{\text{Distance of lens to film}}{\text{Size of image on film}}$$

(Equation 9-A)

Since the ratio on the right side of Equation 9-A deals with factors internal to the camera and is constant for a given camera, the primary variables determining image sizes are the factors shown in the ratio on the left. Since the cross ties shown in Fig. 9-28 are identical in actual size, then the difference in image sizes must be due solely to the difference in the distances of the ties from the lens.

Fig. 9-28. The railroad track example of perspective error.

Perspective errors are greatly magnified when the camera is very close to the object. A possible means of minimizing this error is to use a telephoto lens and place the camera farther away. This procedure is illustrated in Fig. 9-29. Three objects labeled *a, b,* and *c* have identical sizes. Should we look through the camera lens in the situation illustrated on the left, object *b* would appear much larger than objects *a* and *c* due to the fact that this object is closer to the lens than are the other two. The difference in the sizes of these images can be reduced by moving the camera back as shown on the right in Fig. 9-29. Even though the actual distances between the objects remain the same as before, the difference in their distances from the lens has been markedly reduced. Thus their images on the film are more nearly the same.

Linear errors of measurement can also be reduced by photographing an object when it is located directly in front of the lens. The linear dimensions of object *b,* which is directly in front of the lens shown in Fig. 9-29, can be measured with much less error than can those of objects *a* and *c.*

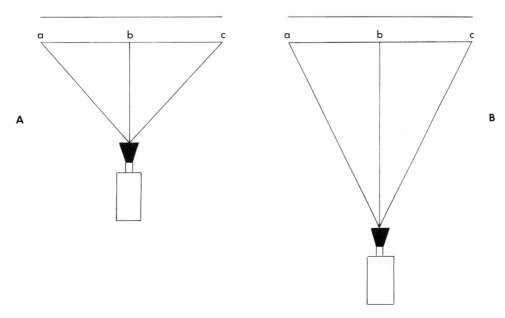

Fig. 9-29. Perspective errors are greatly magnified when the camera is very close to the object **(A).** A possible means of minimizing this error is to use a telephoto lens and place the camera further away as in **B.**

Another fundamental rule of cinematography is to align the objects to be photographed in a plane that is perpendicular to the lens. This was done, for example, with the objects shown in Fig. 9-29. When photographing a movement, be careful to have the camera positioned so its lens is located perpendicular to the plane of the motion.

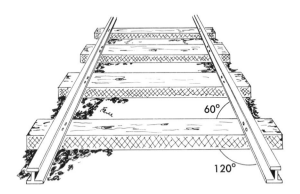

Fig. 9-30. Example of how the parallel rails (which form 90° angles with their cross ties) are altered photographically when the camera lens is not aligned perpendicularly.

When photographing the trajectory of a diver, for example, align the camera lens perpendicular to the line of parabolic flight of the dive. The best position of the camera to photograph the railroad track in Fig. 9-28 would be directly above the track with the lens perpendicular to the plane of the track.

An excellent review of the errors in linear measurement with cinematography is presented by Doolittle (1971). This source is recommended for further information about such errors.

Angular errors of measurement

Fig. 9-30 is a redrawing of the railroad track shown in Fig. 9-28. Note how the parallel rails, which form 90° angles with their cross ties, are altered photographically when the camera lens is not aligned perpendicular to the plane of the track. The true right angle between tie and rail is expanded on the near side of the tie to 120° and correspondingly reduced on the far side. Any angle can be reproduced photographically so its value will vary from 0° to 180° depending upon the alignment of the lens with the plane of the angle (Noss, 1967).

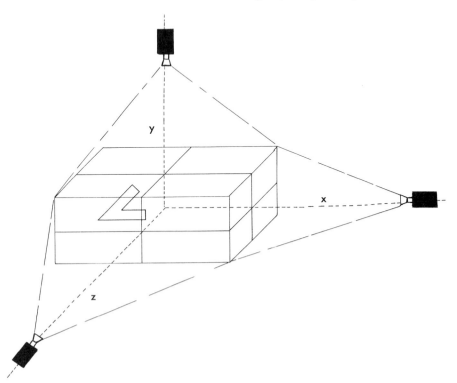

Fig. 9-31. Triplanar arrangement of three cameras used by Noss (1967).

The angular measurement of human movements through cinematography is complicated by the fact that these movements rarely occur in a single plane. A body moving in one direction will present a complex of angles to the camera none of which may exist at right angles to the camera lens. Thus the measurement of these angles could be in great error. The reduction of these errors requires the use of more than one camera plus a three-dimensional analysis approach.

Multicamera cinematography

Many investigators have realized the shortcoming of single-camera methods of data acquisition and have attempted three-dimensional analyses by using two or more cameras. These multicamera procedures involve either a triplanar or a biplanar arrangement of the cameras.

Noss (1967) has proposed a triplanar, triaxial arrangement using three cameras aligned perpendicular to each other and to the three cardinal

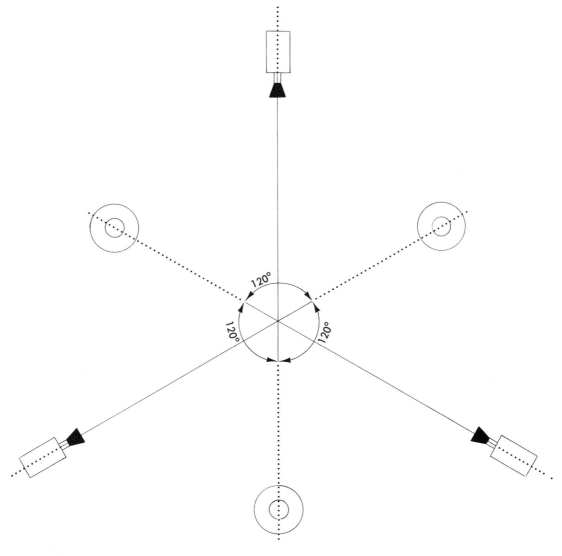

Fig. 9-32. Biplanar arrangement of three cameras used by Miller and Petak (1973).

planes of space parallel with the three cardinal axes. Technically the optical axes of the cameras are aligned parallel with the three axes. An optical axis is defined as a straight line from the geometric center of the lens that is directed toward the center of the projected image. In Noss' triaxial arrangement the three optical axes are assumed to intersect at the origin or the geometric center of their respective fields of view.

When applied in the measurement of angles, triaxial analysis yields three sets of data. Fig. 9-31 indicates how a triaxial analysis might measure a 45° angle. The camera operating on the x axis might record the angle as having a value of 27°; the camera operating on the y axis might read the angle as 72°; the camera operating on the z axis might measure the angle as 36°. While no single camera would accurately recreate the value of the subject angle, the average of all three measurements would equal the true value of the angle as indicated in the diagram.

A biplanar arrangement has been developed by Miller and Petak (1973) in which three cameras are aligned horizontally with their optical axes level, 120° apart, and intersecting as shown in Fig. 9-32. Since the implementation of this system is quite involved, the student should consult the original source for its description.

STUDY GUIDELINES
Three-dimensional analysis of linear quantities
Specific student objectives

1. Identify the three dimensions of space.
2. Differentiate between cardinal planes and secondary planes.
3. Describe the three rectangular planes of space.
4. Describe the anatomical position and its use in defining the reference planes of the human body.
5. Describe the three cardinal directions of linear motion.
6. Describe the rectangular coordinate system.
7. Define a right-handed coordinate system
8. Describe the convention by which negative and positive signs are given to the three rectangular coordinates.

Summary

Points in space are specified by three measurements if the motion description is three dimen-

sional. A two-dimensional analysis requires two measurements to locate a point in space. Measurements are taken from three perpendicular reference planes of the body to locate body positions and record body movements. They are called *cardinal* when laid through the center of gravity of the body and *secondary* when drawn through the center of gravity of a body segment, the center of a joint, or some other point of reference. The three cardinal planes are

1. Sagittal—passing through the body from front to back, dividing it into right and left sections
2. Frontal or lateral—passing through the body from side to side, dividing it into front and back sections
3. Transverse—passing through the body, dividing it into upper and lower sections

The anatomical position is an upright position of the human body with feet parallel, arms at the sides, and the palms of the hands facing forward.

The three cardinal directions of motion are defined as

1. Sideward or x direction—right-to-left or left-to-right movement perpendicular to the y and z directions.
2. Vertical or y direction—upward or downward movement
3. Forward or backward or z direction—near-to-far or far-to-near movement pependicular to the x and y directions.

A right-handed coordinate system is one that when the fingers of the right hand are curled inward (in the direction necessary to rotate the positive x axis toward the positive y axis) the extended thumb will point in the direction of positive z axis.

The rectangular coordinate system is used for recording the position of body parts as follows:

1. Width or x coordinate—to the right of some reference point, *positive;* to the left, *negative*
2. Height or y coordinate—above some reference point, *positive;* below, *negative*
3. Depth or z coordinate—in front of some reference point, *positive;* behind, *negative*

Performance tasks

1. The three spatial dimensions are _____, _____, and _____ (in any order).

2. Planes that pass through the center of gravity of the body are called (cardinal/secondary) _____ planes.
3. Describe the three rectangular planes.
   a. Sagittal _____
   b. Frontal _____
   c. Transverse _____
4. Classify the plane of motion for the following segmental movements:
   a. Flexion-extension _____
   b. Abduction-adduction _____
   c. Rotation _____
5. A human body in upright position, feet parallel, arms at the side, and palms forward is said to be in the (anatomical/cardinal) _____ position.
6. Write out descriptions of the following:
   a. Three rectangular planes of space
   b. Three cardinal directions of linear motion
   c. Rectangular eoordinate system
   d. Right-handed coordinate system
7. Construct a cardboard model of the following:
   a. Rectangular planes of space
   b. Cardinal directions of linear motion
   c. Points in space as represented by the rectangular coordinate system

Three-dimensional analysis of angular space
Specific student objectives

1. Describe an axis.
2. Describe the three axes of angular motion.
3. Describe the three cardinal directions of angular motion.
4. Describe the diagonal motion planes.
5. Describe the polar coordinate system.
6. Describe how the distance between two points in (a) two-dimensional space and (b) three-dimensional space can be determined.
7. State how the three-dimensional movements of body segments can be described.
8. Describe the method of determining the three-dimensional radial and angular coordinates of body positions.

Summary

The axes used in describing angular motion are imaginary straight lines about which circular motion takes place:
1. Longitudinal y—perpendicular to the transverse plane

2. Frontal x—perpendicular to the sagittal plane
3. Sagittal z—perpendicular to the frontal plane

Clockwise movements are negative; counterclockwise movements are positive.

The three cardinal directions of angular motion are defined as follows:
1. Motion in a sagittal plane around a frontal (x) axis; a forward (top spin) or backward (backspin) rotation when viewed from the right side is positive or negative, respectively.
2. Motion in the frontal plane around a sagittal (z) axis; when viewed from the front, rotation to the performer's left is negative; or to the right is positive.
3. Motion in the transverse plane about a longitudinal (y) axis; when viewed from above, rotation to the right is negative; or to the left is positive.

The diagonal motion planes are oblique motion planes located at an angle other than a right angle to one or more of the cardinal or secondary reference planes of the body. These planes are identified as the high and low diagonal planes of the shoulder joint and the diagonal plane of the hip joint.

A point in space can be specified in two dimensions not only in terms of the numbers x and y but also in terms of r and θ, which are the polar coordinates of the point, where r is the distance from the origin to the point in space (i.e., the radius) and θ is the angle that r makes with the positive x axis.

Distance between two points in space, which is r in polar coordinates, is given by the Pythagorean theorem ($r = \sqrt{x^2 + y^2}$).

The three-dimensional addition of rectangular vectors is as follows:
1. Resolve all vectors into their x, y, and z components through either graphical or trigonometric means.
2. Combine the horizontal (x and z) components to obtain a horizontal resultant.
3. Combine the horizontal resultant with the vertical (y) component to obtain the total resultant.

The three-dimensional description of angular segmental movements conceives of the subject inside a globe. The joint movements of the subject are recorded on the surface of the globe. The sphere so recorded can be analyzed in terms of spherical coordinates, which are radius (r) of the sphere, the vertical angle (θ) which r makes with the y axis measured toward the x-z plane, and the horizontal angle (ϕ) which r makes with the z axis.

Performance tasks

1. An imaginary line about which rotation takes place is called an/a (axis/plane)_____ .
2. The three axes of angular motion are
   a. longitudinal, frontal, and lateral
   b. frontal, sagittal, and transverse
   c. linear, polar, and rectangular
   d. frontal, longitudinal, and sagittal
3. Develop a written list and description of the following:
   a. Axes of angular motion
   b. Cardinal directions of angular motion
   c. Polar coordinate system
   d. Diagonal motion planes
   e. Three-dimensional description of angular segmental movements
4. Using angular motion examples, illustrate each of the axes, cardinal directions, and motion planes. For example, flexion and extension of the elbow joint occur in the secondary sagittal plane, around the frontal or x axis, and in positive (extension) and negative (flexion) directions.
5. Describe the meaning of the symbols used in the following formulas:
   a. $r = \sqrt{x^2 + y^2}$
   b. $x = r \cos \theta$
   c. $y = r \sin \theta$
   d. θ = the angle whose tangent is y/x
6. Using angular motion examples, illustrate how the distance between any two points in space can be determined.
7. State how the three-dimensional movement of the arm at the shoulder joint can be described in terms of its
   a. excursion field
   b. spherical coordinates

Three-dimensional cinematography
Specific student objectives

1. Discuss the following aspects of single camera two-dimensional cinematography:

   a. Perspective error and its reduction
   b. $\dfrac{\text{Distance of object to lens}}{\text{Size of object}} = \dfrac{\text{Distance of lens to film}}{\text{Size of image on film}}$
   c. Linear errors
   d. Angular errors
2. Discuss multicamera cinematography:
   a. Triplanar arrangement
   b. Biplanar arrangement

Summary

A single camera produces two-dimensional data about three-dimensional motion. Cinematographic analysis of this type encounters the problem of perspective error, where objects being photographed are of varying distances from the lens.

Linear errors resulting from perspective problems can be reduced by such techniques as filming from long distances, filming directly in front of the camera, and aligning the objects to be filmed in a plane perpendicular to the lens.

Errors in angular measurement can be best reduced through a multicamera approach. Triplanar and biplanar methods have been developed to analyze human motion in three dimensions.

Performance tasks

1. Develop a written description of the advantages and disadvantages of single camera two-dimensional cinematography.
2. Develop a written description of the advantages and disadvantages of multicamera three-dimensional cinematography.
3. Consult two outside references on cinematography for human motion analysis. List and describe the characteristics of two-dimensional and three-dimensional motion picture analysis.

SELF-EVALUATION

Students should use no reference materials for this progress test, and they can check their answers by referring to Appendix A.
1. The three spatial dimensions are
   a. height, width, and depth
   b. frontal, sagittal, and transverse
   c. both a and b
   d. neither a nor b
2. Planes that pass through the center of gravity of a body part are called (cardinal/secondary) _____ .

3. Complete the following rectangular plane descriptions:
   a. Sagittal _____

   _____
   b. Frontal _____

   _____
   c. Transverse _____

   _____
4. Describe the anatomical position.
5. Complete the descriptions for the cardinal directions of motion:
   a. Sideward or x direction _____

   _____
   b. Vertical or y direction _____

   _____
   c. Forward-backward or z direction _____

   _____
6. Describe the location of a point in space using the rectangular coordinate system.
7. An imaginary line about which angular motion takes place is termed an/a (axis/plane)

   _____
8. Complete the descriptions of the three axes for angular motion.
   a. Frontal (x) _____

   _____
   b. Longitudinal (y) _____

   _____
   c. Sagittal (z) _____

   _____

9. Complete the descriptions of the three cardinal directions for angular motion:
   a. Sagittal plane, frontal (x) axis _____

   _____
   b. Frontal plane, sagittal (z) axis _____

   _____
   c. Transverse plane, longitudinal (y) axis

   _____

   _____
10. Identify the axis and plane of angular motion for each of the following:
    a. Knee flexion   e. Dorsal flexion
    b. Shoulder adduction   f. Foot eversion
    c. Trunk flexion   g. Hip abduction
    d. Pronation
11. Identify the Logan and McKinney diagonal motion planes.
12. Describe the polar coordinate system for locating a point in space.
13. List the three steps involved in the three-dimensional addition of rectangular vectors.
14. State how joint segmental movements are described in three dimensions.
15. Describe the following areas of concern in human motion analysis through cinematography:
    a. Reduction of perspective errors, both linear and angular
    b. Multicamera three-dimensional cinematography

Kinematic analysis of fundamental movements

Kinematic analysis of linear motion

Body of lesson
Analysis of linear speed, velocity, and acceleration
 Linear speed and velocity
 Linear acceleration
Analysis of average and instantaneous velocities and accelerations
 Average velocity
 Instantaneous velocity
 Average and instantaneous acceleration
Graphical description of linear motion quantities
 Graphical description of linear velocity
 Graphical description of instantaneous linear velocity
 Graphical description of linear acceleration
 Concept of linear quickness
Study guidelines
Self-evaluation

ANALYSIS OF LINEAR SPEED, VELOCITY, AND ACCELERATION

The rate at which a body moves is called its *speed,* while the rate at which it changes its speed (as in speeding up or slowing down) is called its *acceleration.* Since we have two basic types of motion, linear and angular, we also have two methods of describing the rates of motion. An introduction to the methods of describing linear speed, velocity, and acceleration is presented in this lesson. The extension of the methods of linear kinematics to the study of parabolic motion is presented in Lesson 11. An introduction to the methods of describing the rates of angular motion is presented in Lesson 12.

Linear speed and velocity

A good example of a human motion that follows a linear path is a track man walking on a straight track (Fig. 10-1). The speed of the walker is defined as his rate of change of position on the track. Since by *change of position* we mean the distance he travels, the definition of speed can be written

$$\text{Speed} = \frac{\text{Distance traveled}}{\text{Time}}$$

$$V = \frac{d}{t} \qquad \text{(Equation 10-A)}$$

Therefore the two variables involved in speed are distance traveled (d) and time elapsed (t).

The speed of a body is not, alone, a complete description of its state of motion at any time. The *direction* in which it is moving, as we have already mentioned, is also important. To specify a body's motion more completely, we must include its direction as well as its speed. Such a specification is called *velocity.* Thus speed is a scalar quantity (magnitude only) and velocity is a vector quantity (magnitude plus direction).

Even though the magnitudes of the two quantities are described in the same manner $(V = d/t)$, "speed" is better reserved for the rate at which something covers a distance and "velocity" for its rate of displacement in a given direction (i.e., the more complete description of its motion through space).

Linear acceleration

Whenever the speed of linear motion changes, it is described as an acceleration. An object that is picking up speed has a positive acceleration, while an object that is slowing down has a negative acceleration (also known as deceleration).

More precisely, acceleration is defined as the rate of change of velocity. This means that linear acceleration occurs when either a change in the

173

Fig. 10-1. Linear motion of a walker on a track straightaway.

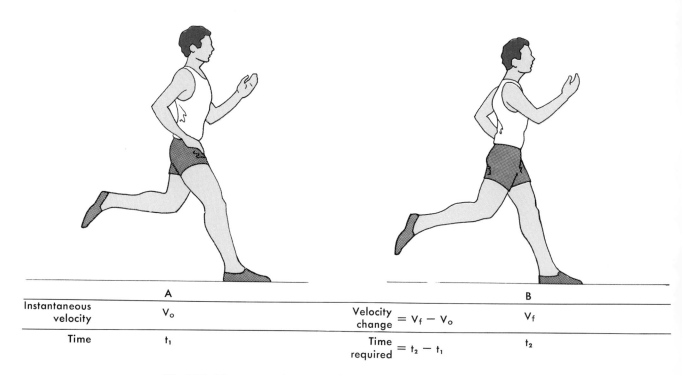

Instantaneous velocity	V_o	Velocity change $= V_f - V_o$	V_f
Time	t_1	Time required $= t_2 - t_1$	t_2

Fig. 10-2. Movements of a runner as an example of linear acceleration.

linear speed or a change in the linear direction of a movement takes place. Acceleration resulting from a change of speed will be discussed in this lesson, while acceleration resulting from a change of direction will be discussed in Lesson 20.

Consider as an example of accelerated motion the sprinter shown in Fig. 10-2. Due to a constantly acting force, exerted by his muscles, he is constantly accelerated as he moves along the straight line *AB*. As he passes *A*, he has a rela-

tively low velocity (V_o, called the original velocity); but farther along the track at point *B* he is moving faster and he has a different velocity (V_f, called the final velocity).

If the time required for the sprinter to go from *A* to *B* is *t*, his acceleration by the definition given above is schematically written:

$$\text{Acceleration} = \frac{\text{Final velocity} - \text{Original velocity}}{\text{Time}}$$

$$a = \frac{V_f - V_o}{t} \qquad \text{(Equation 10-B)}$$

Linear acceleration, like linear velocity, is a vector quantity. It has a direction as well as a magnitude.

Suppose at point *A* in Fig. 10-2 the velocity of the sprinter is 10 ft/sec, at point *B* his velocity has increased to 20 ft/sec, and it takes him 2 seconds to go from *A* to *B*. Thus his acceleration would be $\dfrac{20 \text{ ft/sec} - 10 \text{ ft/sec}}{2 \text{ sec}}$, or 5 ft/sec/sec. This means that his velocity is increasing an average 5 ft/sec during every second of time. The acceleration units of ft/sec/sec can be shortened to ft/sec² because ft/sec/sec is the same as $\dfrac{\text{ft}}{\text{sec}} \div \dfrac{\text{sec}}{1}$ and, inverting and multiplying, we obtain $\dfrac{\text{ft}}{\text{sec}} \times \dfrac{1}{\text{sec}} = \text{ft/sec.}^2$

ANALYSIS OF AVERAGE AND INSTANTANEOUS VELOCITIES AND ACCELERATIONS

In the preceding discussions of this lesson, no distinction has been made between average and instantaneous velocities and accelerations. This distinction should be made now.

Average velocity

When an object is moving with a constant velocity, it is said to be in *uniform motion*. A constant velocity is defined as one in which equal displacements occur in equal intervals of time and the direction of motion remains unchanged. In other words, the distance traveled in any one second is equal to that traveled in any other second.

Except in certain special cases, the velocity of a moving body changes continuously as the motion proceeds. The body is then said to have *variable motion* or an acceleration. A body has a variable velocity when in equal intervals of time its displacements are unequal or its directions of movement are different. Then it is customarily said to have an *average velocity*. Average velocity ($\overline{V}$) is defined in the usual manner as

$$\overline{V} = \frac{d}{t}$$

where *t* is the time required to travel the specified distance *(d)*. The bar above the V signifies an average value.

Consider as an example of average velocity a sprinter who runs a 100-yard dash in 10 seconds flat. His velocity $\left(\dfrac{d}{t}\right)$ would be calculated as $\dfrac{100 \text{ yd}}{10 \text{ sec}} = 10$ yd/sec. Since his velocity is not the same at both the start and the finish of the race, 10 yards/sec is only an average of the many different velocities he achieved over the 100-yard distance.

If the velocity of a body is known, the distance the body will travel in any given interval of time can be calculated by rearranging the velocity formula (V = d/t) to obtain d = Vt. Likewise, the formula can be rearranged to give the time required for a body to travel a certain distance at a given velocity (t = d/V). When we are dealing with a known constant velocity, then, of course, the distance or the time can be predicted precisely. However, when we are dealing with a known average velocity, the distance or the time can only be approximated at any given point.

Instantaneous velocity

The velocity of the sprinter in Fig. 10-2 at one instant of time, or at some point along his path, is called his instantaneous velocity. We witness instantaneous velocity every time we look down at the speedometer of our automobile. Examples of instantaneous velocity can also be drawn from the formula for acceleration: $\dfrac{V_f - V_o}{t}$ Both the final velocity and the original velocity are usually defined as instantaneous.

Average and instantaneous acceleration

Like velocity, acceleration can be either constant or variable. In a *constant acceleration* the velocity changes by equal amounts during each unit of time. A good example of constant acceleration is the acceleration produced by the force of gravity. In a *variable acceleration* the velocity does not change by equal amounts each second.

The *average acceleration* of a moving body is defined as the ratio of the change in velocity to the elapsed time

$$\bar{a} = \frac{V_f - V_o}{t}$$

where $V_f - V_o$ is the total change in velocity and *t* is the total time.

From this definition we can see that the average change in velocity $(V_f - V_o)$ is a product of the acceleration and the time during which the acceleration occurred; i.e., $V_f - V_o = \bar{a}t$. When we are dealing with a constant acceleration such as the acceleration of gravity, we can rearrange the above formula to find the instantaneous final velocity precisely $(V_f = V_o + at)$. Here the final velocity (V_f) of a body is given as the sum of two terms, the initial velocity (V_o) plus the increase in velocity (at). When we are dealing with average acceleration, of course, the predicted final velocity is only an approximation.

The *instantaneous acceleration* of a body is defined as its acceleration at one instant of time or at some point along its path. Instantaneous acceleration, along with instantaneous velocity, will be discussed in greater detail in the next section of this lesson.

GRAPHICAL DESCRIPTION OF LINEAR MOTION QUANTITIES

Human movements can be described not only in terms of the numerical values for speed, velocity, and acceleration but also graphically. A graph is a pictorial representation of the relationship between two quantities.

Graphical description of linear velocity

As an illustration of graphical representation, we can use the linear motion of a sprinter running a 50-yard dash on the straightaway of a track. Say we have marked off 10-yard intervals with chalk (Fig. 10-3) and have stationed five timers with stopwatches one at each 10-yard mark. Along comes the runner, and each timer notes the exact time it takes the runner to travel from the starting line to his marker. The measurements obtained when the runner is allowed to have a

Fig. 10-3. Collection of data for the description of linear motion.

running start might look like those shown in Table 10-1.

To analyze these measurements, a sheet of graph paper is helpful. Along the bottom of the sheet we establish a time scale, perhaps letting each division represent 1 second. On the left-hand side we set up a distance scale, perhaps letting each division represent 15 feet. Next we plot our measurements on the graph, as in Fig. 10-4, where the point is the intersection of the lines which pass through the time and distance measurements on the two scales. Clearly the vari-

ous points form a regular pattern. In fact, if we draw lines to join adjacent points, we see that this pattern is almost a straight line. We may reasonably attribute deviations from perfect straightness to experimental error. We have no way of knowing from the graph itself, of course, whether additional measurements between those we did take would also fall on the same line; but we may suppose that if we had made such additional measurements they too would conform to this line.

When a graph of one quantity versus another results in a straight line, each quantity is *directly proportional* to the other. We can verify this proportionality from the graph of Fig. 10-4, where it can be seen that from point *A* to point *B* doubling the time *(t)* means the distance *(d)* also doubles, tripling *t* means that *d* also triples, and so on. Thus we can write

$$d = Vt$$

where *V* is the constant of proportionality. To determine the value of *V*, we can rearrange the equation to obtain $V = d/t$ and calculate d/t for

Table 10-1. Time and distance measurements of a runner

Total distance (ft)	Elapsed time (sec)
0	0
30	1.4
60	2.7
90	4.1
120	5.4
150	6.7

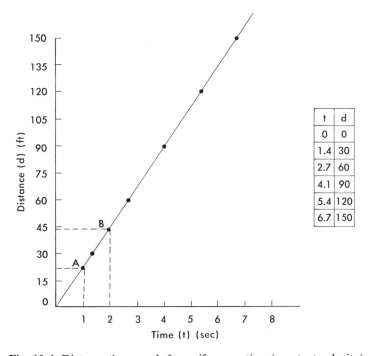

t	d
0	0
1.4	30
2.7	60
4.1	90
5.4	120
6.7	150

Fig. 10-4. Distance-time graph for uniform motion (constant velocity).

Table 10-2. Speeds calculated from the data of Table 10-1

Total distance (ft)	Elapsed time (sec)	Speed (ft/sec)
0	0.0	0.0
30	1.4	21.4
60	2.7	22.2
90	4.1	22.0
120	5.4	22.2
150	6.7	22.4

each of the measurements made. The results are shown in Table 10-2. When the values of V are rounded off, we see that $V = 22$ ft/sec. This is not surprising, since the straight line on the graph indicates a direct proportionality.

The quantity V, as we already know, is called the *velocity* of the runner. It is the rate at which the distance covered by the runner changes with time. Because the distance the runner travels in our example is directly proportional to the elapsed time, V is a constant and the runner is said to move with a *constant* velocity.

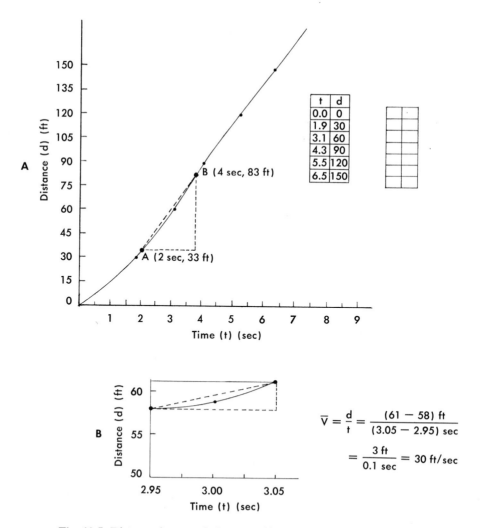

Fig. 10-5. Distance-time graph for nonuniform motion (variable velocity).

Table 10-3. Time and distance measurements of a runner who started from rest at t = 0

Total distance (ft)	Elapsed time (sec)
0	0.0
30	1.9
60	3.1
90	4.3
120	5.5
150	6.5

Graphical description of instantaneous linear velocity

Not all runners, of course, have constant velocities. Another runner, who began running from rest at t = 0, might yield the data shown in Table 10-3. When we plot these data on a graph, as in Fig. 10-5, we find that the line joining the various experimental points is not straight but shows a definite upward curve. In each successive interval of time (as marked off at the bottom of the graph) the runner goes a greater distance; in other words, he is going faster and faster. Because *d* is not directly proportional to *t,* the runner's velocity is not constant.

The *average* velocity ($\overline{V}$) of the runner during any interval of time is given by the slope of a straight line connecting the limits of the interval. For instance, during the interval from t = 2 sec to t = 4 sec the graph reveals that the displacement increases by an amount *(d)* which is 50 feet (83 ft – 33 ft). The time interval *(t)* is 2 sec, so the average velocity (d/t) is 50 ft/2 sec = 25 ft/sec. This is the slope of the dashed triangle in Fig. 10-5, *A.*

The instantaneous velocity can be found by the process of using smaller and smaller time intervals. For instance, in Fig. 10-5, *B,* if the displacement is measured by a time interval from *t* = 2.95 sec to *t* = 3.05 sec (dotted triangle), the displacement will be only 3 feet and the ratio *d/t* can be expected to be closer to the instantaneous velocity at *t* = 3.00 sec than the previous ratio in which a larger time interval was used. By letting the time interval *(t)* become very small, we can have the ratio *d/t* approach the desired instantaneous velocity. This process is called "taking the limit of d/t as t approaches zero."

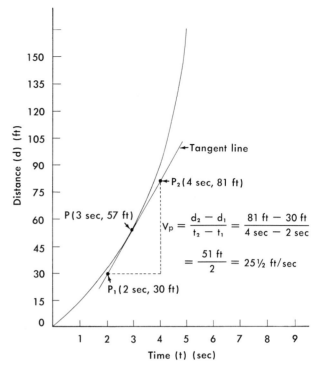

Fig. 10-6. Determination of instantaneous velocity from a distance-time graph by use of the tangent method.

Graphically the direction of the instantaneous velocity is the slope of a tangent drawn on the distance-time graph at the instant in question. A tangent is a straight line which touches the curve at only one point as shown in Fig. 10-6. The tangent line can also be used to calculate the instantaneous velocity at that point. This method is accomplished by selecting any two points on the tangent line and using $V = \dfrac{d_2 - d_1}{t_2 - t_1}$. In Fig. 10-6 this would be $V = \dfrac{81 - 30}{4 - 2} = \dfrac{51 \text{ ft}}{2 \text{ sec}} = 25\frac{1}{2}$ ft/sec. Therefore the estimated instantaneous velocity at point *P* is 25½ ft/sec.

Graphical description of linear acceleration

Since the curve in Fig. 10-6 changes its slope (i.e., the angle between a particular segment of the curve and a horizontal line) continuously, *V* is different at different times. Table 10-4 shows the instantaneous velocity of the runner calculated for the times shown. A graph of this data is shown in Fig. 10-7.

Since acceleration is the time rate change of velocity, we can say that instantaneous acceleration at any time is the slope of the velocity-time curve at the time in question. Therefore instantaneous acceleration is calculated in the same manner as instantaneous velocity, except that $V_f - V_o$ is substituted for $d_2 - d_1$ to obtain $V_f - V_o/t_f - t_o$. Instantaneous acceleration is also determined by applying the tangent method to this velocity-time curve as illustrated in Fig. 10-7 for the instantaneous acceleration of 8 ft/sec^2 at point *P* after an elapsed time of 3 sec.

Table 10-4. Instantaneous speed at 1 sec intervals as determined from Fig. 10-6

Elapsed time (sec)	Instantaneous speed (ft/sec)
0	0
1	16
2	21
3	25
4	26
5	28
6	28

Concept of linear quickness

Quickness is an expression often used in athletics to describe the ability of an athlete to obtain a high rate of movement (velocity) in the shortest time or over the shortest distance. Actually acceleration and quickness are the same. Some coaches, however, use the term quickness to describe the change of velocity per unit of distance. Linear quickness is measured therefore by the minimum amount of displacement (distance) required for the attainment of a given amount of instantaneous velocity. This is usually expressed in terms of the distance required to obtain maximum velocity or as the ratio of velocity obtained to distance required; i.e., velocity/distance or V/d. Linear quickness can also be expressed graphically through the use of a velocity-distance graph.

STUDY GUIDELINES
Analysis of linear speed, velocity, and acceleration
Specific student objectives

1. Describe the following linear motion terms:
   a. Speed   c. Acceleration
   b. Velocity

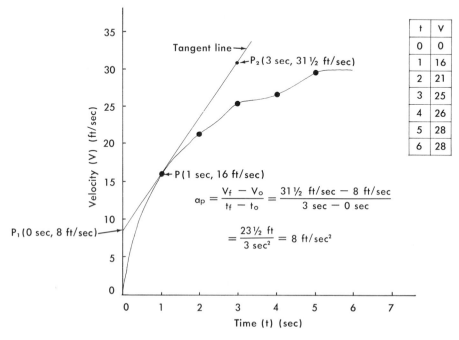

Fig. 10-7. Velocity-time graph for nonuniform motion (variable velocity).

2. Define linear velocity and acceleration in formula and sentence form.

Summary

Speed is the rate of motion. The rate of speed change is called acceleration. Thus speed (V) is defined as distance (d) traveled per unit of time (t), i.e., V = d/t, and is a scalar quantity. Velocity is speed plus the added element of direction and is a vector quantity, i.e., the rate of displacement. The magnitudes of speed and velocity are identical and they are measured through use of the same formula.

Linear acceleration is the time rate change of velocity that occurs when a change of either linear speed or linear direction or both takes place. Quickness is a term synonymous with acceleration.

It can be observed that linear speed is a scalar quantity (magnitude only) while linear velocity and acceleration are both vector quantities (magnitude and direction).

Performance tasks

1. The rate of human motion would be described by its (speed/velocity)_____.
2. Speed plus direction describes_____.
3. A change of velocity per unit of time is called
   a. acceleration
   b. deceleration
   c. both a and b
   d. neither a nor b
4. The formula for speed or velocity is
   a. $V = d/t$
   b. $a = \dfrac{V_f - V_o}{t}$
   c. both a and b
   d. neither a nor b
5. A linear acceleration occurs when there is a
   a. speed increase
   b. speed decrease
   c. change of direction
   d. all of the above
6. Write out the formulas for linear velocity and acceleration. Identify the symbols used in the formulas.
7. Quantities that are measurable are quite common in the environment. Some of these can be described by measurement of magnitude only. This group, called scalars, is exemplified by distance, time, speed, and volume. Other more complex quantities require a measure of magnitude and direction for their complete description. This group, called vectors is illustrated by velocity, displacement, and weight.

Label the following measurable quantities as either a scalar or a vector:

_____ a. Temperature
_____ b. 50 ft west
_____ c. Light
_____ d. 2 gal of water
_____ e. 22 mi/hr
_____ f. Travel from point A to point B
_____ g. 15 mi/hr vertically
_____ h. 4:35 PM
_____ i. 29 cm
_____ j. 50 lb pushing horizontally

8. It is possible to use the speed-velocity formula ($\overline{V}$ = d/t) in three different ways to describe motion along a linear path. $\overline{V}$ = d/t is in a form that is convenient for finding *V* when *d* and *t* are known. This formula can be rearranged (d = $\overline{V} \times$ t) to find *d* if $\overline{V}$ and *t* are known. Likewise, the formula can be rearranged (t = d/$\overline{V}$) to solve for *t*.

Calculate the missing motion quantity in the problems on p. 182.

Analysis of average and instantaneous velocities and accelerations

Specific student objectives

1. Differentiate between average and instantaneous velocities and accelerations.
2. Differentiate between uniform and variable motions.
3. Differentiate between constant and variable accelerations.

Summary

Uniform motion occurs when an object is traveling with a constant velocity, i.e., when equal distances are traveled in equal periods of time. An object traveling with a changing velocity (speed or direction), is said to be in a state of variable motion and to be accelerating. When velocity is variable, unequal distances are traveled in equal periods of time and/or changes in direction occur.

The velocity of an object at a precise instant of

	Distance (d)	Time (t)	Average speed ($\overline{V}$)	Formula
a.	20 mi	4 hr	_____	$\overline{V} = d/t$

Using $\overline{V} = d/t$, we get $V = 20$ mi/4 hr $= 5$ mi/hr.

b.	17 mi	_____	51 mi/hr	$t = d/\overline{V}$

Using $t = d/\overline{V}$, we get $t = \dfrac{17 \text{ mi}}{51 \text{ mi/hr}} = 17 \text{ mi} \times \dfrac{\text{hr}}{51 \text{ mi}} = \frac{1}{3}$ hr.

c.	_____	11 sec	18 ft/sec	$d = \overline{V} \times t$

Using $d = \overline{V} \times t$, we get $d = \dfrac{18 \text{ ft}}{\text{sec}} \times 11 \text{ sec} = 198$ ft.

	Distance (d)	Time (t)	Average speed ($\overline{V}$)	Formula
d.	50 ft	4 sec	_____	_____
e.	300 ft	10 sec	_____ mi/hr	_____
f.	440 yd	_____	20 ft/sec	_____
g.	800 ft	_____	6 ft/sec	_____
h.	_____		24 ft/sec	_____
i.	_____	2 min 8 sec	9 yd/sec	_____
j.	220 yd	24 sec	_____ mi/hr	

time is termed instantaneous velocity. Average velocity is the mathematical average of all the instantaneous velocities for any given interval.

Analogous treatment of acceleration can be carried out. Acceleration may be constant or variable, and an average or instantaneous acceleration can be determined.

Performance tasks

1. Develop a written list of descriptions for:
   a. Average velocity, average acceleration
   b. Instantaneous velocity, instantaneous acceleration
   c. Uniform motion, uniform acceleration
   d. Variable motion, variable acceleration
2. Give some human motion examples illustrating the motion situations listed above.

Graphical description of linear motion quantities

Specific student objectives

1. Describe the graphical description of linear velocity.
2. Describe the graphical description of instantaneous linear velocity.
3. Describe the graphical description of linear acceleration.
4. Describe the concept of linear quickness.
5. Given the raw data of distance and time, construct and label a distance-time graph.
6. Given the raw data of velocity and time, construct and label a velocity-time graph.
7. With the use of a distance-time graph of motion, calculate the instantaneous velocity by use of the *tangent line method* at any selected point on the graph.
8. Using a velocity-time graph of motion, calculate the instantaneous acceleration at any point on the graph by use of the tangent line method.

Summary

Movement can be described in terms of numerical values for speed, velocity, and acceleration. It can be represented graphically to obtain a pictorial view of the motion from which additional information can be derived. Instantaneous velocity and acceleration can also be graphically represented.

Performance tasks

1. Study Fig. 10-8. A graph of distance versus time such as this can be convenient and useful in providing motion information. Since the speed is constant (80 mi/hr), it is represented on the graph as a straight line. The steepness of the line gives us an indication of the magnitude of the speed. A higher rate of speed (as represented by l_1) of 120 mi/hr will result in a steeper line, and a lower rate of speed will produce a less steep line.

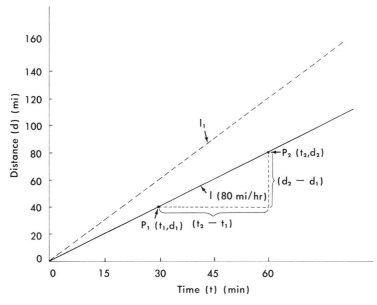

Fig. 10-8. Distance-time graph depicting a constant speed.

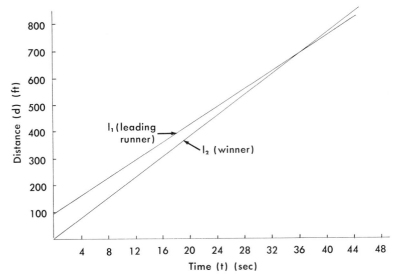

Fig. 10-9. Distance-time graph depicting a race between two runners.

2. Consider the two runners in the final stage of a distance race. The leader is 300 yards from the finish line and traveling at 20 ft/sec. His opponent is 30 yards behind but moving at 23 ft/sec. Will the runner in second place catch the leader before the end of the race?

This question can be answered by construct-ing a distance-time graph (Fig. 10-9). The runner who is ahead at first will lose the race. The graph tells us that the runner represented by l_2 has a steeper graph and will pass l_1 before the end of the race.

3. There is an important relationship between speed and the steepness of the distance-time

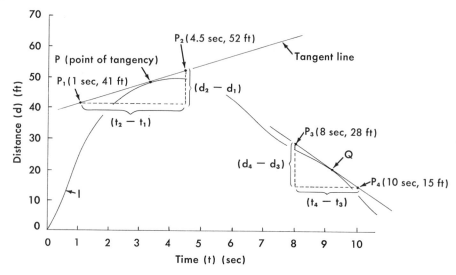

Fig. 10-10. Distance-time graph of nonuniform motion for estimating instantaneous velocity.

graph of Fig. 10-9. We already know that the average speed ($\overline{V}$) is defined as the distance traveled divided by the time elapsed

$$\overline{V} = \frac{\text{Distance traveled}}{\text{Time}} = \frac{d_f - d_o}{t_f - t_o} \text{ or } \frac{d_2 - d_1}{t_2 - t_1}$$

where d_2 is the distance at the end of an interval, d_1 is the distance at the beginning of the interval, t_2 is the time at the end of travel, and t_1 is the time at the beginning of travel.

This can be illustrated by P_1 and P_2 in Fig. 10-8.

$$V = \frac{d_2 - d_1}{t_2 - t_1} = \frac{80 - 40}{60 - 30} = \frac{40 \text{ mi}}{30 \text{ min}} =$$

$$\frac{40 \text{ mi}}{0.5 \text{ hr}} = 80 \text{ mi/hr}$$

In Fig. 10-8 we have, of course, a constant speed of 80 mi/hr. Now let us see what happens when we extend this idea to a distance-time graph with varying speed such as shown in Fig. 10-10. By using the same basic plan, comparing the ratio between the distance traveled and the time elapsed, we can find the *instantaneous* speed of an object. Since variance of speed is quite common in human motion, knowing the instantaneous speed is more useful in a motion analysis than is know-

ing the average speed. Our first step in finding instantaneous speed of an object from a distance-time graph is to draw a line tangent to the graph at the point where we are interested in determining the instant speed. A tangent line is a straight line that touches a curve at only one point as shown in Fig. 10-10. The final step in finding the instantaneous speed (V) is to find the speed from the tangent line by using $V = (d_2 - d_1)/(t_2 - t_1)$.

In Fig. 10-10 this would be

$$V = \frac{52 - 41}{4.5 - 1} = \frac{11}{3.5} = 3.14 \text{ ft/sec}$$

Therefore at point P the instantaneous speed is slightly better than 3 ft/sec.

Calculate the instantaneous speed at point Q in Fig. 10-10.

$$V = \frac{d_4 - d_3}{t_4 - t_3} =$$

The negative sign you will obtain in your answer indicates that the speed is decreasing.

A distance-time graph can be used to obtain a great deal of descriptive information about the movement of an object.

a. If the speed is constant, i.e., the graph is a straight line, we can find the instantaneous

speed (V) where $V = \dfrac{d_f - d_o}{t_f - t_o}$.

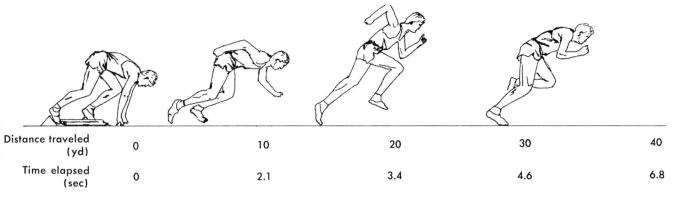

Distance traveled (yd)	0	10	20	30	40
Time elapsed (sec)	0	2.1	3.4	4.6	6.8

Fig. 10-11. Measuring the times for 10-yard intervals of a 40-yard sprint.

b. If the speed varies, i.e., the graph is a curved line, we can find the instantaneous speed (V) by drawing a tangent line at the desired location and use the tangent line to obtain $V = \dfrac{d_f - d_o}{t_f - t_o}$.

4. The measurement of velocity can be illustrated by a sprinter running a 40-yard dash as diagrammed in Fig. 10-11. Imagine that we have stationed four timers with stopwatches along the track represented by the times in the drawing; i.e., there are four timers, one at each 10-yard mark. They will record the time it takes the runner to travel from the start to their yard marker.

Since we know the distance traveled, which is already measured to the yard markers, we can now calculate the *average* velocity for each interval, i.e., the first 10 yards, the first 20 yards, and eventually the whole 40 yards. We are obtaining velocities in this example because the sprint is linear and the direction is determinable (we can assume it is due south).

The different velocities can be calculated as follows:

First 10 yards:
$$\overline{V} = \frac{d}{t} = \frac{10 \text{ yd}}{2.1 \text{ sec}} = 4.76 \text{ yd/sec} = 14.28 \text{ ft/sec}$$

First 20 yards:
$$\overline{V} = \frac{d}{t} = \frac{20 \text{ yd}}{3.4 \text{ sec}} = 5.88 \text{ yd/sec} = 17.64 \text{ ft/sec}$$

First 30 yards:
$$\overline{V} = \frac{d}{t} = \frac{30 \text{ yd}}{4.6 \text{ sec}} = 6.52 \text{ yd/sec} = 19.56 \text{ ft/sec}$$

Total 40 yards (student calculation):
$$\overline{V} = \frac{d}{t} =$$

It is also possible to calculate the average speed during *each* 10-yard interval with the use of the measured quantities.

a. Average velocity from the 10-yard mark to the 20-yard mark: with $\overline{V} = d/t$ (where *d* is $d_f - d_o$, d_f is the final distance mark, d_o is the original distance mark, and *t* is the time taken to traverse the interval $t_f - t_o$), this can be computed as follows:
$$\overline{V} = \frac{d_f - d_o}{t_f - t_o} = \frac{20 \text{ yd} - 10 \text{ yd}}{3.4 \text{ sec} - 2.1 \text{ sec}} =$$
$$\frac{10 \text{ yd}}{1.3 \text{ sec}} = 7.69 \text{ yd/sec} = 23.07 \text{ ft/sec}$$

Thus the sprinter has an *average* velocity of 23.07 ft/sec in a southerly direction during the interval from the 10-yard mark to the 20-yard mark.

b. Average velocity from the 20-yard mark to the 30-yard mark:
$$\overline{V} = \frac{d_f - d_o}{t_f - t_o} = \frac{30 \text{ yd} - 20 \text{ yd}}{4.6 \text{ sec} - 3.4 \text{ sec}} =$$
$$\frac{10 \text{ yd}}{1.2 \text{ sec}} = 8.33 \text{ yd/sec} = 24.99 \text{ ft/sec}$$

Compute the *average* velocities for the intervals from 0 to 10 yards and from 30 to 40 yards. Compare these velocities. They will vary within each interval, which confirms our notion that a spinter exemplifies nonuniform linear motion

(Student calculation) 0 to 10 yards:

30 to 40 yards:

5. Direct determination of average and instantaneous acceleration is more difficult than measuring velocities. Instantaneous acceleration can be found indirectly with the use of the speed-time graph representing the motion instance. Direct measure of acceleration at a given moment can be accomplished only through more sophisticated techniques using such electronic devices as radar. Since average acceleration is completely dependent upon the knowledge of instantaneous velocity, the cal-

culation of this quantity can be carried out only if the instantaneous velocity at two specified instants is known. Remember that average acceleration is the change in velocity ($V_f - V_o$) during any time interval (t). Thus both instantaneous acceleration and average acceleration can be found easily only through indirect methods or with the aid of more complex direct measuring techniques and devices.

Examine the basic descriptive information gathered in the linear motion example depicted in Fig. 10-11. With the known data it is not possible to directly determine either the instantaneous acceleration or the average acceleration. The instantaneous acceleration at desired points, however, can be estimated by graphical methods applied to a speed-time graph. The average acceleration can be found for any desired interval by use of the instantaneous velocities shown at the beginning and end of that interval. These velocities can be found indirectly by use of the distance-time graph or directly with the aid of a measuring device such as an automobile speedometer or radar.

The instantaneous acceleration can be estimated from the speed-time graph by use of the tangent method. Fig. 10-12 shows a speed-time graph for the motion data of the sprinter.

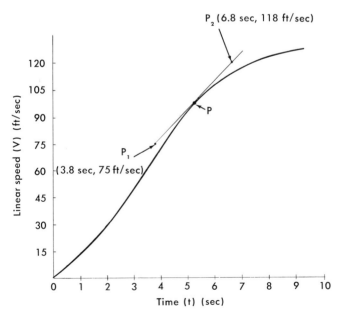

Fig. 10-12. Speed-time graph of a sprinter. (Data are from Fig. 10-11.)

$$a = \frac{V_f - V_o}{t_f - t_o} = \frac{118 \text{ ft/sec} - 75 \text{ ft/sec}}{6.8 \text{ sec} - 3.8 \text{ sec}} = \frac{43 \text{ ft/sec}}{3 \text{ sec}} = 14.33 \text{ ft/sec}^2$$

Place a tangent line at *P* (when, for example, *t* = 5.5 sec) and select two points (*P₁* and *P₂*) on the tangent line. Using any two coordinates obtained from the tangent in Fig. 10-12, we can estimate the instantaneous acceleration at *P* as shown at the top of this page.

6. Using a stopwatch and speedometer to measure elapsed time and instantaneous velocity, conduct an automobile motion experiment similar to that described in Task 10 in the following self-evaluation section. (Use 3-second intervals.)

7. Gather distance and elapsed time data for a 100-yard dash. Determine the time at 20-yard intervals. Carry out the following tasks:
   a. Construct and label a distance-time graph for the motion.
   b. Estimate the instantaneous velocity at the start, at the 50-yard mark, and at the finish line by use of the tangent method applied to the graph.
   c. Calculate the average velocity between the start and the 50-yard mark.
   d. Using the data found in b, construct a speed-time graph of the sprint.
   e. Estimate the instantaneous acceleration at the 20-yard mark. Find the average acceleration between the start and the 20-yard mark.

SELF-EVALUATION

Students should use no reference materials for this progress test, and they can check their answers by referring to Appendix A.

1. The formulas for speed and velocity are (identical/similar)_____.
2. Linear distance traveled per unit of time describes
   a. speed
   b. velocity
   c. both a and b
   d. neither a nor b
3. Linear speed is a (scalar/vector)_____.
4. A change in speed and/or direction produces a/an
   a. acceleration
   b. motion change
   c. both a and b
   d. neither a nor b

5. Describe the following linear motion analysis terms:
   a. Speed _____
   _____
   b. Velocity _____
   _____
   c. Acceleration _____
   _____

6. Define linear velocity and acceleration in sentence form:
   a. $V = d/t$ _____
   _____
   b. $a = \dfrac{V_f - V_o}{t}$ _____
   _____

7. Uniform motion refers to a linear velocity that is (constant/changing)_____.
8. When an object changes speed and/or direction it is said to be in (uniform/variable) _____ motion.
9. Describe the following:
   a. Average velocity; average acceleration
   b. Instantaneous velocity; instantaneous acceleration
   c. Constant acceleration
   d. Variable acceleration
10. Using the instantaneous velocities and elapsed times obtained from a car speedometer and stopwatch during motion of an automobile, solve the following problems:

Location	A	B	C	D	E	F
Elapsed time (t) (sec)	0	1	2	3	4	5
Instantaneous velocity (V) (ft/sec)	0	10	30	60	100	130
Distance traveled (d) (ft)	0	4	26	66	122	180

   a. Determine the average velocity during each interval.

   (A to B) $V = \dfrac{d_f - d_o}{t} = \dfrac{4 \text{ ft} - 0 \text{ ft}}{1 \text{ sec}} =$
   4 ft/sec

   (B to C)_____
   (C to D)_____

188 Kinematic analysis of human motion

(D to E)_____
(E to F)_____

b. Determine the average acceleration during each interval.

$$(A \text{ to } B) \quad \bar{a} = \frac{V_f - V_o}{t} = \frac{10 \text{ ft/sec} - 0 \text{ ft/sec}}{1 \text{ sec}} =$$

$$10 \text{ ft/sec}^2$$

(B to C)_____
(C to D)_____
(D to E)_____
(E to F)_____

c. Which interval will contain the largest instantaneous velocity?

11. Using the data of Fig. 10-11, construct a distance-time graph as shown in Figs. 10-5 and 10-6.
12. Using the data of Fig. 10-11 and the distance-time graph constructed in Task 11, calculate the instantaneous velocity at the end of the first, third, fifth, and seventh seconds (by the tangent method).
13. Using the instantaneous velocities found in Task 12, construct a velocity-time graph as shown in Fig. 10-8.
14. Using the graph constructed in Task 13, calculate the instantaneous acceleration at the end of the first, third, and fifth seconds (by the tangent method).

Kinematic analysis of parabolic motion

Body of lesson

INTRODUCTION

An aerial trajectory is the theoretical path followed by a body as it moves through an air medium when the effects of air friction are neglected. When a ball, for example, is thrown into the air it will follow a predictable path (called its trajectory). In studying the behavior of a projectile, we often neglect air resistance, calculate its theoretical path, and then if necessary make corrections for air friction. As a rule the known factors concerning a projection are V_o, the original speed of projection, and θ (Greek letter theta), the angle of departure (which is usually measured from the horizontal).

The three directions of projection that we shall consider in this lesson are vertical, horizontal, and oblique (pronounced "oh bleak"). The theoretical path followed by a ball thrown into the air at an angle to the horizontal and to the vertical (an oblique projection) is a parabola. A horizontal projection will follow the path represented by half a parabola, while a vertical projection will follow a linear path straight up and straight down.

The factors about a projectile's theoretical behavior that can be calculated are (1) the time of flight, (2) the maximum height reached, and (3) the range attained. The *time of flight* of a projectile moving either obliquely or straight upward will here be defined as the time required for it to rise and then return to the same level from which it was projected. Objects projected either horizontally or downward have a time of flight, of course, measured from the moment of release to the moment of landing. The *maximum*

189

height, called the summit, is defined as the greatest vertical distance reached by a projectile as measured from the horizontal projection plane. The *range* is the horizontal distance from the point of projection to the point where the projectile returns again to the projection plane. Needless to say, vertical projections do not produce ranges and horizontal projections do not produce heights.

ANALYSIS OF VERTICAL PROJECTIONS

The factor that is involved in all projections is the force of gravity, which produces a constant vertical acceleration of approximately 32 ft/sec². In other words, when a body suspended in the air is allowed to fall it immediately undergoes an acceleration caused by the force of gravity, and this acceleration is a constant of approximately 32 ft/sec² or 9.8 m/sec² when the effect of air friction is neglected.

When a body is projected straight upward, its speed rapidly diminishes until at some point it comes momentarily to rest and then falls back

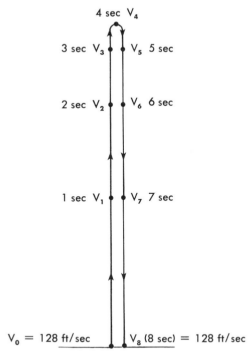

Fig. 11-1. The upward motion of a body has the same velocity sequence as the downward motion but in reverse.

toward earth, acquiring at the end of its flight the same speed it had upon projection. Experiments show that the time taken to rise to the highest point of a trajectory is equal to the time taken to fall from there back. This implies that the upward motions are just the same as the downward motions, but in reverse, and that the time and speed for any point along the path are given by the same equations, which we shall call the *equations for free fall.*

In Fig. 11-1 a ball is shown projected upward with a speed of 128 ft/sec. After each second's time, its speed on the way up is shown to be the same as the speed at the same level on the way down. The acceleration of the ball when it is going up is negative since its speed is decreasing. When the ball is coming down and its speed is increasing, its acceleration is called positive.

Equations for free fall

Fig. 11-2 shows a boy standing on a bridge 100 feet above a river. If the boy should throw a stone vertically upward with an original velocity of 30 ft/sec, how long would it be in the air before it came back to the elevation from which it was thrown? How high would it rise? The answers to these questions, as we shall see later, can be calculated from the equations for free fall, which in turn are made possible by the fact that we are dealing here with a constant acceleration: the acceleration of gravity.

The four equations for free fall are

$$V_f = V_o + gt \qquad \text{(Equation 11-A)}$$
$$d = \frac{(V_f + V_o)\,t}{2} \qquad \text{(Equation 11-B)}$$
$$V_f{}^2 = V_o{}^2 + 2gd \qquad \text{(Equation 11-C)}$$
$$d = V_o t + \tfrac{1}{2}gt^2 \qquad \text{(Equation 11-D)}$$

where V_f is the final velocity, V_o the original velocity, g the acceleration of gravity (32 ft/sec/sec), t the time, and d the distance.

The derivation of these equations is as shown in the boxed material on p. 191.

Calculation of time of flight

Since the time required for a projectile to reach the highest point in its trajectory equals the time required to fall the same distance, Equation 11-A can be used to calculate the time of flight. From

The *first equation* is just a rearrangement of the acceleration formula, $a = (V_f - V_o)/t$. Since we are dealing with free fall situations (in which the constant acceleration is that of gravity), we substitute g for a and rearrange the acceleration formula to find the final velocity (V_f) precisely

$$V_f = V_o + gt$$

Here the final instantaneous velocity (V_f) of a body is given as the sum of the initial velocity (V_o) plus the change in velocity (gt).

Rearranged, this equation assumes the form:

$$g = \frac{V_f - V_o}{t}$$

In the special case where a body starts from rest $(V_o = 0)$ and undergoes a constant acceleration, g is given by the equation $g = V_f/t$. Rearranged, this becomes $V_f = gt$.

The *second equation* is obtained from the concept of constant acceleration. The average velocity of a particle moving with a constant acceleration is given by

$$\overline{V} = \frac{V_f + V_o}{2}$$

The equation $d = Vt$ already given for constant velocity holds for constant acceleration when V is replaced by $\overline{V}$. Thus

$$d = \overline{V}t = \frac{(V_f + V_o)t}{2}$$

For the special case of a body starting from rest $(V_o = 0)$, the equation for the average velocity becomes $\overline{V} = V_f/2$ and the second equation becomes $d = V_f t/2$.

The *third equation* is obtained by combining the first two, i.e., the acceleration equation, $g = (V_f - V_o)/t$, with the distance equation, $d = (V_f + V_o) t/2$. To do this,

we solve each equation for t and then set them equal to each other as follows:

$$t = \frac{V_f - V_o}{g}$$

and

$$t = \frac{2d}{V_f - V_o}$$

Since the right side of both equations is equal to t, the equations equal each other. By setting them equal

$$\frac{V_f - V_o}{g} = \frac{2d}{V_f + V_o}$$

and cross multiplying, we obtain

$$(V_f - V_o)(V_f + V_o) = 2gd$$

By multiplying out, we get $V_f^2 - V_o^2 = 2gd$. Rearranging, we obtain

$$V_f^2 = V_o^2 + 2gd$$

For special cases where a body is initially at rest $(V_o = 0)$, the equation for final velocity becomes $V_f^2 = 2gd$.

The *fourth equation* is also obtained by combining the first two, i.e., the acceleration equation, $g = (V_f - V_o)/t$, with the distance equation, $d = (V_f + V_o)t/2$. To do this, we solve each equation for V_f and then set them equal to each other as follows:

$$V_f = V_o + gt$$

and

$$V_f = \frac{2d}{t} - V_o$$

Since the right side of both equations equals V_f the equations equal each other. Setting them equal $(V_o + gt = 2d/t - V_o)$ and adding V_o to each side gives $2d/t = 2V_o + gt$. Multiplying each side by $t/2$ produces

$$d = V_o t + \tfrac{1}{2}gt^2$$

For special cases where a body is initially at rest $(V_o = 0)$ the equation becomes $d = \tfrac{1}{2}gt^2$.

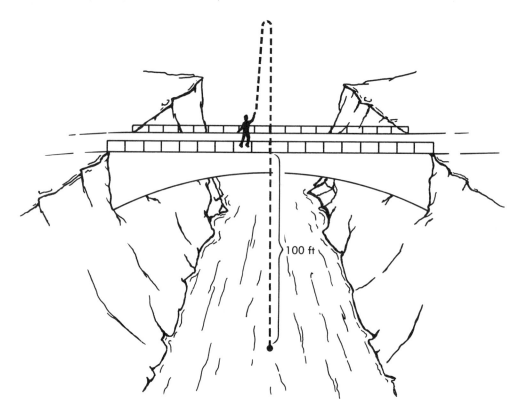

100 ft

Fig. 11-2. A body projected upward rises to a predetermined height and then falls with increasing speed.

this equation we know that an object falling from rest ($V_o = 0$) attains a final velocity equal to the product of the magnitude of the gravitational constant *(g)* and the time *(t)* during which the acceleration occurs; i.e., $V_f = gt$. Since we are dealing with vertical projections in our present discussion, we will use the symbol V_y rather than V_f. Rearranging Equation 11-A to isolate the time *(t)* on the left side of the equation produces $t = V_y/g$. Because *t* is the time taken to rise, or the time to fall, the total time of flight *(T)* of a vertical projection will be twice *t*.

$$T = 2V_y/g$$

Calculation of maximum height

To find the maximum height (h) a projectile will reach, we can use Equation 11-C. This tells us that in a free-fall situation an object will at-

tain a final velocity equal to the square root of the product of the height from which it is dropped (d) and the gravitational constant doubled (2g), i.e., $V_f^2 = 2dg$ and $V_f = \sqrt{2dg}$. In vertical projection situations the letter *d* can be replaced by h to represent height, and the letter V_f can be replaced by V_y to obtain $V_y^2 = 2gh$. Dividing both sides of the equation by *2g* produces

$$h = V_y^2/2g$$

Examples

We are now ready to answer the two questions previously asked in regard to Fig. 11-2, showing a boy on a bridge 100 feet above a river. If he throws a stone vertically upward with an original velocity of 30 ft/sec, how long will it be in the air before it comes back to the elevation from which it was thrown? And how high will it rise?

The first question is concerned with the time of flight and can be answered by the *time of flight* equation (T = $2V_y/g$). Since $V_y = 30$

ft/sec, we can substitute this value into the equation to obtain

$$T = \frac{(2)\ (30\ \text{ft/sec})}{32\ \text{ft/sec}^2} = 1.875\ \text{sec}$$

Thus the stone will be in the air for almost 2 seconds.

The second question is concerned with the maximum height reached by the stone. It can be answered by the *maximum height* equation (h = $V_y{}^2/2g$). Substituting $V_y = 30$ ft/sec gives us

$$h = \frac{(30\ \text{ft/sec})^2}{(2)\ (32\ \text{ft/sec}^2)} = 14.06\ \text{ft}$$

This says the stone will travel about 14 feet into the air before it starts its descent.

If the boy throws the stone vertically downward with the same original velocity, how fast will it be going when it hits the water? We can use Equation 11-C to solve this problem as follows:

$$\begin{aligned} V_f{}^2 &= V_o{}^2 + 2gd \\ &= (30\ \text{ft/sec})^2 + (2)\ (32\ \text{ft/sec}^2)\ (100\ \text{ft}) \\ &= 7300\ \text{ft/sec} \end{aligned}$$

The square root of this number (see Appendix C) gives us approximately 86 ft/sec, which is the velocity with which the stone will strike the water.

ANALYSIS OF HORIZONTAL PROJECTIONS

If one bullet falls freely from rest when another is fired from a rifle horizontally from the same height, the two will strike the ground simultaneously. A conclusion we can draw from this fact is that the downward acceleration of a projectile is the same as the downward acceleration of a freely falling body and takes place independent of its horizontal motion. Furthermore, an experimental measurement of *times* and *distances* shows that the horizontal velocity of projection continues unchanged and is independent of the vertical motion.

In other words, an object projected horizontally carries out two motions independently: (1) a constant horizontal motion with a constant velocity (V_x) and (2) a vertically downward acceleration (g).

Calculation of time of flight

As the bullet in our example falls with an acceleration (g), the vertical distance it falls (d_y) is

proportional to the square of the time and is given by the equation $d_y = \frac{1}{2}gt^2$ (Equation 11-D). Likewise, the vertical time of flight in this situation is found by rearranging Equation 11-D to obtain $t = \sqrt{d/\frac{1}{2}g}$. Remembering that $\frac{1}{2}g = \frac{1}{2}(32) = 16$ and that $d_y = h$ or height when the object is allowed to fall all the way to the ground, we can substitute 16 for $\frac{1}{2}g$ and h for d to obtain $t = \sqrt{h/16}$.

Calculation of range

With an initial horizontal velocity (V_x) the horizontal distance (d_x) traveled by the bullet is proportional to the time (t) and is given by the equation $d_x = V_x t$. Since the horizontal distance traveled is defined as the range (R), we can replace d_x by R and t by $\sqrt{h/16}$. Thus

$$R = V_x\ \sqrt{h/16}$$

Should the boy on the bridge in Fig. 11-2 throw his stone horizontally with the same original velocity as before, its horizontal displacement (range) could be calculated as follows:

$$t = \sqrt{h/16} = \sqrt{100/16} = \sqrt{6.25\ \text{sec}} = 2.5\ \text{sec}$$
$$R = V_x t = (30\ \text{ft/sec})\ (2.5\ \text{sec}) = 75\ \text{ft}$$

ANALYSIS OF OBLIQUE PROJECTIONS

To calculate the time, height, and range of an oblique projection, we need to resolve the initial velocity (V_0) of projection into two components: one vertical and the other horizontal. This is illustrated in Fig. 11-3. With V_0 the speed of projection and θ the elevation angle, the x and y components of the velocity are given by the following trigonometric equations:

$$V_y = V_o \sin \theta$$
$$V_x = V_o \cos \theta$$

The actual trajectory of an oblique projection upward is a combination of two linear motions occurring at the same time. One linear motion is of a body projected *vertically* upward with an initial velocity V_y, and the other is of a body projected *horizontally* with an initial velocity V_x. In other words, an obliquely projected object carries out two motions simultaneously and inde-

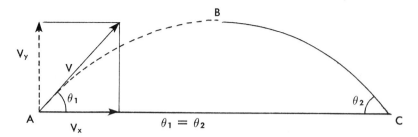

Fig. 11-3. The initial velocity of an oblique projection can be resolved into two components, one vertical *(V_y)* and the other horizontal *(V_x)*. The dashed lines indicate a decrease in velocity until point *B* is reached. The angles of projection *(θ₁)* and landing *(θ₂)* are equal.

pendently. The horizontal motion is characterized by a constant velocity (V_x), and the vertical motion is characterized by a variable velocity (V_y) but a constant vertical acceleration (g).

Thus the kinematic analysis of oblique aerial trajectories is mostly concerned with the combined effects of vertical and horizontal velocity components.

The equations for calculating the time and height of a vertical projection and the range of a horizontal projection can be changed to apply to oblique projections by substituting (V sin θ) for V_y, and (V cos θ) for V_x in these equations as follows:

$$T = 2t = 2V_y/g = 2V \sin \theta/g$$
(Equation 11-E)

$$h = V_y{}^2/2g = (V \sin \theta)^2/2g$$
(Equation 11-F)

$$R = V_x 2t = V_x T = V \cos \theta \frac{(2V \sin \theta)}{g} =$$
$$\frac{2V^2 \sin \theta \cos \theta}{g}$$

This last range formula is put into another and more useful form by using the trigonometric relation that 2 sin θ cos θ = sin 2 θ. When this substitution is made, the equation becomes

$$R = \frac{V^2 \sin 2\theta}{g}$$
(Equation 11-G)

Thus Equations 11-E through 11-G can be used to find *T, h,* and *R* for an oblique projection.

CALCULATION OF TRAJECTORIES WHEN THE ELEVATION OF THE POINT OF RELEASE IS DIFFERENT FROM THAT OF LANDING

In most human movement situations the elevation of the point of release or takeoff of a projectile is different from that of the landing. In basketball goal shooting, for instance, the point of landing of the ball at the goal is above the point of release. In shot-putting the point of landing of the shot on the ground is below the point of release. Determining the time of flight and range in these situations requires a modification of the analytical procedures already discussed.

The illustration in Fig. 11-4 shows an example of an oblique projection in which the point of original projection (takeoff) is above the point of landing. In this example a cat jumps off a table that is 3 ft high. The original velocity of the cat is 10 ft/sec at an angle of 37° above the horizontal. How long is the cat in the air and how far out from the edge of the table does the cat strike the floor?

Calculation of time of flight

To determine the time of flight, we use Equation 11-D (d = Vt + ½gt²). Since the unknown variable in this situation is the time *(t)* and since this unknown appears raised to the second power *(t²)* as well as to the first power *(t)*, it is defined as a quadratic equation. The boxed material on p. 195 shows how this equation can be solved.

By completing the square, we obtain

$$T = \frac{V \sin \theta + \sqrt{(V \sin \theta)^2 + 2gh}}{g}$$
(Equation 11-H)

In a quadratic equation the unknown variable appears raised to the second power (x^2) as well as to the first power (x). By rearranging the terms, we can place any quadratic equation in the form

$$ax^2 + bx + c = 0$$

(Equation 1)

where *a, b,* and *c* are constants.

For example, $c = bx + ax^2$ can be placed in the form of Equation 1 by rearranging so the term involving x^2 is first, the term involving x is second, the constant term is third, and the right side is zero. Thus $-ax^2 - bx + c = 0$.

Likewise, the equation $d = Vt + \frac{1}{2}gt^2$ becomes

$$-\tfrac{1}{2}gt^2 - Vt + d = 0$$

(Equation 2)

If we solve Equation 1 by completing the square, we obtain a formula which is useful and efficient for obtaining the roots of any quadratic equation. The solution of Equation 1 involves the following steps:

1. State the equation.

$$ax^2 + bx + c = 0$$

2. Rearrange *c*.

$$ax^2 + bx = -c$$

3. Divide both sides by *a*.

$$x^2 + \frac{bx}{a} = -\frac{c}{a}$$

4. Add $b^2/4a^2$ to both sides.

$$x^2 + \frac{bx}{a} + \frac{b^2}{4a^2} = \frac{b^2}{4a^2} - \frac{c}{a}$$

NOTE: We can find the number to add to both sides in order to "complete the square" by taking half of the *x* coefficient and squaring (i.e., taking half of *b/a* to get b/2a and squaring to get $b^2/4a^2$).

5. Simplify.

$$\left(x + \frac{b}{2a}\right)^2 = \frac{b^2 - 4ac}{4a^2}$$

6. Take the square root of the two sides.

$$x + \frac{b}{2a} = \frac{\pm \sqrt{b^2 - 4ac}}{2a}$$

7. Solve for *x*.

$$x = -\frac{b}{2a} \pm \frac{\sqrt{b^2 - 4ac}}{2a}$$

Now, since the two denominators on the right side of the last equation are the same, we have

$$x = \frac{-b \pm \sqrt{b^2 - 4ac}}{2a}$$

(Equation 3)

Since Equations 1 and 2 are both quadratic equations, we know that they are the same with only different symbols. The equivalent symbols are

ax^2 and $-\frac{1}{2}gt^2$ (or $a = -\frac{1}{2}g$ and $x^2 = t^2$)
bx and $-Vt$ (or $b = -V$ and $x = t$)
c and d (or $c = d$)

Thus symbols of Equation 2 can be substituted into Equation 3 to obtain

$$t = \frac{-(-V) \pm \sqrt{(-V)^2 - 4(-\frac{1}{2}g)d}}{2(-\frac{1}{2}g)}$$

which (assuming a negative *g*) simplifies to

$$t = \frac{V \pm \sqrt{V^2 - 2gd}}{g}$$

Since we will be using this equation in the study of oblique projections, we can substitute V_y or $V \sin \theta$ for *V* and $-h$ for *d* to obtain

$$T = \frac{V \sin \theta + \sqrt{(V \sin \theta)^2 + 2gh}}{g}$$

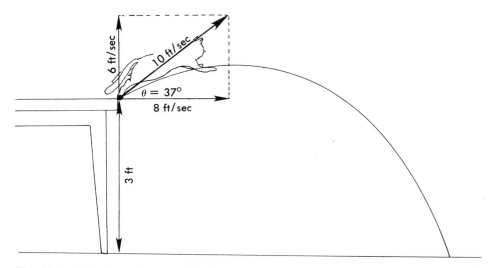

Fig. 11-4. Trajectory of a cat which jumps off a table with a velocity of 10 ft/sec at 37°.

Substituting the data previously given into the above equation, we get

$$T = \frac{V \sin \theta + \sqrt{(V \sin \theta)^2 + 2gh}}{g}$$

$$= \frac{(10 \text{ ft/sec}) (0.6) + \sqrt{((10 \text{ ft/sec}) (0.6))^2 + (2) (32 \text{ ft/sec}^2) (3 \text{ ft})}}{32 \text{ ft/sec}^2}$$

$$= 0.659 \text{ sec}$$

Thus the total time the cat is in the air is about sixth tenths of a second.

Calculation of range

Since $R = V_x T$ and $V_x = V \cos \theta$, we can find the range the cat achieves by substituting the known values from our example into this equation as follows:

$$R = V_x T = V \cos \theta \ T =$$
$$(10 \text{ ft/sec}) (0.8) (0.659) = 5.27 \text{ ft}$$

Thus the cat will land approximately 5 feet away from the edge of the table. Equation 11-H, however, can be combined with the equation which gives us V_x (i.e., $V \cos \theta$) by multiplying the two together as follows:

$$R = V_x T = (V \cos \theta) \left(\frac{V \sin \theta + \sqrt{(V \sin \theta)^2 + 2gh}}{g} \right)$$

which reduces to

$$R = \frac{V^2 \sin \theta \cos \theta + V \cos \theta \sqrt{(V \sin \theta)^2 + 2gh}}{g}$$

(Equation 11-I)

This equation shows us that the four factors which influence the distance a projectile will travel are

V, Original speed of the projection at the point of release and in the direction of the projection

θ, Angle of the projection with the horizontal

g, Acceleration due to gravity (approximately 32 ft/sec²)

h, Height of the projection (the vertical distance from the point of release to the point of landing)

The first three factors (V, θ, and g) are involved in all projections, while the height factor *(h)* is involved only in those situations in which the elevation of the point of release or takeoff is different from that of the landing. If the points of release and landing are at the same elevation so $h = 0$, then Equation 11-I reduces to Equation 11-G (i.e., $R = V^2 \sin 2 \theta/g$). If the point of release is above the point of landing as in shotputting, then the height factor *(h)* is considered to be positive. If on the other hand, the point of release is below the point of landing as in basketball goal shooting, the height factor *(h)* is con-

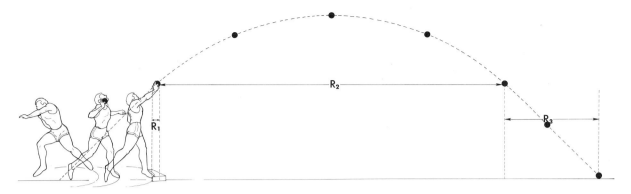

Fig. 11-5. Flight of the shot. $R_1 + R_2 + R_3$ represent the total distances achieved.

sidered to be negative and a negative value of h must be entered into Equation 11-I; this is the only change necessary in these two situations.

Application to shot-putting

According to Bunn (1973) there are three ranges that must be considered in shot-putting (Fig. 11-5):

R_1, the horizontal distance from the inside edge of the toe board to the point in front of the toe board at which the shot was released

R_2, the horizontal distance from the point of release to the point at which the projectile returns to the elevation of the release

R_3, the horizontal distance from the point where the shot falls below the level of release to the point where it hits the ground

These are, of course, the three ranges involved in determining the total horizontal distance traveled by *any* thrown or struck object (neglecting air resistance). The second and third ranges (R_2 and R_3) are calculated through the use of Equation 11-I, where the R is equal to R_2 plus R_3. The distance R_1 is affected by the amount of body lean and length of the shot-putter's arm, and thus it must be measured directly. According to Bunn (1973), Cureton (1936) found from his motion picture studies of the shot put that the usual distance of R_1 is approximately 1 foot.

Application to long jumping

According to Bunn (1973) the long jump introduces additional complications in the calculation of range because it is not possible to assume that the feet of the jumper will follow a parabolic path from the edge of the takeoff board to the spot where they hit the sand pit at the end of the jump. The only part of a jumper's body which follows a parabolic path is his center of gravity.

Calculating the range of a long jump is further complicated by the following facts (Bunn, 1973):

1. Because of a hip lock by the iliofemoral ligament, the knee of the takeoff leg is forced to bend at takeoff. Thus the leg makes a different angle at takeoff from that of the center of gravity. Fig. 11-6 shows the angle of takeoff of the leg to be 30° from the vertical while that of the center of gravity is 25° from the horizontal.

2. At the end of the jump, the thrust of the legs forward to gain distance creates a different angle of the legs at landing from the takeoff angle. This causes the center of gravity to fall below the level at takeoff.

Fig. 11-6 shows the four ranges that must be considered in studying the total horizontal displacement of a long jumper:

R_1, Horizontal distance of the center of gravity in front of the driving foot at the moment of takeoff

R_2, Horizontal distance from the location of the center of gravity at the moment of takeoff to the point where it is located when it returns to the elevation of its takeoff

R_3, Horizontal distance from the point where the center of gravity falls below its level of takeoff to its location when the feet hit the ground (sand pit)

R_4, Horizontal distance the center of gravity is moved from its location when the feet hit the ground to the rear edge of the back foot

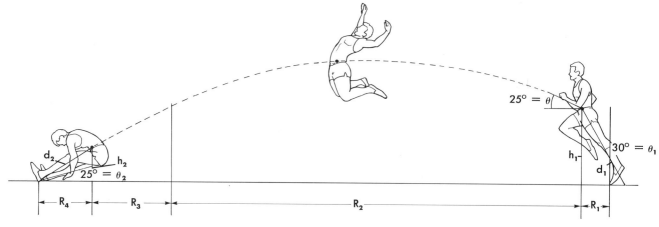

Fig. 11-6. Flight of the long jumper. The distances making up the total length of the jump are shown by R_1, R_2, R_3, and R_4.

There are also three angles that must be considered:

θ, Angle of projection of the center of gravity
θ_1, Angle of lean with the vertical at takeoff
θ_2, Angle of the legs at the moment of alighting with the horizontal

The three heights and two oblique distances that must be measured are

h_1, Vertical distance from center of gravity to ground at takeoff
h_2, Vertical distance from center of gravity to ground at touchdown (when the feet first touch the ground)
H, Height of projection, which is the vertical distance of the center of gravity at takeoff (h_1) minus that at touchdown (h_2); i.e., $H = h_1 - h_2$
d_1, Straight line distance from the center of gravity to the point of foot contact with the ground at takeoff
d_2, Straight line distance from the center of gravity to the point of foot contact with the ground at moment of touchdown

The second and third ranges (R_2 and R_3) are calculated through the use of Equation 11-I, the only modification being the substitution of H for h. Since $H = h_1 - h_2 = (d_1 \cos \theta_1) - (d_2 \sin \theta_2)$, any of these expressions can be substituted into Equation 11-I. Thus Equation 11-I can be used to find the horizontal distance traveled by the center of gravity during the time of its parabolic motion.

The distances due to the leg lean at takeoff (R_1) and at landing (R_4) must be calculated separately, however, and then added to the re-

sults obtained through the use of Equation 11-I. These two distances can be calculated by using trigonometric functions as follows:

$$R_1 = d_1 \sin \theta_1$$
$$R_4 = d_2 \cos \theta_2$$

The total distance of the jump therefore is equal to

$$R_1 + R_2 + R_3 + R_4 =$$
$$\frac{V^2 \sin \theta \; \cos \theta + V \cos \theta \; \sqrt{(V \sin \theta)^2 + 2gH}}{g} +$$
$$d_1 \sin \theta_1 + d_2 \cos \theta_2$$

(Equation 11-J)

OPTIMAL ANGLES OF PROJECTION

As we already know, an oblique projection has both a horizontal and a vertical component of velocity. The vertical component of the projection is what acts against the force of gravity and determines the *height* above the projection point which the projectile will reach and the *time* required for it to reach that point. The effect of this vertical component is identical to the effect of an object projected straight up or straight down with the same speed.

We know that the *time of flight* equals V_y/g and that the *height reached* equals $V_y^2/2g$. Thus the vertical component (V_y) is sometimes called the "time-height" component. Since V_y is equal to ($V \sin \theta$), this component increases with an increase in the angle of projection up to 90°.

Therefore the time of flight of a projectile as well as the height reached varies directly with the angle of projection if we assume that the projection force is constant.

Similarly the horizontal component of a projection velocity is inversely proportional to the angle of projection (i.e., $V_x = V \cos \theta$). As the angle of projection approaches $0°$, that is, as it nears the horizontal or decreases, the horizontal velocity component increases.

The longest theoretical horizontal projection is attained at an angle of projection of $45°$. At this angle of projection, the vertical or time-height component is equal to the horizontal or distance component. Thus this angle results in the best compromise between the time-height component (which operates against the force of gravity) and the distance component (which causes the projectile to move forward during the time available).

The fact that a $45°$ angle gives the greatest theoretical range can be demonstrated through the use of the range equation ($R = V^2 \sin 2\theta/g$). This equation requires the angle to be doubled, which means that the $45°$ angle becomes a $90°$ angle. Since the sine of a $90°$ angle equals 1.00 and since V^2 times 1.00 equals V^2, we can see that the total speed (V) is available for the production of range and thus the maximum possible range will be attained.

Cinematographic studies have determined that the angle of projection of the shot in the case of good shot-putters is $40°$ to $42°$ at the hand and

$45°$ at the ground when maximum distance is obtained. It is evident therefore that the shot actually follows a parabolic path which is continuous from the ground behind the putter to the point of contact with the ground in front of the putter and that the angle at both contacts of the parabola with the ground is $45°$ (Fig. 11-7). Also notice from Fig. 11-7 that the puts are actually intersecting the parabola at various points and that the point of intersection determines the optimal angle at which the shot should be released so its path will coincide with the rest of the parabola.

The following generalizations can be drawn from the facts just discussed:

1. The optimum angle of contact of a projectile with the ground on landing is always $45°$.
2. If an object is released or a body takes off at a height equal to its height of landing and if the purpose of the projection is to attain maximum range, then the optimum angle of release or takeoff is always $45°$ when we neglect air resistance.
3. If an object is released or a body takes off at a height above the point of landing, the optimum angle of projection for the attainment of maximum distance is always less than $45°$.

Hay (1973) suggests these generalizations be further modified as follows:

1. For any given height of release, the greater the *speed* of release the more closely the optimum angle approaches $45°$.
2. For any given speed of release, the greater

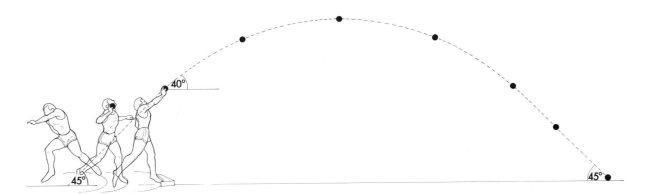

Fig. 11-7. A shot follows a parabolic path that is continuous from the ground behind the putter to the point of contact with the ground in front of the putter. The optimal angle at both contacts is $45°$ when maximum distance is obtained.

the *height* of release the less is the optimum angle.

3. Equal increases in either height of release or speed of release do not yield consistently equal changes in the optimum angle or the resulting range change.

The best angle of projection is determined by the *purpose* of the projectile motion. For example, if speed of horizontal movement is desired (as is sometimes the case with the fastball pitcher in baseball), the angle of projection should be lowered. If maximum distance of projection is desired (as with the home run hitter in baseball), the angle of projection should be about 45°. If maximum height of projection is desired (as with the high jumper in track and field), the angle of projection will have to be as close to 90° as possible.

STUDY GUIDELINES
Introduction
Specific student objectives

1. Define an aerial trajectory.
2. Identify and describe the three directions of projection.
3. Define the three factors about a projectile's theoretical behavior that can be calculated.

Summary

An aerial trajectory is the theoretical path followed by a body as it moves through the air with the effects of air resistance ignored. The known factors concerning a projection are the initial projection speed (V_o) and the angle of projection (θ), usually measured from the horizontal.

The three directions of projection are vertical, horizontal, and oblique (at an angle above horizontal but less than vertical). Vertical projections travel in a linear direction. Horizontal projections trace half a parabola. Oblique projections follow a parabolic path.

Three factors can be found to describe a projectile's behavior:

1. Time of flight. For oblique or straight upward projections this is the time from release until the projectile returns to the original level of projection. For horizontal or downward projections it is the time from release to landing.
2. Maximum height (summit). This is the greatest vertical distance reached above the point of projection.
3. Range. This is the horizontal distance from the point of projection to the point where the projectile again reaches that same level.

Performance tasks

1. Write out descriptions of the following:
   a. Aerial trajectory
   b. Vertical, horizontal, and oblique projections
   c. Maximum height, time of flight, and range of a projection
2. Illustrate the use of the above aerial motion terms in specific motion examples.

Analysis of vertical projections
Specific student objectives

1. Identify the external force involved in all projections.
2. Describe the four equations for free fall.
3. Solve the following free fall motion problems:
   a. Given V_y, find T using $T = 2V_y/g$.
   b. Given V_y, find h using $h = V_y^2/2g$.

Summary

Vertical projections are always affected by a gravitational force, which produces an acceleration (positive or negative) of approximately 32 ft/sec². The upward motion in vertical projection is identical to the downward motion. The equations for free fall are (1) $V_f = V_o + gt$, (2) $d = ((V_f + V_o)\,t/2$, (3) $V_f = V_o^2 + 2gd$, and (4) $d = V_o t + \frac{1}{2}gt$—where V_f is the final velocity, V_o the original velocity, g the gravitational constant (32 ft/sec²), t the time, and d the distance.

The total time of flight (T) can be found from $T = 2t = 2\,V_y/g$. Calculation of maximum height (h) can be found through $h = V_y^2/2g$.

Performance tasks

1. Write out the meanings of the free-fall formulas.
2. State a principle demonstrated by the illustration in Fig. 11-8.
3. You are standing on a bridge 100 feet above a river and you drop a stone straight downward. Find (a) the speed with which it will hit the water (using $V_f^2 = V_o^2 + 2gd$) and (b) the time it will take to hit the water (using

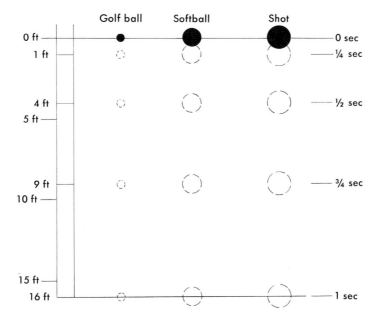

Fig. 11-8. Rates of falling of a golf ball, softball, and lead shot over a distance of 16 ft.

$V_f = V_o + gt$ in the form $t = (V_f - V_o)/g$.

4. A baseball is thrown straight upward with a speed of 80 ft/sec. Find (a) the maximum height (h) it will rise (using $h = V_y^2/2g$) and (b) the time of flight (T) until it reaches the release point again (using $T = V_y/g$, where $g = 32$ ft/sec^2).

Analysis of horizontal projections

Specific student objective

1. Solve the following motion problems:
   a. Given h, find t using $t = \sqrt{h/16}$.
   b. Given V_x and t, find R using $R = (V_x)(t)$.

Summary

Horizontal projections are identical to free-fall projections from a given height with respect to time until impact. In other words, a horizontal projection carries out two motions independently: (1) a constant horizontal velocity (V_x) and (2) a vertically downward acceleration (g) due to the force of gravity.

The time of flight (drop) can be found from the equation $t = \sqrt{h/16}$, where h is the vertical height of fall. The horizontal displacement (range) can be found from $R = (V_x)(t)$, where V_x is the initial horizontal velocity.

Performance tasks

1. Write out the meanings of the horizontal projection equations.
2. An infielder in softball makes a horizontal throw from second to first base with a release speed of 88 ft/sec. The height of the infielder's hand at the time of release is 6 ft. If it is a wild throw that goes past the receiver, find the time of flight (t) and the range (R).

Analysis of oblique projections

Specific student objectives

1. Identify the two rectangular components of an oblique projection.
2. Describe the trigonometric equations used to obtain the x and y components of a velocity.
3. Describe the calculation of time of flight of an oblique projection.
4. Describe the calculation of maximum height reached by an oblique projection.
5. Describe the calculation of range attained by an oblique projection.
6. Solve the following oblique projection problems:
   a. Given V_o and θ, find T using $T = 2 V_o \sin \theta/g$.
   b. Given V_o and θ, find h using $h = (V_o \sin \theta)^2/2g$.

c. Given V_0 and θ, find R using

$$R = \frac{V_0^2 \sin 2\theta}{g}$$

Summary

For oblique projections at an elevation angle θ, the original velocity of projection (V_0) can be resolved into rectangular components: (1) $V_y = V_0 \sin \theta$ and (2) $V_x = V_0 \cos \theta$. The equations for oblique projections will then become (1) $T = 2V_0 \sin \theta/g$, (2) $h = (V_0 \sin \theta)^2/2g$, and (3) $R = V_0^2 \sin 2\theta/g$.

Performance tasks

1. Write out the meaning of all oblique projection equations.
2. An arrow shot with a speed of 200 ft/sec reaches a maximum height of 480 ft. Find (a) the range (R) using $R = V_0^2 \sin 2\theta/g$, (b) the time of flight (T) using $T = 2t = 2V_y/g = 2V \sin \theta/g$, and (c) the angle of projection using $h = (V \sin \theta)^2/2g$.

Calculation of trajectories when the elevation of the point of release is different from that of landing

Specific student objective

1. Solve motion problems of oblique projections when the elevation of the point of release is different from that of landing:
   a. Given V, θ, and h, find T using the formula

$$T = \frac{V \sin \theta + \sqrt{(V \sin \theta)^2 + 2gh}}{g}$$

   b. Given V, θ, and h, find R using the formula

$$R = \frac{V^2 \sin \theta \cos \theta + V \cos \theta \sqrt{(V \sin \theta)^2 + 2gh}}{g}$$

In both problems h is the vertical distance from the point of release to the point of landing.

Summary

Projectile analysis can be carried out when the point of release is at a different height from the point of landing.

In this situation the time of flight (T) can be found using $\quad T = \dfrac{V \sin \theta + \sqrt{(V \sin \theta)^2 + 2gh}}{g}$

while the range (R) can be found by

$$R = \frac{V^2 \sin \theta \cos \theta + V \cos \theta \sqrt{(V \sin \theta)^2 + 2gh}}{g}$$

In both equations V is the original speed of the projectile, θ the angle of projection with the horizontal, g the acceleration due to gravity, and h the height of the projection (positive if the point of release is above the point of landing, negative if it is below the point of landing).

Performance tasks

1. A smooth circular stone is shot from a sling at a speed of 200 ft/sec at an elevation angle of 40°. Calculate its time of flight and range.
2. A baseball is thrown with a speed of 40 m/sec at an elevation angle of 38° and a projection height of 4 ft. Find the range and time of flight of the ball.
3. A basketball is shot with a speed of 15 mi/hr at a projection angle of 60° and a projection height of –10 feet. Find its range and time of flight.
4. Identify the meaning of all symbols used in the analysis equations of this section.

Optimal angles of projection

Specific student objectives

1. Identify the "time-height" and "distance" components of an oblique projection.
2. Describe the relationship between the angle of projection and the magnitudes of "time-height" and the "distance" components.
3. State the three primary purposes of projection.
4. Identify the optimum angle(s) for the three primary projection purposes.
5. State three principles related to the optimal angle of projection when the motion purpose is maximum range.

Summary

Oblique projections have both a horizontal and a vertical component of velocity. The vertical component acts against the force of gravity and determines the height reached and the time required to reach that point; thus this vertical component is termed the *time-height* component and increases directly with the angle of projection (θ) up to 90°. The horizontal force component of a projection velocity, sometimes called the *distance* component is inversely proportional to the angle of projection. The longest theoretical horizontal projection is attained when θ equals 45°. The most desirable angle of projection depends upon the motion purpose, e.g., maximum height, maximum horizontal speed, minimum time, or maximum horizontal distance.

Following are several principles which relate directly to the elevation angle when the primary purpose is maximum range:

1. The optimum theoretical angle of contact of a projectile with the supporting horizontal surface on landing is 45°.
2. If an object is projected from a height above the point of landing, the optimum angle of release is less than 45°.
3. The greater the speed of release, the more closely the optimum angle approaches 45°.
4. For any given speed of release, the greater the height of projection the lower will be the optimum angle.
5. Equal increases in either height or speed of projection do not yield consistently equal changes in the optimum angle or range.

Performance tasks

1. The vertical component of an oblique projection is called (distance/time-height)_____.
2. The distance component of an oblique projection is the component called
   a. horizontal
   b. vertical
   c. both a and b
   d. neither a nor b
3. The relationship between the time-height component and the projection angle is (direct/inverse) _____.
4. The three primary purposes of projection are
   a. length, width, and height
   b. height, speed, and range
   c. time, height, and range
   d. all of the above
5. For maximum horizontal speed the projection angle should be (lowered/raised)_____.
6. For maximum horizontal distance the theoretical projection angle should be near
   a. 0°
   b. 45°
   c. 90°
   d. none of the above
7. Write out the following:
   a. Descriptions of the components of a projection velocity
   b. Three principles relating optimal projection angle to a maximum range purpose

SELF-EVALUATION

Students should use no reference materials for this progress test, and they can check their answers by referring to Appendix A.

1. Describe the following aerial motion terms:
   a. Aerial trajectory_____
   _____
   b. Range_____
   _____
   c. Time of flight_____
   d. Time-height component of an oblique projection_____
   _____
   e. Distance component of an oblique projection_____
   _____
   f. Projectile_____
   g. Acceleration (g) due to gravity_____
   _____
2. The three directions of a projection are
   a. positive, negative, and neutral
   b. low, high, and 45°
   c. speed, velocity, and acceleration
   d. vertical, horizontal, and oblique
3. The four general equations for free fall are listed below. Describe their use and the meaning of all symbols:
   a. $V_f = V_o + gt$
   b. $d = \dfrac{(V_f + V_o)t}{2}$
   c. $V_f^2 = V_o^2 + 2gd$
   d. $d = V_ot + \frac{1}{2}gt^2$
4. Describe the three projection directions:
   a. Horizontal_____
   _____
   b. Vertical_____
   _____
   c. Oblique_____
   _____
5. Calculate R and t for the horizontal projection of a bomb dropped from a plane traveling 200 mi/hr at a height of 300 ft.
6. A javelin is thrown with a speed of 100 ft/sec at an angle of projection of 40°. Find (a) the maximum height from $h = (V \sin \theta)^2/2g$, (b) the range (R) using $R = (V^2 \sin 2\theta)/g$, and (c) the time (T) from $T = 2V \sin \theta/g$.
7. With a specified velocity of projection of 80 ft/sec for a thrown object, calculate the range

(R) at the possible angles of release of 35°, 40°, 45°, 50°, and 55°.

8. Calculate T and R for an object thrown with a velocity of projection of 80 ft/sec at an angle of 35° from a projection height of 4 ft. Use the formulas in the summary section on calculating trajectories (p. 202).

9. Identify a motor task example and optimum projection angle for each of the three primary projection purposes below:
   a. Maximum height
   b. Maximum horizontal speed (least time)
   c. Maximum horizontal distance

10. Oblique projections have horizontal and vertical components of original velocity. They are respectively called the components of
    a. time and height
    b. distance and time
    c. both a and b
    d. neither a nor b

11. State three principles related to the optimum angle of projection when the motion purpose is maximum range.

Kinematic analysis of angular motion

Body of lesson
Study guidelines
Self-evaluation

ANALYSIS OF ANGULAR VELOCITY AND ACCELERATION

The kinematic analysis of angular motion, like that of linear motion, involves a description of velocity and acceleration. The mathematical description of these quantities will be discussed in this section, while their graphical description will be discussed in the next section.

Angular velocity

Angular velocity is symbolized by the lower case Greek letter omega, ω. Just as linear velocity is the rate of linear displacement (d/t), angular velocity is the rate of angular displacement (θ/t). In other words, the angular velocity (ω) (Fig. 12-1) of a motion is defined as the ratio of the angle turned through (θ, in radians) to the elapsed time (t). Expressed in terms of a formula, this definition becomes

$$\text{Angular velocity} = \text{Angle turned through} \div t$$
$$\omega = \theta/t \qquad \text{(Equation 12-A)}$$

The direction factor in angular motion is satisfied by stating the axes and the clockwise or counterclockwise directions of the motion. Clockwise rotation is considered a negative direction; counterclockwise rotation is called positive.

Relationship of angular velocity to linear velocity

From the definition of a radian, it is known that the angle (θ) in radians between any two points on the circumference of a circle is given by d—the length of the arc between the two

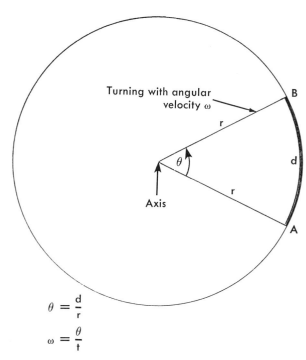

$$\theta = \frac{d}{r}$$

$$\omega = \frac{\theta}{t}$$

Fig. 12-1. A lever turning from point *A* to point *B*, through an angle *(θ)* during a time *(t)*, has an angular velocity *(ω)*.

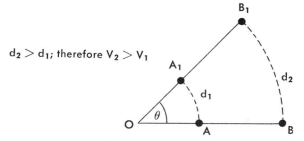

$d_2 > d_1$; therefore $V_2 > V_1$

Fig. 12-2. Two levers of unequal length turning through the same number of radians per unit of time; i.e., $\omega = \theta/t$ is identical for both levers.

points—divided by the radius (r). In other words,

$$\text{Angle (in radians)} = \frac{\text{Arc length}}{\text{Radius}}$$

or, in algebraic symbols,

$$\theta = d/r$$

By rearranging this formula, we can see that the arc length is given by $d = r\theta$. Substituting $r\theta$ for *d* in the linear velocity formula (V = d/t), we obtain $V = r\theta/t$; but since $\omega = \theta/t$, we can replace θ/t with ω to obtain

$$V = r\omega \qquad \text{(Equation 12-B)}$$

Therefore we see that the linear velocity of an object at the end of a lever is the product of the length of the lever (i.e., the radius of rotation) and the angular velocity. Hence the greater the angular velocity and the greater the length of the lever, the greater will be the linear velocity at the end of the lever. This is illustrated in Fig. 12-2. For a given angular velocity the end of lever *B* can be seen to travel a greater distance *(d₂)* per unit of time (t) than does the end of lever *A;* hence the greater linear velocity (d/t) for lever *B*.

Angular acceleration

A rotating body need not have a uniform angular velocity (ω) just as a body in linear motion need not have a uniform velocity (V). If the angular velocity of a body changes from some original value (ω_0) to a new value (ω_f) in a time interval (t), the average angular acceleration (α, lower case Greek letter alpha) of the body is

$$\alpha = \frac{\omega_0 - \omega_f}{t} \qquad \text{(Equation 12-C)}$$

Since angular acceleration is described as the rate of change of angular velocity, it can be either positive (speeding up) or negative (slowing down).

Relationship of angular acceleration to linear acceleration

Remembering that linear velocity (V) equals the product of the radius (r) and the angular velocity (ω), which gives the equation $V = r\omega$, we can obtain the relationship of linear acceleration to angular acceleration by substituting $r\omega$ for *V* in the linear acceleration equation:

$$a = \frac{V_f - V_0}{t} = \frac{r\omega_f - r\omega_0}{t} = \frac{r(\omega_f - \omega_0)}{t} = r\alpha$$
$$\text{(Equation 12-D)}$$

Thus linear acceleration equals the product of the radius (r) and the angular acceleration (α), i.e., $a = r\alpha$. Note the similarity of this relation to the formula depicting the relationship between linear velocity and angular velocity ($V = r\omega$).

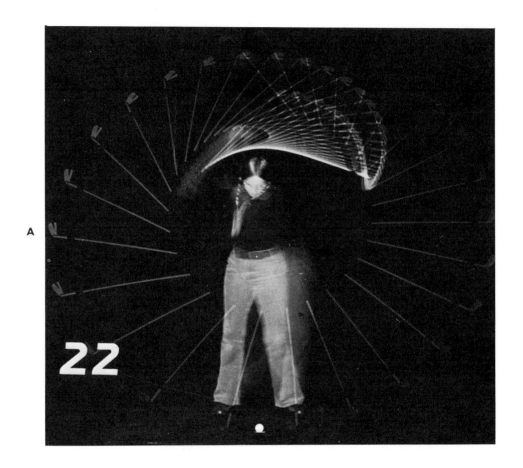

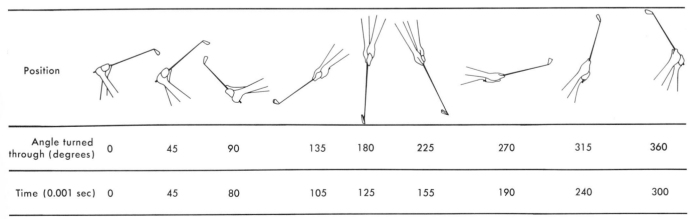

Position								
Angle turned through (degrees)	0	45	90	135	180	225	270	315	360
Time (0.001 sec)	0	45	80	105	125	155	190	240	300

Fig. 12-3. Descriptive data of the angular motion of a golf swing. (Angles shown are for the arms.) (From Cooper, J. M., and Glassow, R. B.: Kinesiology, ed. 4, St. Louis, 1976, The C. V. Mosby Co.)

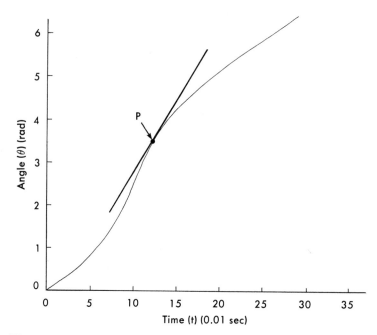

Fig. 12-4. Angle-time graph of the golf swing data described in Fig. 12-3.

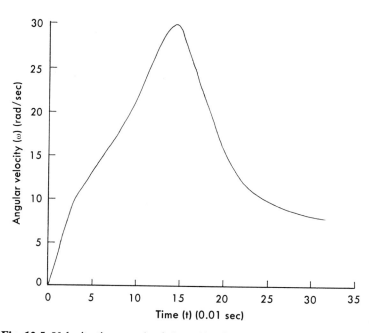

Fig. 12-5. Velocity-time graph of the golf swing data described in Fig. 12-3.

GRAPHICAL DESCRIPTION OF ANGULAR MOTION QUANTITIES

It is possible to graphically describe the angular motion of the golf swing shown in Fig. 12-3 in the same manner as we did the linear motion of the sprinter. With the aid of cinematographic measuring techniques, we can obtain the measures of time and angles shown at the bottom of Fig. 12-3. This descriptive information can be represented graphically in angle-time and velocity-time graphs as shown in Figs. 12-4 and 12-5.

Graphical description of angular velocity

Just as the distance-time graph was used to study linear velocity, so the angle-time graph is used to study angular velocity. Suppose we want to know the instantaneous angular velocity of the golf swing depicted in Fig. 12-3 after the player has swung the club through 200° of rotation. In making this determination, we need to use the angle-time graph shown in Fig. 12-4. A tangent line is drawn on the graph at 3.5 rad (200°) and is labeled *P* (point of tangent). Then selecting any two convenient points on this line, we determine their coordinates by inspection and use this information in the formula

$$\omega = \frac{\theta_f - \theta_o}{t_f - t_o}$$

For example, we can obtain

$$\omega = \frac{5.0 \text{ rad} - 2.2 \text{ rad}}{0.17 \text{ sec} - 0.08 \text{ sec}} = \frac{2.8 \text{ rad}}{0.09 \text{ sec}} = 31 \text{ rad/sec}$$

Thus it should be evident that the angular velocity of the club head after 200° of rotation will be approximately 31 rad/sec. However, remember that this value is merely an estimation whose accuracy greatly depends upon the precision of the graphical techniques. The aforementioned tangent method can also, of course, be used to determine the *angular velocity* occurring between any two angles.

Graphical description of angular acceleration

Angular acceleration can also be examined in a manner analogous to that of linear motion. Thus, just as the linear velocity-time graph was used to study linear acceleration, so the angular velocity-time graph is used to study angular acceleration.

Angular velocities calculated from the golf

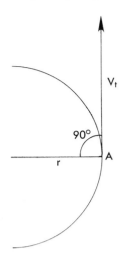

Fig. 12-6. The direction of linear motion of a body in rotation is 90° to the radius (or tangent to the curve) at the intersection with the radius of rotation.

swing data (Fig. 12-3) were used to construct the velocity-time graph shown in Fig. 12-5. The instantaneous angular accelerations at different times during the swing can be found by applying the tangent method to the velocity-time graph according to the techniques already discussed. The velocity-time graph and the tangent method can also be used to find the angular acceleration occurring between any two times during the swing.

Angular quickness is, of course, expressed as the ratio of the change in instantaneous angular velocity to the angular displacement required for this change. Thus the quickest batter in a baseball situation is the one who can obtain the greatest angular velocity of the bat in the shortest amount of angular displacement, i.e., the fewest degrees turned through. Angular quickness can also be expressed graphically through the use of a velocity-angle graph.

DIRECTION OF LINEAR MOTION OF A BODY IN ROTATION

The direction of linear motion of a body in rotation is 90° to the radius of rotation or tangent to the curve of the rotation. This direction is shown in Fig. 12-6. Since linear motion is always tangent to the curve, it is frequently called simply tangential (tan-jen'shal) motion. Thus the terms tangential motion and linear motion are synonymous when used in regard to a body in rotation.

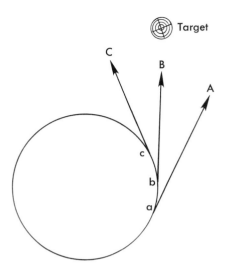

Fig. 12-7. The point of release of a projection is crucial in accuracy events.

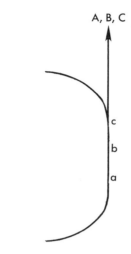

Fig. 12-8. Accuracy can be increased by flattening the arc of the swing.

Effect of the release point on direction

Since an object continues to move in the direction it was moving at release until acted upon by some other force, its direction of movement at release obviously is crucial in projection events that involve accuracy. Fig. 12-7 shows an object in rotation with three release points identified as *a, b,* and *c.* If the purpose of the release is to project the object to the target shown, the best tangent to the arc will be the one which arises from release point *b.* Release points *a* and *c* have tangents which are directed either too far right or too far left to even come close to the target.

The importance of the release point is soon learned by most athletes involved in sports which require objects to be projected at targets. The football quarterback, for instance, soon learns that if he releases the ball too soon on a short pass it will go over the receiver's head. Likewise, if he releases the ball too late it will end up in the turf short of the receiver. This problem of accuracy is, of course, just another application of the laws affecting projectiles.

Effect of the curvature of the arc on direction control

The effect of the curvature of the arc on the control of direction is illustrated in Fig. 12-8.

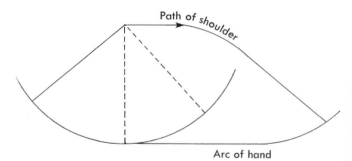

Fig. 12-9. The arc of an underhand throw can be flattened by moving the shoulder forward.

This illustration shows that the more nearly the arc approaches a straight line the less divergent will be the tangents at various points on it. Therefore the flatter the arc through which the object travels, the greater is the possibility of accuracy.

Broer (1973) suggests that the arc of the throwing hand can be flattened by moving the center of the arc (i.e., the shoulder) in the direction of the movement and in the direction of the circumference of the arc.

1. In the underhand throw the shoulder moves forward and downward as shown in Fig. 12-9. This is accomplished by transferring the weight to a bent-forward leg and following through in a forward direction.

2. In the overhand throw there is a sequential extension of the joints of the arm as the shoulder moves forward due to the rotation of the spine and the transference of weight.

3. In the sidearm throw the rotation of the trunk moves the shoulder forward and out to the side of the throwing hand. Weight transfer also acts to flatten the arc.

EFFECTS OF VARYING THE LENGTH OF THE RADIUS OF ROTATION ON MOTION QUANTITIES

The following two principles are extremely important in the analysis of human motion.

When the angular velocity of a body in rotation is constant, its linear velocity is *directly* proportional to the length of its radius of rotation.

$$V = r\omega$$

When the linear velocity of a body in rotation is constant, its angular velocity is *inversely* proportional to the length of its radius of rotation.

$$\omega = V/r$$

Situations in which a lengthening of the radius is desired

When the maximum angular velocity of a body part has been attained, its linear velocity can be increased by lengthening its radius of rotation. A maximum angular velocity of a body part is attained when it no longer is undergoing an angular acceleration ($\alpha = 0$); during the continuation of its movement, the angular velocity is constant.

For example, in the velocity-time graph of Fig. 12-10 the body is being accelerated between points *A* and *B*. Beyond *B* it is moving with a constant velocity. If the object moving in this situation is the leg of a kicker and if the knee of the kicker is bent, then we can see that the linear velocity of the foot will be increased through a straightening of the knee (which will produce a longer radius of rotation).

Likewise, a golfer must have arms straight at the moment the club strikes the ball, and a tennis player's serving arm must be straight as the racket strikes the ball. There are multitudes of examples of the applications of this principle, and students should try to identify as many of these as they can.

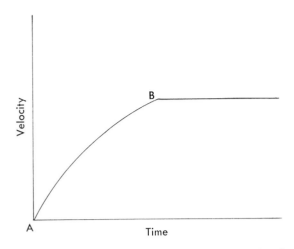

Fig. 12-10. A velocity-time graph showing acceleration between points *A* and *B*.

Situations in which a shortening of the radius is desired

When the maximum linear velocity of a body part has been attained, its angular velocity can be increased by shortening its radius of rotation.

For example, in somersaults from the trampoline or diving board, the linear speed of the rotation is determined when the body leaves the trampoline or the springboard. If a faster turn is desired to complete the stunt properly, the radius of rotation can be shortened by assuming a tucked position. This decrease in the radius of rotation increases the angular velocity of the turn. Likewise, if the body is turning too fast the radius can be lengthened by assuming a layout position. The increase in the radius of rotation will decrease the angular velocity of the turn.

In the same manner the angular speed with which a batter swings the bat can be increased by choking up on the bat thus decreasing the radius of rotation. A kicker or runner can increase the angular speed of leg movement by bending the knee. Students should try to generate other examples of this principle.

Situations in which an alternate lengthening and shortening of the radius is desired

In vertical swinging activties, e.g., the high bar, parallel bars, or flying rings in gymnastics, shortening the radius of rotation (distance between

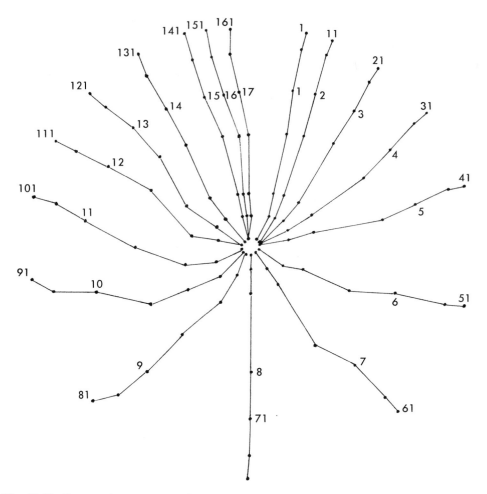

Fig. 12-11. Segmental movements of a gymnast doing a giant swing. The dots represent the centers of the wrist, elbow, shoulder, hip, knee, and ankle joints and the center of gravity of the feet. Every tenth frame is plotted. The camera speed was 64 f/s.

the center of gravity and the center of rotation) on the upswing will accelerate the movement. Lengthening the radius on the downswing will increase the linear velocity of the center of gravity at the bottom of the swing. This alternate lengthening and shortening of the radius during a swinging activity is illustrated in Fig. 12-11.

STUDY GUIDELINES
Analysis of angular velocity and acceleration
Specific student objectives

1. Define angular velocity (in both sentence and formula form).
2. Describe the relationship of angular velocity to linear velocity.
3. Define angular acceleration (in both sentence and formula form).
4. Describe the relationship of angular acceleration to linear acceleration.

Summary

Angular velocity (ω) is defined as the rate of angular displacement, i.e., the ratio of the angle turned through (θ) in radians to the elapsed time (t). Thus we see that $\omega = \theta/t$. Clockwise rotation is called a negative direction; counterclockwise rotation is termed positive. The relationship between linear velocity (V) and angular velocity (ω) is expressed by the formula $V = r\omega$.

Angular acceleration, the rate of change of angular velocity, can be an increase in angular speed (positive) or a decrease in angular speed (negative); or there can be a change in the direction of rotation. Linear acceleration (a) and angular acceleration (α) are related by the formula a = rα.

Performance tasks

1. State the meaning of d = rθ, V = rω, and a = rα.
2. The angle turned through per unit of time is called angular
   a. speed
   b. velocity
   c. V = d/t
   d. V = t/d
3. The difference between angular speed and velocity is the element of (acceleration/direction)_____.

4. Write out the description of the following:
   a. Angular speed, velocity, and acceleration
   b. Relations between angular and linear velocity and angular and linear acceleration
5. The formula for angular velocity is
   a. $\omega = \theta/t$
   b. $\omega = t/\theta$
   c. $\omega = d/t$
   d. all of the above
6. Using V = rω, describe the relationship between V and ω with the aid of a motor performance example.
7. The angular velocity formula can be used to analyize motion problems. By rearrangement, should the magnitudes of the remaining two quantities be known we can find the magitude of a missing quantity.
   Select the correct form of the angular velocity formula and calculate the missing quantity in the problems below:

Forms of the angular velocity formula

$$\omega = \frac{\theta}{t} \qquad t = \frac{\theta}{\omega} \qquad \theta = t \times \omega$$

Angle turned through (θ)	Time (t)	Angular velocity (ω)	Formula
a. 35 rad	7 sec	_____	$\omega = \frac{\theta}{t}$

Using $\omega = \frac{\theta}{t}$, we get $\bar{\omega} = \frac{35 \text{ rad}}{7 \text{ sec}} = \frac{5 \text{ rad}}{\text{sec}}$.

This is 5/6 of a revolution per second.

b. 27 rad	_____	16 rad/sec	$t = \frac{\theta}{\omega}$

Using $t = \frac{\theta}{\omega}$, we get $t = \frac{27 \text{ rad}}{16 \text{ rad/sec}} = \frac{27 \text{ rad} \times \text{sec}}{16 \text{ rad}} = 1.69$ sec

c. _____	14 min	2 rev/sec	$\theta = t \times \bar{\omega}$

Using $\theta = t \times \bar{\omega}$ and changing $\frac{2 \text{ rev}}{\text{sec}}$ to $\frac{4 \pi \text{ rad}}{\text{sec}} = \frac{12.56 \text{ rad}}{\text{sec}}$,

we obtain $\theta = 840 \text{ sec} \times \frac{12.56 \text{ rad}}{\text{sec}} = 10{,}550.40$ rad (approx. 1680 rev).

d. 12 rad	4 sec	_____	_____
e. 11 rev	5 sec	_____	_____
f. 13.3 rad	_____	9 rad/sec	_____
g. 92 rad	_____	4 rev/sec	_____
h. _____	26 sec	7 rad/sec	_____
i. _____	2 min	4 rad/sec	_____
j. 540°	3.1 sec	_____	_____

Graphical description of angular motion quantities

Specific student objectives

1. Identify the graphical description of angular velocity.
2. Identify the graphical description of angular acceleration.
3. Describe the use of the Greek letters theta (θ), omega (ω), and alpha (α) as angular motion symbols.
4. Given the raw data of angles and time, construct and label an angle-time graph.
5. Given the raw data of angular velocity and time, construct and label an angular velocity-time graph.
6. With the use of an angle-time graph of angular motion, calculate the approximate instantaneous angular velocity by the tangent-line method at any selected point on the graph.
7. With the use of an angular velocity-time graph of angular motion, calculate the approximate instantaneous acceleration at any point on the graph by the tangent-line method.
8. Describe angular quickness.

Summary

Uniform angular motion occurs when an object is rotating with a constant velocity. An object traveling with a changing velocity is in a state of acceleration or variable motion. A variable velocity is one in which unequal angles are turned through in equal periods of time (or vice versa) and/or the direction of rotation changes. The angular velocity of a turning object at a precise instant of time is termed instantaneous velocity. Average velocity is the mathematical average of all the instantaneous velocities for any given interval. Angular acceleration, likewise, can be constant or variable, and average and instantaneous accelerations can be determined. Angular quickness is the ratio of instantaneous angular velocity to angular displacement.

Rotational movement can be described in numerical values for angular speed, velocity, and acceleration. It can also be represented graphically to obtain a pictorial view of the motion from which additional information can be derived, such as instantaneous angular velocities and accelerations.

Performance tasks

1. Write out descriptions of the following angular motion terms:
   a. The use of alpha (α), theta (θ), and omega (ω)
   b. Uniform and variable angular velocity
   c. Uniform and variable angular acceleration
   d. Average and instantaneous angular velocity and acceleration
2. Identify the steps in graphing the following angular motion situations:
   a. Angle-time graph
   b. Velocity-time graph
3. It is possible to examine the angular velocity of the golf swing shown in Fig. 12-3 as we did the linear motion of a sprinter. With the aid of cinematographic measuring techniques, we can obtain the measures of time and angles shown in Fig. 12-3.

   The average angular velocity during selected intervals may be found using the obtained measures as shown below:

First 45°:

$$\omega = \frac{\theta}{t}$$

where θ is 45° or $\pi/4$ rad and t is 0.045 sec.

$$\omega = \frac{0.785 \text{ rad}}{0.045 \text{ sec}} = 17.44 \text{ rad/sec}$$

First 90°:

$$\omega = \frac{\theta}{t} = \frac{1.57 \text{ rad}}{0.08 \text{ sec}} = 19.63 \text{ rad/sec}$$

The student should calculate the average angular velocity for the first 135° and the first 180° for comparison purposes.

It is also possible to determine the average angular velocity during specified intervals. Average angular velocity from 45° to 90°:

$$\bar{\omega} = \frac{\theta_f - \theta_o}{t_f - t_o} = \frac{1.57 \text{ rad} - 0.785 \text{ rad}}{0.08 \text{ sec} - 0.045 \text{ sec}} =$$

$$\frac{0.785 \text{ rad}}{0.035 \text{ sec}} = 22.14 \text{ rad/sec}$$

Average angular velocity from 90° to 135°:

(Student calculation):

$$\omega = \frac{\theta_f - \theta_o}{t_f - t_o} =$$

4. Using the appropriate formula for acceleration, calculate the average acceleration from the given information.

$$\bar{\alpha} = \frac{\omega_t - \omega_o}{t}$$

	ω_t	ω_o	t	$\bar{\alpha}$
a.	20 rad/sec	10 rad/sec	4 sec	_____
b.	38 rad/min	28 rad/min	2 sec	_____
c.	6 rev/sec	0 rev/sec	3 sec	_____
d.	0 rev/sec	11 rev/sec	2.2 sec	_____
e.	4π rad/sec	π rad/sec	1.7 sec	_____
f.	$3\pi/4$ rad/sec	8 rad/sec	3.8 sec	_____

5. Since $V = r \times \omega$ is the relationship between linear velocity and angular velocity, the golf swing might also be used to illustrate this relation in concrete terms. If it is assumed that the radius of rotation *(r)* is approximately 5 ft for the example cited in Fig. 12-3 and that *V* and ω at specified locations have been determined graphically, then the relationship between *V* and ω would be

Angle (rad)	ω (rad/sec)	V (ft/sec)
$\pi/4$	14	70
$\pi/2$	18	90
$3\pi/4$	26	130
$\pi*$	31	155
$5\pi/4$	18	90
$3\pi/2$	12	60

A *direct* relationship is shown by this comparison between linear velocity and angular velocity. Thus, with a constant radius, the linear velocity increases or decreases in direct proportion to the increase or decrease in angular velocity. This tells us that the linear velocity of the club head at impact with the ball is a direct function of the angular velocity of the arms and club acting in combination.

6. Using $a = r\alpha$, list and describe the existing proportionality between pairs of terms in the formula. Illustrate each proportionality with a human motion example. An example would be

a. If "r" is held constant, "a" and "α" are directly proportional; i.e., as one doubles the other will do the same.

b. An illustration of this situation is that of

*Indicating instant of impact.

swinging a "bat." Assume you will always use the same bat and it has a constant "r"; i.e., you grip it in the same place each time. Then, assuming you have the necessary strength, the more you increase the torque to cause an increase in the rotation of the bat (α) the greater will be the linear acceleration (a) of the end of the bat.

7. The descriptive information presented on the golf swing in Fig. 12-3 can be represented graphically in both angle-time and velocity-time graphs. This graphical representation is presented in Figs. 12-4 and 12-5, respectively. The information obtainable from the angle-time graph is as follows:

a. If the angular velocity is constant, i.e., the graph is a straight line, then the average velocity $\bar{\omega}$ can be found directly from the graph by

$$\omega = \frac{\theta_t - \theta_o}{t_t - t_o}$$

b. If angular velocity varies, i.e., the graph is a curved line, the instantaneous angular velocity can be found by use of the tangent line method and

$$\omega = \frac{\theta_t - \theta_o}{t_t - t_o}$$

With the angle-time graph shown in Fig. 12-4, the instantaneous velocity after half a revolution (at impact) can be found. For example, at 180° (3.14 rad) we can construct a tangent line on the angle-time graph. Then, selecting any two convenient points on this line, we can determine their coordinates by inspection and use the information obtained in the formula

$$\omega = \frac{\theta_t - \theta_o}{t_t - t_o}$$

For example we can obtain

$$\omega = \frac{5 \text{ rad} - 2.2 \text{ rad}}{0.17 \text{ sec} - 0.08 \text{ sec}} = \frac{2.8 \text{ rad}}{0.09 \text{ sec}} =$$

31 rad/sec at the moment of impact

It must be realized that this is an estimate whose accuracy depends upon the precision of the graphical techniques. The student should use the tangent method to estimate the instantaneous angular velocity at several other points in the golf swing.

8. Angular acceleration can be examined in a manner analogous to that of linear motion. The data of Fig. 12-3 will be used to illustrate this point.

Instantaneous angular acceleration can be found using cinematographic measuring techniques or applying the tangent method to a velocity-time graph of the data. The student should construct this graph and verify the instantaneous velocity estimations at 90°, 180°, 270°, and 360°. These estimates are as follows:

Angle (θ)	Angular velocity (ω)
90° or $\pi/2$	18 rad/sec
180° or π	31 rad/sec
270° or $3\pi/2$	12 rad/sec
360° or 2π	8 rad/sec

The average rotatory acceleration ($\bar{\alpha}$) can be obtained either directly from advanced measuring techniques (radar) or indirectly from a velocity-time graph. Between 180° and 270° the average acceleration obtained from the data above is

$$\bar{\alpha} = \frac{\omega_t - \omega_o}{t - t} = \frac{12 \text{ rad/sec} - 31 \text{ rad/sec}}{0.190 \text{ sec} - 0.125 \text{ sec}} =$$

$$\frac{-19 \text{ rad/sec}}{0.065 \text{ sec}} = -292.3 \text{ rad/sec}^2$$

This means that the golf swing is decelerating at the rate of 292.3 rad per second during every second.

Direction of linear motion of a body in rotation
Specific student objectives

1. Describe the direction of tangential motion.
2. Describe the effect of the release point on the direction of motion of a projectile.
3. Describe the effect of the arc of curvature on the direction control of projectiles.
4. Identify a specific technique used to flatten the arc in an angular motion situation.

Summary

During movement about an axis, the direction of linear motion upon release is 90° to the radius

of rotation, i.e., tangent to the curve at the point of release. Tangential motion is the term used to describe this effect.

Thus it can be shown that the release point during angular motion will precisely determine the linear direction of the object released according to the tangential effect.

When changing from angular motion to linear motion, the performer can flatten the arc in order to make the tangents less divergent and, in turn, improve accuracy. This can be accomplished by such movement techniques as combining linear motion with the angular motion just prior to release.

Performance tasks

1. Write descriptions of each of the following:
   a. Tangent
   b. The direction of linear motion of a rotating object
   c. Tangential motion
   d. The effect of the release point on the direction of linear motion
   e. The effect of the arc of curvature on linear direction control
2. Select four angular motion situations that result in linear motion upon release. Specify exact techniques you would recommend for increasing task accuracy by flattening the arc of curvature.

Effects of varying the length of the radius of rotation on motion quantities
Specific student objective

1. Identify and describe with illustrative motion examples the principles relating radius of rotation to either angular velocity or linear velocity.

Summary

The two principles describing the effects of changing the length of the radius of rotation in angular motion situations are

1. When the angular velocity of a body in rotation is constant, the linear velocity of the body is *directly* proportional to the length of its radius of rotation. In formula form this relationship is expressed as $V = r\omega$.
2. When the linear velocity of a body in rota-

tion is constant, the angular velocity of the body is *inversely* proportional to the length of its radius of rotation. In formula form this relationship is expressed as $\omega = V/r$.

The length of the radius should be increased in situations where maximum angular velocity is attained (i.e., angular acceleration is 0) and increased linear velocity is desired.

A shortening of the radius should be carried out in angular motion situations when maximum linear velocity has been reached and increased angular velocity is desired.

Combinations of the aforementioned angular motion situations sometimes occur and call for alternate lengthening and shortening of the radius.

Performance tasks

1. Give three sports performance examples of how increasing the radius length when angular acceleration is 0 will increase the linear velocity.
2. Give three sports performance examples of how decreasing the radius length when linear acceleration is 0 will increase angular velocity.

SELF-EVALUATION

Students should use no reference materials for this progress test, and they can check their answers by referring to Appendix A.

1. The difference between angular speed and angular velocity is the element of
   a. distance
   b. time
   c. both 1 and 2
   d. neither 1 nor 2
2. The formula for angular velocity is
   a. $\omega = \theta/t$
   b. $\omega = t/\theta$
   c. $V = d/t$
   d. $V = t/d$
3. Define angular velocity (in sentence form).

   _____
   _____

4. Using $V = r\omega$, describe the relationship between V and ω with the aid of a motor performance example. _____
   _____
   _____
   _____

5. Calculate $\bar{a}$ or $\bar{\alpha}$ for the following motion problems:
   a. $V_f = 20$ ft/sec, $V_o = 50$ ft/sec, t = 3 sec
   b. $V_f = 10$ mi/hr, $V_o = 0$ mi/hr, t = 0.6 sec
   c. $\omega_f = 5$ rad/sec, $\omega_o = 22$ rad/sec, t = ¼ sec
   d. $\omega_f = 0$ rad/sec, $\omega_o = 1$ rad/sec, t = 3 sec
6. Define angular acceleration (in both formula and sentence form).

   _____
   _____

7. Velocity (and therefore acceleration) changes when the following element(s) varies (vary):
   a. Magnitude
   b. Direction
   c. Both a and b
   d. Neither a nor b
8. Angular acceleration is a (scalar/vector) quantity _____ .
9. If $a = r\alpha$, describe the relationship between r and a if α is constant.
10. Insert the word "equal" or "unequal" in each blank below:
    a. Uniform motion takes place when angles are traversed in_____time periods and no direction change occurs.
    b. Nonuniform motion results when angles are traversed in_____time periods and/or direction changes occur.
11. Define instantaneous angular velocity.

    _____
    _____

12. Average angular velocity is the mathematical average of all_____angular velocities over an interval.
13. Using the data of Fig. 12-12, depicting the motion of a baseball batter, carry out the following exercises:
    a. Construct and label an angle-time graph.
    b. Using the angle-time graph and the tangent method, estimate instantaneous angular velocities at 0.05 sec intervals (to 0.30 sec).
    c. Using the instantaneous velocities found in (b), construct and label an angular velocity-time graph.

Angle turned through (degrees)	0	45	90	135	180	225	270	315	360
Time (0.01 sec)	0	5	7	10	12	15	19	23	28

Fig. 12-12. Descriptive data of the angular motion of baseball batting.

d. Using the angular velocity-time graph and the tangent method, calculate instantaneous accelerations when t = 0.10 and 0.20.

14. Describe the following angular motion terms:
    a. Angular quickness_____
    _____
    b. Direction of tangential motion_____
    _____
    c. Effect of the release point on the direction of linear motion of a projectile_____
    _____
    d. Effect of the arc of curvature on direction control of a projectile_____
    _____

15. The technique(s) used to flatten the arc for possible accuracy increase would be
    a. weight transfer
    b. radius length change
    c. both a and b
    d. neither a nor b

16. Illustrate each of the following motion principles with movement examples:
    a. Increasing the radius length when angular acceleration is zero and increased linear velocity is desired
    b. Decreasing the radius when maximum linear velocity has been obtained and increased angular velocity is needed
    c. Combinations of a and b

Kinematic analysis of force and torque

Body of lesson

KINEMATIC ANALYSIS OF FORCE

One of the key concepts of mechanics is that of force. Force is simply any influence tending to cause or change the movement and/or shape of a body. Therefore it is both the instigator and the modifier of motion, and as such it is the focal point in the causal analysis of motion, which we have already identified as *kinetics*. The kinetic analysis of forces, however, is based upon an accurate qualitative and quantitative description of these forces and of their vector components. Thus a *kinematic* analysis of forces must be made *before* a kinetic analysis is even possible.

Classification of forces

Even though a qualitative analysis consists of both nominal and evaluative types, our present consideration will be limited to the classification of forces—which forms the basis of a nominal analysis. The evaluative analysis of force appli-

cations in terms of their effectiveness and efficiency will be reserved for the lesson on kinetics.

Forces can be classified according to their effects and according to their sources.

The two possible motion *effects* are propulsion and resistance. The effect of a propelling force is, of course, to cause motion. The effect of a resistive force is to oppose motion. Thus, when we see an object in motion we know that it is moving because a propelling force has acted upon it and that this propelling force was greater than the object's resistive force.

The forces which cause or modify human motion may come from a variety of *sources*. Therefore we sometimes consider the human body and its parts as constituting a "system," and we classify the forces originating outside this system (e.g., gravity, surface friction, air resistance, and the pushes and pulls exerted by other people) as external while the forces generated from within

Fig. 13-1. A bowling ball and the pins that it strikes are considered to be a closed system, and the forces they exert on one another are internal forces.

the body (e.g., through muscle contractions and friction between moving parts) are internal.

Actually any set of interacting objects can be considered to constitute a system, and we normally refer to the forces that they exert on one another as *internal forces* and the forces arising outside the system as *external forces*. For example, consider the bowling ball and pins illustrated in Fig. 13-1. When the ball strikes the pins, it exerts a force on each of the pins with which it makes contact. In turn, all the pins exert forces on each other as they make contact. When we consider the ball and pins as constituting a system, then the forces they exert on each other are considered to be internal forces. When a pin strikes the end cushion of the alley, however, the

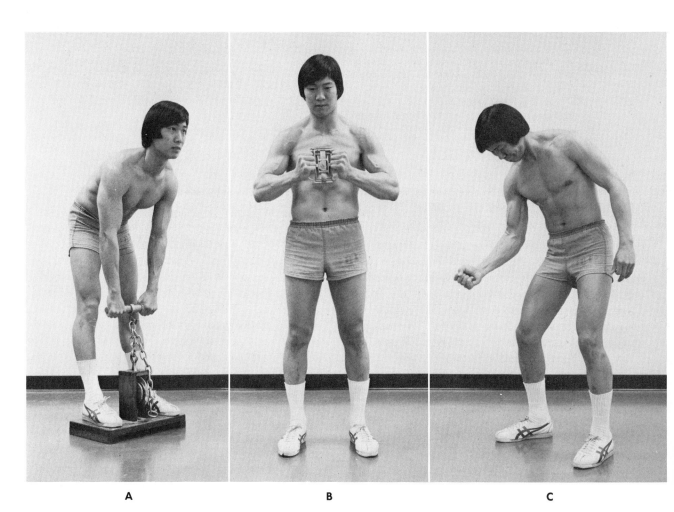

A   B   C

Fig. 13-2. Selected mechanical devices used in the measurement of force.

resulting force of impact then is considered to be an external force since the end cushion is not a constituent of the system. Actually the pin and cushion form a new system of interacting objects.

The criteria for the inclusion of objects within a system are (1) their relevancy to the questions being studied, (2) their practicality from the standpoint of the measurement problems involved, and (3) the significance of their contributions to the behavior of the total system. Because the end cushion does not meet these criteria, it is excluded from the system being considered.

Direct measurement of force

Forces can be measured either directly through the use of mechanical and electronic devices or indirectly through an analysis of the movements and movement changes they produce. Our present consideration will be restricted to the direct approaches. The lessons on kinetics will deal with the indirect approaches.

Some mechanical devices used to measure muscle-produced forces are illustrated in Fig. 13-2. The force a muscle group can exert against one of these devices in a single maximal effort is called its strength. Since there are two basic types of muscle contraction, isometric and iso-

tonic, there are two basic approaches to the measurement of muscle strength. The definition of the isometric and isotonic types of contraction as well as the definition of the different types of isotonic contractions are as follows:

isometric contraction This is a contraction in which the propulsive force produced by a muscle group is not sufficient to overcome the resistive forces of the object acted on by the muscle group. Thus the motion state of the object and the length of the muscles do not change.

isotonic contraction This is a contraction during which the object acted on by a muscle group does move and the length of the muscles does change. The two types of isotonic contractions are

concentric contraction In which the propulsive force produced by a muscle group is greater than the resistive forces of the object acted on; thus the muscles are allowed to shorten and produce motion

eccentric contraction In which the resistive force produced by a muscle group is not sufficient to prevent motion of the object acted on; thus the muscles lengthen during the resulting motion

NOTE: A concentric or eccentric contraction during which the speed of contraction is held constant is called an isokinetic contraction.

Since any time a force acts through a distance mechanical work is performed, the isotonic types

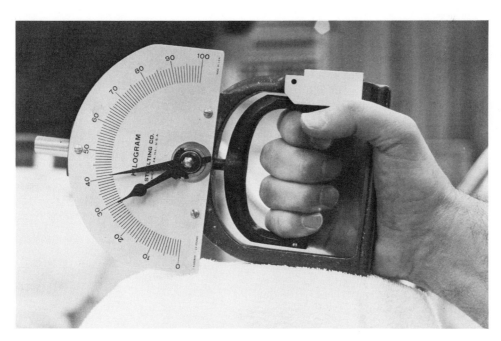

Fig. 13-3. Smedley dynamometer.

of strength measures will be discussed in the next lesson, which focuses on work and power. Thus the present discussion is further restricted to a consideration of static forces. A static force is defined as one acting in a situation where the propulsive and resistive forces are balanced and the force is not allowed to undergo a displacement.

A device that measures force is called a dynamometer (from Greek *dynamis,* power; *metrein,* to measure) and a self-registering dynamometer is called a dynamograph (*graphein,* to write). The two most common types of mechanical dynamometer (di'-nah-mom'-eter) are the spring steel type and the tensiometer. The spring steel type is illustrated in Fig. 13-3, which shows the Smedley dynamometer. All spring steel dynamometers are based upon the same principle: deformation of a piece of steel (in the form of a ring, ellipse, or coil) is proportional to the force applied. This is actually a description of Hooke's law (to be explained in Lesson 19). The cable tensiometer is illustrated in Fig. 13-4. Cable tension is determined by measuring the force applied to a riser causing an offset in a cable

stretched taut between two sectors. The tension is then converted into pounds or kilograms on the calibration chart of the instrument.

The spring steel dynamometer and tensiometer have both been used to successfully measure maximum isometric forces. However, in recent years emphasis has shifted from merely assessing maximum force to examining the interaction of force and time during a muscle contraction. This change of emphasis has necessitated the development of electronic dynamometers through which force and time can be continually measured.

The three most common types of transducers used in these systems are the strain gauge, the linear variable differential transformer (LVDT), and piezoelectric crystals.

The Smedley hand grip dynamometer (Fig. 13-3) can be used as either a mechanical or an electronic force-measuring device. When used as an electronic measuring device, it becomes a *strain gauge transducer.* When force is applied to the dynamometer, its electrical resistance is changed due to the mechanical strain produced by the force. This altered resistance results in changes in the output voltage which are moni-

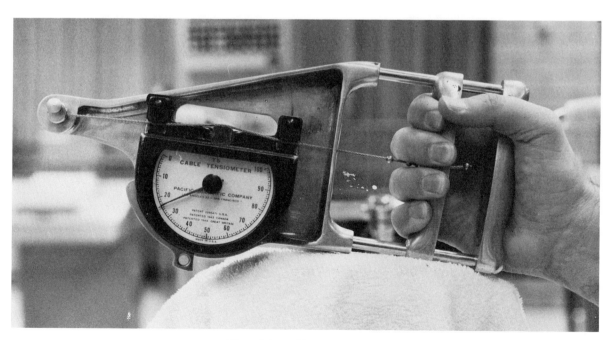

Fig. 13-4. Cable tensiometer.

tored and recorded. Calibration procedures provide direct conversion values from voltage to pounds or kilograms.

The *linear variable differential transformer* (LVDT) is a device which continuously translates the displacements of a magnetic iron core into a linear AC voltage (Fig. 13-5). Completely stepless resolution, combined with extreme sensitivity, allows displacements of one millionth of an inch or less to be detected readily. The output signal voltage is precisely proportional to the mechanical displacement of the sensing probe on which the force is applied, and these displacements are directly proportional to the applied force.

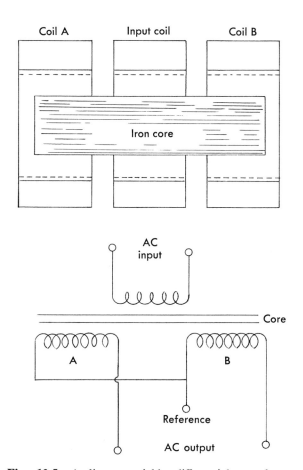

Fig. 13-5. A linear variable differential transformer (LVDT) is an electromechanical device which continuously translates displacement or position change of an iron core into linear AC voltage.

Piezoelectric transducers (Greek *piezein,* to press or squeeze) use the piezoelectric effect (discovered in 1880 by the Curie brothers): an electrical charge appears on the surfaces of certain crystals when they are subjected to a mechanical load. Of the numerous piezoelectric materials, quartz is by far the most suitable for force measurement because it is a very stable material with constant properties. The use of quartz crystals in multicomponent force measurement will be discussed later in this lesson.

KINEMATIC ANALYSIS OF TORQUE

When a single force acting on a body tends to produce angular motion, it is said to exert a torque. As we already know, angular movement always occurs around a center of rotation, called an axis or fulcrum. To make anything rotate (i.e., to cause angular motion), it is always necessary to apply an *off-center* force.

Every force has a line of action. In Lesson 7, line of action was defined as the line that extends through the force vector indefinitely in both directions (Fig. 13-6). In order for a force to have a twisting or rotating effect (called torque), it must have a line of action which does *not* pass through the center of rotation or axis. In other words, the force must be applied *eccentrically* to cause rotation.

Only linear motion can be produced by a force whose line of action passes through the center of rotation. Simultaneous linear and angular motion, however, results from an eccentric force. This effect can be seen in Fig. 13-7, which shows that F_1 has a tendency to move the body downward as well as produce counterclockwise rotation while F_2 has a tendency to move the body upward and to the right as well as to produce clockwise rotation. In a *force couple* situation (Fig. 13-8), i.e., two equal and opposite parallel eccentric forces, the linear tendencies are neutralized by the opposite linear directions of the forces but since

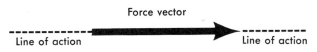

Fig. 13-6. The line of action of a force extends through the force vector indefinitely in both directions.

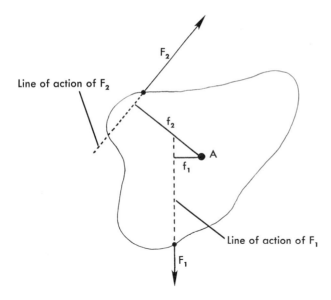

Fig. 13-7. The torque of a force about an axis is the product of the force and its lever arm.

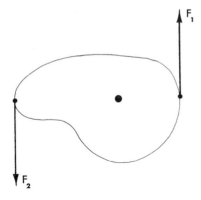

Fig. 13-8. A force couple consists of two equal and opposite parallel eccentric forces.

the rotation tendencies are in the same angular direction (clockwise) they are added together.

As with forces the two possible effects of a torque are propulsion and resistance. The effect of a propelling torque is, of course, to cause rotation while that of a resistive torque is to oppose rotation. Torques can also be classified as being either internal or external depending upon the source of the forces.

The directional effect produced on a body by a torque of any given magnitude depends on the direction of its line of action in relation to the

body's axis of rotation. Thus in Fig. 13-7, the effect of F_1 is to produce counterclockwise rotation about the axis while the effect of F_2 is to produce clockwise rotation. To distinguish between these directions of rotation, we shall adopt the convention that counterclockwise torques are positive and clockwise torques are negative. Hence the torque (T) for F_1 is positive while T for F_2 is negative. Rotation will take place in the direction of the greater torque, and no rotation will take place when the opposite torques are equal (i.e., when the algebraic sum of the torques is zero).

Measurement of torque

The term torque is synonymous with force-moment or moment of force and is defined as the product of the force multiplied by the perpendicular distance from the line of action of the force to the fulcrum. The perpendicular distance from the line of action of the force to the fulcrum is the *moment* or *lever arm* of that force and is symbolized by the lower case letter f. Thus torque (T) equals the product of the force *(F)* multiplied by the length of its lever arm *(f)* (Fig. 13-7).

$$T = F \times f \qquad \text{(Equation 13-A)}$$

Torque is expressed in weight-distance units such as pound-feet (lb-ft).

Another method of calculating torque involves breaking the force into its two components (which act at right angles to each other). These two components are shown in Fig. 13-9, where they are identified as the radial *(F_r)* component and the tangential *(F_t)* component. One component acts along the radius of the rotation *(F_r)* and the other at right angles to the radius *(F_t)*. As is evident from the diagram, the component acting along the radius has a zero torque since its line of action passes directly through the point of rotation. Therefore we need be concerned only with the torque produced by the component acting at right angles to the radius *(F_t)*.

If we consider the distance from the center of rotation or axis to the point of application of the force to be the radius of rotation (r), then torque can be found by multiplying the tangential component (F_t) by r; i.e., $T = F_t r$. Since the tangential component, as we shall see later, is found by multiplying the force (F) by the sine of the angle

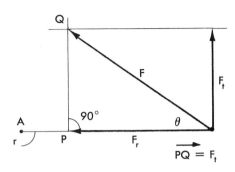

Fig. 13-9. A force acting on a radius can be resolved into its radial *(F_r)* and tangential *(F_t)* components.

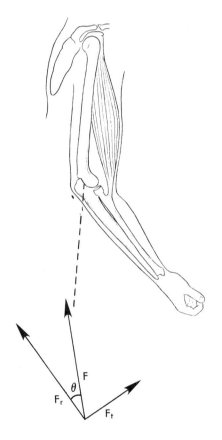

Fig. 13-10. The biceps brachii applies a force that tends to rotate the forearm at the elbow joint. (From Krause, J. V., and Barham, J. N.: The mechanical foundations of human motion: a programmed text, St. Louis, 1975, The C. V. Mosby Co.)

$(F_t = F \sin \theta)$, then the second method of calculating torque can be expressed as

$$T = F_t r = (F \sin \theta) r$$

(Equation 13-B)

The measurement of torque through the use of Equation 13-B will be further discussed in the next section of this lesson.

ANALYSIS OF FORCE COMPONENTS OPERATING IN ANGULAR MOTION

An example of the measurement of torque through the resolution of the applied force into its components is illustrated in Fig. 13-10. This illustration shows the biceps brachii muscle, located on the anterior surface of the arm. When a muscle contracts, it applies a force on all bones to which it is attached. Let us assume that the angle of pull of the biceps brachii on the forearm is 30°. After a glance at the illustration, you should be able to see that the action of the muscle will be to cause rotation of the forearm at the elbow. However, the action will also cause the forearm to be pulled into the elbow joint, thus tending to stabilize this joint. The force of biceps brachii contraction therefore has two components: one causing rotation and the other causing stabilization. These components are calculated in the same manner as were the horizontal and vertical components discussed in Lesson 8.

Concept of radial and tangential components

The vector diagram of the biceps brachii muscle force is given at the bottom of Fig. 13-10. One component of the force acts parallel with the forearm, which is the radius of rotation, while the other component acts perpendicular to the forearm. The component acting parallel with the forearm or radius is called the radial component (F_r); the component acting perpendicular or tangent to the forearm or radius is called the tangential component (F_t).

Concept of centripetal and centrifugal forces

Radial components can be classified as being either centripetal or centrifugal according to the direction in which they act. Centripetal components act *toward* the axis of rotation (Latin *petere*, to seek), while centrifugal forces act *away* from the center of rotation (Latin *fugere,* to flee). The radial component illustrated in Fig. 13-10 is a centripetal component because it is directed inward toward the center of rotation. The centripetal component therefore does not have a rotatory tendency, but it does have a stabilizing tendency for it pulls the forearm more tightly into the elbow. Thus more force is required to dislocate

this joint. Similarly a centrifugal force component has a dislocating tendency on a joint.

Graphical determination of force components

The magnitudes of the radial and tangential components are determined graphically by the

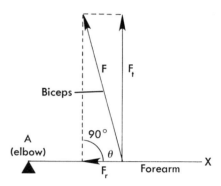

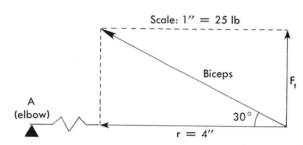

Fig. 13-11. Analog drawing of the biceps brachii force acting on the forearm.

Fig. 13-12. The tangential component (F_t) is the side opposite in a right triangle, while the radial component (F_r) is the side adjacent.

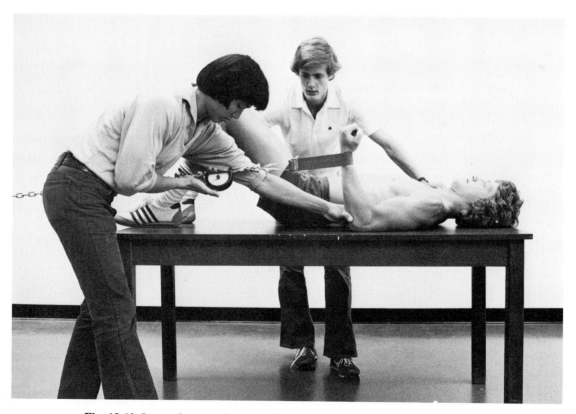

Fig. 13-13. Isometric strength test of the elbow flexors using the cable tensiometer.

same methods as were discussed in Lesson 8. The two steps in this method are (1) draw the vector to scale on a sheet of paper and (2) draw perpendicular lines from the head of the arrow vector to the radial and tangential axes (located at the tail of the vector). The magnitudes of the radial component (F_r) and the tangential component (F_t) are then measured directly from the diagram. The drawing shown in Fig. 13-11 is an analog enlargement effect of the force created by the biceps brachii muscle acting on the forearm and causing it to rotate around the elbow joint.

Trigonometric determination of force components

The trigonometric method of determining force components can also be applied to problems concerned with the resolution of force vectors into

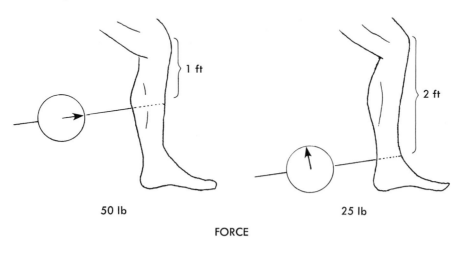

FORCE

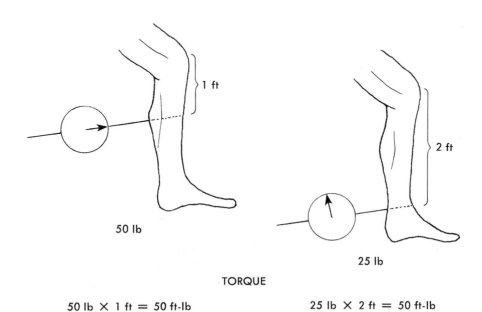

TORQUE

50 lb × 1 ft = 50 ft-lb   25 lb × 2 ft = 50 ft-lb

Fig. 13-14. The force, measured by a dynamometer, is dependent upon its distance from the joint axis, while the torque is constant, regardless of distance, throughout the entire length of the segment.

radial and tangential components. Shown in Fig. 13-12 is a force vector acting upon a body. Here it can be seen that the tangential component is the side opposite in a right triangle while the radial component is the side adjacent. Thus the tangential component (F_t) is found by multiplying the force by the sine of the angle ($F_t = F \sin \theta$), and the radial component (F_r) is found by multiplying the force by the cosine of the angle (or $F_r = F \cos \theta$).

Torque analysis of muscle strength tests

An isometric muscle strength test is shown in Fig. 13-13. The force applied by the subject's forearm registers on the face of the dynamometer. This is the resistive force, which opposes the force produced by the subject's elbow flexors. It can be converted into a torque measure by multiplying it by the resistive moment arm of the subject. Here the resistive moment arm of the subject equals the longitudinal axis of his forearm. Since the subject's elbow is in a state of angular equilibrium, obviously the measured resistive torque will be equal to the muscle-produced torque. Thus muscle strength tests are actually muscle torque tests since the muscle force is not measured directly.

The effect of varying the location of the dynamometer on knee extension strength is shown

at the top of Fig. 13-14. When the dynamometer is located 1 ft from the center of the knee, the force registered is 50 lb; when it is located 2 ft from the center of the knee joint, the force is 25 lb. Thus the force obtained in a strength test is dependent upon the distance of the dynamometer from the joint axis. The lower part of the illustration shows that the torques calculated in the two situations, however, are identical.

THREE-DIMENSIONAL ANALYSIS OF FORCES AND TORQUES

A device that measures forces in the x, y, and z directions and the torques about each of these axes is usually called a *force platform* or *force plate*. The sensing devices can be strain gauges, LVDTs, or piezoelectric transducers. Even though strain gauges and LVDTs are used in the construction of force platforms, only the piezoelectric transducers will be discussed in this section.

As mentioned before, the primary piezoelectric material used to make piezoelectric transducers is the quartz crystal. Of the three piezoelectric effects which occur in quartz, i.e., the longitudinal, shear, and transverse effects, only the longitudinal

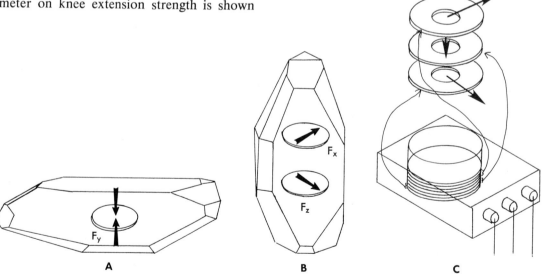

Fig. 13-15. Depending upon the position in which they are cut out of quartz, discs are obtained which are sensitive to pressure only (**A**) (longitudinal effect) or to shear in one direction only (**B**). A single three-component force transducer (**C**) contains two pairs of shear-sensitive discs to measure F_x and F_z and one pair of pressure-sensitive discs to measure F_y. Using them in pairs doubles the sensitivity and permits simple electrical contact by a central electrode.

and shear effects are used in multicomponent force measurements.

In its crystalline form quartz is anisotropic; this means its material properties are not identical in all directions. Depending on how they are cut out of the crystal, discs are obtained which are sensitive only to perpendicular pressure (longitudinal pressure effect) or sensitive only to shear in a particular direction (shear effect) as shown in Fig. 13-15. The pressure-sensitive discs measure the vertical force component (F_y), while the shear-sensitive discs measure the horizontal force components $(F_x$ and $F_z)$ as well as the torque around the three cardinal axes (M_x, M_y, M_z)* (Fig. 13-16).

*M (which means *moment*) is the same as T (torque).

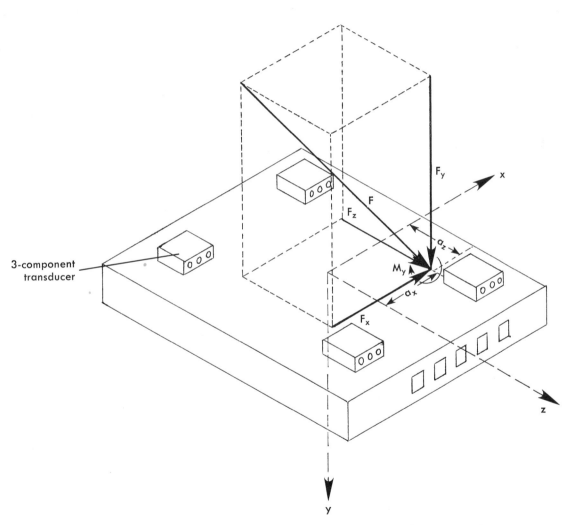

Fig. 13-16. Four three-component transducers are mounted, under high prestress between a base and a top plate, to form a force platform. For simple three-component force measurement, the respective x, y, and z channels of the four transducers are paralleled. This is achieved directly through switches built into the platform. The three components are measured independently of the point of force application. Determining the location of the point of force application (coordinates a_x and a_z) and the moment about the y axis (M_y) requires the individual signals from the four transducers to be processed according to mechanics formulae by a digital computer (see Fig. 3-11).

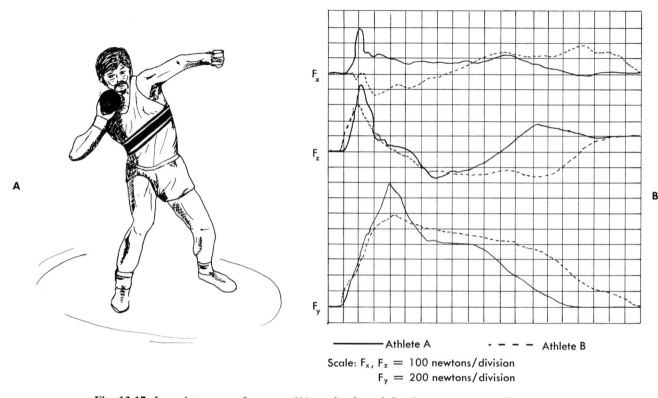

Scale: F_x, F_z = 100 newtons/division
F_y = 200 newtons/division

Fig. 13-17. In a shot put performance **(A)**, evaluation of the three components F_x, F_z, and F_y in the thrusting leg is important. Shown in **B** is the evolution of the three components of the force exerted by the right foot of two shot-putters. *Athlete A*, weighing 105 kg, puts the shot 18 m. *Athlete B*, weighing 86 kg, attains 14.5 m. Note the difference in the corresponding force components.

Design of quartz multicomponent dynamometers

The utilization of quartz discs in the design of three-component dynamometers is described and illustrated in Fig. 13-15. The design of a multicomponent force platform is described and illustrated in Fig. 13-16.

As can be seen in Fig. 13-17, the piezoelectric force platform can measure the forces exerted by an athlete's feet in three components. The point of instantaneous force application as well as the torque about a vertical axis can be determined. The diagrams in Fig. 13-17, *B,* show the evolution of the three components of the force exerted by the right foot of two shot-putters. *Athlete A* weighs 105 kg and puts the shot 18.00 m while *Athlete B,* weighing 86 kg, attains 14.50 m. Note the difference in the corresponding force

components. Usually such force measurements are supplemented by cinematographic and other recordings.

STUDY GUIDELINES
Kinematic analysis of force
Specific student objectives

1. Define force and static force.
2. Differentiate between propulsion and resistance forces.
3. Differentiate between internal and external forces.
4. Identify and describe the types of muscle contractions.
5. Define muscle strength.
6. Describe the following force analysis terms:
   a. Dynamometer   c. Transducer
   b. Dynamograph

Summary

Force is defined as any influence tending to distort and to cause motion or to change the movements of a body. The motion effects of a force are propulsion (causing motion) and resistance (opposing motion).

Forces that affect human motion can be classified as external (originating outside the body), such as gravity and friction, and internal (generated within the body) such as muscle forces.

The two basic types of muscle contraction are isometric (no movement or change in muscle length) and isotonic (muscle length change occurs). The types of isotonic contraction are concentric (muscle shortening) and eccentric (muscle lengthening). Isokinetic contractions are isotonic contractions during which the speed of contraction is held constant.

Force can be measured directly with mechanical or electronic devices. Strength is described as the static force (one maximal effort) a muscle can exert against one of the measuring devices.

A device that measures force is called a dynamometer. A self-registering dynamometer is called a dynamograph. The two most common types are the spring steel and the tensiometer. Electronic dynamometers are called transducers.

Performance tasks

1. Any influence tending to distort and to cause motion or tending to change motion of a body is called a (force/resistance)————————.
2. Develop a written list of descriptions for the following:
   a. Static force
   b. Propulsive and resistive forces
   c. Internal and external forces
   d. The types of muscle contractions
   e. Dynamometers
   f. Transducers
   g. Muscle strength
3. Identify specific motor performance examples of the following forces:
   a. Propulsive
   b. Internal
   c. Resistive
   d. External
4. Identify specific motor performance examples of the basic types of muscle contraction.
5. Write out a summary of mechanical and electronic means for direct measurement of force.

Kinematic analysis of torque

Specific student objectives

1. Define the concept of torque (in both sentence and formula form).
2. Define "axis of rotation" and "line of action" of a force.
3. Define "lever arm" of a force.
4. Differentiate between positive and negative torques.
5. Differentiate between radial and tangential components of a force.
6. Describe an eccentric force.
7. Differentiate between propulsive and resistive and between internal and external torques.
8. Identify the two methods of measuring torque.
9. Given a force (F) and its lever arm (f), calculate the produced torque.
10. Given a tangential force (F_t) and radius of rotation (r), calculate the produced torque.
11. Given a torque-producing force (F), resolve it into radial and tangential components.

Summary

A turning or twisting effect, called torque, is produced when an off-center eccentric force is applied at a distance from the axis of rotation (fulcrum). Torque (T) is synonymous with moment of force and is defined as the product of the *force* (F) and the *perpendicular distance* (f) from the line of action of the force to the fulcrum; thus T = F × f. This distance is called the "lever arm" of the force. Torques are also either propulsive or resistive and external or internal. Clockwise torques are negative while counterclockwise torques are designated as positive.

The force producing a torque, as a vector, can be resolved into two components: one acting perpendicular to the radius (called tangential, F_t) and the other acting parallel with the radius (called radial, F_r); thus $\vec{F} = \vec{F}_r + \vec{F}_t$ (arrows signify vector quantities). The magnitude of these components can be determined by the approximation method of graphing or by the more precise trigonometric methods described in Lesson 8, where $F_t = F \sin \theta$ and $F_r = F \cos \theta$.

Torque can thus be measured by $T = F \times f$ or by $T = F_t \times r = (F \sin \theta)r$.

Performance tasks

1. Develop written descriptions for the following:
   a. Torque
   b. Axis of rotation
   c. Line of action
   d. Lever arm
   e. Positive and negative torques
   f. Radial and tangential components of a force
   g. Eccentric force
   h. Propulsive and resistive; internal and external torques
2. A twisting effect is called (force/torque)____.
3. The two variables that makeup torque are
   a. force and axis
   b. propulsive and resistive
   c. internal and external
   d. force and force arm
4. Calculate the torques (force moments) in the following situations: Use torque (T) = force (F) × lever arm (f).

	Force	Lever arm	Torque
a.	20 lb	2 ft	?

(Example: $T = F \times f = 20$ lb $\times 2$ ft $= 40$ ft-lb)

	Force	Lever arm	Torque
b.	4 lb	7 yd	_____ ft-lb
c.	57 oz	3 in	_____ ft-lb
d.	4200 oz	4 yd	_____ ft-lb
e.	13 lb	9 ft	_____ ft-lb

NOTE: Principles for changing units of measure are presented in Appendix D.

5. Identify two specific examples each of propulsive and resistive torques in human motion situations.
6. Identify one example of both an internal and an external torque in a human motion situation.
7. Resolve the following torque-producing forces (F) into their radial and tangential compononets: $F_r = F \cos \theta$; $F_t = F \sin \theta$

	F	θ	r	F_r	F_t
a.	100 lb	45°	3 in	_____	_____
b.	150 lb	10°	2 ft	_____	_____
c.	41 lb	5°	6 in	_____	_____
d.	56 lb	85°	1 ft	_____	_____
e.	90 lb	30°	18 in	_____	_____

8. Determine the torques produced by the forces acting through the angles in Task 7. Use $T = F_t \times r$ or $T = F \times f = (F \sin \theta) \times r$.

Analysis of force components operating in angular motion

Specific student objectives

1. Describe the graphical determination of force components.
2. Differentiate between centripetal and centrifugal forces.
3. Describe the role of torque measures in strength testing.

Summary

A torque-producing force can be analyzed into its radial component (parallel with the radius of rotation) and tangential component (tangent to the arc of rotation and perpendicular to the radius).

Radial components can be resolved into either centripetal components (acting toward the axis) or centrifugal components (acting away from the axis of rotation). Centripetal components of a muscle force tend to stabilize, while centrifugal components tend to dislocate, joints.

The magnitudes of the above components can be found by the vector analysis methods of resolution by graphical and trigonometric methods (Lesson 8).

Strength tests must be analyzed to consider the measure of torque as well as force.

Performance tasks

1. Develop a written list of descriptions for each of the following:
   a. Radial and tangential force components
   b. Centripetal and centrifugal components of a radial force
2. Carry out at least two strength-testing measures and determine both static force and torque.
3. Force acting toward the center refers to (centripetal/centrifugal)_____.
4. Torque measures are usually expressed in
   a. pound-feet
   b. mi/hr
   c. foot-pounds
   d. feet/sec

Three-dimensional analysis of forces and torques

Specific student objective

1. Describe a force platform.

Summary

A force platform is a mechanical-electrical device that measures forces and torques in three dimensions, i.e. forces in the x, y, and z directions and torques about each of these axes. The sensing devices are called transducers. These are three-component dynamometers which can produce data that are usually supplemented by other techniques such as cinematographic analysis.

Performance tasks

1. Develop a written description of a force platform.
2. Consult one outside reference concerning three-dimensional force and torque analysis and develop a written summary of the use of a force platform in human motion analysis.

SELF-EVALUATION

Students should use no reference materials for this progress test, and they can check their answers by referring to Appendix A.

1. A tendency to cause motion or change the state of motion is called a
   a. force
   b. machine
   c. both a and b
   d. neither a nor b
2. When propulsive and resistive forces are balanced, the force is called
   a. static
   b. external
   c. internal
   d. torque
3. The two types of force and torque classified according to their effects are
   a. internal and external
   b. propulsive and resistive
   c. both a and b
   d. neither a nor b
4. Muscle force would be an example of an (external/internal)_____ force.
5. Complete the descriptions of the following types of muscle contractions:

a. Isometric_____
   _____
b. Isotonic_____
   _____
   (1) Concentric_____
   _____
   (2) Eccentric_____
   _____
   (3) Isokinetic_____
   _____

6. The force a muscle group can exert against a measuring device is called the muscle
   a. strength
   b. torque
   c. both a and b
   d. neither a nor b
7. A device that measures force is called a (dynamometer/goniometer)_____.
8. An electronic force-measuring device is called a (tensiometer/transducer)_____.
9. Torque can be defined as follows:
   a. $T = d/t$
   b. $T = gt$
   c. $T = F \times f$
   d. $T = f_r \times f_t$
10. Define the following terms:
    a. Lever arm
    b. Line of action
    c. Positive torque
    d. Resistive torque
    e. Radial and tangential force components
11. A radial force component that has a dislocating effect away from the fulcrum is called (centripetal/centrifugal)_____.
12. A force producing a torque is termed (eccentric/static)_____.
13. A torque can be changed by varying the
    a. causal force
    b. lever arm
    c. both a and b
    d. neither a nor b
14. A three-dimensional force-torque analysis device is a
    a. dynamometer
    b. force platform
    c. graph
    d. strength test
15. Identify an example of
    a. propulsive and resistive force and torque
    b. internal and external force and torque

16. Strength testing should include measurements of
    a. force
    b. torque
    c. both a and b
    d. neither a nor b

17. Calculate the torques in the following motion problems:

	Force (F)	Lever arm (f)	Torque (T)
a.	11 lb	3 ft	_____
b.	35 oz	5 in	_____
c.	100 lb	1½ in	_____
d.	300 lb	2 in	_____
e.	84 lb	¼ in	_____

18. Resolve the following torque-producing forces into radial and tangential components:

	F	θ	r	F_r	F_t
a.	27 lb	15°	3 in	____	____
b.	92 lb	8°	4 in	____	____
c.	100 lb	30°	9 in	____	____
d.	210 lb	3°	7 ft	____	____
e.	136 lb	73°	2 ft	____	____

19. Determine the torques produced by the forces acting through the angles given in Question 18. Use $T = F_t \times r$.

Kinematic analysis of energy and power

Body of lesson

CONCEPTS OF ENERGY AND WORK

When a physical event or change occurs in the universe, it does so because of the redistribution of some form of energy. Therefore one of the most fundamental concepts in all of nature is that of energy. But just what is this phenomenon? How can the concept of energy be defined?

We know from experience that energy is not a "thing" which can be described in such terms as size, shape, or mass. It must therefore be defined in dynamic terms and in context with the environmental changes it produces. Indeed, the presence of energy is revealed only when change is occurring. Accordingly, we shall define energy as the ability to produce change or, more precisely, the ability to perform work. Work is therefore a measure of energy, and the magnitude of work can be measured in terms of the forces acting upon matter and the resulting displacement of these forces. In order to make this notion definite, the physical quantity called work is defined as follows: Work equals force times the distance through which the force acts.

$$\text{Work} = W = F \times d$$

(Equation 14-A)

We have all had the experience of applying forces against objects whose resistance is greater than we have the strength to overcome. Pushing against a brick wall might be an example of a situation where we can apply a force and nothing happens. The force is applied, but the wall does not yield and no effects of the force on the wall are shown. On the other hand, when we apply exactly the same amount of force to a ball, the ball flies through the air for some distance. In the first case, where we pushed against the wall, our hand did not move; the force remained in place. In the second case, where we threw the ball, our hand *did* move while the force was being applied and before the ball left our hand. The *motion* of the force was what made the difference.

If we think carefully along these lines, we will see that whenever a force acts so as to produce motion in a body the force itself undergoes a displacement. For this reason work was defined as the product of force and the distance through which the force is applied. The definition of work is a great help in clarifying the effects of forces. Unless a force acts through a distance, no work is done, no matter how great the force. And even

237

if a body moves through a distance, no work is done unless a force is acting upon it or it exerts a force on something else.

Different forms of energy and work

Such diversified phenomena as the flight of a golf ball, the lifting of a book, the fall of a leaf, the flashing of a light, the charging of a battery, and the processes of life in general involve the emission, absorption, and redistribution of energy. Energy manifests itself in such obvious forms as mechanical energy, heat energy, electrical energy, sound energy, and light or radiant energy. Apparently distinct as these forms of energy may appear, each possesses in common with every other form of energy the feature that under certain conditions it can be made to do work.

For example, we know that the combustion of a gasoline air mixture is a heat-producing process. If the combustion takes place very rapidly within a confined space such as the cylinder of a gasoline engine, it is called an explosion. This explosion sets countless millions of tiny high-speed gas molecules in motion; these are hurled against a piston, and the piston moves. By means of the motion of tiny particles (molecules) (felt as heat), a force has acted through a distance and work has been done. The force which causes the movement of the tiny particles is the source of *heat energy.*

The common dry cell or flashlight battery produces a flow of electrons (an electric current) in a piece of wire, e.g., in the filament of a flashlight bulb, as a result of a chemical reaction. Even though electrons, which are constituents of metallic atoms of the filament, are very tiny, they do have mass and work must be done to make them move. The chemical reaction which has supplied the force to cause their movement has been a source of *chemical energy.*

When an electric current flows through a wire lying between the poles of a magnet, the wire is subject to a force which makes it move. Hence we speak of *electric energy.*

A rapidly vibrating violin string sets in motion air molecules, which have mass. These molecules in turn transmit their motion to others and eventually impinge on human eardrums, making the eardrums move. Thus sound has exerted force and caused objects to move. Hence *sound* qualifies as a form of energy.

Likewise, we have *light energy,* which is capable of causing the movement of molecules in the retina of the eye. The movement of the retinal molecules in turn causes nerve impulses to be transmitted into the brain to give us visual sensations.

Since there are many forms of energy and work, it is important that we pinpoint the matter upon which a force is being applied. For example, when we are pushing on a brick wall, no work is being performed with respect to the wall but a great amount of work is being performed by our skeletal muscles. This is true because of the large amount of molecular motion that occurs inside the muscles during their contractions.

Law of energy conservation

One of the fundamental laws of nature is the law of the conservation of energy. The key concept in this law is that energy can never be created or destroyed and, therefore, the total of all the energy in the universe remains constant. Thus in changing from one form to another, energy is always conserved. In other words, for any quantity of one form of energy that disappears, an equivalent amount of other forms is produced.

Should we drop, say, a book on the floor, what happens to the mechanical energy? According to the concept of energy conservation, the mechanical energy of the book is converted into heat and sound. The conversion of energy from one form into another can be demonstrated quite easily. All we need to do is rub one thing against another: two blocks of wood, perhaps, or even our hands. When this is done, we find that the objects are warmer than before. This is a quite general conclusion. We might do work against *viscosity,* which is friction in a fluid, by stirring some water vigorously and again find that it becomes warmer. Therefore we have a conversion of mechanical energy into heat energy.

Einstein's discovery of the interchangeability of mass and energy has compelled the extension of the law of conservation of energy to include situations involving the conservation of mass to energy and vice versa.

CONCEPT OF MECHANICAL ENERGY

Energy may be thought of as the ability to do work. When we say that something has energy, we mean it is capable of exerting a force on something else and doing work. On the other hand, when we do work on something, we have added an amount of energy equal to the work done. The units of energy are the same as those of work—the foot-pound in the British system.

What properties can a body have that may be converted into work? In other words, what forms does energy take? We shall consider in this lesson two broad categories of mechanical energy: *kinetic energy,* which is energy of motion, and *potential energy,* which is energy of position or condition.

We all know from experience that a moving body is capable of doing work upon objects with which it may collide. A bowling ball, for example, is able to do work upon the pins it strikes; i.e., when it collides with the pins, it exerts a force on them through a distance and therefore does work. Likewise, a baseball bat striking a baseball does work upon the ball; a football fullback does work on the opposing lineman with whom he collides, etc. The energy of a moving body is called kinetic energy. The concept and the quantifications of kinetic energy will be further considered in Lesson 18.

The other type of mechanical energy is called potential energy. The key idea here is that a body, by virtue of its position or condition, is able to do work. A ball held at a certain height (h) above the ground is capable of doing work when it is released and allowed to fall. The work it does is equal to the product of its weight (w) and the height (h) at which it was originally held above the ground, i.e., w × h. This type of energy is called *gravitational potential energy.* Another type of potential energy is contained in a bent bow in archery. The bow of an archer is capable of doing work upon an arrow because of the elastic restoring force contained in the materials making up the bow. Thus when the archer releases the string on a bent bow, this elastic restoring force acts upon the arrow through a distance. The potential energy imparted to the bow by bending is called *elastic potential energy.* The gravitational and elastic types of potential energy will be further analyzed in Lessons 18 and 19, respectively.

KINEMATIC ANALYSIS OF WORK

Work is done only when a force is exerted on a body and moves the body in such a way that there is a force component along the line of motion of the point of application of the force. If the component of the force is in the *same direction* as the displacement, the work (W) is *positive.* If the component is *opposite* the displacement, the work is *negative.* If the force is at *right angles* to the displacement, it has no component in the direction of the displacement and the work is *zero.*

Thus when a body is lifted, the work of the lifting force is positive; when a spring is stretched, the work of the stretching force is positive; when a gas is compressed in a cylinder, again the work of the compressing force is positive. However, the work of the gravitational force acting on a body being lifted is negative since the downward gravitational force is opposite the upward displacement. Also, though it would be considered "hard work" to hold a heavy object stationary at arm's length, no work would be done on the *object* thus held because the forces applied would not be undergoing any displacement. The movement of muscle components, however, does give us considerable chemical or physiological work, as evidenced by the heat produced.

Even if we were to walk along a level floor while carrying a heavy object, no vertical work would be done since the vertical supporting force would have no component in the direction of the horizontal motion. Similarly, the work of the reaction force of gravity exerted on a body by a surface on which it was moving would be zero, as would the work of the centripetal force acting on a body moving in a circle.

Analysis of work components

In the preceding section, note was made of the fact that in defining positive and negative work the vector displacement (d) must be parallel with the direction of the force (F). If d and F are not parallel, we must take into account the angle (θ) between them. For example, in analyzing the movements of a runner during a single stride in sprinting we know that he produces two force components simultaneously: a horizontal force component (F_x), which acts to overcome the forces of air resistance and surface friction, and a vertical force component (F_y), which acts primarily against the force of gravity. We also know that these components can be found by using $F_y = F \sin \theta$ and $F_x = F \cos \theta$. Thus the total work done by the total force equals the vector sum of its two components, i.e., $\vec{W} = \vec{W_x} + \vec{W_y}$ = $(\overline{Fd \cos \theta}) + (\overline{Fd \sin \theta})$.

In the analysis of horizontal work ($W_x = Fd \cos \theta$), when the force and its horizontal motion are parallel, the angle (θ) = 0°; and since the cos of 0° = 1, we have the maximum production of work. When the force and its horizontal motion are perpendicular to each other, θ = 90°; and since the cos of 90° = 0, we can see that no work is done, even though the force has moved. An example of this is motion parallel with the earth's surface, where the force of gravity (pulling downward) is perpendicular to all horizontal displacements. Thus the work done against gravity is zero. The primary resistance forces which must be overcome in this example are surface friction and air resistance. When we lift an object, however, work is done against gravity due to the fact that the motion and the force of gravity are not perpendicular.

The vertical and horizontal forces in effect during a single stride of a runner can be measured through the use of the force platform mentioned in the last lesson. The distance through which these forces act can be measured by cinematographic techniques as we have already discussed. Thus the work done during a single stride can be calculated. The amount of work produced per stride is multiplied by the number of strides in an event (which can be counted through the use of cinematography) to obtain the total work done. Since this procedure for the measurement of work is quite involved, many kinesiologists have found it more convenient to measure the work of walking and running through the use of a treadmill (Fig. 14-1).

A treadmill consists of a conveyor belt operated by an electric motor. The speed of a walk or run is determined by the speed of the belt movement. Since the subject walks or runs in place with no horizontal displacement, we know the horizontal work produced by the subject is

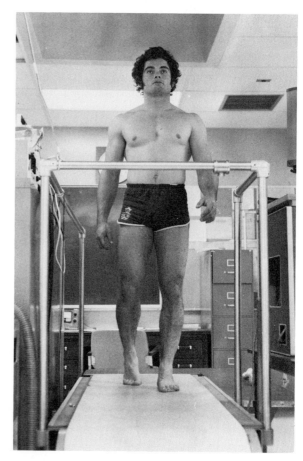

Fig. 14-1. A treadmill can be used to measure the work of walking or running.

zero. Thus the work of treadmill performance can be calculated only when a treadmill grade is established; i.e., work can be measured only when the treadmill is positioned so the subject walks or runs up an incline (grade) and thereby does vertical work.

Usually the slope of incline of a treadmill is reported as percent grade rather than degrees. Percent grade can be defined as the units of rise or vertical displacement per hundred horizontal units. For example, a +4% grade is interpreted as a 4-meter vertical displacement for every 100 meters of belt movement. The percent grade is equal to the tangent of the angle of inclination multiplied by 100.

$$\text{Percent grade} = \text{Tan } \theta \times 100$$

(Equation 14-B)

Likewise, the tangent of an angle of inclination is equal to the percent grade divided by 100. The angle which is equal to this tangent can be found by consulting a table of trigometric functions (Appendix B).

It is easy to compute the work done in lifting a body against gravity. The force of gravity acting on a body is given by the body's weight. Hence, if we lift any simple object to a height (h) we have $F = w$ and $d = h$ so the work done is $W = Fd = wh$.

Work output on a treadmill is also equal to the weight of the subject times the vertical distance (h) he would have raised himself in walking or running up the incline of the treadmill. The height (h) is obtained through use of the linear velocity equation ($V = d/t$). We substitute into this equation h for *d* and V_y for *V* to obtain $V_y = h/t$. The equation is then rearranged to obtain $h = V_y t$; and when we substitute $V \sin \theta$ for V_y we obtain

$$h = (V \sin \theta)\, t$$

(Equation 14-C)

Thus the height *(h)* is the product of the speed of belt movement, the sine of the angle of inclination, and the duration of the movement.

For example, if a subject walks for 30 minutes (0.5 hour) on a treadmill set at 3 mi/hr at an angle of 2°, his vertical displacement is determined as $h = (V \sin \theta)\,(t) = (3 \text{ mi/hr})(0.035)(0.5 \text{ hr}) = 0.0525$ mile. This height can be changed to feet by multiplying by the number of feet in a mile (5280) to obtain 277.2 feet. If the weight of the subject is 160 pounds, then the work he does during this exercise is determined as

$$W = wh = (160 \text{ lb})(277.2 \text{ ft}) = 44,052 \text{ ft-lb}$$

Perhaps it should be pointed out here that to lift a body of weight (w) to a height (h) above its starting point requires that we perform the amount of work (wh) and that the particular route taken by the body in being lifted to such height is not involved. Therefore, just as much work must be done to climb a flight of stairs as to go up in an elevator to the same floor (though not by the same agent).

By way of review, remember that a force which is applied at an angle to the horizontal can in-

deed produce two types of work on a body simultaneously. There is the work of lifting the body and the work of moving the body forward. Thus the total work performed on a body can be divided into two components: a vertical work component and a horizontal work component. The vertical work component would be determined by the equation

$$F_y = Fd (\sin \theta)$$

(Equation 14-D)

and the horizontal component by the equation

$$F_x = Fd (\cos \theta)$$

(Equation 14-E)

Analysis of angular work

Angular work is calculated in the same way as linear work except that we substitute either $r\theta$

Fig. 14-2. A bicycle ergometer can be used to measure work.

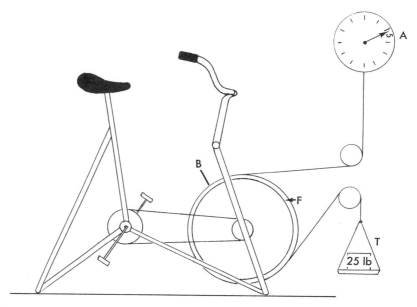

Fig. 14-3. Friction-type bicycle ergometer. When no frictional resistance is being applied, the spring scale *(A)* shows a measure equal to the weight in the tray *(T)*. When the subject pedals the bicycle, the flywheel *(F)* causes friction against the belt *(B)*. The reading on the spring scale will be reduced by an amount equal to the frictional resistance. In this type of ergometer, the work output equals the weight in the tray (e.g., 25 lb) minus the reading on the spring scale (15 lb). Thus in our example the work output is equal to 10 lb.

(where θ is in radians) or $2\pi rn$ (where *n* equals the number of revolutions) for *d* in the work equation (14-A) to obtain:

$$W = Fd = Fr\theta = F2\pi rn$$

(Equation 14-F)

If angular displacement is measured in terms of angles, then $r\theta$ is substituted for *d* in Equation 14-A. If, on the other hand, angular displacement is measured in numbers of revolutions, then $2\pi rn$ is substituted for *d*.

A common laboratory angular work–measuring device is the bicycle ergometer (Fig. 14-2). When work is measured through the use of this device, the equation for determining the amount of work done is $W = F2\pi rn$. Thus the variables in this case are (1) the resistive force *(F)*, which is the frictional resistance of the wheel, (2) the radius of the wheel *(r)*, and (3) the number of wheel revolutions *(n)*.

The friction-type bicycle shown in Fig. 14-3 was the most common work-measuring device employed by the early kinesiologists, many of whom constructed their own. The front wheel

of a bicycle is converted into a 30 to 40 lb flywheel by welding around it a heavy metal rim. Placed around the flywheel is a leather belt. One end of this belt is attached to a tray which holds the weights that apply resistance to the wheel. The other end of the belt is attached to a scale. When the bicycle is not being used, the reading on the spring scale equals the weight on the tray. When the subject begins to pedal, however, the leather belt causes friction against the flywheel. The reading on the spring scale is reduced by an amount equal to the imposed resistance. Thus the frictional resistance used in the work equation will equal the weight in the tray minus the spring scale reading. In our example, the weight on the tray is 25 lb and the reading on the scales is 15 lb. Therefore the frictional resistance acting on the wheel equals 10 lb. If the radius of the flywheel is 2 ft and the number of flywheel revolutions is 1000, then the work done is determined as:

$$F2\pi rn = (10 \text{ lb})(2)(3.1416)(2 \text{ ft})(1000)$$
$$= 125,664 \text{ ft-lb}$$

KINEMATIC ANALYSIS OF POWER

The time element is not involved in the definition of work. The same amount of work is done in raising a given weight through a given height whether the work is done in 1 second, 1 hour, or one year. In many instances, however, it is necessary to consider the rate at which work is done by a working agent, called the power developed by that agent.

If a quantity of work (W) is done in a time interval (t), the average power ($\bar{P}$) is defined as

$$\text{Average power} = \frac{\text{Work done}}{\text{Time interval}}$$
$$\bar{P} = W/t \text{ (or } Fd/t) \qquad \text{(Equation 14-G)}$$

If the rate of doing work is not uniform, the power at any instant is the ratio of the work done to the time interval when time is extremely small.

In our treadmill example discussed in the last section, we had a subject walking on the treadmill for 30 minutes at 3 mi/hr at an angle of 2°. Thus the subject in his 30 minutes of walking did 44,052 ft-lb of work. We can now use Equation 14-G to determine the average power output of the subject:

$$\text{Power} = \frac{W}{t} = \frac{44,052 \text{ ft-lb}}{30 \text{ min}} = 1468.4 \text{ ft-lb/min}$$

This value can be converted into foot-pounds per second by dividing it by the number of seconds in a minute; i.e., 1468.4 ft-lb/60 sec = 24.47 ft-lb/sec.

In the British system, where work is expressed in foot-pounds and time in seconds, the unit of power is 1 foot-pound per second. Since this unit is inconveniently small, a larger unit called the *horsepower* (hp) is in common use. Horsepower is defined as follows:

$$1 \text{ hp} = 550 \text{ ft-lb/sec} = 33,000 \text{ ft-lb/min}$$

Thus a 1 hp motor running at full load is doing 33,000 ft-lb of work every minute it runs. The number of horsepower produced by our treadmill performer can now be determined as

$$\text{hp} = \frac{\text{ft-lb/sec power produced}}{550 \text{ ft-lb/sec}} =$$
$$\frac{24.47 \text{ ft-lb/sec}}{550 \text{ ft-lb/sec}} = 0.044$$

Relationship of power to velocity

Suppose a constant force (F) is exerted on a body while the body undergoes a displacement (d) in the direction of the force. The work done is given by W = Fd, and the average power developed is given by $\bar{P} = \frac{W}{t} = F\frac{d}{t}$. Since d/t is the average velocity, the symbol $\bar{V}$ can be substituted for it in the average power equation to obtain $P = FV$. If the time interval is made extremely short, then this equation reduces to $P = FV$, where *P* and *V* are instantaneous values.

For example, the instantaneous power of a shot-putter at the time he releases the shot is equal to the weight component of the shot multiplied by the vertical component of the instantaneous velocity of the shot at the moment it is released.

Measurement of work and power

The concept of power is extremely important in the study of human movement because it combines all three of the fundamental quantities of motion—force, displacement, and time. Since all movements involve matter moving through space and time, we should now be able to see that every human movement can be described in units of power. The only nonpower events are those of the static type in which no displacements occur when a force application is made. Thus isometric strength tests are considered to be nonpower events.

To be able to jump for maximum distances, to run sprints, to throw objects for maximum distances, and to execute fast starts (as required of football players) are a few examples of the athlete's production of power. We should now easily see that the ability to develop considerable power is a prime factor in many types of athletic success. Thus there is a need to qualitatively and quantitatively analyze this important movement parameter.

The first requirement for the analysis of power is the measurement of work. A device used to measure human work output is called an ergometer (Greek *ergon,* work). An ergometer (ergom'eter) that also graphically records the work done is called an ergograph (er'go-graf), and the graphic record is called an ergogram. The most common types of ergometers found in college

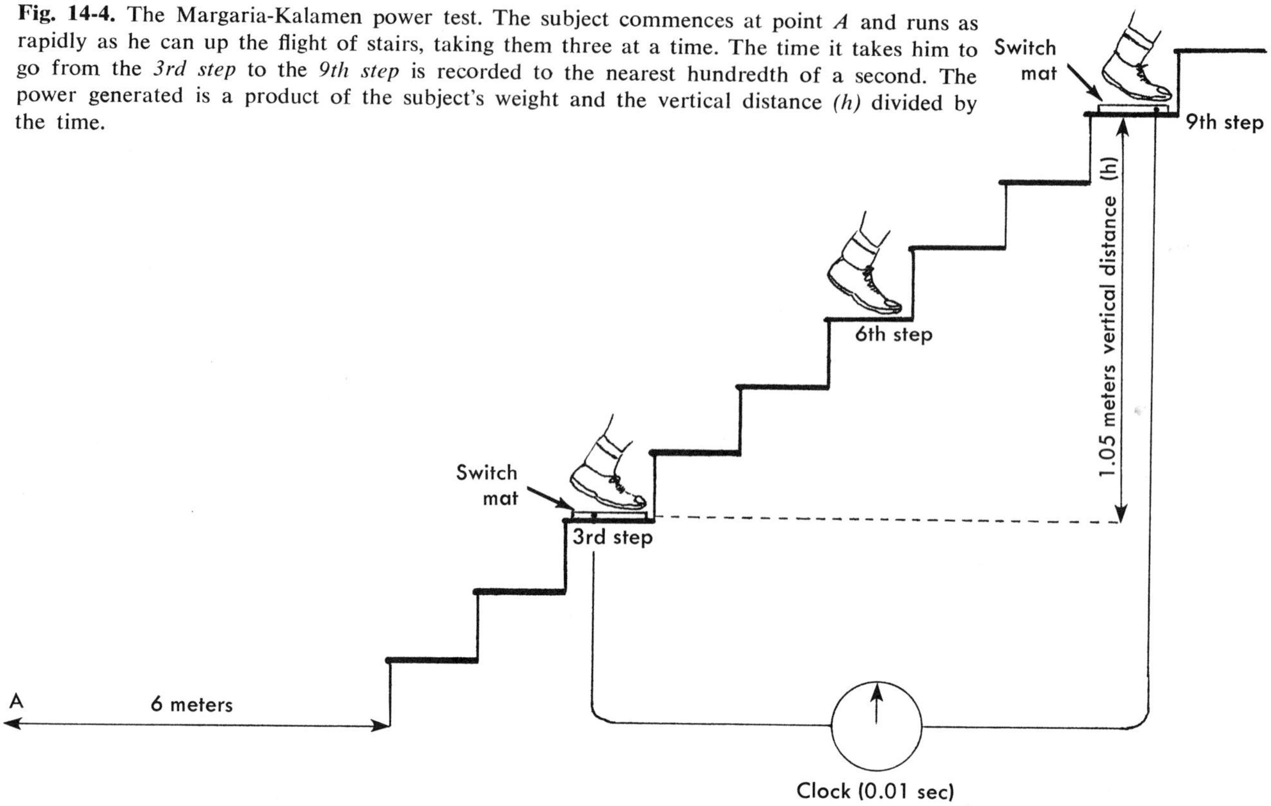

Fig. 14-4. The Margaria-Kalamen power test. The subject commences at point *A* and runs as rapidly as he can up the flight of stairs, taking them three at a time. The time it takes him to go from the *3rd step* to the *9th step* is recorded to the nearest hundredth of a second. The power generated is a product of the subject's weight and the vertical distance *(h)* divided by the time.

kinesiology laboratories are the treadmill and the bicycle ergometer already discussed. Expensive equipment is not needed for the precise measurement of work and power.

An inexpensive yet excellent test of power is that developed by Margaria and modified by Kalamen. In the *Margaria-Kalamen* power test (Fig. 14-4) the subject stands 6 m in front of a staircase. When he is ready, he runs up the stairs as rapidly as possible, taking three steps at a time. A switchmat is placed on the third and ninth stair. A clock starts when the person steps on the first switchmat and stops when he steps on the second. Time is recorded to a hundredth of a second. The vertical displacement of the subject is determined by multiplying the height of an average step by the number of steps. If an average step is about 174 mm high and there are six steps, then the vertical displacement is 1.044 m.

The two variables in this test are (1) the weight of the subject and (2) the time recorded. As an example, let us assume that the subject weighs 75 kg and that the time measured is 0.49 sec. The average power output of the subject, therefore is

$$\overline{P} = \frac{Fd}{t} = \frac{(75\ kg)\ (1.05\ m)}{0.49\ sec} = \frac{78.75\ kg\text{-}m}{0.49\ sec} = 160.71\ kg\text{-}m/sec$$

As reported by Mathews and Fox (1971), Kalamen obtained a high coefficient of relationship between the time of running the 50-yard dash with a 15-yard running start and the Margaria-Kalamen power test (r = 0.974). Thus if electronic timing equipment is available, the Margaria-Kalamen test may be used; if not, the 50-yard dash from a running start may be used.

CONCEPT OF MACHINES

A machine is defined as a device that helps directly in the performance of work. The specific function of a machine is to transmit mechanical energy from one location to another, often in a different form. There are many types of machines which transform energy, but only simple

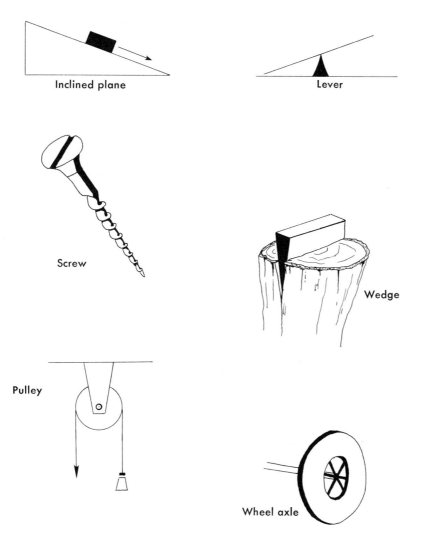

Fig. 14-5. The six types of simple machines.

machines carry out this transformation directly. A golf club is an example of a simple machine (lever) that greatly increases our ability to do work upon a golf ball directly.

The six types of simple machines are the lever, wheel-axle, pulley, wedge, screw, and inclined plane (Fig. 14-5). The three types of simple "machines" represented in the human body (to be described in the next lesson) are the lever, wheel-axle, and pulley.

We will center our attention in this text upon the three simple machines found in the human body. We will do this even though human movement is also accomplished using many combina-

tions of simple machines, which we shall call compound machines. The automobile is an example of a compound machine that is a combination of simple machines.

STUDY GUIDELINES
Concepts of energy and work
Specific student objectives

1. Define the concept of energy.
2. Define work (in both sentence and formula form.
3. Identify the different forms of energy and work.
4. State the law of energy conservation.

Summary

Energy is the ability to produce change or the ability to perform work. Work is defined as the product of any force times the distance through which the force acts, i.e., W = F × d (where *W* stands for work, *F* for applied force, and *d* for the distance). Mechanical work implies the presence of both an applied force and the displacement of that force.

Energy occurs in many forms. For example:
1. Mechanical—the energy of position, state, and motion
2. Heat—energy of molecular motion
3. Chemical—energy from chemical reactions
4. Electric—the energy of electron movement
5. Sound—vibration waves
6. Light—the energy of light waves
7. Atomic and nuclear

The law of the conservation of energy states that energy can never be created or destroyed (just changed in form); i.e., the total of all the energy in the universe remains constant.

Performance tasks

1. Write out definitions of the following:
   a. Energy
   b. Work
2. The formula for work is
   a. $W = F \times d$
   b. $T = F \times f$
   c. $V = d/t$
   d. $W = T \cos \theta$
3. List five different forms of energy.
4. Write out your description of the law of energy conservation. Illustrate its application with three examples.
5. Discuss the statement "Energy is the most important concept in nature."

Concept of mechanical energy
Specific student objectives

1. Describe the concept of mechanical energy.
2. Differentiate between kinetic energy and potential energy.
3. Differentiate between gravitational potential energy and elastic potential energy.
4. Identify the two types of mechanical energy.
5. Identify the most common unit of measure for work and energy in the British system.

Summary

Mechanical energy, the energy of material objects caused by their position, state, or motion, is categorized into two types:
1. Kinetic energy—the energy possessed by objects because of their movement
2. Potential energy—the energy possessed by objects because of their state or position
   a. Gravitational potential energy—the energy possessed by an object because of its position, i.e., an object at a height and with the force of gravity acting upon it
   b. Elastic potential energy—the energy possessed by an object because of its deformation and its elastic restoring forces

Energy and work are expressed in distance-weight units: foot-pounds, meter-kilograms.

Performance tasks

1. Write out descriptions of the following:
   a. Mechanical energy
   b. Kinetic and potential mechanical energy
   c. Gravitational and elastic potential energy
2. The British system unit of measure for work and energy is
   a. ft/sec   c. meters
   b. lb/ft   d. ft-lb
3. Illustrate the different forms of mechanical energy with motor performance examples.
4. The two types of mechanical energy are
   a. gravitational and elastic
   b. kinetic and potential
   c. both a and b
   d. neither a nor b

Kinematic analysis of work
Specific student objectives

1. Differentiate between positive, negative, and zero work.
2. Describe the vertical and horizontal work components.
3. Identify the work formula when the F and d vectors are not parallel.
4. Describe work determination when the applied force is acting in an oblique direction.
5. Describe the following aspects of treadmills:
   a. Basic description of the device
   b. Percent grades
   c. Calculation of work during use of the device

6. Identify and describe the two angular work formulas.
7. Describe the following:
   a. Bicycle ergometer and work calculations with the device
   b. Friction bicycle and work calculations with the device
8. Calculate work using the appropriate equations.

Summary

Work is done only when a force is exerted on a body while the body at the same time moves in such a way that the force has a component along the line of motion of its point of application. If the component is in the same direction as the line of displacement, the work is positive. If the component is opposite the displacement, the work is negative. If the force is at right angles to the displacement, the work is zero.

The total work done on a body can be divided into two components: a vertical work component (determined by the product of the force F, the height of vertical displacement h, and the sine of the angle of force application θ, $W_y = Fh \sin \theta$) and a horizontal work component ($W_x = Fd \cos \theta$, where d is the horizontal displacement and θ is the oblique angle that the applied force makes with the horizontal).

A force platform and cinematographic techniques can be used to measure the total work done in the human motion task of running (or walking). A treadmill (a conveyor belt operated by an electric motor) can also be used to measure work. The percent grade is defined as units of rise per 100 horizontal units. With a treadmill, percent grade is found as the product of the tangent of the angle of inclination times 100; i.e., percent grade = $(\tan \theta) (100)$. The work output on a treadmill is found by $W = wh$ and $h = (V \sin \theta) (t)$, where V is the treadmill speed, t is the time, h is the vertical displacement, and w is the subject's weight.

Angular work is calculated similarly by using $W = Fd = F (r\theta) = F (2\pi rn)$, where r is the radius, θ the angle turned through, and n the number of revolutions. A common device used to measure angular work is the bicycle ergometer. With this ergometer, work can be found by $W = F2\pi rn$ (where F is the resistive force of the wheel, r the radius of the wheel, and n the number of wheel revolutions).

Performance tasks

1. Develop a written list of descriptions for each of the following:
   a. Positive, negative, and zero work
   b. Vertical and horizontal components of work
   c. Treadmill
   d. Percent grade of a treadmill
   e. Calculation of work on a treadmill
   f. Angular work
   g. Bicycle ergometer
2. If a force component acts in the same direction as the displacement, it is said to do (positive/negative)_____work.
3. The total work done on a body is given by
   a. vertical work component
   b. horizontal work component
   c. both a and b
   d. neither a nor b
4. A horizontal work component can be found by
   a. $W_x = Fd \cos \theta$ c. both a and b
   b. $W_y = Fh \cos \theta$ d. neither a nor b
5. List two devices for measuring work other than a force platform and cinematographic devices and techniques.
6. A device with a conveyor belt powered by an electric motor is called a (force platform/treadmill)_____.
7. Calculate the work accomplished in the examples that follow. Use $W = F \times d$ and express results in ft-lb.

	F	d	W
a.	22 lb	3 ft	_____
b.	48 oz	30 in	_____
c.	65 lb	3 in	_____
d.	240 lb	8 ft	_____
e.	100 lb	7.7 ft	_____

8. Calculate work done on a treadmill in the examples that follow:

	Angle of inclination (θ)	Velocity (V)	Time (t)	Weight (w)
a.	5°	2 mi/hr	15 min	200 lb
b.	3°	3 mi/hr	20 min	170 lb
c.	1°	5 mi/hr	40 min	180 lb
d.	2°	6 mi/hr	30 min	105 lb
e.	10°	4 mi/hr	10 min	90 lb

9. Calculate angular work done on a bicycle ergometer in the following examples:

Wheel force (F)	Radius of wheel (r)	Revolutions (n)
a. 10 lb	18 in	100
b. 20 lb	18 in	150
c. 5 lb	12 in	300
d. 30 lb	2 ft	500
e. 15 lb	2 ft	1000

Kinematic analysis of power
Specific student objectives

1. Define average power (in both sentence and formula form).
2. Identify the units of power in the British system.
3. Calculate power using the appropriate equation.
4. State the relationship of power to velocity.
5. Describe an ergometer and an ergograph.
6. Describe the Margaria-Kalamen power test.

Summary

Power is described as the rate of doing work. Average power ($\overline{P}$) is the work done during any time interval, i.e., $\overline{P} = W/t = Fd/t$. Power can also be described as $P = F \times \overline{V}$. The units of power are foot-pound per second (ft-lb/sec) and horsepower (hp), where 1 hp = 550 ft-lb/sec. Power can be expressed in average and instantaneous terms. It can be measured with the aid of an ergometer (device used to measure work) and a time recorder.

Performance tasks

1. Develop a written list of descriptions for each of the following:
   a. Average power
   b. The formula for average power (both forms)
   c. The relationship of power to velocity
   d. Ergometer and ergograph
   e. Margaria-Kalamen power test
2. Differentiate between muscle strength and muscle power.
3. The units of power in the British system of measurement are
   a. ft-lb/sec
   b. hp
   c. both a and b
   d. neither a nor b

4. Calculate the average power generated in the motion example given in Task 8 (p. 247). Express results in horsepower units.

Concept of machines
Specific student objectives

1. Define a machine.
2. Identify the six types of simple machines.
3. Differentiate between simple machines and compound machines.
4. Identify the three types of simple machines found in the human body.

Summary

A machine is defined as any device that assists in performing work, i.e., to transmit mechanical energy directly from one location to another. Simple machines carry this transformation out directly. The six simple machines are the wedge, screw, inclined plane, lever, wheel-axle, and pully. The latter three are represented in the human body. A compound machine is a combination of simple machines.

Performance tasks

1. Develop written descripitons for each of the following:
   a. Machine
   b. Simple machine
   c. Compound machine
2. The three types of simple machines found in the human body are
   a. lever, wheel-axle, and pulley
   b. lever, screw, and wedge
   c. pulley, wedge, and inclined plane
   d. pulley, lever, and wedge
3. List examples of the six types of simple machines.

SELF-EVALUATION

Students should use no reference materials for this progress test, and they can check their answers by referring to Appendix A.

1. The ability to produce change or the ability to perform work is a description of (energy/power)_____.

2. Define work (in both sentence and formula form).

3. "Energy can never be created or destroyed"

(just changed in form) is a statement of the law of conservation of

a. energy
b. momentum
c. power
d. work

4. Define energy.
5. Identify five different forms of energy.
6. Illustrate the law of energy conservation with an example of conversion from one form to another.
7. The energy of material objects caused by forces in motion is called (chemical/mechanical)_____.
8. List and describe the two types of mechanical energy.
9. Identify the units of measurement for
   a. work and energy
   b. power
10. Differentiate between work and power.
11. The two types of potential mechanical energy are
    a. chemical and nuclear
    b. elastic and gravitational
    c. both a and b
    d. neither a nor b
12. Work in which the applied force tends to cause a displacement in a direction opposite that of the actual displacement is called (positive/negative/zero)_____work.
13. When a force is applied in an oblique direction, the horizontal and vertical components of the total work can be found by the formula $\overrightarrow{W_T} = \overrightarrow{W_x} + \overrightarrow{W_y}$, where $W_x =$_____and $W_y =$_____.
14. Describe an ergometer.
15. Provide written descriptions concerning the following work-measuring devices:
    a. treadmill, percent grade, calculation formula for work
    b. bicycle ergometer, work calculation formula

16. The angular work formula is
    a. $W = Fd$
    b. $W = F(r\theta)$
    c. $\overrightarrow{W} = \overrightarrow{W_x} + \overrightarrow{W_y}$
    d. $W = wh$
17. Define a machine.
18. The three types of simple machines found in the human body are
    a. wedge, wheel-axle, and pulley
    b. inclined plane, lever, and screw
    c. lever, wedge, and screw
    d. lever, wheel-axle, and pulley
19. Calculate work done in the following motion situations. Express results in foot-pounds.

	F	d	W
a.	57 lb	6 ft	_____
b.	500 lb	8 ft	_____
c.	28 lb	6 in	_____
d.	40 oz	4 in	_____

20. Calculate work done on a treadmill in the following examples.

	θ	V	t	w	W	P
a.	6°	4 mi/hr	20 min	150 lb	____	____
b.	4°	6 mi/hr	25 min	125 lb	____	____
c.	7°	5 mi/hr	18 min	110 lb	____	____
d.	2°	8 mi/hr	7 min	195 lb	____	____

21. Calculate the average generated power in Problem 20. Express the results in horsepower units.
22. Calculate angular work accomplished on a bicycle ergometer. Express the results in foot-pounds.

	F	r	n	W
a.	35 lb	2 ft	200	_____
b.	20 lb	3 ft	400	_____
c.	10 lb	1 ft	1200	_____
d.	8 lb	2 ft	1400	_____

LESSON 15

Kinematic analysis of machine components and functions

CONCEPT OF A MECHANICAL MACHINE SYSTEM

In describing the movements of body segments, we often think of the bones of the segments, the muscles that act upon the bones, and the joints at which the segments are free to rotate as collectively comprising a mechanical machine system. The purpose of this lesson is to present the basic concepts and nomenclatures involved in describing the machine components of a movement performance.

The three major components of a machine system are (1) a source of input energy, called the *mover*, (2) a *machine* that changes the energy into a more useful form, and (3) a *resistor* that serves as the load upon which the system acts.

The skeletal muscles are, of course, the *movers* of this mechanical system; i.e., they are the source of input energy for the machine system of the human body. The function of muscles in this respect can be likened to that of the electric motor, water turbine, or gasoline engine in other machine systems. Thus skeletal muscles may be conceptualized, as was mentioned in Part I, as being the "motors" of the human body. The word motor is a term often used in kinesiology to describe muscle-produced movements—motor performance, motor activity, and motor task.

Motor performances like running, jumping, lifting, pulling, and throwing can be taken as examples of the mechanical capabilities of the musculoskeletal system. These movements are produced by skeletal muscles, which apply force to the bones (i.e., the *machines*) operating at the joints of the body.

Skeletal muscles, of course, can apply their forces only to the bones to which they are attached. Thus one important role that the bones of the body play is changing the energy of muscle

250

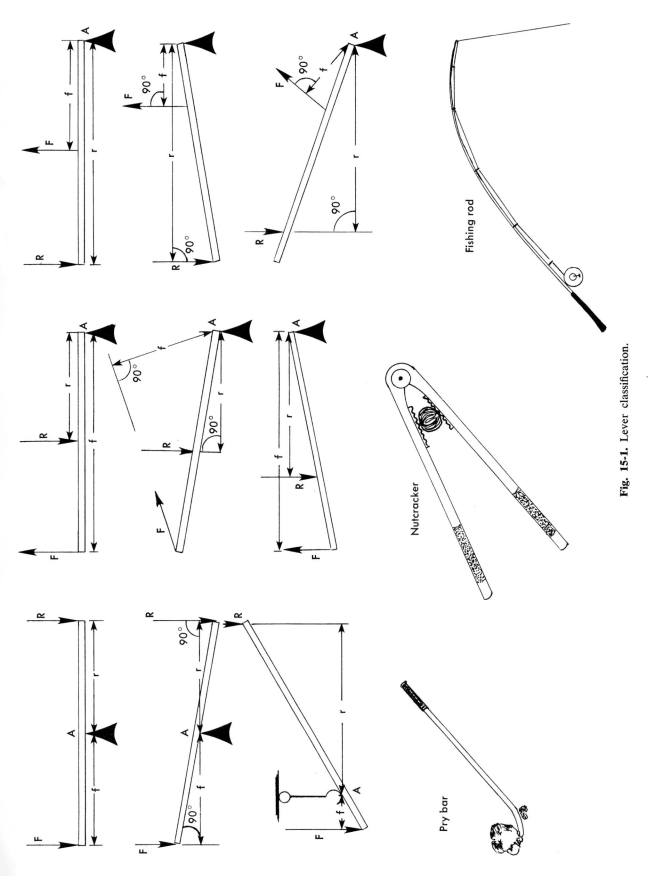

Fig. 15-1. Lever classification.

contraction into a more useful form. The bones of the body in their function as machines may be compared to the combination of levers, wheels, and gears of an automobile.

The various segments of the body involved in a throwing movement, for example, may be considered to be the *resistors* which oppose the action of the movers. The function of the body segments as resistors may be compared with that of an airplane propeller, the wheels of a car, or the hands of a clock.

The machine system of the human body can be summarized as follows: the coordinated action of muscles (movers) acting upon bones (machines) to cause the coordinated motion of body segments (resistors) resulting in motor performances.

The three types of simple machines found in the human musculoskeletal system are the lever, the wheel-axle, and the pulley.

ANALYSIS OF LEVER COMPONENTS AND FUNCTIONS

The lever is one of the simplest of all mechanical devices that may rightfully be called a machine. In fact, any rigid object which is free to turn about a center of rotation when a force is applied to it can be classified as a lever. When the object turns, it overcomes a resistance and thereby produces mechanical work. The resistance consists of the weight of the object, friction between the moving parts, plus any external load that may be added to it.

Classification of levers

The important points on a lever are (1) the point about which it turns, (2) the point at which force is applied to it, and (3) the point at which the resistance to its movement is concentrated. These three points along with the classes of levers are shown in their simplest form in Fig. 15-1. In each illustration the point of the axis *(A)* is the pivot or fulcrum about which the lever is made to turn. *R* is the point at which the resistance force or load to be lifted is concentrated, and *F* is the point at which the applied force is acting. The arrangement of these three points provides the basis for the classification of levers.

Since there are three points on a lever, there are three possible arrangements of these points.

Any one of the three may be situated between the other two. In a first-class lever Fig. 15-1, *A,* the axis is located at any point between the force and the resistance. In a second-class lever (Fig. 15-1, *B*) the resistance is located anywhere between the axis and the force. In a third class lever (Fig. 15-1, *C*) the force is located at any point between the axis and the resistance.

In Fig. 15-1 the distance *r* is called the lever arm of the resistance or load, and the distance *f* is the lever arm of the applied force. These lever arms are commonly called the resistance arm and the force arm of the lever, respectively. The lever arm of any force, including the resistance force, was defined in Lesson 13 as the perpendicular distance from the line of action of the force to the fulcrum.

Machine functions of levers

As discussed before, a bony lever is a machine capable of changing the mechanical force of muscle contraction into a more useful form. The four functions which levers are capable of performing in this regard are (1) the *magnification* of a force, which is possible when the force arm is longer than the resistance arm, (2) an *increase in the speed and range* of motion through which the end of a lever may move, as is possible when the resistance arm is longer than the force arm, (3) the *balancing* of equal forces, which is possible when the lengths of the force and resistance arms are equal, and (4) *changing the direction* of the input force. This last function is the characteristic of a fixed pulley, but it is also one of the characteristics of a first-class lever.

Examples of levers

There are numerous examples of the lever functions and of the three types of levers as used in everyday life. Diagrams of three such devices are shown in Fig. 15-2.

In the seesaw or teeter-totter (Fig. 15-2, *A*) the straight line *FAR* represents a lever of the first class—with *A* as the axis, *R* the weight of one rider, and *F* the weight of the other rider. Since the relative lengths of the force and resistance arms can vary in a first-class lever depending upon the location of the axis, this type of lever can be used to perform all four functions of a lever. In other words, the force arm of a first-class lever

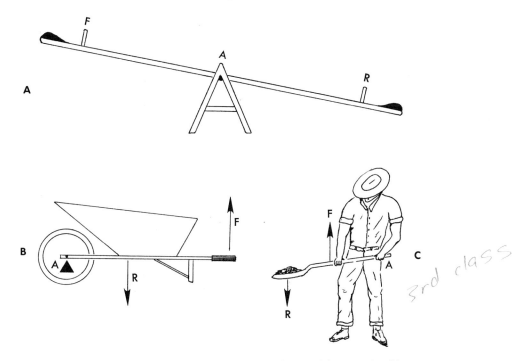

Fig. 15-2. The three classes of levers as commonly found in everyday life.

may be shorter than, equal to, or longer than the resistance arm. You can also see that the first-class lever changes the direction of the input force *(F)*. The direction of *F* is opposite the direction of the output force acting to move the resistance *(R)*; i.e., the applied force *(F)* acts downward while the output force acts upward to move the resistance *(R)*.

The wheelbarrow in Fig. 15-2, *B,* employs a lever of the second class with the axis *(A)* located at the wheel. The heavily loaded barrow is lifted to a rolling position by the application of a smaller force *(F)* at the handles, which illustrates that a second-class lever is more often designed for the magnification of force.

The loaded shovel in Fig. 15-2, *C,* illustrates a lever of the third class, the fulcrum being located in the man's left hand. This class of lever is almost always designed to increase the speed and range of motion of the lever.

Every bone in the body acts as a lever at one time or another in the production of bodily movements. The bones may act either alone or in combination to form the lever systems of the musculoskeletal system. In fact, the body as a whole acts like one complex series of third-class

Fig. 15-3. A tennis player using the body as a lever.

levers in performing, for example, a correctly executed tennis serve (Fig. 15-3). The right-handed server pivots over his left foot *(A)* and transfers his body weight *(F)* into the serve, which through a stabilized wrist overcomes the resistance of the racket *(R)*. This lever system greatly increases the speed of the racket in striking the ball.

The vast majority of bony levers of the human body are of the third class, which means that humans are built for speed more than for strength. The first-class levers are few in number but are still more numerous than the second-class levers.

ANALYSIS OF THE COMPONENTS AND FUNCTIONS OF WHEEL-AXLE AND PULLEY ARRANGEMENTS

The wheel-axle arrangement is a simple machine involving the basic principles of the lever. It consists of a wheel attached to, and rotating about, a central axle. Note the distinction between *axle* and the point of rotation, called *axis*. The force causing the rotation can be applied either to the rim of the wheel (type I arrangement), as in the case of the steering wheel of an automobile, or to the axle (type II arrangement), as in the case of the rear wheel of a bicycle. A type I wheel-axle arrangement is illustrated in Fig. 15-4.

Machine functions of a wheel-axle arrangement

In a type I wheel-axle arrangement, with the force being applied at the rim of the wheel (e.g., a steering wheel), the force is magnified as in a second-class lever at the expense of speed and distance. The turning effect of the wheel exerts a torque that is the product of the force *(F)* and the radius of the wheel (r_w); i.e., $T = F \times r_w$

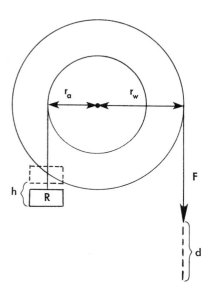

Fig. 15-4. Wheel-axle machine.

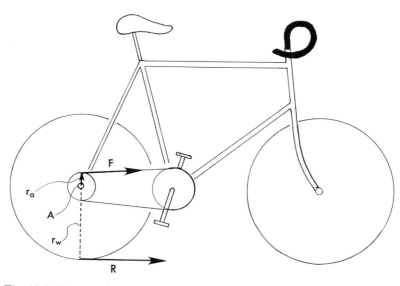

Fig. 15-5. The rear wheel of a bicycle forms a type II wheel-axle arrangement.

(Fig. 15-4). The radius of the wheel in this case corresponds to the force arm of a lever. Thus the larger the diameter of the wheel, the greater will be the magnification of the force.

In a type II wheel-axle arrangement, with the force being applied at the axle (e.g., the rear wheel of a bicycle), speed and range of motion (displacement) are magnified—as in a third-class lever—at the expense of force. The turning effect of the axle exerts a torque that is the product of the force *(F)* and the radius of the axle *(r_a)*; i.e., $T = F \times r_a$ (Fig. 15-5). The radius of the wheel in this case corresponds to the resistance arm of a lever. Thus the larger the diameter of the wheel, the greater will be the magnification of the speed and the displacement. The secondary function of a machine, to produce a change of direction of the input force, does not apply to the wheel-axle arrangement in the human body.

Examples of wheel-axle arrangements in the body

Most of the wheel-axle arrangements in the body, like anatomic levers, are for gaining distance and speed at the expense of force. This happens when the movement force is applied at the axle. Both kinds, however, are represented in the body.

Most twisting or rotation movements in the body involve a wheel-axle type of arrangement. Shaking the head, twisting the trunk, and inward and outward rotations of the arm and thigh are good examples of movements produced by wheel-axle arrangements. A cross section of the rib cage, for instance, shows that the ribs serve as a wheel with the vertebral column serving as the axle in a wheel and axle system. Some muscles (deep posterior muscles) act upon the axle, represented by the vertebrae, in producing rotation of the trunk (Fig. 15-6). A cross section of the arm or the thigh, likewise, presents the characteristics of a wheel-axle arrangement. Here the longitudinal axis of the long bones serves as the axle, and the surrounding tissues as the wheels. The rotation of both the trunk and the limbs around their longitudinal axes constitutes the movements of a wheel-axle arrangement.

Machine function of a fixed pulley

Of the several types of pulleys, only the fixed pulley is represented in human anatomy. The machine function of a single fixed pulley is the same as the secondary function of a lever having equal lever arms (Fig. 15-7). This function is to change the direction of the line of action of a force.

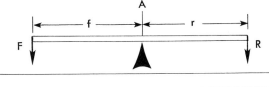

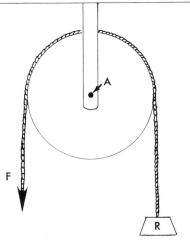

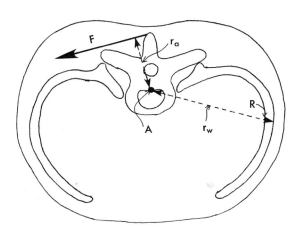

Fig. 15-6. The spinal muscles acting upon the vertebrae operate as a wheel-axle machine.

Fig. 15-7. A fixed pulley and its analog as a first-class lever with equal lever arms.

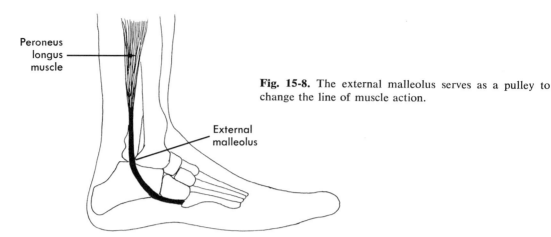

Peroneus
longus
muscle

External
malleolus

Fig. 15-8. The external malleolus serves as a pulley to change the line of muscle action.

Fig. 15-8 shows the lateral malleolus of the ankle serving as a pulley to change the direction of the line of pull of the peroneus longus muscle. This pulley arrangement allows the peroneus longus to perform a more useful function than would be possible otherwise.

KINEMATIC ANALYSIS OF THE BODY'S MACHINERY

The three primary problems involved in the kinematic analysis of the body's machinery are locating the joint centers, describing the longitudinal axes of body segments, and locating the points of force applications.

Locating the joint centers

The joint centers are generally located through a manual analysis process called tracing. Determining the joint centers introduces a subjective error which varies with the tracer's knowledge of anatomy. Aids recommended by Plagenhoef (1971) for locating the joint centers from external landmarks are illustrated in Figs. 15-9 and 15-10. Plagenhoef recommends that a skin pencil be used to draw the plane of the joint center so the joint center can be located properly when the segment moves out of the vertical plane as shown in Fig. 15-11.

The importance of locating the joint centers precisely can be realized when we remember that the cartesian coordinates of these points are what serve as the only input in many computerized analysis systems.

Describing the longitudinal axes of body segments

The lengths of the longitudinal axes of body segments can be measured directly from the subjects, or they can be calculated from a knowledge of their segmental end points (proximal and distal joint centers). The x and y coordinates of two segmental end points are illustrated in Fig. 15-12. These coordinates can be used to find the vertical and horizontal distances between the segmental end points which form the sides opposite and adjacent in a right triangle. Thus the segmental lengths (longitudinal axes) are equal to the hypotenuses of these right triangles and can be calculated through the use of the Pythagorean theorem ($h = \sqrt{a^2 + b^2}$). This calculation can be done automatically in a computerized analysis system.

Locating the points of force application

The two points of force application that must be determined in the kinematic analysis of the body's machinery are (1) the point at which the line of action of the resistance force intersects the longitudinal axis of the segment and (2) the point at which the line of action of the applied force intersects the axis. Since gravity is the primary resistive force acting in human movement situations, methods of locating its point of application (center of gravity) in a segment will be discussed in Lesson 26. An approximation method is presented in Table 15-1 and illustrated in Fig. 15-13. Problems involved in locating the second

Text continued on p. 265.

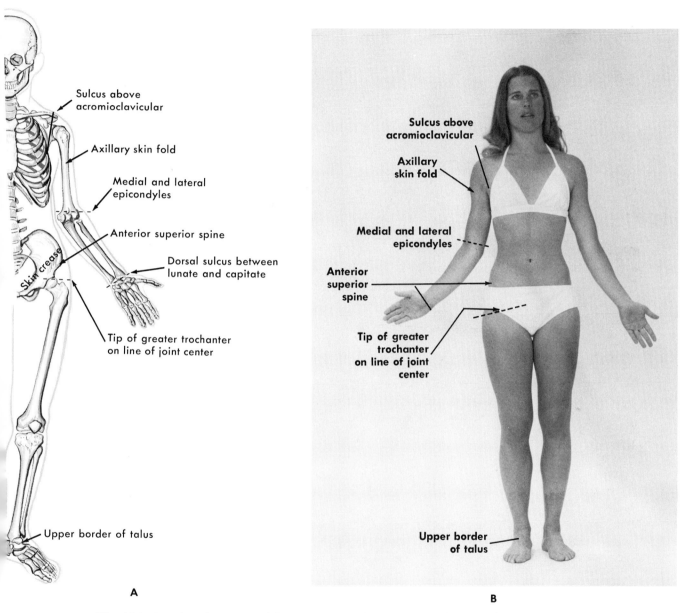

Fig. 15-9. Anterior view of the joint centers and planes. (**A** modified from Anthony, C. P., and Kolthoff, N. J.: Textbook of anatomy and physiology, ed. 9, St. Louis, 1975, The C. V. Mosby Co.)

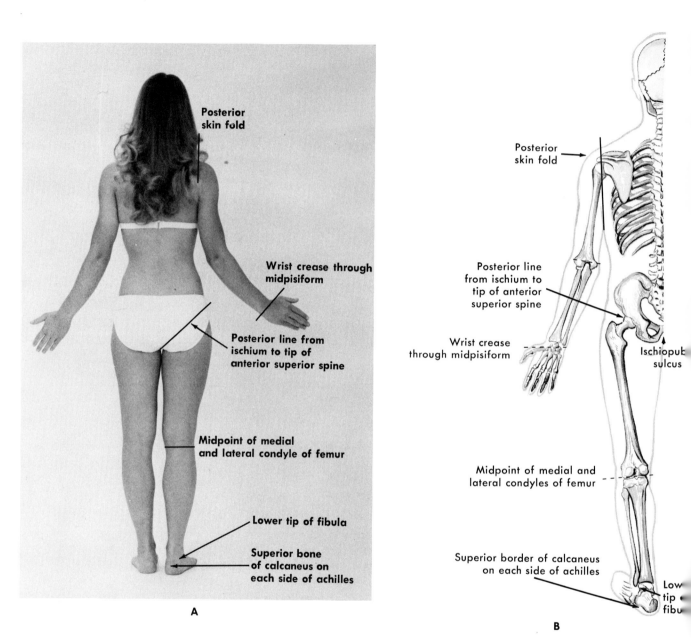

Fig. 15-10. Posterior view of the joint centers and planes. (**B** modified from Anthony, C. P., and Kolthoff, N. J.: Textbook of anatomy and physiology, ed. 9, St. Louis, 1975, The C. V. Mosby Co.)

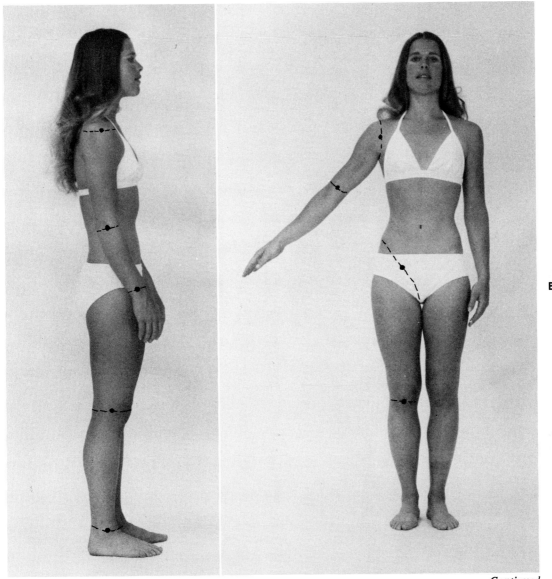

Continued.

Fig. 15-11. Skin creases and bony landmarks help locate the plane of the center so the position of the center can be located. If a skin pencil is used to draw the plane of the joint center, the center must be located properly when the segment movement is out of the vertical plane as shown in **E, F,** and **G.** (Modified from Plagenhoef, S. C.: Patterns of human movement: a cinematographic analysis, Englewood Cliffs, N.J., 1971, Prentice-Hall, Inc., p. 14.)

C

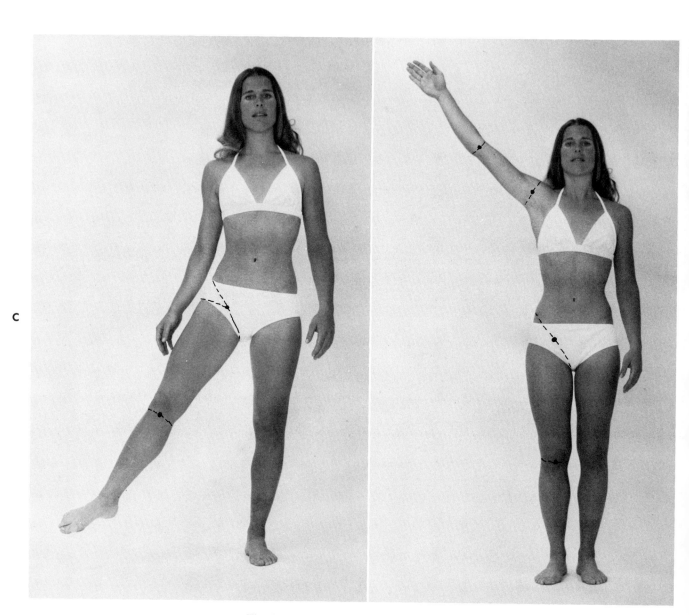

Fig. 15-11, cont'd. For legend see p. 259.

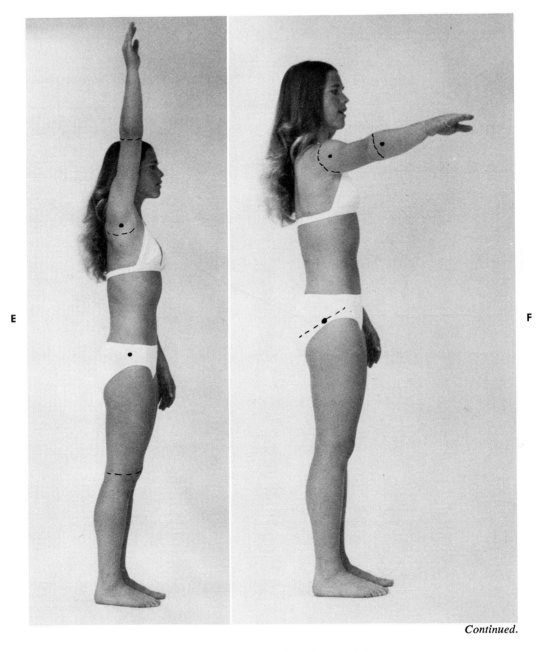

E F

Continued.

Fig. 15-11, cont'd. For legend see p. 259.

G

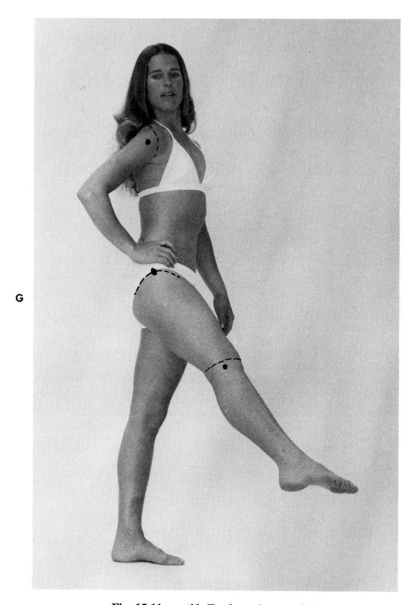

Fig. 15-11, cont'd. For legend see p. 259.

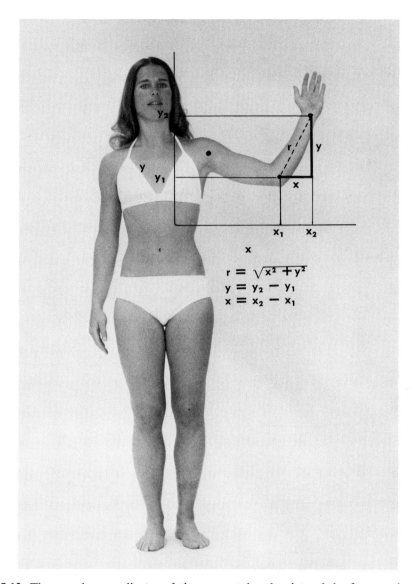

$$r = \sqrt{x^2 + y^2}$$
$$y = y_2 - y_1$$
$$x = x_2 - x_1$$

Fig. 15-12. The x and y coordinates of the segmental end points of the forearm (centers of elbow and wrist joints) can be used to find the vertical and horizontal distances between these points, which form two sides of a right triangle. The segmental length is found by use of the Pthagorean theorem.

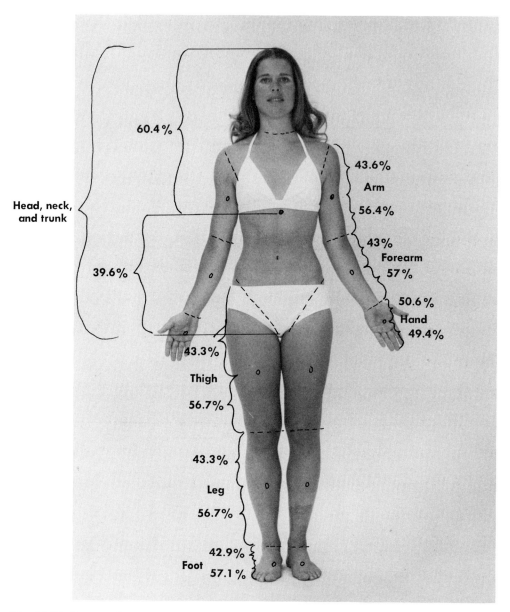

Fig. 15-13. Segmental end points and percentage distances of the centers of gravity from these points.

Table 15-1. Location of segmental centers of gravity*

Segment	Percent distance to proximal end	Percent distance to distal end
Head-neck and trunk	60.4	39.6
Arm	43.6	56.4
Forearm	43.0	57.0
Hand	50.6	49.4
Thigh	43.3	56.7
Leg	43.3	56.7
Foot	42.9	57.1

*Modified from Dempster, W.: Space requirements of the seated operator. WADC technical report 55-159, July, 1955, Office of Technical Services, U.S. Department of Commerce, Washington, D.C.

point, which is primarily determined through the vector addition of muscle forces, is discussed in Lesson 25.

STUDY GUIDELINES
Concept of a mechanical machine system
Specific student objectives

1. Describe a mechanical machine system.
2. Identify the three components of a machine system.
3. Describe the machine components found in the human body.
4. Identify the three simple machines found in the human body.

Summary

A machine is any device that converts mechanical energy into a more useful form. Mechanical energy tends to cause or change the motion state of an object. A mechanical machine system consists of a mover that acts as a source of input energy, a machine, and a resistor which serves as the source of output energy. The three machines present in the human body are the lever, the wheel-axle, and the simple pulley. In the human body the muscle is the mover, the bone is the machine, and the body segment is the resistor.

Performance tasks

1. Write the description of a mechanical machine system and its components.
2. A machine system is composed of a mover, the machine, and its
   a. pulley

b. machine component
c. both a and b
d. neither a nor b

3. Identify a structural part of the human body corresponding to each of the following:
   a. Mover _____
   b. Machine _____
   c. Resistor _____

Analysis of lever components and functions
Specific student objectives

1. Describe a lever.
2. Identify the points on a lever.
3. Describe the three classes of levers.
4. Differentiate between a force arm and a resistance arm.
5. Describe the four machine functions of levers.
6. Identify examples of the lever functions and of the three types of levers.

Summary

A lever is a rigid object that can turn about an axis of rotation (fulcrum) when a force is applied to it. The turning lever overcomes a resistance and is said to produce work. The essential elements of a lever are the pivot or fulcrum, or axis (A), the resistance (R), and the applied force (F).

Levers are classified according to the relative location of A, R, and F. First-class, second-class, and third-class levers are represented when A, R, or F, respectively, is located between the remaining two elements. In the human body, levers carry out three primary machine functions: magnifying a force, increasing displacement, and balancing forces. First-class levers are also capable of the secondary function of a machine: changing the direction of the input force.

The perpendicular distance from the axis of rotation to the line of action of a force is called the lever arm of that force. The lever arms for the applied force and the resistance are respectively termed the force arm (f) and the resistance arm (r).

Performance tasks

1. Develop written descriptions of each of the following:
   a. A lever
   b. The three primary points on a lever

c. The three classes of levers
d. A moment or lever arm, a force arm, and a resistance arm
e. The three primary machine functions and the one secondary function
2. The most common type of lever found in the human body is of the (first/second/third) _____class.
3. Levers are classified according to the location of the (left side/middle/right side)_____ element.
4. Identify examples of each lever classification and identify its primary machine function.

Analysis of the components and functions of wheel-axle and pulley arrangements
Specific student objectives
1. Describe wheel-axle and pulley machines.
2. List the machine functions of a wheel-axle and of a pulley.
3. Describe the two classes of wheel-axle machines.
4. Identify two examples each of wheel-axle and pulley arrangements in the human body.

Summary

A machine composed of a rotating axle and a larger wheel surrounding the axle can produce angular motion in the human body. The applied force that tends to move this wheel-axle arrangement may be applied either at the axle or on the wheel. When the force causing the rotation is applied to the rim of the wheel, the arrangement is classified as type I—in which the machine function is magnifying force at the expense of speed and distance. A type II arrangement has the force applied at the axle and possesses the machine function of magnifying speed and distance at the expense of force.

The single fixed pulley is a mechanical device used to change the direction of an applied force. The tendons in the body act around bone structure or other tissue to produce a pulley-like effect. This type of pulley is the same as that of a first-class lever having equal lever arms; thus the pulley balances forces and also changes the direction of the applied force, a secondary function.

Performance tasks
1. Write out descriptions for each of the following:

a. A wheel-axle machine
b. A pulley machine
c. The two classes of wheel-axles
2. Differentiate between axis and axle.
3. The primary machine function of a type I wheel-axle is to
a. magnify force
b. magnify speed
c. balance forces
d. change direction
4. The secondary machine function of a pulley is to
a. magnify force
b. magnify speed
c. balance forces
d. change direction
5. Identify human body examples of both types of wheel-axle arrangements and the fixed pulley.

Kinematic analysis of the body's machinery
Specific student objectives
1. Identify and describe the three problems involved in the kinematic analysis of body machinery.
2. Carry out an experimental analysis of a selected segmental movement by
a. locating the center of the joint axis
b. locating the longitudinal axes of both levers terminating at the joint
c. locating the point of application of the resistance force

Summary

Locating the joint centers, identifying the longitudinal axes of body segments, and determining the point of resistance force application are the three primary tasks of kinematic analysis of the body's machinery.

Joint centers (axes) are located by tracing or with aids recommended by Plagenhoef. Longitudinal axes are located directly or calculated from the proximal and distal joint centers. The point of resistance force application is located by center of gravity determinations and vector analysis of forces.

Performance tasks
1. Write out a step-by-step procedure for the following:
a. Locating joint centers (anatomical axes)

b. Identifying longitudinal axes of body segments

c. Determining the resistance force point of application

2. Select a segmental movement in each of the three main divisions of the skeleton and carry out an estimated kinematic analysis by locating the

a. center of the joint axis

b. longitudinal axis of involved levers

c. point of application of resistance force

SELF-EVALUATION

Students should use no reference materials for this progress test, and they can check their answers by referring to Appendix A.

1. Identify and describe the three components of a mechanical machine system.

2. Illustrate the machine components by identifying within the human body an example of each of the following:

a. Mover_____

b. Machine_____

c. Resistor_____

3. A, R, and F are the three primary (arms/points)_____ on a lever.

4. Describe the three classes of levers:

a. First (class I) _____

b. Second (class II)_____

c. Third (class III)_____

5. The two types of lever arms are

a. force and resistance

b. speed and distance

c. both a and b

d. neither a nor b

6. List and describe the three primary and one secondary machine functions.

7. Identify human body examples of third-class and first-class levers and determine their machine functions.

8. The perpendicular distance from the axis to the line of action of a force is called

a. device

b. speed

c. resistance

d. lever arm

9. A machine composed of a rotating axle and a larger wheel surrounding it is called a

a. lever

b. wheel-axle

c. pulley

d. all of the above

10. Wheel-axle machines have (one/two/three) _____classifications.

11. Describe a pulley machine.

12. The primary machine function possessed by a pulley is

a. favor force

b. favor speed and distance

c. both a and b

d. neither a nor b

13. Select a human body example of both types of wheel-axle and one pulley example. Identify the machine function(s) for each.

14. Kinematic analysis of body machinery involves location of:

a. joint centers and longitudinal axes

b. point of resistance force application

c. both a and b

d. neither a nor b

Kinematic analysis of machine mechanical advantages and efficiencies

Body of lesson
  Introduction
    The law of the lever
    Mechanical advantage
  Concepts of theoretical and actual mechanical advantages
    Mechanical efficiency of a machine
    Theoretical mechanical advantage and machine functions
  Conservation of energy and machine functions
  Mechanical advantages of wheel-axle and fixed pulley arrangements
  Analysis of motor efficiency
    Heat as the common denominator of all the energy released in the body
    Energy units
    Measurement of energy
    Basal metabolism
    Internal and external work
    Efficiency of human movement
    Techniques of measuring motor efficiency
Study guidelines
Self-evaluation

INTRODUCTION

Machines are able to assist their users in the performance of work and to perform their machine functions because of the mechanical advantages they have. Fundamental to an understanding of mechanical advantage, however, is the law of the lever. This law describes the condition of angular equilibrium.

The law of the lever

A lever is said to be in a state of equilibrium when the algebraic sum of all the torques acting upon it equals zero. In other words, a lever is in equilibrium when the sum of all the torques tending to produce clockwise rotation equals the sum of all the torques tending to produce counter-clockwise rotation. This is known as the law of the lever.

The law of the lever is of value because it enables us to calculate the amount of force needed to balance a known resistance by means of a known lever or to calculate where to place the fulcrum in order to balance a known resistance with a given force. If we remember that the resistance torque equals the product of the resistance force (R) and the length of the resistance arm (r) and that the force torque equals the product of the applied force (F) and the length of the force arm (f), then we can use the following formula to express the law of the lever:

$$F \times f = R \times r$$

(Equation 16-A)

If any three of the four values are known, the remaining one can be calculated through the use of the formula.

Mechanical advantage

The mechanical advantage of a machine may be thought of as the ratio of the output force delivered by the machine to the amount of input force applied by the mover. With a simple machine like the lever, the output force is represented by the resistance force (R) while the input force is represented by the applied force (F). The mechanical advantage (M.A.) of a lever therefore is given by dividing the applied force into the resistance force; i.e., M.A. = R/F. By rearranging the law of the lever (F × f = R × r), we can also express the mechanical advantage as the ratio of the lever arms

$$\text{M.A.} = \frac{R}{F} = \frac{f}{r}$$

CONCEPTS OF THEORETICAL AND ACTUAL MECHANICAL ADVANTAGE

Perhaps it should be pointed out here that the two expressions of mechanical advantage (R/F and f/r) are equal only in theory. In actuality the mechanical advantage expressed in terms of the output and input forces (R/F) is always less than the mechanical advantage expressed in terms of the lever arms (f/r) because of the opposing force of friction. For this reason, the ratio of the lever arms is called the *theoretical* mechanical advantage while the ratio of the forces is called the *actual* mechanical advantage.

Assume, for example, that in a certain lever application an applied force of 10 lb is able to move a 100 lb resistance. The actual mechanical advantage of this lever arrangement is R/F = 100/10 = 10. In practice, however, the length of the force arm in the lever application must be slightly more than ten times the length of the resistance arm in order to overcome the additional resistance due to friction. Thus the actual and theoretical mechanical advantages are equal to each other only in theory. In reality, the *R* in the actual mechanical advantage formula should be considered as the total resistance to a movement. In other words, it should be considered the weight of the load being moved *plus* all other forces which are opposing the movement (e.g., friction).

Since R is usually thought of as just the weight of the object being moved by a lever, we should be able to see that this weight will always be less than the true total resistance. In practice, there-

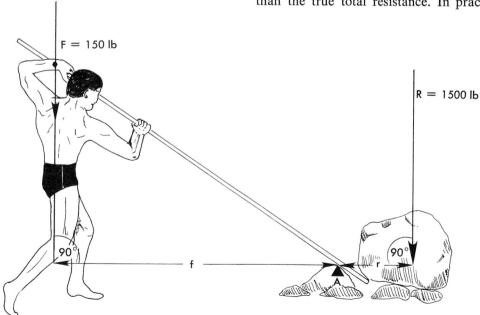

Fig. 16-1. Lever with a mechanical advantage greater than unity (>1).

fore, the theoretical mechanical advantage (f/r) will always be greater than the actual mechanical advantage (R/F).

Mechanical efficiency of a machine

The mechanical efficiency of a machine may be found by comparing the actual mechanical advantage (A.M.A.) to the theoretical mechanical advantage (T.M.A.):

$$\text{Mechanical efficiency} = \text{M.E.} = \frac{\text{A.M.A.}}{\text{T.M.A.}}$$

(Equation 16-B)

Illustrative of this would be a machine with an actual mechanical advantage of 8 and a theoretical mechanical advantage of 10, resulting in an efficiency rating of 0.80 or 80%. As will be seen in the next section of this lesson, mechanical efficiency measured by the ratio of the A.M.A. to the T.M.A. is the same as that determined by measuring the ratio of the *actual work* done by a machine to its *theoretical work* potential.

Theoretical mechanical advantage and machine functions

The concept of theoretical mechanical advantage (f/r) is closely related to the primary machine functions of a lever. When the machine function is to *balance forces* and the magnitudes of the forces are equal, then the length of the lever arms must also be equal, which results in a theoretical mechanical advantage of 1 (unity). Likewise, when the machine function of a lever is to *magnify a force,* the force arm must be longer than the resistance arm, which results in a theoretical mechanical advantage greater than unity (>1) (Fig. 16-1). The remaining possibility, a theoretical mechanical advantage less than unity, occurs when the resistance arm is greater than the force arm. Therefore a lever with a theoretical mechanical advantage <1 has the function of increasing the speed and range of motion of an object at the expense of force.

CONSERVATION OF ENERGY AND MACHINE FUNCTIONS

We have found so far in this lesson that levers are used for three primary purposes: (1) to gain force at the expense of range of motion, which is referred to as the *magnification of force,* (2) to

gain speed and range of motion at the expense of force which is referred to as the *magnification of displacement,* and (3) to balance forces.

In the second-class lever the force arm usually coincides with the total lever; hence the force arm is usually longer than the resistance arm and the lever usually has the effect of *magnifying force.* In the third-class lever the resistance arm usually coincides with the total lever; hence the resistance arm is usually longer than the force arm and the lever usually has the effect of *magnifying displacement.* In the first-class lever, however, the arms may be of equal length or either the force arm or the resistance arm may be longer depending upon the relative position of the fulcrum.

As we know, the law of conservation of energy states that in the transformation of energy from one form to another, energy is always conserved. Even though a lever is capable of magnifying either force or displacement, energy relations show that the user of the lever is not getting "something for nothing." The law of conservation of energy requires that the work done *by* the machine be no greater than the work done *on* the machine.

Mechanical work was defined in Lesson 14 as the product of force and the distance through which the force is applied (W = F × d). Thus using the terms "force" and "resistance" once again, we can define the work done *on* a machine as being equal to the product of the input force *(F)* and the distance *(d)* through which the force acts (F × d) while the work done *by* the machine is equal to the product of the resistance *(R)* and the distance or height *(h)* that the resistance is moved (R × h). By the law of conservation of energy, we know that the output work *(R × h)* theoretically equals the input work *(F × d)* (Fig. 16-2).

By rearranging the law of conservation of energy (R × h = F × d), we obtain R/F = d/h; this equation tells us that the mechanical advantage, defined as the ratio of the forces *(R/F)*, is also given by the inverse ratio of the distances through which the forces operate *(d/h)*. The latter is called the reduction or the magnification of the displacements. If *d* is greater than *h*—there is a reduction in the displacements, the force arm *(f)* is greater than the resistance arm *(r),* the mechanical advantage *(R/F)* is greater than

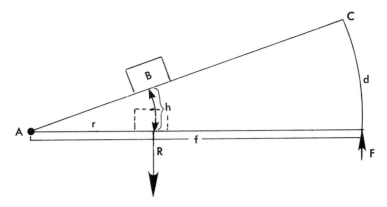

Fig. 16-2. According to the law of conservation of energy, the output work *(R × h)* is theoretically equal to the input work *(F × d).*

unity, and the effect of the force is magnified. If *d* is smaller than *h*—there is a magnification in the displacement, the force arm *(f)* is less than the resistance arm *(r),* and the mechanical advantage is less than unity.

Combining the equation M.A. = R/F = f/r with the equation R/F = d/h demonstrates the relation between lever arms and displacement (f/r = d/h). This effect is illustrated in Fig. 16-2. The two triangles *ARB* and *AFC* are similar, and their corresponding sides are proportional.

Since *R,* as mentioned in the last section, is usually thought of as just the weight of the object being moved by a machine and not the total resistance opposing motion, the statement that output work (actual work done) equals input work (potential work capability) is true only in theory. Thus mechanical efficiency can be determined by comparing not only a machine's T.M.A. and A.M.A., as discussed in the last section, but also its input and output works (energies) as follows:

$$\text{Mechanical efficiency} = \frac{\text{Work output}}{\text{Work input}} = \frac{R \times h}{F \times d}$$

(Equation 16-C)

MECHANICAL ADVANTAGES OF WHEEL-AXLE AND FIXED PULLEY ARRANGEMENTS

The law of the lever also applies to a wheel-axle arrangement and becomes $F \times r_w = R \times r_a$ in a type I situation (where the force is applied to the wheel) and $F \times r_a = R \times r_w$ in a type II situation (force applied to the axle). Thus a wheel-axle machine is in equilibrium when the applied torque equals the resistance torque.

The theoretical mechanical advantage of a wheel-axle machine is given by the ratio of the applied force arm to the resistance arm; i.e., M.A. = r_w/r_a or r_a/r_w depending upon which radius is acting as the force arm. As in the lever, the actual mechanical advantage is given by the ratio of the resistance overcome by the machine to the force applied (R/F). The mechanical efficiency of a wheel-axle machine is given by the ratio of the actual mechanical advantage to the theoretical mechanical advantage.

Although an apparent theoretical mechanical advantage of unity obtained with a fixed pulley may be looked upon as no mechanical advantage at all, it often happens that a change in a muscle's line of pull (such as increasing the angle of its action on a bone) will increase the muscle's effectiveness as a mover. This is illustrated in Fig. 16-3. Note how the position of the kneecap (acting as a pulley) increases the angle of pull of the quadriceps muscle. The example reveals that the kneecap actually increases the mechanical advantage of the tibia, to which the quadriceps muscle is attached, indirectly by improving the line of action of the muscle. In turn, the muscle increases the length of the force arm of the lever and the tangential component ($F_t = F \sin \theta$) of the force.

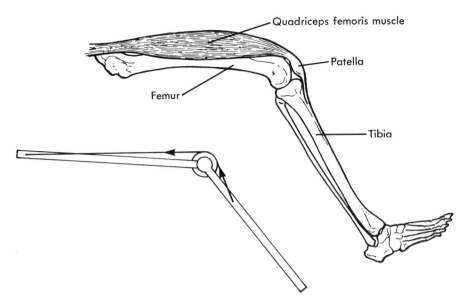

Fig. 16-3. The patella, acting as a pulley, increases the angle of pull of the quadriceps femoris muscle.

ANALYSIS OF MOTOR EFFICIENCY

In Lesson 15, muscles were defined as being the motors of the body and an analogy was drawn between muscles and other types of motors (e.g., the gasoline engine). Actually the human body may be considered as a heat engine or a thermodynamic machine which consumes food and oxygen as fuel, does external work by muscular activity, and gives off heat and carbon dioxide as the principal waste products. The physiological term applied to such transformations in the body is *metabolism.* While the process of metabolism is chiefly one of chemistry, the principal result is the production of heat and mechanical work.

Heat as the common denominator of all the energy released in the body

The metabolism of the body is simply all the chemical reactions in all the cells of the body, and the *metabolic rate* is normally expressed in terms of the rate of heat liberation during the chemical reactions.

The body's energy is obtained through the oxidation of foods—carbohydrates, fats, and proteins—in the cells, and by this process large amounts of energy are released. The same foods can be burned with pure oxygen outside the body in an actual fire, again releasing large amounts of energy, but then the energy is released suddenly all in the form of heat. The energy needed by the physiological processes of the cells, however, is not heat but *chemical energy* to perform cellular functions.

To provide this energy, the chemical reactions must be "coupled" with the systems responsible for these physiological functions. The coupling is done by special cellular enzyme systems for transferring energy. Because of the inefficiency of the transfer mechanisms, not all the energy in foods is transferred by the functional systems of the cells. In fact, only about 25% of it is finally utilized by the functional systems. The remaining 75% is given off as heat.

Even though 25% of the energy finally reaches the functional systems of the cells, the major proportion of this also becomes heat. For example, consider the energy used for muscle activity. Much of this energy simply overcomes the viscosity of the muscles themselves or of the tissues so the limbs can move. The viscous movement, in turn, causes friction within the tissues, and the friction generates heat. For similar reasons the physiological functioning of the other body systems finally ends up in the form of heat. Therefore we can say that essentially all the energy expended by the body is converted into heat.

The only real exception to the preceding generalization occurs when the muscles elevate an object to a height or carry the person's body up steps. A type of potential energy is thus created by raising a mass against gravity. Also, if muscular energy is used to turn a flywheel so kinetic energy is developed in the flywheel, this, too, is an external expenditure of energy. When external expenditure of energy is not taking place, however, it is safe to consider all the energy released by the metabolic processes as eventually becoming body heat.

For example, the readers of this text undoubtedly have an external expenditure of energy close to zero as they sit wherever they are and read since they are doing almost no external work at all (unless they have very unusual study habits!).

Energy units

To discuss the metabolic rate and related subjects intelligently, we must use some unit for expressing the quantity of energy released from the different foods or expended by the various functional processes of the body. Among the units commonly employed in measuring energy are foot-pounds, ergs, joules, kilowatt-hours, watt-seconds, electron volts, British thermal units, and calories. Since the common denominator of all energy released in the body is heat, the unit generally employed in human metabolism studies is a unit of heat called the kilocalorie (Kcal), also known as the kilogram calorie or the large calorie. In terms of heat, 1 Kcal equals approximately the energy necessary to raise 1 kg of water (or 1 liter of water) 1° centigrade. A kilocalorie is equal to 1000 "common" or "small" calories. The "calories" referred to in standard human nutrition tables listing the energy content of various foods are actually kilocalories.

Measurement of energy

The measurement of heat energy is called calorimetry. The two basic types of calorimetry are direct and indirect.

In *direct* calorimetry the metabolic rate is measured by actually determining the amount of heat produced and given off by the body in a known period of time. To do this, human subjects are placed in a large chamber called a *human calorimeter*. The chamber is cooled by water flowing through a radiator system. The heat given off from the body is picked up by the cooling system and measured by an appropriate physical apparatus.

An *indirect* method for measuring the metabolic rate is based on the amount of oxygen burned by the body in a given period of time. From this the rate of energy release can be calculated. The amount of energy released when 1 liter of oxygen burns with carbohydrates, proteins, or fats is essentially the same regardless of which of the three foods is being used for energy. Therefore it is reasonable to use an approximate average of the different values (4.825 Kcal) as the amount of energy released in the body every time a liter of oxygen is burned. When the metabolic rate is calculated this way, the value obtained is never in error by more than 4% even though a great excess of one type of food or another might be used for metabolism momentarily; and 4% is far less than the error of measurement anyway. Therefore, to determine the amount of energy being generated in the body, you need only determine the amount of oxygen being used.

Basal metabolism

During rest periods in which a subject lies motionless in a warm comfortable environment, a steady rate of heat production takes place. While this rate, called basal metabolism, varies with height, weight, age, sex, and state of health, it is a valuable measure for the interpretation of energy dynamics.

When a person is standing and working, energy over and above the basal metabolic rate is being transformed through muscular activity into mechanical work. Basal metabolism is therefore comparable to the idling of a car motor, when a certain amount of fuel is consumed in keeping the motor running. The body at work is like the running of an automobile along a level road or up a hill.

Internal and external work

In the study of energy dynamics, it is important to remember the difference between internal and external work. In Lesson 14 it was mentioned that when a subject holds a weight horizontally at arm's length he is doing no mechanical work since the weight is stationary. The physiological work of the subject, however, has increased significantly since this task has increased the subject's oxygen consumption, heat production, metabolic rate, etc. Thus, in discussions of "work," the *external* mechanical force-times-distance work must be carefully distinguished from the *internal* or physiological rate of free energy expenditure.

Efficiency of human movement

The efficiency of the body considered as a mechanical machine fueled by chemical energy can be calculated from the ratio of external work to internal energy conversion rate. The two types of efficiency are gross and net.

gross efficiency The rate of external work to total internal energy conversion rate

net efficiency The ratio of external work to internal energy necessary to accomplish the work

Thus a person who begins to exercise by doing external work at the rate of 20 Kcal per hour might show a steady state metabolic rate increase from 80 Kcal per hour just before exercise to 180 Kcal per hour during the exercise. This would give a gross efficiency of $20/180 = 11\%$ and a net efficiency of $20/(180 - 80) = 20\%$.

Under optimal conditions the net efficiency of the body as a machine is about 25%. This is greater than the efficiency of a steam engine, about equal to the efficiency of an internal combustion engine, and greatly inferior to the efficiency of a well designed electric motor. The body's efficiency in exercise is less than that of isolated muscle since in exercise energy expenditure is increased for circulation, respiration, and other internal supportive activities not present in isolated muscle preparations.

In typical circumstances, the mechanical efficiency of the body is considerably less than 25%. For example, a person sitting quietly watching television has an efficiency at that time of close to 0% with nearly 100% of his energy going into heat.

Techniques of measuring motor efficiency

In Part I it was stated that a motor performance can be evaluated in terms of effectiveness and efficiency. Motor effectiveness was defined as the extent that the performance goals are achieved, while motor efficiency was defined as the ratio of external work achieved to energy expended. Thus it was conceptualized that performers who were equal in their effectiveness could vary in their efficiency, with the most efficient performer requiring less energy for the achievement of a given amount of work.

The techniques involved in measuring motor efficiency are those involved in (1) measuring the external work accomplished by the individual (through the use of such ergometers as the treadmill and bicycle) and (2) measuring the gross and net energy expenditures of the individual (through use of indirect calorimetry).

The units of external and internal work are interconvertible as follows:

1 foot-pound = 0.13825 kilogram-meter =
0.000324 kilocalorie

1 kilogram-meter = 7.23 foot-pounds =
0.002343 kilocalorie

1 kilocalorie = 3086 foot-pounds =
426.4 kilogram-meters

STUDY GUIDELINES
Introduction
Specific student objectives
1. State the law of the lever.
2. Describe mechanical advantage.

Summary

Machines are used to perform work and carry out machine functions. Mechanical advantage in a machine may be thought of as the effect of a machine on the input force when the machine is used. More precisely the mechanical advantage (M.A.) of a machine is the ratio of the output force (R) delivered by the machine to the input force (F) applied by the mover; i.e., M.A. = R/F.

The condition for angular equilibrium is called the law of the lever. $F \times f = R \times r$ is the formula statement of this law, where f and r are the lever arms for the input force and output force, respectively. This state of angular balance exists when there is zero angular acceleration.

Performance tasks
1. Write out descriptions for the following:
   a. Mechanical advantage
   b. F, R, f, and r
   c. The law of the lever
2. Illustrate the concept of mechanical advantage with an example related to human motion.

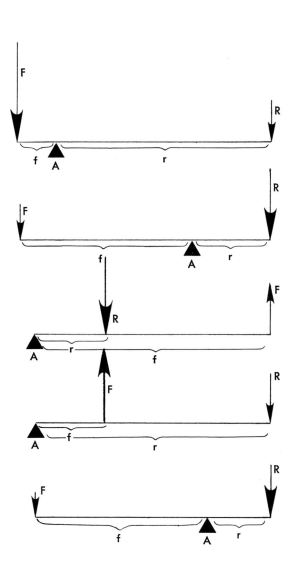

3. Using the law of the lever ($F \times f = R \times r$) for a lever in a state of angular equilibrium, calculate the missing element in the problems at the bottom of this page and illustrated in the diagrams on the left. Note that $F \times f = R \times \dot{r}$ can be rearranged to obtain:

$$F = \frac{R \times r}{f}, \quad R = \frac{F \times f}{r}, \quad f = \frac{R \times r}{F},$$

$$r = \frac{F \times f}{R}$$

Principles for the rearrangement of formulas are presented in Appendix D.

Concepts of theoretical and actual mechanical advantages

Specific student objectives

1. Differentiate between theoretical and actual mechanical advantages.
2. Describe the mechanical efficiency of a machine.
3. Describe the relationship of machine functions to the concept of theoretical mechanical advantage.
4. Given the input and output forces, calculate mechanical advantage.
5. Conduct a lever analysis.

Summary

The mechanical advantage (M.A.) of a lever is a comparison between output force and input force or between force and resistance arm:

$$\text{M.A.} = \frac{R}{F} = \frac{f}{r}$$

	Force (F)	Force arm (f)	Resistance (R)	Resistance arm (r)
a.	_____	3 in	120 lb	12 in

$$(\text{Example } F = \frac{R \times r}{f} = \frac{120 \text{ lb} \times 12 \text{ in}}{3 \text{ in}} = \frac{120 \text{ lb} \times 12}{3} = 480 \text{ lb.})$$

	Force (F)	Force arm (f)	Resistance (R)	Resistance arm (r)
b.	220 lb	8 in	_____	4 in
c.	82 lb	_____	219 lb	18 in
d.	17 lb	2 in	4 lb	_____
e.	_____	6 ft	98 lb	2 ft

A distinction is made between theoretical mechanical advantage (f/r) and actual mechanical advantage (R/F). In practice the former is always greater than the latter. The efficiency of a machine is the ratio of the actual mechanical advantage to the theoretical mechanical advantage; i.e., Efficiency = Actual M.A./Theoretical M.A.

Theoretical mechanical advantage (T.M.A.) is closely related to the machine function of a lever; when the function is to *balance* forces, T.M.A. = 1; when the function is to *magnify* forces, T.M.A. > 1; finally, when the function is to *increase the speed and range* of motion of an object at the expense of force, T.M.A. < 1.

Performance tasks

1. Develop a list of written descriptions for the following:
   a. Theoretical mechanical advantage (T.M.A.)
   b. Actual mechanical advantage (A.M.A.)
   c. The relationship between A.M.A. and T.M.A.
   d. Mechanical efficiency (M.E.)
   e. The relationship between T.M.A. and the primary machine functions
2. The range of values for mechanical advantage is
   a. greater than zero
   b. any positive number
   c. both a and b
   d. neither a nor b
3. Mechanical efficiency is usually expressed as a
   a. decimal
   b. percent
   c. both a and b
   d. neither a nor b
4. The possible range of values for mechanical efficiency is from
   a. zero up
   b. 0% – 100%
   c. any positive number
   d. all of the above
5. With the information given at the bottom of p. 275, calculate the A.M.A. and T.M.A. for each lever and list its mechanical function. Ignore the resistive force of friction in each case. Use

$$\text{M.A.} = \frac{F}{R} = \frac{f}{r}$$

6. Calculate the A.M.A. for the following machine examples:

	Input force (F)	Output force (R)	A.M.A.
a.	8 lb	72 lb	_____
b.	25 newtons (n)	120 n	_____
c.	18 n	18 n	_____
d.	80 lb	20 lb	_____
e.	10 dynes	2 dynes	_____

7. Using the machine examples and information of Tasks 6 and 11, calculate the mechanical efficiency for Problems a through e in each task. Express each as a percent.
8. It is possible to utilize many of the basic concepts of levers in the analysis of bony machines. To illustrate this, consider the brachialis muscle, which acts upon the forearm as shown in Fig. 16-4.

The scaled diagram of Fig. 16-4 has been drawn to depict the following:
a. Muscle force (F) = 120 lb
b. Force arm (f) = 2 in
c. Resistance (R) = ?
d. Resistance arm (r) = 16 in

The forearm lever in this illustration is a

						Function
a.	$\text{M.A.} = \dfrac{R}{F} = \dfrac{120 \text{ lb}}{480 \text{ lb}} = \dfrac{1}{4}$ or $\text{M.A.} = \dfrac{f}{r} = \dfrac{3 \text{ in}}{12 \text{ in}} = \dfrac{1}{4}$					Magnification of displacement
b.	M.A. =					
c.	M.A. =					
d.	M.A. =					
e.	M.A. =					

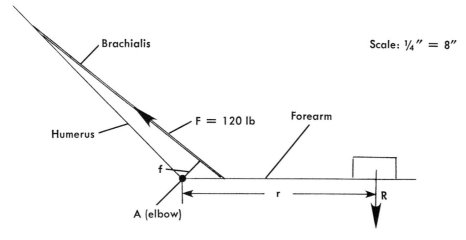

Fig. 16-4. Analog diagram of the brachialis muscle acting upon the forearm lever.

third-class lever because the point of application of the muscle force is located on the forearm between the axis of rotation (the elbow joint) and the weight held in the hand.

In the estimation of T.M.A.,

$$f/r = 2 \text{ in}/16 \text{ in} = \frac{1}{8}$$

Thus the machine function of this lever is to favor speed and range of motion. From the law of the lever, we can use $R = F \times f/r$ to demonstrate that a muscle force of 120 lb will balance a resistance force (R) of 15 lb with the forearm in the position shown.

$$R = \frac{120 \text{ lb} \times 2 \text{ in}}{16 \text{ in}} = 15 \text{ lb}$$

NOTE: The weight of the lever itself, which also must be moved, is ignored in the above illustration and the effects of friction are also ignored.

Finally, both positive and negative torques can be found with the lever in a balanced position as follows:

$$+T = F \times f = 120 \text{ lb} \times 2 \text{ in} = 240 \text{ lb-in} = 20 \text{ lb-ft}$$
$$-T = R \times r = 15 \text{ lb} \times 16 \text{ in} = 240 \text{ lb-in} = 20 \text{ lb-ft}$$

9. A similar analysis can be made for the tibia

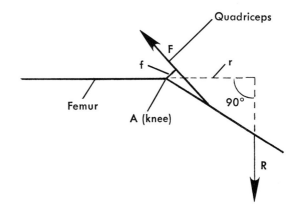

Fig. 16-5. Analog diagram of the quadriceps femoris muscle acting upon the lower leg lever.

of the leg acting as a lever moved by the quadriceps femoris muscle. The scaled diagram of Fig. 16-5 has been drawn to depict the following:

a. Muscle force (F) = ?
b. Force arm (f) = 4 in
c. Resistance (R), the center of weight of the leg = 8 lb.
d. Resistance arm (r), the "right angle" distance from the axis (A) to the line of action (R) = 20 in

The lever is of the third class; i.e., it is de-

278 *Kinematic analysis of human motion*

signed to increase speed through a fuller range of motion at the expense of force.

The mechanical advantage of this lever

$$\text{T.M.A.} = \frac{f}{r} = \frac{4 \text{ in}}{20 \text{ in}} = \frac{1}{5}$$

indicates that five times as much input force must be developed to balance a given resistance.

Using $R = 8$ lb and $R \times r = F \times f$, we can obtain

$$F = \frac{R \times r}{f} = \frac{8 \text{ lb} \times 20 \text{ in}}{4 \text{ in}} = 40 \text{ lb}$$

This means that a 40 lb muscle force is necessary to (balance/lift) _____ the leg as shown.

Finally, both positive and negative torques can be found as follows:

$$+T = F \times f = 40 \text{ lb} \times 4 \text{ in} = 160 \text{ lb-in}$$
$$-T = R \times r = 8 \text{ lb} \times 20 \text{ in} = 160 \text{ lb-in}$$

10. Select three other levers of the body and conduct a lever analysis; i.e., given a lever of the human body (with its position and action), develop a scaled diagram to show the following:
    a. Muscle force (F) and its line of action
    b. Lever arm of the force (f)
    c. Fulcrum (A)
    d. Resistance (R) and its line of action
    e. Lever arm of the resistance (r)
    Then do each of the following tasks:
    f. Classify the lever.
    g. Estimate its T.M.A. in that position.
    h. State the machine function of the lever.
11. Calculate the T.M.A. for the following machine examples. Also list the primary machine function in the spaces provided:

Conservation of energy and machine functions
Specific student objectives

1. State the law of energy conservation in terms of mechanical work.
2. Describe the relationship of machine functions to the concept energy.

Summary

Work done on a machine (F × d) equals work done by a machine (R × h) according to the law of conservation of energy; i.e., R/F = d/h.

Thus, from the law of the lever, we know that f/r = d/h and we can see that, according to the energy conservation law, work output equals work input when a mechanical machine is used.

Performance tasks

1. Illustrate the law of energy conservation with a human motion example of "work output equals work input."
2. Work done on a machine is found by
    a. F × d
    b. R × h
    c. both a and b
    d. neither a nor b
3. Even though machines may be used to "magnify force," according to the energy conservation law they do so at the expense of
    a. balancing forces
    b. range of motion
    c. both a and b
    d. neither a nor b

Mechanical advantages of wheel-axle and fixed pulley arrangements
Specific student objectives

1. Identify the two types of wheel-axle machines.
2. Apply the law of the lever to the two types of wheel-axle arrangements and a pulley.
3. Describe the A.M.A. and T.M.A. for a wheel-axle and pulley.

	Applied force arm (f)	Resistance arm (r)	T.M.A.	Function
a.	10 in	6 in	_____	_____
b.	48 cm	18 cm	_____	_____
c.	1.5 cm	1 cm	_____	_____
d.	2 ft	4 ft	_____	_____
e.	20 mm	5 mm	_____	_____

Summary

The two wheel-axle arrangements occur when the force is applied at the wheel (type I) or at the axle (type II). The law of the lever, i.e., the statement of angular equilibrium, also applies to a type I ($F \times r_w = R \times r_a$) or type II ($F \times r_a = R \times r_w$) arrangement.

The T.M.A. for a wheel-axle is given by the ratio of the applied force arm to the resistance arm; i.e., T.M.A. $= r_w/r_a$ (type I) or r_a/r_w (type II). The A.M.A. is found by comparing the resistance overcome to the force applied (R/F). Again, mechanical efficiency (M.E.) is expressed as a decimal or percent and is found by M.E. $=$ A.M.A./T.M.A.

For a pulley, T.M.A. $= 1$ (unity), A.M.A. $=$ R/F, and M.E. $=$ A.M.A./T.M.A. The mechanical advantage of unity for a pulley indicates neither favoring force nor displacement. A pulley serves the primary function of balancing forces when its T.M.A. $= 1$. A pulley also possesses the secondary machine function of directional change, which can, in effect, improve the effective mechanical advantage of the machine.

Performance tasks

1. State the two primary functions of a wheel-axle.
2. Cite examples of wheel-axle arrangements in the body and state their primary functions by estimating their theoretical mechanical advantages.
3. The deep spinal muscles can act upon the vertebral axle, as shown in Fig. 16-6, to rotate the ribs (wheel). The following information is given: $r_a = 2$ in, $r_w = 12$ in.

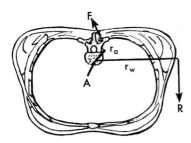

Fig. 16-6. The deep spinal muscles can act upon the vertebral axle to rotate the ribs (wheel) in a wheel-axle arrangement.

The mechanical advantage of the machine is the ratio of r_a to r_w. Thus T.M.A. $= 2$ in/12 in. This mechanical advantage (with a value less than unity) is characteristic of a wheel-axle in which the force is applied at the axle and the resistance at the wheel, and it indicates a machine function of increasing (speed and range of motion at the expense of force/force at the expense of speed and range of motion)_____.

4. The other possible function of a wheel-axle machine occurs if the force is applied at the rim of the wheel and the resistance occurs at the axle. This situation is exemplified by the oblique abdominal muscles acting on the rib cage. Using the same information as given in the previous example, we obtain T.M.A. $= 12$ in/2 in. Thus when the force is applied at the wheel, the mechanical advantage is greater than unity and the function is of magnifying (force/displacement)_____ at the expense of (force/displacement)_____.
5. The T.M.A. for a single fixed pulley is
   a. less than one
   b. equal to one
   c. greater than one
   d. all of the above
6. A pulley possesses the machine function of
   a. magnifying forces
   b. balancing forces
   c. favoring speed and distance
   d. none of the above
7. Identify examples of a wheel-axle and a pulley. Estimate their T.M.A. and state their machine function(s).
8. List examples of human body pulleys and state their function.
9. A simple fixed pulley is a mechanical device designed to change the direction of an applied force. The line of action of a force may be changed as shown in Fig. 16-7. No mechanical advantage occurs, but the applied force acting downward causes the weight to move upward through the utilization of the pulley.

   The change of direction may aid *or* hinder motion capabilities in the body. Fig. 16-8 depicts a situation whereby the pulley arrangement decreases the angle of pull of a muscle by its arrangement. On the flexor side of the finger joints, the tendons of the flexor digi-

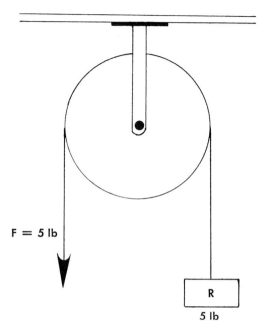

Fig 16-7. A single fixed pulley changes the direction of the applied force.

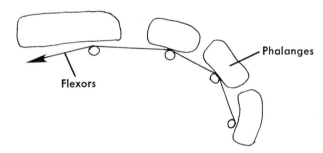

Fig. 16-8. Tendons of the flexor digitorum profundus and flexor digitorum superficialis deflected by pulleys.

torum profundus and flexor digitorum superficialis are deflected and limits are placed upon their motion by a pulley-like loop.

Analysis of motor efficiency

Specific student objectives

1. Identify the products produced by the human body when it is used as a thermodynamic machine.
2. Describe metabolism and metabolic rate.
3. Identify the common denominator of all energy released in the body.
4. Describe calorimetry and a calorie.
5. Identify and describe one direct and one indirect method of calorimetry.
6. Differentiate between external and internal work.
7. Define gross and net efficiency of the body.
8. Describe techniques employed in motor efficiency measurements.

Summary

The human body can be considered as a thermodynamic machine with the principal result being heat production and mechanical work.

Heat is the common denominator of all energy released in the body. Metabolism consists of all the chemical reactions that take place within the body; the metabolic rate is the quantitative measure of the heat liberated during the chemical reactions. Only about 25% (maximum) of all energy from food is utilized by the functional systems of the body. The remainder becomes heat. The heat energy unit (kilocalorie) is the quantity of energy needed to raise 1 liter of water 1° centigrade. The measuring of heat energy is termed calorimetry. The two types of calorimetry are direct (measuring the heat) and indirect (measuring the oxygen used).

Basal metabolism is the rate of chemical reactions when a person lies motionless in a warm comfortable environment.

Internal work is the physiological rate of free energy expenditure. External work is mechanical work. Body efficiency is defined as the ratio of external work to internal energy conversion. Gross efficiency uses total internal energy, while net efficiency utilizes only the internal energy necessary to do the work. Body efficiency will commonly vary from 0% to 25%. Motor efficiency can be measured by (1) measuring external work with ergometers and (2) measuring gross and net energy expenditures through the use of calorimetry.

Performance tasks

1. Develop a written list of descriptions for each of the following:
   a. Thermodynamic machine
   b. Metabolism, basal metabolism, and metabolic rate

c. Calorimetry and calorie
d. External and internal work
e. Motor efficiency, gross and net
f. Indirect and direct calorimetric methods
2. Metabolism of the body usually results in the production of
a. heat
b. mechanical work
c. both a and b
d. neither a nor b
3. When exercising, we produce mechanical energy in motion; but a temperature rise in and around muscles indicates the production of
a. heat
b. mechanical work
c. both a and b
d. neither a nor b
4. The sum of the chemical reactions of all the body cells is called (metabolism/thermodynamics) _____ .
5. The rate of heat liberation in the body is a measure of
a. efficiency
b. effectiveness
c. metabolic rate
d. work
6. The oxidation of carbohydrates, fats, and proteins in the body results in a transformation of chemical energy to heat energy and the energy of motion called _____ .
7. Most chemical energy of the body is transformed into
a. heat   c. both a and b
b. work   d. neither a nor b
8. Consult any textbook concerned with the physiological aspects of exercise and examine the physiological processes which result in chemical energy being transformed into heat and mechanical energy.
9. An energy unit for human metabolism studies is
a. calorie   c. both a and b
b. kilocalorie   d. neither a nor b
10. The amount of energy necessary to raise one kilogram of water by one degree centigrade is called a
a. calorie
b. kilocalorie
c. both a and b

d. neither a nor b
11. Measurement of heat energy is called (calorimetry/thermodynamics)_____ .
12. Direct calorimetry measures determine the production of heat while indirect methods measure
a. oxygen   c. both a and b
b. work   d. neither a nor b
13. Using a human calorimeter or an oxygen measuring device, set up a human motion experiment to measure the amount of heat energy given off during a specified motor task.
14. Set up an experiment to determine your basal metabolism.
15. "Lifting" and moving an object would be an example of (external/internal)_____ work.
16. Physiological tasks indicated by oxygen consumption and heat production are evidence of (external/internal)_____ work.
17. How well a task is done is the question addressed by the topic of motor (effectiveness/efficiency) _____ .
18. Comparing the roles of external work to total internal conversion rate is called
a. gross efficiency   c. efficiency
b. net efficiency   d. all of the above
19. Calculate gross and net efficiency in the following thermodynamic motion situations:

	External work	Resting metabolic rate	Exercising metabolic rate
a.	30 Kcal/hr	60 Kcal/hr	150 Kcal/hr
b.	15 Kcal/hr	90 Kcal/hr	200 Kcal/hr
c.	20 Kcal/hr	70 Kcal/hr	190 Kcal/hr
d.	40 Kcal/hr	120 Kcal/hr	280 Kcal/hr
e.	5 Kcal/hr	60 Kcal/hr	160 Kcal/hr

20. Set up and carry out an efficiency experiment for a human motion task. Measure external work with an ergometer and gross and net energy expenditure through indirect calorimetry.

SELF-EVALUATION

Students should use no reference materials for this progress test, and they can check their answers by referring to Appendix A.
1. The law of the lever (in formula form) is
a. $T = F \times f$

b. V = d/t

c. A.M.A. = f/r

d. F × f = R × r

2. The effect of a machine upon input force is a description of mechanical
   a. advantage
   b. effectiveness
   c. efficiency
   d. work

3. Describe each symbol used in the formula Ff = Rr.
   a. F _____
   b. f _____
   c. R _____
 • d. r _____

4. Comparing the actual forces involved in using a machine, i.e., comparing the output resistance force (R) to the input applied force (F), would result in a description of
   a. A.M.A.
   b. M.A.
   c. M.E.
   d. T.M.A.

5. Comparing the lever arms of a machine, i.e., comparing the applied force arm (f) to the resistance force arm (r), would give
   a. A.M.A.
   b. M.A.
   c. M.E.
   d. T.M.A.

6. Comparing the way a machine actually operates to the way it should work is a measure called
   a. A.M.A.
   b. M.A.
   c. M.E.
   d. T.M.A.

7. Calculate A.M.A., T.M.A., and M.E. in the following machine examples. Also identify the primary machine functions.

8. "Work input equals work output" is a statement of the law of
   a. effectiveness
   b. efficiency
   c. energy conservation
   d. energy dissipation

9. A machine that "magnifies force" does so at the expense of
   a. speed and distance
   b. work
   c. both a and b
   d. neither a nor b

10. For a type I wheel-axle machine, the law of the lever would be modified to become
    a. $F \times r_a = R \times r_w$
    b. $F \times r_w = R \times r_a$
    c. both a and b
    d. neither a nor b

11. With a single fixed pulley, T.M.A. has a value of
    a. <1
    b. $= 1$
    c. >1
    d. none of the above

12. A pulley has the secondary machine function of
    a. favoring force
    b. balancing forces
    c. favoring speed and distance
    d. changing direction

13. Human motion results in the production of
    a. heat   c. both a and b
    b. work   d. neither a nor b

14. Chemical reactions in the body are called (metabolism/thermodynamics)_____.

15. The measure of heat energy is termed (calorimetry/goniometry)_____.

16. Indirect calorimetry measures oxygen while direct calorimetry measures
    a. heat   c. both a and b
    b. work   d. neither a nor b

	f	r	F	R	A.M.A.	T.M.A.	M.E.	Function
a.	20 cm	4 cm	10 kg	40 kg	____	____	____	____
b.	50 cm	50 cm	176 kg	180 kg	____	____	____	____
c.	$r_w = 1$ m	$r_a = 0.2$ m	480 g	160 oz	____	____	____	____
d.	3 in	18 in	120 lb	15 lb	____	____	____	____
e.	2 in	20 in	140 lb	10 lb	____	____	____	____

17. The rate of chemical reactions within the body when a person lies motionless in a neutral environment would be
    a. basal metabolism
    b. external work
    c. motor efficiency
    d. torque production
18. Describe motor efficiency.
19. Comparing the rate of external work to total internal conversion rate is called
    a. gross efficiency
    b. net efficiency
    c. efficiency
    d. all of the above
20. Identify the appropriate measurement techniques to use in determining motor efficiency for a specified motor task.
    a. External work
    b. Gross and net energy expenditure
21. Calculate gross and net efficiency from the data in the following experimental human motion situations.

22. Analyze the body levers listed below. Develop a scaled diagram. Classify the lever, estimate the theoretical mechanical advantage, and list the primary machine function of the lever.
    a. The biceps brachii muscle acting upon the radius-ulna with the forearm in a horizontal position and the arm in a vertical position; the resistance is the weight of the forearm (4 lb)
    b. The triceps as the mover of the forearm; the forearm is above the head making a 45° angle with the horizontal humerus (as in shooting a basketball); consider the resistance to be the weight of the ball only (20 oz)
    c. The leg in a 45° angle measured from vertical, the thigh partially flexed, and the body in anatomical position; analyze the action of the hamstrings on the leg (which weighs 4 kg)

	External work	Steady state metabolic rate	Exercise metabolic rate	Gross	Net
a.	25 Kcal/hr	150 Kcal/hr	250 Kcal/hr	_____	_____
b.	10 Kcal/hr	120 Kcal/hr	200 Kcal/hr	_____	_____
c.	8 Kcal/hr	100 Kcal/hr	140 Kcal/hr	_____	_____

PART III

KINETIC ANALYSIS OF HUMAN MOTION

SECTION ONE

Kinetic analysis of linear motion

LESSON 17

Introduction to kinetics

CONCEPT OF KINETICS

In preceding lessons the discussions have been centered on *how* motion and forces are described mathematically. However, the answer to the question of *why* an object moves the way it does has not been given. This is the central question of that branch of mechanics called kinetics, which deals with the causes for the states of motion of an object.

Rest and motion

The key concepts in kinetics are those concerned with the states of "at rest" and "in motion." An object is said to be *in motion* when its position with respect to some point, line, or surface is changing. Likewise, an object is said to be *at rest* when its position with respect to some point, line, or surface remains unchanged. For example, a person may sit quietly in a jet airliner and be at rest with respect to the airplane but be in a great state of motion with respect to the ground. Thus motion and rest are relative terms. They are also contrasting terms because a body at any given time in relation to a given point of

reference *must* by definition be either at rest or in motion.

To Issac Newton goes the credit of first discovering the laws which govern all states of rest and motion. These laws constitute the fundamental principles of kinetics.

NEWTON'S LAWS OF MOTION

Newton's first law, also known as the law of *inertia,* deals with the resistance of an object to any change in its state of motion. This law says that an object at rest has a tendency to remain at rest and that an object in motion tends to remain in motion and to travel in a straight line with uniform speed unless acted on by some net external force.

Newton's second law, also known as the law of *acceleration* because it deals with the factors affecting the acceleration of an object, says that when a body is acted upon by a net force the resulting acceleration is directly proportional to the net force, and inversely proportional to the mass, and takes place in the direction of the acting net force. This law can be expressed in formula form:

$$a = \frac{net\ F}{m}$$

(Equation 17-A)

Newton's third law, known as the law of *reaction,* says that for every action force there is an equal and opposite reaction force. This law emphasizes that when a force is applied upon an object the object pushes back on the source of the effort with a force equal to and opposite the original force.

CONCEPT OF NET FORCE

According to Newton's first law, an object moves because a force greater than its resistance has been applied to it. Thus many ambiguities are cleared up about the concept of force, which is simply any influence that causes a body to be either positively or negatively accelerated.

This law introduces the concept of *net force.* The net force acting on an object is the resultant of all the forces affecting the object. A body at rest, for example, can have many forces acting upon it, but then their magnitudes and directions will cancel one another out to leave no net force. In other words, in the parlance of vector addition, the resultant or vector sum of all the forces acting upon the object equals zero, a condition known as linear equilibrium or zero acceleration. However, when one or more forces begin to act upon a body at rest, and their vector sum is not zero, the body will be set into motion. Under such conditions an unbalanced force is acting and this force alone accounts for the motion. As defined in Lesson 2, the branch of kinetics dealing with bodies in a state of equilibrium (a condition brought about by balanced net forces) is called *statics.* On the other hand, *dynamics* is the branch of kinetics dealing with changes in motion brought about by unbalanced net forces.

Effects of net forces

The term resistance refers to the actual forces exerted in opposition to applied forces. Thus a net force is the difference between the applied force (F) and the resistive force (R),

$$Net\ F = F - R$$

(Equation 17-B)

From this equation it is evident that a positive net force exists and both positive acceleration and positive work are produced when the applied forces are greater than the resistive forces. Likewise, both negative acceleration and negative work are produced when the applied forces are less than the resistive forces, and zero work is produced when the applied and resistive forces are equal.

Concept of surface friction as a resistive force

To clearly understand the characteristic properties of resistive and applied forces, consider what happens when we attempt to slide a packing case across a level floor (Fig. 17-1). At first the packing case is stationary with no horizontal forces whatever acting on it. As we begin to push, the packing case does not move because the floor exerts a force on its bottom which opposes the force we apply. This opposition force is surface friction which arises from the nature of the contact between the floor and packing case. As we push harder, the frictional force also increases to match our efforts, until finally we are able to exceed the frictional force and the packing case begins to move.

Evidently the opposing frictional force has a maximum value beyond which it cannot go, and when we apply a force greater than this maximum the packing case experiences a net force. Because this force acting on the packing case is the force of our push *minus* the force of friction, it is always less than the force we apply. The net force may even be zero, as we have seen when no horizontal forces are being applied or when the forces are balanced.

Concept of gravity as a resistive force

Exactly the same type of situation as above exists when we attempt to lift a suitcase, except the resistance force is now the force of gravity. For example, if the weight of the suitcase is 50 lb and we apply an upward force of 10 lb, then the magnitude of the resistance force is also 10 lb according to Newton's third law and the net force is zero. Thus the suitcase will continue to be at rest. Fig. 17-2 reveals that the resistance force of gravity can be increased up to a maximum (which in this instance is 50 lb). Any applied force greater than the maximum resistance

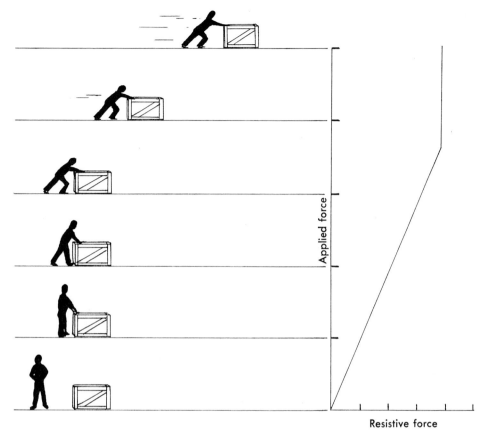

Applied force

Resistive force

Fig. 17-1. A horizontal force is applied to a packing case on a level floor. The frictional resistive force increases to a certain maximum and then remains constant until movement is initiated; after movement is initiated, it decreases to a new value.

will produce a net force, and an upward motion (acceleration) will be imparted to the suitcase. For example, if 65 lb of force is applied upward, then the net force will be 15 lb and this is the force which is entered into the equation for Newton's second law to determine the amount of vertical acceleration that will be imparted to the suitcase.

Newton's laws and the effects of net tangential and radial forces

We can express the laws of motion in terms of the effects of net radial and tangential components of a force. In Lesson 13 we learned that any net force not acting on a line through the axis of an object will impart angular motion to the object. In other words, an object with a *fixed axis* of rotation must have a force acting perpendicular to its radius of rotation in order for it to

have angular motion. This perpendicular component was identified as the tangential force (F_t). The tangential direction was also identified as being the *linear path* of a body in rotation. The linear path of a body in rotation is located at a tangent to the curved path of the motion or perpendicular to the radius of rotation.

A body in motion that does not have a *fixed axis of* rotation must have a force acting perpendicular to its linear path in order to have a change in the direction of that path. This force was identified as the normal or radial force (F_r). Thus the two components that a force can have are:

1. Tangential force: acting parallel with the linear path of a body, perpendicular to a possible radius of rotation
2. Radial force: acting perpendicular to the linear path of a body, parallel with a possible radius of rotation

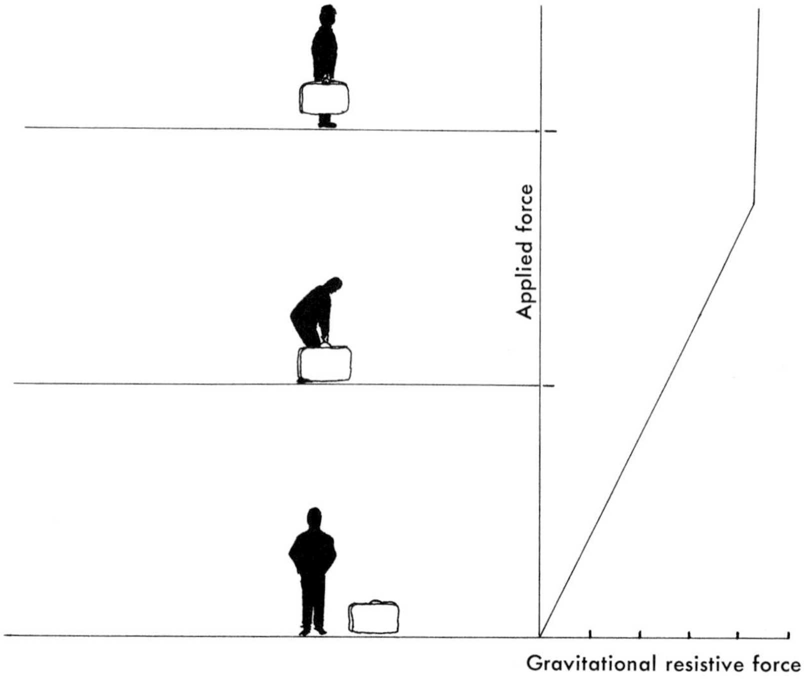

Fig. 17-2. As vertical force is applied to a suitcase, the gravitational resistive force increases to a certain maximum (equal to the weight of the suitcase) and then remains constant.

Therefore it can be seen that when a body is in linear motion the speed may be increased or decreased by the presence of a net tangential force and that the direction can be changed by the presence of a net radial force.

Newton's laws can be stated in terms of tangential and radial forces as follows:

NEWTON I. A body in motion will travel with a constant speed unless acted upon by a net tangential force, and it will travel in a straight line unless acted upon by a net radial force.

NEWTON II. The rate of change in speed of a body in motion is directly proportional to the magnitude of the net tangential force and inversely proportional to its mass.

The rate of change in direction of a body in motion is directly proportional to the magnitude of the net radial force and inversely proportional to its mass.

NEWTON III. For every tangential and radial force component there is an equal and opposite reaction tangential and radial force component.

CONCEPTS OF INERTIA AND MASS

Newton's first law of motion imputes to material bodies the property of tending to resist changes in their states of rest or uniform motion, a property known as inertia. We can specify inertia in a precise way with the help of the second law of motion. Since the response of a body to a net force (F) is an acceleration (a) proportional to F, we can write the equation

$$\text{Net F} = \text{ma}$$

(Equation 17-C)

In this equation m is a constant of proportionality characteristic of the particular body being subjected to the force. The larger the value of m, the smaller will be the acceleration produced by a given net force. Thus m is a measure of the inertia of a body.

Upon what properties of a body does m depend? From everyday experiences we know that such external characteristics of an object as color, shape, and texture are not involved in m but that what we might loosely call the *amount of matter* in the object is involved. The quantity m is called the *mass* of the body.

Measurement of mass

The second law of motion provides a method for measuring the mass of a body. All that is necessary is to apply the same force in turn to a standard mass and to an unknown mass and then to compare their accelerations. This comparison can be made anywhere in the universe with the same results. No matter where we are, we can always tell a baseball from a lead ball of the same size by throwing them. The difference in the inertia or masses of the two balls would make the balls feel very different to us even though we might be isolated far out in interstellar space where sensations of weight would be absent.

On the earth's surface, however, the inertia or mass of an object can be determined through the use of gravitational quantities. In this case we use as a standard the *force* of gravity and the *acceleration* which results when a suspended object is allowed to fall. At a given place on the earth, we measure the acceleration (g) resulting from the earth's gravity at that point. This acceleration has been found to average around 32 ft/sec². We might also measure the force of gravity acting on any object at that place, i.e., the weight of the body (w).

According to Newton's second law of motion, $F = ma$; therefore with $F = w$ and $a = g$, we have $w = mg$, and by rearranging we get

$$\text{Mass} = m = \frac{\text{Weight}}{\text{Acceleration of gravity}} = \frac{w}{g}$$

(Equation 17-D)

In other words, the mass of a body expressed in terms of gravitational quantities is equal to the weight of the body divided by the acceleration of gravity. Thus the mass of a person weighing 160 lb is determined by substituting 160 lb for w and 32 ft/sec² for g in the equation $m = w/g$ to obtain $m = 160$ lb/32 ft/sec² $= 5$ lb/ft/sec². Since 5 lb/ft/sec² is such a large and cumbersome unit, the unit of 5 *slugs* is substituted for it in the British system. Thus in the British system with weight (w) expressed in pounds and the acceleration of gravity averaging 32 ft/sec², mass (m) is expressed in slugs. The weight of a 1-slug mass is 32 lb, and the mass of a 1 lb weight is 1/32 slug.

There is no need for any confusion between the concepts of mass and weight: the *mass* of a body is a measure of its *inertia;* the *weight* of a body is the gravitational *force* with which it is attracted toward the center of the earth.

A very significant empirical fact is that the mass and weight of a body are proportional to each other. We might ask why when we double the weight of a body we do not increase the acceleration of the body during free fall. The answer is simply that the increase in the weight of the body is offset by an equal increase in the mass of the body. A significant difference between the mass and weight of a body, however, is that its mass is the same anywhere in the universe while its weight can be determined only when it is located within a gravitational field. Another significant difference between weight and mass is the fact that weight is a measure of force and is therefore a vector quantity, with both magnitude and direction. Mass, on the other hand, is a scalar quantity, with only magnitude.

As an example of the scalar nature of mass, imagine a body being located at rest out in space far removed from the gravitational field of the earth or that of any other planet. The body in this situation would have a weight of zero, but it would still have a mass. In what direction, however, would our imagination say the mass is acting? The answer, of course, would be that the mass of the body in such a situation has no direction and thus it qualifies as a scalar quantity, with only a magnitude.

Difference between inertia and resistive forces

Resistive forces are of many kinds but all act to impede motion. The effects of resistance must be distinguished from the effects of inertia. The term inertia merely refers to the fact that bodies maintain their original states of rest or motion in the absence of net forces acting on them; even a relatively minute net force is sufficient to accelerate a body despite its inertia. The term resistance, on the other hand, refers to the actual forces which are exerted in opposition to applied forces.

It can be seen from this discussion that the difference between resistance and inertia is due to the difference in the effects of these two quantities. The magnitude of the resistance, for example, affects the magnitude of the net force while the magnitude of the inertia affects the magnitude of the acceleration produced by the net force. Because a person weighs less on the moon than on the earth, he or she can jump higher on the moon. This is due to the lesser resistive force of gravity on the moon and the resulting greater net force produced by a given amount of muscle force.

The inertia of a body, however, is the same on the moon as it is on the earth.

INTRODUCTION TO THE PRINCIPLES OF STATICS

Newton's first law, from the standpoint of statics, can be turned around to say that any object remaining at rest (or moving with uniform motion) is in equilibrium and the resultant of all forces acting upon it is zero. Thus an object in motion can also be in equilibrium.

Some students might feel uneasy with Newton's first law. Although their everyday experience does indicate that bodies in motion *tend* to remain in motion along a straight line at a constant speed, they know that sooner or later these bodies invariably come to a stop and often deviate from a straight path as well, A golf ball, for instance, rolling along a smooth perfectly level fairway will not continue forever owing to the resistance of air and to friction between it and the grass. However, try to imagine what would happen should the ball be moving in outer space far away from the gravitational field of the earth. In this case, what would *cause* the ball to change its state of motion? The answer according to Newton's first law is that since no force would be acting on the ball it would continue to move in a straight line with uniform speed.

On the earth, however, we do have air resistance and we do have frictional forces which tend to change the velocity of moving objects. Therefore in order to maintain uniform velocity, we must have a constantly acting propelling force to counteract the resistive forces.

Newton's third law is also involved in problems of equilibrium. In Fig. 17-3 a 2 lb cat is lying stationary on a table, pressing down on the table with a force of 2 lb. The table pushes upward on the cat with a reaction force of 2 lb. Why does the cat not fly upward into the air? The answer is that the upward force of 2 lb acting on the cat merely balances the cat's weight of 2 lb, which acts downward. If the table were not there to cancel out the latter 2 lb force, the cat would, of course, be accelerated downward. This situation also exists when we try to push or lift an object heavier than we have the strength to move. Suppose we apply 100 lb of force against an object. Since the object pushes back against us with

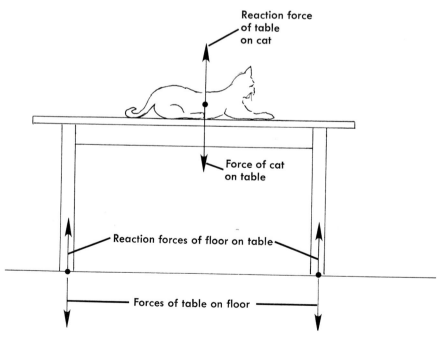

Fig. 17-3. Action-reaction forces between a cat and a table and between the table and the floor.

an equal and opposite reaction force of 100 lb, the resultant or vector sum of the two forces equals zero and the object remains in equilibrium at rest.

As another example of equilibrium, consider the game of tug-of-war (Fig. 17-4) and assume that two opposing teams are pulling on the ends of a rope with equal and opposite forces of 500 lb. One team pulls with a force of 500 lb, and the other pulls in the opposite direction with a force of 500 lb. It should be noted here that the tension in the rope is 500 lb and not 1000 lb. This apparent paradox can be explained by supposing that one team becomes tired and ties its end of the rope to a post. The other team, still pulling with 500 lb, maintains the same equilibrium condition as before and in so doing maintains the tension of 500 lb. One team can be looked upon as holding the rope so the other team can pull. Therefore the force of one team is offset by the reaction force of the other team. If the two forces, however, become unequal, as might occur when one team begins to fatigue, equilibrium will no longer exist and the rope will move in the direction of the greater force.

The principles of statics can be summarized by stating the two conditions of equilibrium:

1. In order for an object to be in *linear equilibrium,* the sum of all the rectangular components of the forces, acting upon it must equal zero; i.e., the x, y, and z components must total zero ($\Sigma F_x = 0$, $\Sigma F_y = 0$, and $\Sigma F_z = 0$).

2. In order for an object to be in *rotatory equilibrium,* the sum of all the torques tending to produce clockwise rotation around each of the three axes must be counterbalanced by the sum of all the torques tending to produce counterclockwise rotation around these axes; i.e., the sum of all the torques acting upon the object must equal zero.

INTRODUCTION TO THE PRINCIPLES OF DYNAMICS

As mentioned before, Newton's first law defines force as any influence causing a body to be accelerated. When an object at rest, for instance, is set into motion, its speed is changed from zero to a positive value and it is therefore accelerated. Likewise, when the velocity of an object already in motion is either increased or decreased, or changed in direction, that object is, by definition, being accelerated.

Newton's second law provides a method for

Fig. 17-4. A tug-of-war can be used as an example of linear equilibrium when the forces produced by two teams are equal and opposite.

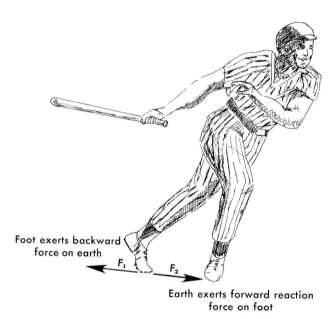

Foot exerts backward
force on earth

F_1 F_2

Earth exerts forward reaction
force on foot

Fig. 17-5. When a runner pushes backward with one foot, the earth pushes forward on the runner. This reaction force produces the forward motion of the runner.

Fig. 17-6. The force applied to a snowball by a person is equal to the reaction force applied by the snowball to the person.

analyzing and comparing forces in terms of the accelerations they produce. Thus a force which produces twice the acceleration that another force produces must be twice as great. Newton's first law is clearly a special case of his second law: when the net applied force (resultant or vector sum of all the forces acting upon a body) is zero, the acceleration is also zero.

The nature of the net applied force itself is given by Newton's third law. In the act of running, for example, we push backward with one foot and the earth pushes forward on us. The forward reaction force exerted by the earth causes us to move forward; at the same time the backward force of our push causes the earth to move backward (Fig. 17-5). Because of the earth's tremendous mass, however, its motion cannot be detected practically but the backward action is there. In analyzing action-reaction forces, we must remember that every example of such an interaction is an example of forces acting on *different* bodies. We push on the earth, and the earth pushes back on us. If the reaction force is greater than our resistance, we proceed to undergo acceleration.

We might conceivably find ourselves on a frozen lake with a perfectly smooth surface. We could not walk because the absence of friction would prevent us from exerting a backward force on the ice to produce a forward force on us. We could throw a snowball forward, however, by applying forces on it. The snowball would press back on us at the same time with an identical force but in the opposite direction. As a consequence of this reaction force, we would find ourselves being moved (accelerated) backward (Fig. 17-6) but at a slower speed because of our greater mass.

STUDY GUIDELINES
Concept of kinetics
Specific student objectives

1. Describe kinetics.
2. Differentiate between the states of rest and motion.

Summary

Kinetics is the area of mechanics in which the forces causing motion are examined. Thus kinetics deals with the causes for the states of motion of an object.

An object is either at rest or in motion with respect to some reference. An object is said to be in motion when its position in relation to a reference point, line, or surface changes. Similarly it is said to be at rest when its posiiton in relation to some reference remains unchanged.

Performance tasks

1. Write out descriptions of kinetics and the states of rest and motion.
2. Identify three examples of objects "at rest" and three examples of objects "in motion" during sports activities.

Newton's laws of motion
Specific student objectives
1. Describe Newton's three laws of motion.
2. Illustrate Newton's three laws of motion with linear motion examples.

Summary

Isaac Newton formulated the fundamental laws governing all states of rest and motion. These laws focus upon the causal analysis of motion:
1. *Law of inertia:* An object at rest or in uniform motion has a tendency to remain either at rest or in uniform motion unless an external net force is applied to it.
2. *Law of acceleration:* When a body is acted upon by a new force (F), its resulting linear acceleration (a) is directly proportional to the net force and inversely proportional to its mass and takes place in the direction of the acting net force. In formula form this law is a = Net F/m.
3. *Law of reaction:* Whenever one body exerts an action force on another body, the second body exerts a reaction force on the first body. These forces are equal in magnitude and opposite in direction.

Performance tasks

1. State Newton's laws of linear motion in your own words.
2. Write out motor performance examples of the applications of Newton's three laws for linear motion as found in sports.

Concept of net force
Specific student objectives

1. Describe the concept of net force.
2. Differentiate between statics and dynamics.
3. Distinguish between states of linear equilibrium and disequilibrium.
4. Describe the effects of net forces.
5. Differentiate between tangential and radial forces in terms of both their directions and their effects.
6. State Newton's laws in terms of net tangential and net radial forces.

Summary

The net force acting on an object is the resultant of all forces affecting the object. If the net force is 0, the object is said to be in equilibrium. This condition is brought about by balanced forces, and its study is called statics. Linear equilibrium exists when an object is at rest or in linear motion with a constant speed. Dynamics is the branch of kinetics which examines changes in motion experienced by objects in disequilibrium, i.e., accelerations caused by unbalanced net forces.

A net force (net F) is the difference between an applied force (F) and a resistance force (R); i.e., Net F = F – R. Positive work is produced when F is greater than R, and either negative work or no work is produced when F is less than or equal to R.

A body in linear motion can be affected by a tangential force (the component of a force acting parallel with the linear path of the body), which changes its speed, and/or by a radial force (the component acting perpendicular to the linear path of motion), which changes its direction of linear motion.

Following are Newton's laws stated in terms of tangential and radial forces:

1. A body in motion will travel with a constant speed unless acted upon by a net tangential force, and in a straight line unless acted upon by a net radial force.
2. The rate of change in speed of a body in linear motion is directly proportional to the magnitude of the net tangential force and inversely proportional to its mass.
   The rate of change of direction of a body in linear motion is directly proportional to the magnitude of the net radial force and inversely proportional to its mass.
3. For every tangential and radial force component there is an equal and opposite reactive tangential and radial force component.

Performance tasks

1. The resultant or vector sum of all forces acting upon an object is called
   a. propulsion
   b. resistance
   c. weight
   d. none of the above
2. Illustrate states of equilibrium and disequilibrium with two human motion examples.
3. Zero acceleration or linear equilibrium occurs
   a. at rest
   b. in motion
   c. both a and b
   d. neither a nor b
4. Match the conditions at the left with the appropriate descriptions on the right:
   a. Equilibrium   Balanced forces
   b. Disequilibrium   Dynamics
                          Unbalanced forces
                          Statics
5. The two branches of kinetics are
   a. mass and weight
   b. kinematics and statics
   c. statics and dynamics
   d. none of the above
6. Write out descriptions for
   a. statics
   b. dynamics
   c. net force
   d. linear equilibrium
   e. kinetics
   f. zero acceleration
7. An applied force greater than the resistive force will produce (positive/negative)_____ _____ work.
8. A force component acting parallel with the linear path of a body is called
   a. radial
   b. resultant
   c. tangential
   d. all of the above
9. A tangential force component will change the (direction/speed)_____ of linear motion.

10. Describe radial and tangential force components.
11. Illustrate Newton's three laws of motion with a human motion example which includes radial and tangential force components.

Concepts of inertia and mass
Specific student objectives

1. Describe inertia.
2. Describe mass.
3. Identify a method of measuring mass.
4. Identify the British system unit for mass.
5. Differentiate between mass and weight.
6. Differentiate between inertia and a resistive force.

Summary

The resistance of an object to a change in its state of motion (acceleration) is called inertia. The quantity m in the equation net $F = ma$ is called mass or inertia and is described as a measure of the amount of matter comprising a body. With $g = 32$ ft/sec² it can be shown that $m = w/g$, where w equals the weight of an object. Mass is a measure of inertia while weight is the gravitational force with which an object is attracted toward the earth. In the British system weight is measured in pounds, and mass is measured in slugs or lb/(ft/sec²).

Mass and weight are proportional to each other. However, the mass of an object stays relatively the same while the weight, a changing quantity, depends upon the location of the object within a gravitational field. Mass is a scalar quantity; weight is a vector quantity.

Inertia is the property of an object which resists motion change. It must be separated from the concept of resistive forces, which are actual forces exerted in opposition to applied forces.

Performance tasks

1. The resistance of an object to changes in its state of motion is called
   a. acceleration
   b. inertia
   c. weight
   d. velocity
2. Weight is a (scalar/vector)_____ quantity.
3. The measure of inertia is the quantity called
   a. mass
   b. weight
   c. both a and b
   d. neither a nor b
4. The force of attraction between an object and the earth is its
   a. acceleration
   b. mass
   c. matter
   d. weight
5. Mass and weight are (directly/inversely) _____ proportional to each other.
6. Mass is a quantity that is (changing/constant)_____.
7. Actual forces exerted in opposition to applied forces are called
   a. propulsive
   b. resistive
   c. both a and b
   d. neither a nor b
8. Identify two human motion examples of resistive forces.
9. The British system unit for mass is
   a. dyne
   b. newton
   c. pound
   d. slug
10. Weight depends on the gravitational field an object is located in as well as on its (mass/position/mass and position)_____ within that field.
11. A person's weight would be (different/the same)_____ on the earth and on the moon.
12. Differentiate between inertia and a resistive force.
13. Develop a written list of descriptions for the following:
    a. Inertia
    b. Mass
    c. Weight
    d. Resistive force
    e. Net force
    f. Acceleration due to gravity
    g. Units of measure for mass and weight

Introduction to the principles of statics
Specific student objectives

1. Describe Newton's first and third laws in terms of the principles of statics.
2. State the two conditions of equilibrium.

Summary

Statics, the branch of kinetics that deals with states of balanced forces, is concerned with the conditions of equilibrium or zero acceleration.

Newton's first and third laws can be stated in static terms as follows:
1. Any object remaining at rest, or moving with uniform motion, is in linear equilibrium and all forces acting on it have a vector sum of zero (Newton I).
2. When an object is in equilibrium, every action force applied on it is balanced by an equal and opposite reaction force applied by the object (Newton III).

The two conditions of equilibrium are *linear* (the sum of all rectangular components of forces acting upon an object equals zero) and *rotatory* (the vector sum of all torques acting upon the object equals zero).

Performance tasks

1. State two examples of objects related to human motion in a state of zero acceleration. Use Newton's laws to explain these static states.
2. Explain Newton's law of reaction by applying it to a classroom desk that is in equilibrium.
3. The two conditions of equilibrium are
   a. action-reaction
   b. angular-linear
   c. both a and b
   d. neither a nor b
4. Describe linear and angular equilibrium.

Introduction to the principles of dynamics
Specific student objectives

1. Describe Newton's three laws in terms of the principles of dynamics.
2. Differentiate between balanced and unbalanced forces.

Summary

Dynamics is that branch of kinetics dealing with objects in a state of disequilibrium. This condition exists when net forces and torques acting upon an object are not equal to zero. Newton's second law describes the linear acceleration resulting from these unbalanced forces. Newton's first law is a special case of the second law when the net force is zero resulting in zero acceleration.

Finally, Newton's third law describes the action-reaction force pairs that produce linear acceleration.

Forces are balanced when their algebraic sum is equal to zero and they are unbalanced when their algebraic sum is not equal to zero.

Performance tasks

1. The state of disequilibrium brought about by unbalanced forces is examined in
   a. dynamics
   b. kinetics
   c. statics
   d. all of the above
2. Unbalanced forces cause
   a. accelerations
   b. equilibrium
   c. both a and b
   d. neither a nor b
3. Show that Newton's first law is a special case of his second law, net F = ma.
4. Explain and illustrate how action-reaction force pairs can produce accelerations.
5. Forces with an algebraic sum of zero are said to be (balanced/unbalanced)_____.

SELF-EVALUATION

Students should use no reference materials for this progress test, and they can check their answers by referring to Appendix A.

1. The causal analysis of movements involved in a motor performance is called
   a. dynamics
   b. kinetics
   c. kinematics
   d. statics
2. The two branches of kinetics are
   a. dynamics-statics
   b. kinematics-mechanics
   c. both a and b
   d. neither a nor b
3. States of "at rest" and "in motion" are (mutually exclusive/overlapping)_____.
4. The tendency of an object to resist changes in its existing state of motion is called
   a. equilibrium
   b. inertia
   c. resistive force
   d. weight
5. "An object at rest or in uniform linear mo-

tion will continue in its existing state unless acted upon by an external net force." This is a statement of Newton's law of (acceleration/inertia/reaction)_____.

6. Illustrate each of Newton's three laws of motion with three specific motion examples.

7. An object is in equilibrium when all net forces and torques are
   a. less than zero
   b. equal to zero
   c. greater than zero
   d. none of the above

8. List the two conditions for equilibrium.

9. The causal analysis of situations in which there are balanced forces is the concern of
   a. dynamics
   b. kinematics
   c. both a and b
   d. neither a nor b

10. State Newton's three laws in your own words.

11. Mass is a (scalar/vector)_____quantity.

12. The measure of gravitational force acting on a body is the quantity called_____
   a. mass
   b. weight
   c. both a and b
   d. neither a nor b

13. Weight is a (constant/changing)_____ quantity.

14. Differentiate between resistance and inertia.

15. The forces of friction and gravity are examples of
   a. propulsion
   b. resistance
   c. both a and b
   d. neither a nor b

16. The British unit for weight is (pound/slug) _____ .

17. Describe a net force.

18. An object at rest or in linear motion at a constant or unchanging velocity is in a state of
   a. linear equilibrium
   b. zero acceleration
   c. both a and b
   d. neither a nor b

19. Unbalanced forces cause linear
   a. acceleration
   b. equilibrium
   c. both a and b
   d. neither a nor b

20. A force component acting perpendicular to the linear path of a body is termed
   a. radial
   b. resultant
   c. tangential
   d. all of the above

21. A radial force component will change the (direction/speed)_____of linear motion.

22. State Newton's three laws of motion in terms of radial and tangential force components.

23. Complete the following conceptual diagram:

KINETICS
Causal analysis of motion

Statics		Dynamics

LESSON 18

Kinetic analysis of energy and momentum

Body of lesson

KINETIC ANALYSIS OF WORK AND MECHANICAL ENERGY

In Lesson 14 the concepts of work, energy, and power were discussed. Energy was defined as the ability to do work. When we say that something has energy, we mean it is capable of exerting a force and doing work on something else. Remember that when we do work on something we have added to it an amount of energy equal to the work done. The units of energy are the same as those of work: the foot-pound in the British system and the erg and joule in the metric system, as will be discussed at the end of this lesson.

What properties can a body have that may be converted into work? In other words, what forms does energy take? We shall consider in this lesson two broad categories of mechanical energy: kinetic energy, which is the energy of motion, and potential energy, which is the energy of condition or position. As discussed in Lesson 14, there are other varieties of energy—heat, electrical, magnetic, chemical, and nuclear—as well as mass energy, the type possessed by a body due to its mass alone. Our present discussion, however, will be limited to external or mechanical energy.

Analysis of kinetic energy

When we perform work on a ball by throwing it, what becomes of this work? Suppose we apply a uniform force (of magnitude F) to the ball for a distance (d) before it leaves our hand (Fig. 18-1). If the ball's mass is m, the second law of motion (F = ma) tells us that the acceleration of the ball while it is in our hand is a = F/m. We have already learned (Lesson 11) that when a body starting from rest undergoes an acceleration (a) through a distance (d) its final velocity (V) is related to a and d by the formula $V^2 = 2ad$. Inserting for a the value F/m from Newton's second law, we obtain $V^2 = 2Fd/m$. Rearranging this equation, we obtain

$$Fd = \tfrac{1}{2}mV^2$$

(Equation 18-A)

The quantity on the left, *Fd,* is the work our hand has done in throwing the ball. The quantity on the right, $\tfrac{1}{2}mV^2$, must therefore be the energy acquired by the ball as a result of the work we did on it. Accordingly we define kinetic energy of a moving body as $\tfrac{1}{2}mV^2$; or, in other words, kinetic energy is equal to half the product of the mass of the body and the square of its

300

Fig. 18-1. The work done on a ball being thrown is the product of the acting force and the distance through which the force is applied while in the hand of the thrower. (Modified from Cooper, J. M., and Glassow, R. B.: Kinesiology, ed. 4, St. Louis, 1976, The C. V. Mosby Co.)

velocity. A moving body is able to perform an amount of work equal to $\frac{1}{2}mV^2$ in the course of being stopped. Thus the more massive a running back in football and the faster he is moving, the more difficult he is to tackle through his center of gravity. Remember, in this example (and in this lesson), we are dealing only with linear motion situations. The effects of tackling the runner at his ankles and thereby creating a force couple (Fig. 13-8), in which his forward kinetic energy actually works to the advantage of the tackler, will be considered in the lessons on angular kinetics.

We all know, however, that objects with great mass and moving with great speed hit harder than do objects with less mass and moving slower. Thus the force of impact or the force with which one object can strike another is affected by the kinetic energy of the striking object. This force of impact can be determined through the use of Equation 18-A. By moving the displacement (d) to the right side, we obtain:

$$F = \frac{\frac{1}{2}mV^2}{d}$$

(Equation 18-B)

This equation tells us that the force of impact a moving object can exert in an impact situation is directly proportional to its kinetic energy and inversely proportional to the distance through which it is applied (reduced or absorbed in this situation). We have probably all had the experi-

ence of trying to catch a fast-moving baseball with our hand. We know that the faster the baseball is moving the greater its kinetic energy is and the harder it will hit our hand, and that in order to prevent damage to our hand we must *give* with the ball. By giving with the ball, we increase the distance through which its kinetic energy is absorbed and thereby reduce the force of impact.

This topic is so important it will be studied in greater detail in Lesson 30, which is devoted to a kinetic analysis of impact and rebound forces.

Analysis of potential energy

In Lesson 14 we saw that there are actually several forms of potential energy. Elastic potential energy and gravitational potential energy (Fig. 18-2) are the types with which we are most concerned in this text.

The stretching of a trampoline, for instance, imparts to the trampoline the *elastic* potentiality of doing work once the stretching force has been removed, as we shall see in Lesson 19.

When we lift a body above the ground, we give it the *gravitational* potentiality of doing work once the lifting or holding force is removed and the body is allowed to fall. The force acting in this case is the force of gravity, which is equal to the weight of the body (w); and, as we saw in the last lesson, w = mg. The work done on the falling body is equal to the gravitational potential energy (G.P.E.) and is the product of the body's weight and the distance it falls (h):

$$\text{Work} = \text{Fd} = \text{wh} = \text{mgh} = \text{P.E.}$$

(Equation 18-C)

The amount of work the body can do when it strikes the ground after dropping from height h also equals *mgh*. Thus we can see that the body at its original position *(h)* above the ground has the capacity to do work even though it is stationary at the time. It must only be released and allowed to drop. This capacity of a body to do work by virtue of its position has been defined as its *gravitational potential energy* and is equal to *mgh;* i.e., G.P.E. = P.E. = mgh.

Conservation of mechanical energy

We have seen thus far that we can do work on something and thereby give it kinetic or potential energy, which, in turn, gives it the ability to do work. When we throw a ball into the air, the

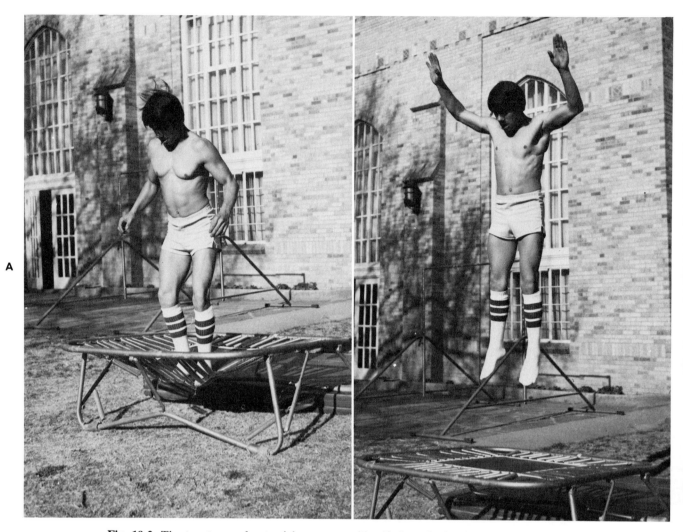

Fig. 18-2. The two types of potential energy are identified as elastic **(A)** and gravitational **(B)**. The stretching of a trampoline bed imparts to the bed the elastic potential of doing work upon returning to the nonstretched position. Likewise, when a body is located a certain height above a surface it has the gravitational potential of doing work when it falls back to the surface.

work we do appears first as kinetic energy; and as the ball rises, this kinetic energy gradually becomes potential energy. At its highest point, the ball has potential energy exclusively. As it begins to fall, the potential energy changes back into kinetic energy; and when the ball strikes the ground, the kinetic energy does work.

Therefore the ball has 100% kinetic energy when first released at ground level. The kinetic energy is converted into 100% potential energy at the highest point of its rise, and this potential energy is converted again into 100% kinetic energy when the ball strikes the ground. The po-

tential energy at the top of the trajectory equals the kinetic energy at the bottom; that is, P.E. (at top) = K.E. (at bottom) or Fd or mgh at top = $\frac{1}{2}mV^2$ at bottom.

When the falling ball is a quarter of the way down during its fall, it still has three fourths of its original potential energy while the other one fourth has been converted into kinetic energy. When the ball is three quarters of the way down, only a fourth of its original potential energy remains, the other three fourths having been converted into kinetic energy. When the ball reaches the ground, 100% of its energy is again kinetic.

At the instant it reaches the ground, it is suddenly stopped and all its kinetic energy is quickly transformed into an equal amount of heat. Thus, as we have already discussed, the total mechanical energy contained in a body is the sum of its kinetic and potential energies.

$$\text{M.E.} = \text{K.E.} + \text{P.E.} = \tfrac{1}{2}mV^2 + mgh = Fd = \text{Work}$$

(Equation 18-D)

KINETIC ANALYSIS OF MOMENTUM AND IMPULSE

The quantity or amount of motion in a moving body is called momentum. The concept of momentum can be derived from a knowledge of mass and Newton's second law.

Remembering that acceleration is the rate of change of velocity and is written algebraically as $a = (V_f - V_o)/t$, we can replace *a* by $(V_f - V_o)/t$ in the force equation ($F = ma$) and obtain $F = m(V_f - V_o)/t$. Multiplying out, we get $F = (mV_f - mV_o)/t$. This equation introduces us to the concept of momentum. Momentum is defined as the "quantity of motion" and is the product of mass times velocity (mV).

$$\text{Momentum} = M = mV$$

(Equation 18-E)

According to this definition all moving bodies have momentum.

By using momentum rather than velocity, we can change Newton's second law to say the rate of change of momentum of a body is proportional to the applied net force and takes place in the direction of that force. The relationship $mV_f - mV_o$ is defined as the change in momentum. By rearranging the force equation expressed in terms of momentum and by moving time *(t)* to the left side of the equation, we obtain

$$Ft = mV_f - mV_o = M_f - M_o$$

(Equation 18-F)

This is called the impulse equation, with *Ft* being the impulse and $mV_f - mV_o$ being the change in momentum produced by the impulse. Since impulse and momentum are vector quantities, they require the description of both magnitude and direction in order to be completely specified.

From Equation 18-F we can see that the change in the momentum of an object is a product of the force and the time during which the force is applied. Therefore the greater the force and the greater the time during which the force acts, the greater will be the change in momentum. The importance of this principle becomes evident when we realize that nearly all sport activities involve the constant changing of momentum of the body or of external implements such as rackets, bats, and balls. In some situations the goal of the performance is to *increase* momentum (e.g., sprinting and shot-putting), while in others the goal is to *reduce* momentum (e.g., landing from a fall or catching a baseball). In either case the principle is the same: the magnitude of the momentum change is directly proportional to the magnitude of the applied impulse (Ft). This important principle will be further discussed in many of the remaining lessons of the text.

Conservation of momentum

When two or more bodies exert forces on each other and we wish to consider them as a single system, the forces acting between the bodies are defined as *internal* forces. No matter what interactions take place between the forces which operate within a system, the total momentum of the system never varies. This theorem is known as the conservation of linear momentum. More formally the theorem states:

> When the vector sum of the external forces acting upon a system of particles equals zero, the total linear momentum of the system remains constant.

Consider a specific example readily treated with the help of the conservation of momentum theorem. Suppose we have a hand grenade (Fig. 18-3) of mass *m*, initially at rest, which suddenly explodes into many particles of masses m_1, m_2, m_3 . . . m_n that fly apart. The forces acting on the hand grenade causing it to break up are internal forces; no external force is present. Since the hand grenade has an initial momentum of zero, the final momentums of the many smaller masses, when added together, must also be zero. In other words, momentum as a vector quantity has directional values, which in this instance cancel out to leave a net momentum of zero.

One of the principles underlying rocket flight is the conservation of momentum. The total momentum of a rocket on its launching pad is zero. When it is fired, exhaust gases shoot downward at high speed and the rocket moves upward to

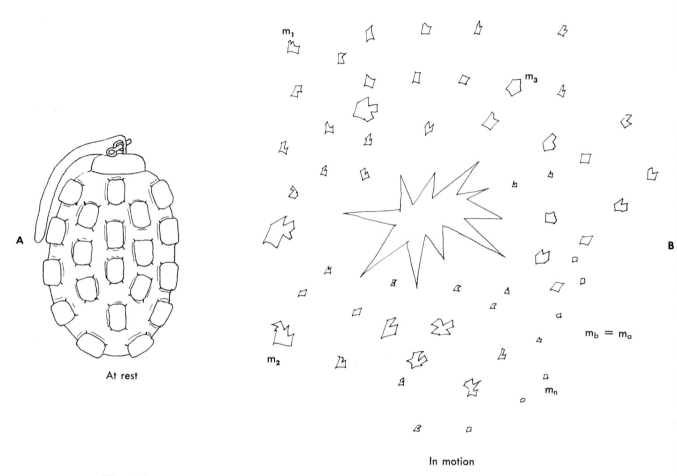

Fig. 18-3. A hand grenade before explosion **(A)** has an initial momentum of zero. The final momentums of the small masses after explosion **(B)** when added together are also equal to zero. Thus the momentum (and the mass) before explosion (m_b) is equal to the momentum (and mass) after explosion (m_a).

balance the momentum of the gases. The total momentum of the system of rocket plus exhaust remains at the initial value of zero. Rockets do not operate by "pushing" against their launching pads, the air, or anything else; in fact, they perform best in space, where there is no atmosphere to impede their motion. The energy of the rocket and its exhaust comes from the chemical energy stored in its fuel.

Analysis of collision situations

When two or more bodies collide with each other, momentum is conserved. Since no external forces act on the bodies, the law of conservation of momentum applies to all collision phenomena and says that the total momentum before impact equals the total momentum after impact. The essential effect of the collision is to redistribute the total momentum of the particles.

An example of a collision problem might be helpful. Consider the head-on encounter of a bat and a ball as shown in Fig. 18-4. Before impact the bat (with mass m_1) is moving with a velocity V_1 and has a momentum m_1V_1 while the ball (with m_2) is moving with a velocity V_2 and has a momentum m_2V_2. The total momentum before impact is therefore equal to the sum of the two, $m_1V_1 + m_2V_2$.

By similar reasoning it is clear that after impact, m_1 and m_2 with their new velocities V_3 and V_4 have a total momentum $m_1V_3 + m_2V_4$. The law of conservation of momentum requires that

Fig. 18-4. The sum of the momentums of a bat and ball before impact $(m_1V_1 + m_2V_2)$ is equal to the sum of their momentums after impact $(m_1V_3 + m_2V_4)$.

$$m_1 V_1 + m_2V_2 = m_1V_3 + m_2V_4$$

Momentum before impact = Momentum after impact

(Equation 18-G)

In a bat-and-ball impact situation the momentum of the ball (actually the velocity) is usually increased, while that of the bat is decreased. In order for the sum of the momentums of the bat and ball after impact to equal the sum of their momentums before impact, obviously the increase in the momentum of the ball must be exactly equal to the decrease in momentum of the bat.

You might wonder how elasticity of the bodies impacting affects the law of conservation of momentum. For example, what happens in an impact situation when two lumps of clay collide? Before analyzing this situation, keep in mind the directional character of momentum. In adding up the individual momentums (mV), you must consider the directions of the velocities and not merely add them algebraically. Sometimes the problem under consideration involves motion along a straight line, as in the example above; but in general it will be in two or three dimensions, and you must be sure to take this into account by a vector calculation.

We might consider what effect elasticity has on the law of conservation of momentum. Suppose that a 5-slug lump of clay is moving with a velocity of 10 ft/sec to the left and strikes a 6-slug lump of clay moving with a velocity of 12 ft/sec to the right and that the two lumps stick together after the collision (Fig. 18-5). We will call the

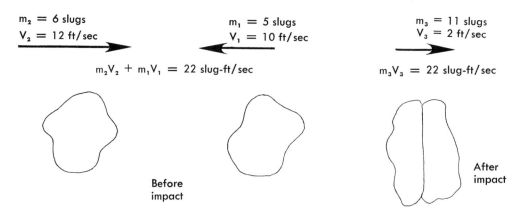

$m_2 = 6$ slugs
$V_2 = 12$ ft/sec

$m_1 = 5$ slugs
$V_1 = 10$ ft/sec

$m_3 = 11$ slugs
$V_3 = 2$ ft/sec

$m_2V_2 + m_1V_1 = 22$ slug-ft/sec

$m_3V_3 = 22$ slug-ft/sec

Before
impact

After
impact

Fig. 18-5. The sum of momentums of two lumps of clay before impact is equal to the sum of their momentums after impact.

mass of the final body m and its velocity V. The momentum before collision $(m_2V_2 + m_1V_1)$ must equal the momentum after collision (m_3V_3), so we proceed to determine the two momentums.

As mentioned in Lesson 9, the convention has been adopted that motion to the left is negative and motion to the right is positive. Thus

$$m_1V_1 = (5 \text{ slugs}) (\text{-}10 \text{ ft/sec}) = \text{-}50 \text{ slug-ft/sec}$$
$$m_2V_2 = (6 \text{ slugs}) (\ 12 \text{ ft/sec}) = \underline{\ \ 72 \text{ slug-ft/sec}}$$
$$22 \text{ slug-ft/sec}$$

Combining these two, we get 22 slug-ft/sec.

In the determination of mV, quite obviously m times V will equal 11 slugs times $(-10 + 12 =)$ 2 ft/sec, which comes to 22 slug-ft/sec also. Thus elasticity has no effect on the conservation of momentum since momentum after impact equals momentum before impact.

Relationship of energy and momentum

Energy and momentum are entirely independent concepts. In the above problem the lumps of clay before collision had the kinetic energies

$$\text{K.E.}_1 = \tfrac{1}{2}m_1V_1^2 = \tfrac{1}{2}(5 \text{ slugs}) (\text{-}10 \text{ ft/sec})^2 =$$
$$250 \text{ slug-ft}^2/\text{sec}^2 = 250 \text{ ft-lb} = \text{Work}$$

$$\text{K.E.}_2 = \tfrac{1}{2}m_2V_2^2 = \tfrac{1}{2}(6 \text{ slugs}) (12 \text{ ft/sec})^2 =$$
$$432 \text{ slug-ft}^2/\text{sec}^2 = 432 \text{ ft-lb}$$

After the collision the new lump of clay had the kinetic energy

$$\text{K.E.}_3 = \tfrac{1}{2}m_3V_3^2 = \tfrac{1}{2}(11 \text{ slugs}) (2 \text{ ft/sec})^2 =$$
$$22 \text{ slug-ft}^2/\text{sec}^2 = 22 \text{ ft-lb}$$

The total kinetic energy before the collision was $432 + 250 = 682$ ft-lb while afterward it was only 22 ft-lb. The difference of 660 was dissipated largely as heat energy in the collision, with some probably being lost to sound energy as well.

While conservation of *energy* is an excellent guide in situations that do not involve the dissipation of mechanical energy into other forms, conservation of *momentum* must be employed whenever there are situations involving several bodies interacting with one another. In many situations, however, both laws must be used together in order to understand certain processes, as we will discuss in Lesson 30, which deals with the kinetic analysis of impact and rebound forces.

UNITS OF LINEAR MEASURE

Now that we have introduced Newton's laws and the concepts of energy and work, it is important to see how these laws and concepts are used to define the units of linear measure in the British and metric systems.

Units of force, mass, and acceleration

In the British or *foot-pound-second* (fps) system of units, in which forces are expressed in pounds and acceleration in feet per second squared, the unit of mass is the *slug*, equal to 1 lb/ft/sec².

In the *meter-kilogram-second* (mks) system of units, the standard kilogram is chosen to be the unit of mass. The unit of force in this system is that which gives a standard kilogram an acceleration of 1 m/sec.² This force is called 1 *newton* (n) and is approximately equal to ¼ pound of force (more precisely 0.22481 lb).

Table 18-2. Units of kinetic energy

System of units	Expressed in mass units $(\frac{1}{2}mV^2)$	Expressed in force units $\left(\dfrac{\frac{1}{2}wV^2}{g}\right)$	Expressed in work units
British (fps)	1 slug $\dfrac{ft^2}{sec^2}$	1 $\dfrac{lb}{ft/sec^2}\dfrac{ft^2}{sec^2}$	1 ft-lb
mks	1 kg $\dfrac{m^2}{sec^2}$	1 $\dfrac{n}{m/sec^2}\dfrac{m^2}{sec^2}$	1 joule
cgs	1 gm $\dfrac{cm^2}{sec^2}$	1 $\dfrac{dyne}{cm/sec^2}\dfrac{cm^2}{sec^2}$	1 erg

Table 18-1. Units of force, mass, and acceleration

System of units	Force	Mass	Acceleration
British (fps)	Pound (lb)	Slug	ft/sec^2
mks	Newton (n)	Kilogram (km)	m/sec^2
cgs	Dyne	Gram (gm)	cm/sec^2

Table 18-3. Conversion tables for the units of work and energy

1 foot-pound $= 3.2389 \times 10^{-4}$ Kcal	1 kilogram-meter $= 2.3427 \times 10^{-3}$ Kcal
1 foot-pound $= 1.35582 \times 10^7$ ergs	1 kilogram-meter $= 9.8066 \times 10^7$ ergs
1 foot-pound $= 1.3558$ joules	1 kilogram-meter $= 9.8066$ joules
1 foot-pound $= 1.2854 \times 10^{-3}$ BTU	1 kilogram-meter $= 9.2967 \times 10^3$ BTU
1 foot-pound $= 0.13825$ kpm	1 kilogram-meter $= 7.2330$ ft-lb
1 foot-pound $= 5.0505 \times 10^{-7}$ hp-hr	1 kilogram-meter $= 3.6529 \times 10^{-6}$ hp-hr
1 joule $= 2.3889 \times 10^{-4}$ Kcal	1 kilocalorie $= 4.186 \times 10^{10}$ ergs
1 joule $= 1 \times 10^7$ ergs	1 kilocalorie $= 4186$ joules
1 joule $= 9.4805 \times 10^{-4}$ BTU	1 kilocalorie $= 3.9680$ BTU
1 joule $= 0.73756$ ft-lb	1 kilocalorie $= 3087.4$ ft-lb
1 joule $= 0.10197$ kpm	1 kilocalorie $= 426.85$ kpm
1 joule $= 3.7251 \times 10^{-7}$ hp-hr	1 kilocalorie $= 1.5593 \times 10^{-3}$ hp-hr
1 erg $= 2.3889 \times 10^{-11}$ Kcal	1 BTU $= 0.25198$ Kcal
1 erg $= 1 \times 10^{-7}$ joule	1 BTU $= 1.0548 \times 10^{10}$ ergs
1 erg $= 9.4805 \times 10^{-14}$ BTU	1 BTU $= 1054.8$ joules
1 erg $= 7.3756 \times 10^{-8}$ ft-lb	1 BTU $= 777.98$ ft-lb
1 erg $= 1.0197 \times 10^{-8}$ kpm	1 BTU $= 107.56$ kpm
1 erg $= 3.7251 \times 10^{-14}$ hp-hr	1 BTU $= 3.9292 \times 10^{-4}$ hp-hr

In the *centimeter-gram-second* (cgs) system the mass unit is 1 gram, which is equal to 1/1000 kilogram. The force unit in this system is that which gives a body of 1 gm mass an acceleration of 1 cm/sec^2; it is called 1 *dyne*.

The units of force, mass, and acceleration in the three systems are summarized in Table 18-1.

Units of work and energy

In the British (fps) system, work (Fd) is measured in foot-pounds. One foot-pound may be defined as the work done when a constant force of 1 lb is exerted on a body and moves the body a distance of 1 ft in the same direction as the force.

In the mks system, where forces are expressed in newtons and distances in meters, the unit of work is the *newton-meter* (n-m). One newton-meter is called a *joule*.

In the cgs system the unit of work is the *dyne-centimeter* (dyne-cm). One dyne-centimeter is called an *erg*.

Another unit commonly used in the measurement of mechanical work is the kilopond-meter (kpm): 1 kilopond (kp) is the force acting on the mass of one kilogram at normal acceleration of gravity.

$$1 \text{ kp} = 9.80665 \text{ newtons (n)}$$
$$1 \text{ kpm} = 9.80665 \text{ joules (j)}$$

Heat energy is measured in calories; 1 calorie is the amount of heat required at a pressure of

Table 18-4. Conversion tables for the units of power

1 watt	=	0.001 kilowatt
1 watt	=	0.73756 ft-lb/sec
1 watt	=	1×10^7 ergs/sec
1 watt	=	0.056884 BTU/min = 3.41304 BTU/hr
1 watt	=	0.01433 Kcal/min
1 watt	=	1.341×10^{-3} hp
1 watt	=	1 joule/sec
1 watt	=	6.12 kpm/min

1 atmosphere (760 mm Hg) to raise the temperature of 1 gm of water 1 degree centigrade (from 15° to 16°). This unit is also called a small calorie. One kilocalorie (Kcal), large calorie, equals 1000 small calories. A British thermal unit (BTU) is equal to 252 Kcal.

Kinetic energy in any system is equal to the unit of work in that system, and thus it is customarily expressed in foot-pounds, joules, or ergs. The relationships between these measures are shown in Table 18-2.

The conversion tables for the units of work and energy are given in Table 18-3.

Units of power

In the British (fps) system, power is measured in foot-pounds per second and in horsepower.

In the metric system, with work measured in joules or ergs, power is expressed in either joules per second or ergs per second. One joule per second is a watt. The kilowatt, another unit of power, is equal to 1000 watts.

The conversions for the units of power are given in Table 18-4.

STUDY GUIDELINES
Kinetic analysis of work and mechanical energy
Specific student objectives

1. Define energy.
2. Identify the measurement units for work and energy in the British system.
3. Identify and describe the two types of external mechanical energy.
4. Identify and define the formula for kinetic energy.
5. Identify the two variables that determine kinetic energy. State their relationship to kinetic energy.
6. Identify the two types of potential energy.
7. Define gravitational potential energy (in sentence and formula form).
8. Describe the law of energy conservation applied to mechanical energy.

Summary

Energy, the ability to do work, is measured in the British system in foot-pounds. Work is accomplished by a force (F) moving an object through a distance (d); i.e., W = F × d. The relation between the kinetic energy of a moving body produced when a force moves the body through a distance is shown by the equation $F \times d = \frac{1}{2}mV^2$. Thus kinetic energy (K.E.) equals half the product of the object's inertia or mass and the square of its final velocity.

Potential energy (P.E.) is described as the ability of an object to do work by virtue of its state or position. Gravitational potential energy, caused by the force of gravity acting on an object, is equal to mgh = wh, where *w* is the weight of the object and *h* is the height of the object above the earth's surface. Elastic potential energy is the mechanical energy possessed by an object because of its elastic restoring forces and depends upon the nature of the material making up the object.

Mechanical energy also obeys the law of the conservation of energy, which states that energy is neither created nor destroyed but only transformed to another form. Thus the total mechanical energy (K.E. + P.E.) in a motion situation remains constant unless it is changed into another form; i.e., $M.E. = \frac{1}{2}mV^2 + mgh$.

Performance tasks

1. The mechanical energy of an object in motion depends upon the variables of
   a. force and distance
   b. mass and velocity
   c. force and velocity
   d. force and acceleration
2. The ability to do work is
   a. acceleration
   b. energy
   c. inertia
   d. torque
3. The measurement units for work and energy in the British system are
   a. foot-pounds

b. meter-kilograms
c. both a and b
d. neither a nor b

4. The formula for kinetic energy (K.E.) is
   a. $\frac{1}{2} mV^2$
   b. mgh
   c. F × t
   d. A.M.A./T.M.A.

5. Differentiate between elastic and gravitational potential energy and illustrate each with an example.

6. Kinetic energy depends directly upon an object's
   a. mass divided by two
   b. velocity squared
   c. both a and b
   d. neither a nor b

7. The two types of potential mechanical energy are
   a. kinetic and static
   b. gravitational and elastic
   c. action and reaction
   d. positive and negative

8. State the meaning of P.E. = mgh.

9. Gravitational potential energy depends upon
   a. height and weight
   b. mass and velocity
   c. both a and b
   d. neither a nor b

10. Apply the law of mechanical energy conservation, using M.E. = K.E. + P.E. to the human motor task of high jumping.

Kinetic analysis of momentum and impulse

Specific student objectives

1. Define momentum.
2. State Newton's second law in terms of momentum.
3. Define impulse (in both sentence and formula form).
4. State the conservation of linear momentum law.
5. Identify the external variables influencing change in momentum.
6. State the relationship between energy and momentum.

Summary

Momentum (M) is defined as "the quantity of motion" of an object with mass m and velocity V; i.e., $M = mV$. An expression of Newton's second law in momentum terms is $F = (mV_f - mV_o)/t$. Thus the rate of change of momentum is proportional to the net force and takes place in the direction of that force. This formula can be rearranged to obtain $Ft = M_f - M_o$, where *Ft* is the impulse causing a change in the momentum *($M_f - M_o$)* of an object. The change in momentum is directly dependent upon the net force and the time during which it is applied. Impulse and momentum are both vector quantities.

The law of conservation of linear momentum states that "when the vector sum of the external forces acting upon a system of particles equals zero the total linear momentum of the system remains constant"; i.e., linear momentum is neither created nor destroyed but only transferred during collisions. The total momentum before impact equals the total momentum after impact.

Performance tasks

1. Identify an object in motion and state how its momentum is described.

2. The two variables that make up momentum (M) are
   a. V and t
   b. m and w
   c. m and V
   d. V and a

3. Newton's second law of motion (net F = ma) can be stated in momentum terms as
   a. a = net F/m
   b. $Ft = M_f - M_o$
   c. M = mV
   d. None of the above

4. Change in momentum is caused by
   a. impulse
   b. mass
   c. velocity
   d. weight

5. Illustrate Newton's law of acceleration stated in momentum terms with a specific human motion example.

6. Illustrate impulse as the causal agent for momentum change.

7. Change in momentum is related (directly/inversely)_____ to the applied force (F) and the time (t) during which it is applied.

8. Impulse and momentum are (scalar/vector) _____ quantities.
9. Illustrate the conservation of momentum law with a specific impact situation as found in sports.
10. Energy and momentum are related as (dependent/independent)_____quantities.

Units of linear measure
Specific student objectives

1. Identify the units of force, mass, and acceleration in the British and metric systems of measurement.
2. Identify the units of work, energy, and power in the British and metric systems of measurement.

Summary

The basic units of force, mass, and acceleration are summarized below.

| | Metric system | | British system |
	(mks)	(cgs)	(fps)
Force	Newton (n)	Dyne	Pound (lb)
Mass	Kilogram (kg)	Gram	Slug
Acceleration	m/sec²	cm/sec²	ft/sec²

The basic units of work, energy, and power are summarized at the bottom of this page.

Performance tasks

1. In the British system, force is measured in pounds. The metric system uses the
   a. dyne and newton
   b. dyne and kilogram
   c. gram and kilogram
   d. gram and slug
2. Acceleration is always measured with one distance unit and (one/two/three)_____time unit(s).

3. In the British system, work is measured in
   a. calories   c. foot-pounds
   b. ergs   d. newton-meters
4. The metric units for work are
   a. calories and kilocalories
   b. ergs/sec and joules/sec
   c. newton-meter and dyne-meter
   d. newton-meter and dyne-centimeter
5. Power is measured in units of
   a. horsepower
   b. kilowatts
   c. watts
   d. all of the above
6. The units for heat energy are
   a. calorie
   b. BTU
   c. both a and b
   d. neither a nor b
7. Develop a written table of the basic units of measure in the British and metric measurement systems. Include the quantities of acceleration, force, heat energy, kinetic energy, mass, power, and work.

SELF-EVALUATION

Students should use no reference materials for this progress test, and they can check their answers by referring to Appendix A.

1. Energy is the ability to do (power/work) _____.
2. Units of measurement for work and energy are
   a. dyne-cm (erg)
   b. ft-lb
   c. newton-meter (joule)
   d all of the above
3. The two types of external mechanical energy are
   a. chemical and heat

| | Metric system | | British system |
	(mks)	(cgs)	(fps)
Work	Newton-meter (n-m) (1 n-m = 1 joule)	Dyne-centimeter (dyne-cm) (1 dyne-cm = 1 erg)	Foot-pound (ft-lb)
Heat energy	Kilocalorie	Calorie	British thermal unit (BTU)
Kinetic energy	Joule	Erg	Foot-pound
Power	Joule/sec (1 joule/sec = 1 watt)	Erg/sec	Foot-pounds/sec or horsepower (1 hp = 550 ft-lb/sec)

   b. kinetic and potential
   c. gravitational and elastic
   d. all of the above

4. The formula for kinetic energy is
   a. mgh   c. ma
   b. $M_f - M_o$ d. $\frac{1}{2}mV^2$

5. Describe the formula for kinetic energy and the variables that determine K.E.

6. Measurement units for power are
   a. horsepower   c. both a and b
   b. kilowatts   d. neither a nor b

7. The relationship between the kinetic energy of an object and its mass and velocity would be called (direct/inverse)_____ .

8. The two types of potential energy are
   a. elastic and gravitational
   b. kinetic and static
   c. both a and b
   d. neither a nor b

9. Define gravitational potential energy (in formula and sentence form).

10. M.E. = K.E. + P.E. is a formula statement for the law of conservation of
    a. energy   c. both a and b
    b. momentum   d. neither a nor b

11. The product of an object's mass (m) and velocity (V) is its
    a. kinetic energy   c. potential energy
    b. momentum   d. work

12. Newton's second law (acceleration) stated in terms of momentum is
   a. net F = ma
   b. net $F = (mV_f - mV_o)/t$
   c. M = mV
   d. P.E. = mgh

13. Impulse is the (causal/resistive)_____ agent involved in momentum changes.

14. Define impulse (in both formula and sentence form).

15. Impulse and momentum are (scalar/vector) _____ quantities.

16. "When the vector sum of the external forces acting upon a system of particles equals zero, the total linear momentum of the system remains constant" is a statement of the conservation law for
   a. energy   c. both a and b
   b. forces   d. neither a nor b

17. Complete the "unit of measure" table below:

	Metric (mks)	(cgs)	British (fps)
Force	_____	_____	_____
Mass	_____	_____	_____
Work	_____	_____	_____
Heat	_____	_____	_____
Power	_____	_____	_____

LESSON 19

Kinetic analysis of elastic harmonic motion

Body of lesson

INTRODUCTION

Shown in Fig. 19-1 is a spring with a certain weight suspended which causes the spring to be stretched or put under tension. If the spring is pulled out farther and then released, the weight bounces up and down. The importance of this kind of motion lies in the fact that it is produced whenever the distortion of an elastic body is released. This special motion is called vibratory or harmonic motion. The terms vibratory and harmonic are synonyms and refer to any back-and-forth motion that generally follows the same repetitive path. The motion of a dribbled basketball is an example of this type of motion.

The two types of harmonic motion are linear and angular. Linear harmonic motion occurs along a straight line, as seen in the bouncing spring illustrated in Fig. 19-1. Angular harmonic motion, on the other hand, occurs along the arc of a circle, as seen in the motion of a clock pendulum or a child swinging on a swing (Fig. 19-2). The topic of this lesson will be the analysis of linear harmonic motion. The topic of angular harmonic motion will be taken up in Lesson 24.

Concept of elastic harmonic motion

Most linear types of harmonic motion result from the elastic nature of the materials involved and their behavior when subjected to an alternate application and release of force. All matter is distorted more or less by the application of force and is furthermore characterized by a relative tendency to recover from such distortion with a given speed following the release of the force. This property of a body is called elasticity. In other words, the resistance of a body to distortion and the tendency of a body to restore itself after being distorted are called elasticity.

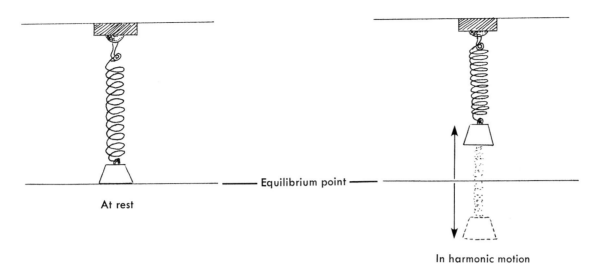

Fig. 19-1. A bouncing spring is an example of linear harmonic motion.

Fig. 19-2. A child swinging is an example of angular harmonic motion.

Fig. 19-3. For every distorting force tending to change the shape of an object, there is an equal and opposite restoring force.

MEASUREMENT OF ELASTICITY

When a force of any magnitude is applied to a solid body, the body becomes distorted. Whether the distortion is large or small, some portion of the body is moved with respect to a neighboring portion. As a result of this displacement, atomic forces of attraction set up restoring forces. These forces, in turn, resist the alteration and tend to restore the body to its original shape. The restoring forces are equal in magnitude to the distorting forces; but like all action and reaction forces, as predicted by Newton's third law, they act in opposite directions. Thus (as can be seen in Fig. 19-3) if a distorting force (F_1) acts downward on an object, it will create within the object an equal restoring force (F_2) that will be directed upward.

It is common practice in kinesiology to describe either the distorting force or the restoring force acting on a body as a *stress* and to give to this term the quantitative definition of force per unit of area. The actual deformation of the body produced by an applied force involves a change in geometrical form called *strain*. Thus strain is defined as a quantitative measure of deformation.

Definition of Hooke's law

If a vertically suspended rod, wire, or spring is supported rigidly at its upper end and weights are added to its lower end, the amount by which it is stretched is found to be proportional to the weights applied. This is known as Hooke's law.

The stretching of a spring is illustrated in Fig. 19-4. Due to an added weight *(w)* the spring is stretched a distance *(d)*. If a second equal weight is added, the total distance stretched will be twice that for the first weight. If a third weight is added, the total distance stretched will be three times that for the first weight, etc. This is illustrated by the graph shown at the right in Fig. 19-4. Each value of *d* is plotted vertically on the graph and the corresponding loads are plotted horizontally.

More specifically, when the first 10 gm weight is added, the stretch or elongation is 2 cm. With two 10 gm weights the total elongation is 4 cm. With three weights $d = 6$ cm, etc. A continuation of this shows, as does the graph at the right in Fig. 19-4, that each 10 gm weight produces an added elongation of 2 cm. To make an equation of this, we write w = Kd (where *K* is a constant, which in this experiment is equal to 5). Thus each value of *d* multiplied by 5 gives the corresponding weight *(w)*. When the spring in Fig. 19-4 is stetched a distance *d,* it exerts an upward restoring force (F_r) equal but opposite in direction to *w.*

$$\text{Restoring force} = F_r = Kd$$

(Equation 19-A)

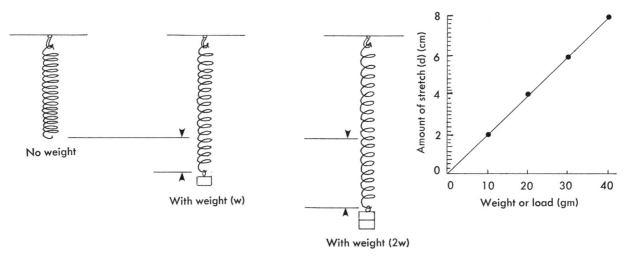

Fig. 19-4. When a vertically suspended spring is supported rigidly at its upper end and weights are added to its lower end, the amount by which it is stretched is found to be proportional to the weight applied. This is known as Hooke's law.

This equation is the formula for Hooke's law.

Hooke's law, as just described for the stretching of a spring, applies equally well to other types of deformation. In general, the law states that the deformation of a body is proportional to the applied force per unit of area or, in other words, the strain produced in a body is directly proportional to the applied stress.

Concept of elastic limit

The term elastic limit refers to the maximum deformation an object can undergo as the result of stress forces without being permanently altered. When this limit is exceeded, the object may not be far from breaking. Shown in Fig. 19-5 is a graph which relates the restoring force (F_r) to the displacement (d). The curve of the graph between points A and B is practically a straight line. Point B is often called the elastic limit of the spring. Between the two points A and B the restoring force developed in a body is directly proportional to its deformation (d). Beyond the straight line determined by A and B, the relation between F_r and d is not one of a simple direct proportion; i.e., the elastic limit is exceeded and a material breakdown occurs.

Within the elastic limit of a spring, the graph of F_r versus d is a straight line, indicating that the relationship of F_r and d is a direct proportion. Beyond the elastic limit the graph of F_r versus d is a curved line (the relation between F_r and d is not a direct proportion). Resilience is the ability of an object to undergo a deformation or compression without permanent deformation. In other words, an object is resilient only within its elastic limits.

Coefficient of elasticity

As already defined, the resistance of a body to deformation (or the speed of its resiliency) is called its elasticity. We might also call this elastic and resistive force the *force of restitution* or, in an impact situation, the *force of rebound* since the greater the elasticity of a body the greater will be its tendency when deformed to return to its original shape once the deforming force is removed. Upon returning to its original shape, the material of the body exerts a force that according to Newton's second law (F = ma) determines the velocities with which the bodies separate following collision.

The number that expresses the ratio of the velocity with which two bodies separate after collision to the velocity of their approach before collision is defined as the coefficient of elasticity or restitution (r). Thus

$$r = \frac{\text{Velocity of separation}}{\text{Velocity of approach}}$$

or

$$\frac{V_s}{V_a}$$

The r for a body is determined through use of this equation by dropping the body upon a hard object such as a steel anvil.

Rather than measure these velocities directly, we often make use of the laws of falling bodies. A collision is created by vertically dropping one body upon another. The two variables in this situation are the height (h) to which the body will rebound and the height (H) from which it is dropped.

Since $V^2 = 2gh$ for falling bodies, $\sqrt{2gH}$ can be written for V_a, and $\sqrt{2gh}$ can be written for V_s

$$r = \frac{V_s}{V_a} = \frac{\sqrt{2gh}}{\sqrt{2gH}} = \sqrt{\frac{h}{H}}$$

(Equation 19-B)

Therefore the coefficient of restitution (r) can also be stated as the square root of the ratio of the height (h) to which a body will rebound after collision to the height (H) from which it is dropped.

The coefficient of elasticity (r) for any given material varies somewhat with the velocity of

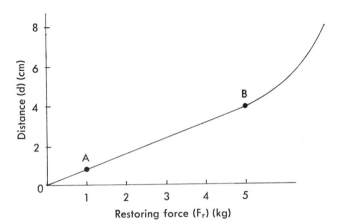

Fig. 19-5. A displacement versus restoring-force graph.

impact but is always between 0 and 1. The coefficient is highest at low velocities. All liquids and gases are highly elastic. Air is used in many inflated balls for games and sports and can cause the ball to become highly elastic. Liquid is used in the best golf balls as a core because of its high elasticity.

The elasticity of a basketball can be varied, of course, by increasing or decreasing the amount of air in it. Decreasing the amount of air in it will make it soft and easily deformed and will decrease the height to which it will rebound. Increasing the amount of air will, of course, have the opposite effect; both elasticity and speed of rebound will be increased.

Now that we understand the concept of elasticity, we will turn our attention to an analysis of the harmonic motion which is produced when the distortion of an elastic body is released.

ANALYSIS OF ELASTIC HARMONIC MOTION

Perhaps the analysis of elastic harmonic motion can best be understood by starting our discussion with an example. To fix our ideas, suppose a metal arrow is clamped vertically in a vise and a small body is attached to its upper tip as shown in Fig. 19-6. When we pull the top of the arrow to the right a distance (d), we establish an elastic restoring force in the arrow which is directed toward the left; i.e., we have established the condition necessary for the initiation of elastic harmonic motion.

When we release the arrow, it will be acted on by the restoring force which accelerates the body toward the equilibrium position (shown in Fig. 19-6 as the center of the vibration and labeled *0*). Thus it will move in toward the center with increasing speed. The rate of increase in speed (i.e., the acceleration) will not be constant since the accelerating force becomes smaller as the body approaches the center. When the body reaches the center, the restoring force has decreased to zero; but because of the velocity that has been acquired, the body overshoots the equilibrium position and continues to move toward the left. As soon as the equilibrium position is passed, the restoring force again comes into play directed now toward the right.

As the body moves past the equilibrium posi-

tion, the restoring force has the effect of decelerating the body at a rate which increases with increasing distance from the center. The body will be brought to rest at some point to the left of center and repeat its motion in the opposite direction. If there were no loss of energy by friction, the back-and-forth movement of the arrow would continue indefinitely once started.

Harmonic motion and the conservation of energy

The motions of the metal arrow occurring in our example represent the continuous conversion of the energy of position (potential energy) to the energy of motion (kinetic energy) and back. As a body in harmonic motion moves past its equilibrium position, it loses kinetic energy and gains potential energy. Likewise, as it moves toward its equilibrium position it gains kinetic energy and loses potential energy.

Measurement of work performed during harmonic motion

The amount of work done in deforming a body that obeys Hooke's law can be calculated. As we learned in Lesson 14, the work done by a force is the product of the force and the distance through which it acts; i.e., $W = Fd$. In harmonic motion, however, the force used to deform the body is not constant but is proportional to the deformation (d) at each point in the deforming process. The average force $(\overline{F})$ applied while a body is deformed from its normal shape by an amount (d) is $(F_o + F_f)/2$, where F_o and F_f are the original and final forces, respectively. Since the initial force is zero and the final force is Kd (from $F = Kd$, Hooke's law), the equation for average force becomes

$$\overline{F} = \frac{F_o + F_f}{2} = \frac{0 + Kd}{2} = \tfrac{1}{2} Kd$$

(Equation 19-C)

Thus the work done (Fd) is the product of the average force $\overline{(F)}$ and the total displacement *(d)*, so

$$\overline{W} = \overline{F}d = (\tfrac{1}{2}Kd)d = \tfrac{1}{2} Kd^2$$

(Equation 19-D)

Therefore we can see that to bend the arrow shown in Fig. 19-6, whose force constant is K, by an amount *(d)* from its normal position re-

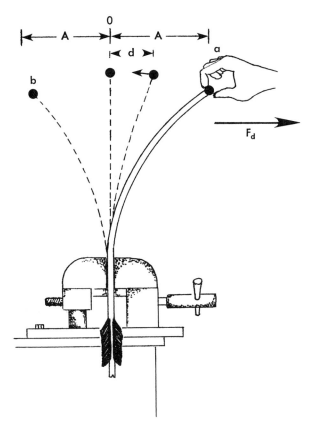

Fig. 19-6. The harmonic motion of a metal arrow clamped vertically in a vice with a small body attached to its upper tip.

quires $\frac{1}{2}Kd^2$ of work to be done. This work goes into elastic potential energy.

When the arrow is released, its potential energy ($\frac{1}{2}Kd^2$) is transformed into kinetic energy (remember that K.E. $= \frac{1}{2}mV^2$) or into work done on something else such as something it might strike. Work done against frictional forces within the arrow itself always absorbs some fraction of the available mechanical energy. Thus we see that during harmonic motion the work done by a deforming force ($W = \frac{1}{2}Kd^2$) goes into elastic potential energy, which in turn is transformed into kinetic energy. We also can note that the amount of potential energy created depends upon the force constant (K) and the total displacement (d).

Newton's laws and elastic harmonic motion

Newton's laws of motion can be applied to an analysis of harmonic motion. For example, his second law of motion (acceleration) was previously expressed as F = ma or a = F/m. In harmonic motion the force producing acceleration is the restoring force (F_r). Therefore F = ma becomes F_r = ma; and, since F_r = Kd, Newton II becomes Kd = ma and a = Kd/m.

Newton's three laws of elastic harmonic motion are as follows:

> NEWTON I. A body with a given shape will retain that shape and a body at rest will remain at rest unless a net distortion force acts upon it. If a body is in harmonic motion, it will remain in harmonic motion with equal displacements unless an unbalanced force acts upon it.
>
> NEWTON II. The acceleration of a body in returning to its equilibrium position or shape following distortion is directly proportional to the product of the force constant (K) of the body and its displacement (d) and is inversely proportional to its mass.
>
> $$a = Kd/m$$
> (Equation 19-E)
>
> NEWTON III. For every force tending to distort a body within its elastic limits, there is an equal and opposite restoring force produced in the body that will tend to return the body to its original shape or equilibrium position. In other words, for every action-distorting force acting upon a body there is created within the body an equal and opposite reaction-restoring force.

Analysis of harmonic motion quantities

Any sort of harmonic motion which repeats itself is also called periodic, and if the motion is back and forth over the same path, it is also called oscillatory. Periodic oscillatory motion such as that shown in Fig. 19-6 can be described through measurement of the quantities derived in the boxed material and listed below.

> The *periodic time* or simply the *period* of the motion (represented by *T*) is the time required for one complete oscillation. A complete oscillation is one round trip, say from *a* to *b* and back to *a* as shown in Fig. 19-6 or from *0* to *b* to *0* to *a* and back to *0*.
>
> $$T = 2\pi \sqrt{m/K}$$
> (Equation 19-1)
>
> The *frequency (f)* is the number of complete oscillations per unit of time. Evidently the frequency is the reciprocal of the period, or
>
> $$f = \frac{1}{T} \text{ or } f = \frac{1}{2\pi} \sqrt{\frac{K}{m}}$$
> (Equation 19-G)

Following are the definitions of symbols used in the drawing on the right:

A, Axis of rotation
r, Radius of rotation
θ, Angular displacement
Q, A point revolving in a circle
P, A point on the horizontal diameter directly below or above Q
d, Horizontal distance of P to A

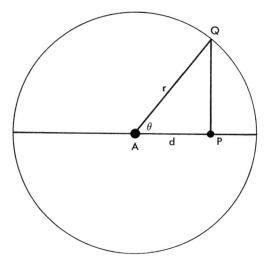

As can be seen in the figure, d is the side adjacent to θ in a right triangle and thus is given by

$$d = r \cos \theta$$

(Equation 1)

If point *Q,* revolving about *A* with a constant angular velocity (ω), is at the extreme right end of the diameter (i.e., at t = 0), the angle θ is given by

$$\theta = \omega t$$

(Equation 2)

This is, of course, a rearrangement of the angular velocity equation ($\omega = \theta/t$).

Equation 2 can be substituted into Equation 1 to obtain

$$d = r \cos \omega t$$

(Equation 3)

Since θ equals the number of radians in a movement and there are 2π radians in one revolution, $\theta = 2\pi n$, where *n* is the number of revolutions. Substituting $2\pi n$ for θ in the angular velocity equation gives $\omega = 2\pi n/t$. Since n/t is equal to the frequency (f), we can substitute f to obtain

$$\omega = 2\pi f$$

(Equation 4)

Substituting Equation 4 into Equation 3 gives

$$d = r \cos 2\pi ft$$

(Equation 5)

This equation gives the displacement of point *P* at any time (t) after the start of the motion and thus corresponds to Equation 11-D ($d = V_0 t + \frac{1}{2}at^2$) for a body moving with constant acceleration. Note that Equa-

tion 5 gives the distance from the center of the path and not the distance from the starting point.

The instantaneous tangential velocity of point *Q* is equal to $r\omega$ or (according to Equation 4) $r2\pi f$.

The instantaneous velocity of point *P* is equal to the x component of V_t of point *Q;* i.e., $V = V_t \sin \theta$ or

$$V = 2\pi fr \sin \theta$$

(Equation 6)

Since $\theta = \omega t$ and $\omega = 2\pi f$ (Equation 4), we can substitute $2\pi f$ for ω in the angular displacement equation ($\theta = \omega t$) to obtain

$$\theta = 2\pi ft$$

(Equation 7)

Substituting from Equation 7 into Equation 6, we obtain

$$V = 2\pi fr \sin 2\pi ft$$

(Equation 8)

Since $\sin \theta = \sqrt{1 - \cos^2 \theta}$ and $\cos \theta$ is equal to the ratio of the side adjacent *(d)* to the hypotenuse *(r)*, i.e., $\cos \theta = d/r$, Equation 8 can be rewritten as

$$V = \pm 2\pi fr \sqrt{1 - \frac{d^2}{r^2}} \quad \text{or}$$

$$V = \pm 2\pi f \sqrt{r^2 - d^2}$$

(Equation 9)

The acceleration of point *P* is equal to the x component of the centripetal acceleration of *Q,* which is equal to $r\omega^2$. Substituting $2\pi f$ for ω (Equation 4) gives

$$a_r = r\omega^2 = r(4\pi^2 f^2)$$

(Equation 10)

The x component (a) is obtained from a_r as

$$a = a_r \cos \theta$$

(Equation 11)

Substituting $r(4\pi^2 f^2)$ for a_r from Equation 10 and $2\pi ft$ for θ from Equation 7 gives

$$a = r(4\pi^2 f^2) \cos 2\pi ft$$

(Equation 12)

Since $r \cos 2\pi ft$ equals d from Equation 5, Equation 12 becomes

$$a = 4\pi^2 f^2 d$$

(Equation 13)

In harmonic motion the instantaneous force acting in a situation is symbolized with the letters Kd, where *K* is a force constant and *d* is the horizontal displacement. Thus Newton's second law for harmonic motion is

$$a = \frac{F}{m} = \frac{Kd}{m} = 4\pi^2 f^2 d$$

Since the *d*'s cancel on each side of the equal sign, the equation becomes

$$a = \frac{F}{m} = \frac{K}{m} = 4\pi^2 f^2$$

(Equation 14)

The frequency of harmonic motion is found by rearranging Equation 14 as follows:

$$f = \frac{1}{2\pi} \sqrt{K/m}$$

(Equation 15)

Since period T is equal to the reciprocal of the frequency, Equation 15 can also be written

$$T = 2\pi \sqrt{m/K}$$

(Equation 16)

The displacement *(d)* at any instant is the horizontal distance of the object away from the equilibrium position or center of the path of oscillation at that instant.

The amplitude *(A)* is the maximum displacement.

The range *(R)* is the total displacement of the motion and is, therefore, equal to twice the amplitude; i.e., *2A.*

ANALYSIS OF RESONANCE IN ELASTIC HARMONIC MOTION

In general, whenever a body is acted on by a periodic series of impulses having a frequency equal to one of the natural frequencies of vibration of the body, it is set into vibration with a relatively large amplitude. This phenomenon is called resonance, and the body is said to resonate with the applied impulses.

All bodies have natural periods or frequencies of vibration depending upon their masses, their geometric characteristics, and the maner in which they are set into vibration. A body can be set into vibration very readily when a force acts upon it periodically with its natural frequency or period.

Resonance is defined as the prolongation and intensification of vibrations. The production of resonance is a common occurrence in motor skills. Children, for example, readily learn how to pump a swing; i.e., they learn that they can give the swing a vibration of considerable amplitude if they properly time their impulses in accord with the natural period or frequency of the swing. This type of pendular vibration is further discussed in Lesson 24. The springboard diver also learns how to take advantage of the natural period or frequency of the diving board to increase his or her projection velocity by bouncing at the right time. This is another example of resonance.

Examples of "in-phase" and "out-of-phase" actions

Two vibrations with the same frequency are said to be "in phase" with each other if both start out together. Vibrations are "out of phase" if either has a different frequency or they do not start out together. Resonance occurs when two similar vibrations are in phase. Thus a trampolinist in phase with the trampoline will have the greatest conservation of force, while a trampolinist out-of-phase with the trampoline will have the greatest dissipation of force. The in-phase trampolinist will be projected higher than will the out-of-phase trampolinist. The performer who wishes to stop the trampoline from vibrating can do so by executing a movement that is out of phase.

Examples of undesirable resonance

Resonance effects are sometimes undesirable and are then to be avoided. Frequently it is observed that at certain speeds a car vibrates more readily than at other speeds. This is due to a coincidence of the vibration frequencies produced by irregularities in the road and the vibration frequencies produced by the rotating motor. We would want to decrease resonance through the production of out-of-phase vibrations.

A bridge or, for that matter, any structure is capable of vibrating with certain natural frequencies. If the regular footsteps of a column of soldiers were to have a frequency equal to one of the natural frequencies of a bridge which the soldiers were crossing, a vibration of dangerously large amplitude might result. Therefore in crossing a bridge a column of soldiers is ordered to break step. These out-of-phase steps or vibrations will decrease resonance.

Center of percussion

We know that when we hit a baseball on certain parts of our bat we feel more "sting" (i.e., we get more vibrations or resonance) than when we hit the ball on other points. Some baseball players give a special name to the point on their bat that produces the fewest vibrations upon

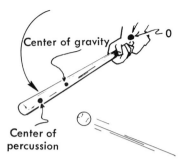

Fig. 19-7. The center of gravity and center of percussion of a baseball bat are located at different points. (From Krause, J. V., and Barham, J. N.: The mechanical foundations of human motion: a programmed text, St. Louis, 1975, The C. V. Mosby Co.)

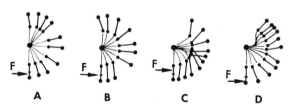

Fig. 19-8. A series of drawings made from multiflash photographs shows the motion of a body suspended by a cord when the body is struck a horizontal blow. The center of gravity is marked by a black band. In **A** the body is struck at its center of percussion, while in **B** it is struck at its center of gravity. In **C** the object is struck above, and in **D** below, its center of gravity. (From Krause, J. V., and Barham, J. N.: The mechanical foundations of human motion: a programmed text, St. Louis, 1975, The C. V. Mosby Co.)

hitting a ball. They call this point the "sweet spot." The technical name for this point is *center of percussion* (Fig. 19-7). All striking implements have a center of percussion.

A ball hit at the center of percussion of a bat will have the greatest force imparted to it because the less the vibration of the bat the greater are the conservation and transfer of force to the ball. In other words, the force of impact is dissipated by the vibrations produced in a bat through resonance.

Shown in Fig. 19-8 are a series of drawings made from multiflash photographs illustrating the motion of a body suspended by a cord when the body is struck a horizontal blow. The center of gravity is marked by a black band. In *A* the body is struck at its center of percussion relative to a pivot at the upper end of the cord, and it starts to swing smoothly about this pivot. In *B* the body is struck at its center of gravity. Note that it does not start to rotate about the pivot but that its initial motion is one of pure linear movement; i.e., no torque is produced by the blow. Thus the center of percussion does not coincide with the center of gravity. In *C* the body is struck above, and in *D* below, its center of percussion. These last two diagrams illustrate the dissipation of force.

STUDY GUIDELINES
Introduction
Specific student objectives

1. Describe harmonic motion.
2. Identify the two types of harmonic motion.
3. Describe elasticity and elastic harmonic motion.

Summary

Harmonic or vibratory motion is a back-and-forth motion that generally follows the same repetitive path. Linear harmonic motion occurs along a straight line path; angular harmonic motion traces a circular path.

Elasticity is the resistance of a body to distortion and the tendency of that body to restore itself quickly after being distorted. Elastic objects which accept a distorting force can produce harmonic motion, i.e., elastic harmonic motion.

Performance tasks

1. Develop a written list of descriptions for
   a. harmonic motion
   b. linear and angular harmonic motion
   c. elasticity and resiliency
   d. elastic harmonic motion
2. Illustrate linear and angular harmonic motion with specific motion examples.
3. Back-and-forth motions that generally follows the same repetitive path can be called
   a. harmonic motions   c. both a and b
   b. vibratory motions   d. neither a nor b
4. An object that accepts a distorting force and has a tendency to restore itself quickly possesses the characteristic of
   a. harmonic motion   c. both a and b
   b. flexibility   d. neither a nor b

5. Identify four examples of elastic objects.

Measurement of elasticity
Specific student objectives

1. Describe mechanical stress and strain.
2. State Hooke's law.
3. Describe elastic limit.
4. Define the coefficient of elasticity or restitution.
5. Identify and describe the two formulas for determining the coefficient of elasticity.
6. List five considerations involved in elasticity.
7. Given V_a and V_s or h and H, determine r.

Summary

In an impact situation, stress is described as the distorting or restoring forces acting on a body per unit of its area. The actual deformation produced by the applied force is called strain, i.e., the quantitative measure of deformation.

Hooke's law states that the deformation of a body is proportional to the applied force per unit of area, or strain is proportional to stress.

The maximum deformation an object can undergo as the result of stress forces without being permanently altered is its elastic limit. This is its "breaking point," where the object no longer obeys Hooke's law. Resiliency is the ability of an object to become distorted and to restore itself, i.e., to stay within its elastic limits.

The coefficient of elasticity or restitution is a numerical value that is a quantitative measure of an object's elasticity. This coefficient of elasticity (r) has a value characteristic of the makeup and material of the elastic object. The two formulas for this coefficient are

1. $r = V_s/V_a$, where V_s is velocity of separation and V_a is velocity of approach.
2. $r = \sqrt{h/H}$, where H is the height from which an object is dropped and h is the height to which a body will rebound after collision.

Several considerations involved in elasticity are listed as follows:

1. r varies with the velocity of impact.
2. r is highest at low velocities of collision.
3. r has a range of values between 0 and 1.
4. Liquids and gases have high elasticity and for this reason are used inside balls.
5. Air is used in inflated objects to increase r.

Performance tasks

1. Hooke's law is concerned with
   a. elasticity   c. momentum
   b. energy   d. power
2. The destorting forces acting per unit of area of a body in a collision situation are called
   a. stresses   c. both a and b
   b. strains   d. neither a nor b
3. A term synonymous with elasticity is (resiliency/restitution)_____
4. The ability of an object to accept an applied force during impact and restore itself after deformation is termed_____.
5. Speed of resiliency could be used as a description for
   a. elasticity   c. modules
   b. Hooke's law   d. all of the above
6. The point at which material "breaks down" in an object upon impact is its
   a. elasticity   c. both a and b
   b. resiliency   d. neither a nor b
7. List five considerations involved in restitution.
8. Describe in written form the meaning of the following:
   a. Resiliency   e. r
   b. Elasticity   f. $r = V_s/V_a$
   c. Hooke's law   g. $r = \sqrt{h/H}$
   d. Elastic limit   h. Mechanical stress
9. With the information given below for collision situations, determine the coefficient of restitution using $r = V_s/V_a$ or $r = \sqrt{h/H}$.
   a. $V_a = 50$ ft/sec, $V_s = 40$ ft/sec
   b. $V_a = 33$ m/sec, $V_s = 29$ m/sec
   c. $V_a = 110$ cm/sec, $V_s = 87$ cm/sec
   d. H $= 6$ ft, h $= 4$ ft
   e. H $= 6$ ft, h $= 2$ ft
   f. H $= 21$ m, h $= 18$ m

Analysis of elastic harmonic motion
Specific student objectives

1. Describe the work equation for elastic potential energy situations.
2. State Newton's three laws for elastic harmonic motion.
3. Describe the following harmonic motion terms:
   a. Periodic   e. Displacement (d)
   b. Oscillatory   f. Amplitude (A)
   c. Period (T)   g. Range (R)
   d. Frequency (f)

Summary

Harmonic motion exemplifies the law of energy conservation as found in the conservation of elastic potential energy to kinetic energy and vice-versa. Work done during elastic harmonic motion can be found by $W = \frac{1}{2}Kd^2$, where K is the force constant of the material and d is the deformation distance.

Newton's laws applied to elastic harmonic motion are as follows:

1. A body with a given shape will tend to retain that shape, and a body at rest will tend to remain at rest unless a net distortion force acts upon it. If a body is in harmonic motion, it will remain in harmonic motion with equal displacements unless an unbalanced force acts upon it.
2. The acceleration of an object in returning to its equilibrium position or shape following distortion is directly proportional to the product of the force constant (K) of the object and its displacement (d) and is inversely proportional to its mass; i.e., a = Kd/m.
3. For every force tending to distort a body within its elastic limits, there is an equal and opposite restoring force produced in the body that will tend to return it to its original position or shape.

Harmonic motion quantities can be analyzed by use of the equations developed in the boxed material on pp. 318 and 319. These quantities are (1) period, T, the time required for one complete oscillation, (2) frequency, f, the number of complete oscillations per unit of time, (3) displacement, d, the horizontal distance of the object from the equilibrium position, (4) amplitude, A, the maximum displacement, and (5) range, R, the total displacement of the harmonic motion (i.e., 2A).

Performance tasks

1. During elastic harmonic motion, the work done can be found by $W = \frac{1}{2}Kd^2$, where K is the force constant and d is the _____.
2. Develop a written description for the following terms:
   a. Periodic   d. Frequency
   b. Oscillatory   e. Displacement
   c. Period   f. Amplitude
                       g. Range

3. Illustrate Newton's three laws of motion for elastic harmonic motion with specific motion examples.
4. State Newton's three laws of elastic harmonic motion.

Analysis of resonance in elastic harmonic motion

Specific student objectives

1. Describe resonance.
2. Describe "in-phase" and "out-of-phase" vibrations.
3. Describe the center of percussion of an object.

Summary

Resonance is defined as the prolongation and intensification of vibrations. It occurs when an object is acted upon by a periodic series of impulses having a frequency equal to one of its natural frequencies of vibration.

Two vibrations with the same frequency are "in phase" with each other if they occur together or "out of phase" if they have different frequencies or do not occur together. Resonance occurs when two similar vibrations are "in phase."

The center of percussion of an object is that point which when struck produces the least vibration and the greatest conservation of force.

Performance tasks

1. Identify two motion examples of objects with either "in-phase" or "out-of-phase" vibrations.
2. Write out descriptions of the following:
   a. Resonance
   b. In-phase vibrations
   c. Out-of-phase vibrations
   d. Center of percussion

SELF-EVALUATION

Students should use no reference materials for this progress test, and they can check their answers by referring to Appendix A.

1. Harmonic motion is sometimes called (vibratory/linear)_____ motion.
2. The two types of harmonic motion are
   a. linear and angular
   b. resonance and vibratory
   c. both a and b
   d. neither a nor b
3. The ability of an object to be compressed and restore itself is called
   a. elasticity   c. both a and b
   b. resiliency   d. neither a nor b
4. Speed of resiliency is termed
   a. elasticity   c. both a and b
   b. restitution   d. neither a nor b
5. Back-and-forth motion that generally follows the same repetitive path is
   a. harmonic motion   c. both a and b
   b. displacement   d. neither a nor b
6. The restoring force acting per unit of area in an impact situation is
   a. elasticity   c. stress
   b. resilience   d. strain
7. State Hooke's law.
8. The maximum deformation an object can undergo as a result of stress without being permanently altered is its
   a. elastic limits   c. elasticity
   b. strain   d. restitution
9. The coefficient of elasticity for an object depends upon its
   a. makeup   c. both a and b
   b. materials   d. neither a nor b
10. State five considerations involved in elasticity.
11. Given the following information for impact situations, determine the coefficient of elasticity:
    a. H = 11 m, h = 10 m
    b. H = 8 ft, h = 6 ft
    c. H = 196 cm, h = 180 cm
    d. V_a = 90 ft/sec, V_s = 80 ft/sec
    e. V_a = 42 ft/sec, V_s = 37 ft/sec
12. Describe the meaning of $W = \frac{1}{2}Kd^2$.
13. The time required for one complete oscillation in harmonic motion is the
    a. amplitude   c. period
    b. displacement   d. range
14. The maximum displacement during vibratory motion is
    a. amplitude   c. period
    b. frequency   d. range
15. State Newton's three laws for elastic harmonic motion and illustrate each with a motion example.
16. The prolongation and intensification of vibrations is called
    a. amplitude

b. resonance
c. restitution
d. frequency

17. In-phase vibrations produce
    a. elasticity
    b. resonance
    c. restitution
    d. resiliency

18. The point on an object which, upon impact, produces the least vibration is the center of
    a. gravity
    b. mass
    c. percussion
    d. vibration

SECTION TWO

Kinetic analysis of angular motion

Kinetic analysis of uniform angular motion

Body of lesson
  Introduction
  Analysis of radial acceleration and force
    Analysis of radial acceleration
    Analysis of radial force
  Determining the inward angle of lean of a runner
    Example
Study guidelines
Self-evaluation

INTRODUCTION

According to Newton's first law, the basic path of a body in motion is a straight line and any deviation from this path must be produced by a net force. In Lesson 17 this force, being perpendicular to the linear path of a moving body and acting to change the direction of that path, was identified as the radial force (F_r).

Since a body in curvilinear motion is constantly being deflected toward the inside of a curve, there must be an inward force acting upon it to keep it moving along its curved path. This radial force is given the special name of *centripetal force,* which means, literally, "force seeking the center." Without this force, curvilinear motion cannot occur.

The crucial role of centripetal force in curvilinear motion is illustrated in Fig. 20-1. A man is whirling a ball at the end of a string horizontally above his head. As the ball swings around, he must continually exert a centripetal force by means of the string. Should the man let go of the string, the ball would resume its normal straight line path (as Newton's first law predicts) and fly off at a tangent to its original circular path. The tangent is, of course, 90° to the radius

of the circle. Once released, the ball will, as everyone knows, be acted upon immediately by the centripetal force of gravity, which will cause it to curve toward the ground.

The above example illustrates two types of curvilinear motion. First, we have the circular motion produced by the man whirling the ball horizontally above his head. Second, there is the curved path of a projectile followed by the ball after the man releases it. Consideration of the first case, which involves uniform circular motion, allows us to introduce the concepts of radial (centripetal) acceleration and force.

ANALYSIS OF RADIAL ACCELERATION AND FORCE

The rate of change of velocity was defined in Lesson 10 as acceleration. Acceleration (a) is given by $a = (V_f - V_o)/t$, where $V_f - V_o$ is the change in velocity (ΔV). Since velocity is a vector and has both magnitude and direction, this change can involve (1) a change in magnitude only, (2) a change in direction only, or (3) a change in both magnitude and direction. In Lessons 13 and 17 we identified the force that produces the change in direction of a velocity as a radial force

Fig. 20-1. Whirling a ball at the end of a string illustrates the crucial role of centripetal force in curvilinear motion.

and the force that produces the change in speed as a tangential force. Thus the accelerations produced by these forces are likewise called the radial and tangential accelerations.

We have already had experience with the use of the acceleration formula in the description of tangential acceleration resulting from a change in only the magnitude of a linear or tangential velocity (Lesson 10). The present discussion, to begin with, is restricted to a description of radial acceleration which results from a change in only the direction of a velocity. The more usual case of acceleration resulting from a change in both

magnitude and direction of a velocity will be discussed in the next lesson.

Analysis of radial acceleration

Fig. 20-2 shows the instantaneous velocity of a particle in rotation at two points, A and B. The velocity, as indicated by the vectors V_f and V_o, is seen to be changing in direction but not in magnitude (i.e., the vectors have the same length but different directions).

To find the change in velocity $(V_f - V_o)$ of the particle in moving from point A to point B, we need to use vector subtraction. This is done

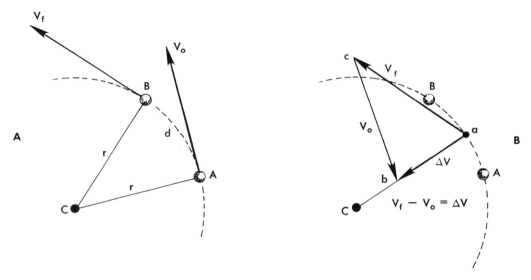

Fig. 20-2. Subtraction of the original velocity from the final velocity of an object in **uniform** rotation to find the radial change of velocity (ΔV_r).

simply by attaching a negative sign to V_o (which reverses its direction) and then adding the two vectors in the usual manner of vector addition, i.e., $V_f + (-V_o)$.

Fig. 20-2, *B* is a velocity diagram showing ΔV as the change in velocity that occurs in going from *A* to *B*. The construction of the velocity diagram follows the steps below:

1. Draw V_f to the proper scale and in the proper direction starting from the midpoint of the interval between *A* and *B*. The midpoint is used to construct the velocity diagram because we are interested here in the average change in velocity over the time interval (t) that it takes the particle to move from *A* to *B*. This has been done in Fig. 20-2, *B*, where the vector V_f is transferred to point *a*, midway between *A* and *B*.
2. Reverse the direction of V_o and attach its tail to the head of V_f.
3. Draw a line attaching the tail of V_f to the head of V_o. This line describes the change in velocity of the two vectors (ΔV).

We see that the vector difference (ΔV) points toward the center of the circle parallel with the radius and hence the average acceleration ($\Delta V/t$ or $(V_f - V_o)/t$) points in this direction also.

The magnitude of ΔV can be determined by using the two triangles shown in **Fig. 20-2.** Triangles *ABC* (on the left) and *abc* (on the right) are proportional since both are isosceles triangles and their long sides are mutually perpendicular. The proportionality of the triangles means the ratio of the base of the triangle on the right *(ΔV)* to either of its two sides is equal to the ratio of the base of the triangle on the left *(d)* to either of its two sides. Hence, if we let V represent the magnitudes of either V_f or V_o, we have

$$\frac{\Delta V}{V} = \frac{d}{r}$$

The distance the particle actually covers in going from *A* to *B* is the arc joining these points *(d)*, whose length is Vt; i.e., d = Vt, or linear displacement *(d)* is the product of the speed *(V)* and the time of motion *(t)* as discussed in Lesson 10. Hence the equation $\Delta V/V = d/r$ can be changed by substituting Vt for *d* to obtain $\Delta V/V = Vt/r$. This equation can be rearranged to obtain $\Delta V/t = V^2/r$. Remembering that ΔV equals ($V_f - V_o$) and that $(V_f - V_o)/t$ equals acceleration, we can see that the radial acceleration (a_r) is directly proportional to the square of the velocity and inversely proportional to the radius of the circle.

$$a_r = \frac{V^2}{r}$$

(Equation 20-A)

Radial acceleration (Equation 20-A) can also be expressed in terms of the angular velocity (ω). Since $V = r\omega$, we can substitute $r\omega$ for V in the radial acceleration formula to obtain

$$a_r = \frac{(r\omega)^2}{r} = \frac{r^2\omega^2}{r} = r\omega^2$$

(Equation 20-B)

Analysis of radial force

Having obtained an expression for the radial acceleration of a particle revolving in a circle, we can now use Newton's second law ($F = ma$) to find the radial (centripetal) force acting on the particle to produce its curved path. Since the magnitude of the centripetal acceleration equals V^2/r and its direction is parallel with the radius of the circle, the magnitude of the radial force (F_r) acting on a particle of mass m is

$$F_r = \frac{mV^2}{r}$$

(Equation 20-C)

or in angular quantities

$$F_r = mr\omega^2$$

(Equation 20-D)

Like the radial acceleration, the radial force is directed toward the center of rotation (which is the same as saying it is parallel with the radius of the circle).

DETERMINING THE INWARD ANGLE OF LEAN OF A RUNNER

We can use the problem of a runner going around the curve of a track as an example of curvilinear motion. Since radial force is given by the equation $F_r = mV^2/r$, it can be seen that the magnitude of the radial force a runner must exert in order to go around the curve is directly proportional to his mass and the square of his velocity, and inversely proportional to the radius of the track. Thus the heavier the runner and the greater his velocity and the shorter the radius of the track, the more acute will be his problem in rounding the curve.

The radial force of the runner is supplied either by his inward angle of lean or by the slope of the track. Fig. 20-3 shows a runner with an inward lean of 18°, which gives him the radial force to make the turn. We know from Lesson 8, however, that his inward lean actually gives the runner

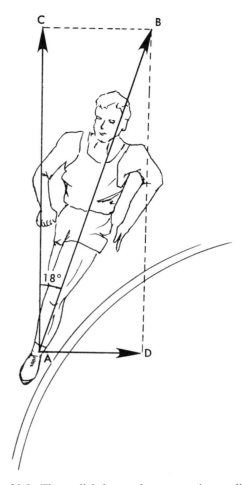

Fig. 20-3. The radial force of a runner is supplied by his inward angle of lean or by the slope of the track.

two force components: a vertical component and a horizontal component. Fig. 20-3 shows that the vertical component is represented by the "side adjacent" in the right triangle and that the horizontal component is represented by the "side opposite." This reversal of normal relationships is due to the fact that the angle of 18° is measured from the positive y coordinate rather than from the positive x coordinate as is the usual case. The magnitudes of the vertical and horizontal components can be determined as follows:

1. From the information presented in Lesson 8, we know that the "side opposite" of a right triangle is equal to the hypotenuse multiplied by the sine of the angle. Since the horizontal component of the runner equals the "side opposite" and the total

force acting is equal to the hypotenuse, we can see that $F_x = F \sin \theta$.

2. Likewise, it should now be evident that $F_y = F \cos \theta$.

3. Since the horizontal component is directed toward the center of the curve, this component provides the radial force mV^2/r. Thus the horizontal component ($F \sin \theta$) equals the radial force mV^2/r, or $F \sin \theta = mV^2/r$.

4. Since there is no vertical acceleration, the vertical component ($F \cos \theta$) equals the weight of the runner (mg) or $F \cos \theta = mg$.

The angle of inward lean required for a runner to go around a given curve can be determined as follows:

1. In Lesson 8 we learned that the tangent of an angle equals the ratio of the "side opposite" to the "side adjacent," which in our present case is the ratio of the "horizontal component" to the "vertical component."

2. When we divide the "horizontal component" by the "vertical component," we obtain the tangent of the angle of inward lean of a runner required to take him through a given turn.

$$\text{Tan } \theta = \frac{\text{Horizontal component}}{\text{Vertical component}} = \frac{F_x}{F_y} = \frac{mV^2/r}{mg} = \frac{V^2}{rg}$$

(Equation 20-E)

Thus the greater the linear velocity of the runner, and the shorter the radius of the turn, the greater is the inward angle of lean that is required.

The angle of lean can be provided for by simply banking the track over which the runner is moving. Thus it should now be evident that the tangent of the angle of banking of a track, or the angle of inward lean of a runner, is directly proportional to the square of the velocity and

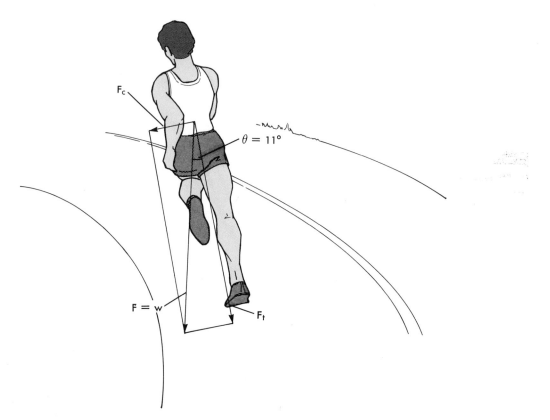

Fig. 20-4. Back view of a runner going into the curve of a track. The radial force required to make the turn is supplied by the inward angle of lean of the runner (θ).

inversely proportional to the length of the radius. It should also be evident that for a given radius no one angle is correct for all velocities. Hence in the design of highways, railroads, and race-tracks, curves are banked for the average velocity of the traffic over them.

The same considerations apply to the correct banking angle of a plane when it makes a turn in level flight.

Example

As an example of a problem involving the application of centripetal force and an inward lean on the part of a runner, consider the runner shown in Fig. 20-4. He must go around the curve on a cinder track. Most outdoor quarter-mile cinder tracks have a radius of about 100 ft. Assume that the runner is traveling 25 ft/sec and that he weighs 160 pounds. What force tends to pull him out of his lane at the turn or, in other words, how much force is needed to permit him to follow the turn of the track? This force can be determined through the use of Equation 20-C:

$$F_r = \frac{mV^2}{r} = \frac{(160 \text{ lb})/(32 \text{ ft/sec}^2)\ (25 \text{ ft/sec})^2}{100 \text{ ft}} =$$

$$\frac{3125 \text{ lb/ft}}{100 \text{ ft}} = 31.25 \text{ lb}$$

The amount of inward lean required of the runner in order to furnish this radial force can be determined through use of Equation 20-E:

$$\text{Tan } \theta = \frac{V^2}{rg} = \frac{(25 \text{ ft/sec})^2}{(100 \text{ ft})\ (32 \text{ ft/sec}^2)} =$$

$$\frac{625 \text{ ft}^2/\text{sec}^2}{3200 \text{ ft}^2/\text{sec}^2} = 0.195$$

When we look this tangent up in the table of trigonometric functions in Appendix B, we find that a tangent of 0.195 is equal to an angle of approximately 11°.

STUDY GUIDELINES
Introduction
Specific student objectives

1. Describe radial force (F_r) and define centripetal force.
2. Describe a tangent.
3. Describe radial (centripetal) acceleration.

Summary

A force which acts perpendicular to the path of a motion is called a radial force (F_r), and it tends to change the direction of this motion. A special type of radial force directed toward the center of rotation is termed centripetal; thus it creates angular motion in a body without a fixed axis of rotation. The acceleration caused by a centripetal force (a_r) is toward the axis and is called a radial or centripetal acceleration.

When the net centripetal force acting upon a moving object becomes zero, the object will resume a linear path at a tangent to its former circular motion; i.e., at the point of release, the object will begin moving off in a tangential direction, at right angles to the radius.

Performance tasks

1. Describe and illustrate the following terms:
   a. Radial force
   b. Centripetal force
   c. Radial acceleration
   d. Centripetal acceleration
   e. Tangent to a circle
   f. Radius of a circle
2. The pure or natural state of motion of a body when no external net force is acting on it is (angular/linear) _____.

Analysis of radial acceleration and force
Specific student objectives

1. Identify the two ways to change linear velocity.
2. Differentiate between radial and tangential forces; radial and tangential accelerations.
3. Describe the meaning of $a_r = V^2/r = r\omega^2$.
4. Describe the meaning of $F_r = mV^2/r = mr\omega^2$.
5. Given r and V or ω, calculate a_r.
6. Given m, r, and V or ω, calculate F_r.

Summary

In angular motion the force that acts at right angles to the linear path of motion and changes the direction of that path is called the radial force (F_r). For a body in curvilinear motion this "inward" directed force is termed centripetal (toward the center). If the centripetal force causing curved motion is removed, the object will begin to travel in a straight line tangent (90° to the circle's radius) to the original circular path. Radial forces in circular motion produce radial accelerations (a_r), and the quantity a_r has been found to be directly proportional to the square of the velocity and inversely proportional to the radius of the circle; i.e. $a_r = V^2/r$. Also $a_r = (r\omega)^2/r = r\omega^2$. Using Newton's second law of acceleration ($F = ma$), we can obtain $F_r = m(V^2/r) = mr\omega^2$, where the direction of the radial force is parallel with the radius of the circle.

Performance tasks

1. Velocity may be changed by changing the motion's
   a. speed only
   b. direction only
   c. speed and direction
   d. all of the above
2. Radial forces produce changes in a motion's
   a. speed
   b. direction
   c. both a and b
   d. neither a nor b
3. A change in an object's speed is produced by
   a. radial force
   b. tangential force
   c. both a and b
   d. neither a nor b
4. Describe radial and tangential accelerations.
5. Describe the meaning of $a_r = V^2/r = r\omega^2$.
6. The radial force F_r is inversely proportional to
   a. mass
   b. velocity squared
   c. both a and b
   d. neither a nor b
7. Illustrate the relationships of $a_r = V^2/r$ and $F_r = mV^2/r$ with specific motion examples.
8. F_r and a_r are (scalar/vector) _____ quantities.
9. Using the formulas for radial force and acceleration, find a_r and F_r in the following angular motion situations (NOTE: $m = w/g$):

	r	m	V	ω
a.	6 m	6 kg	_____	3 rad/sec
b.	52 cm	176 g	42 cm/sec	_____
c.	2 ft	10 slugs	_____	50 rev/min
d.	8 ft	3 slugs	44 ft/sec	_____

Determining the inward angle of lean of a runner

Specific student objectives

1. Describe the meaning of $\tan \theta = V^2/rg$.
2. Describe the angle of lean.
3. Identify the variables that determine the angle of lean.
4. State two principles related to the angle of lean of a runner.
5. Given V and r, determine θ (the angle of lean).

Summary

The radial force (F_r) necessary to keep a runner on a curved path or track can be supplied by an inward angle of lean of the runner or by the slope of the track.

The tangent of this angle equals the square of the velocity of the runner divided by the product of the radius and the gravitational constant (g). Thus $\tan \theta = V^2/rg$ shows that the angle of lean is directly proportional to the velocity squared and inversely proportional to the length of the radius.

Performance tasks

1. The angle of lean of a runner produces the
   a. radial force
   b. tangential force
   c. both a and b
   d. neither a nor b
2. Describe the angle of lean of a runner.
3. Identify and describe all symbols in the equation $\tan \theta = V^2/rg$.
4. The angle of lean is related to the "velocity squared" of a runner (directly/inversely) _____.
5. The angle of lean depends upon the radius of the curve in a (direct/inverse) _____ proportional relationship.
6. Determine the angle of lean to fit the following situations:

θ (radius of curve)	Maximum velocity
a. _____90 ft	20 ft/sec
b. _____30 m	10 m/sec
c. _____40 m	10 m/sec

SELF-EVALUATION

Students should use no reference materials for this progress test, and they can check their answers by referring to Appendix A.

1. A centripetal force is a special type of
   a. radial force
   b. tangential force
   c. torque
   d. weight
2. Centripetal is a word that means (away from/toward) _____ the center.
3. A linear acceleration has been described as any change in linear velocity. State the three possible ways that a velocity change may occur.
4. A "pure" radial acceleration is produced by a change in an object's motion
   a. direction
   b. magnitude

c. both a and b

d. neither a nor b

5. A "pure" tangential acceleration is produced by a change in an object's motion

   a. direction   c. both a and b

   b. magnitude   d. neither a nor b

6. State the relationship between a radial force and a tangential force in angular motion.

7. Using the formulas for radial force ($F_r = mV^2/r = mr\omega^2$) and radial acceleration ($a_r = V^2/r = r\omega^2$), find F_r and a_r in the follow-lowing situations (NOTE: $m = w/g$, $g = 32$ ft/sec^2, 32 ft/sec^2 = 22 mi/hr^2, and 1 rev = 2π rad.)

	r	w	V	ω
a.	2 ft	5 kg	_____	2 rad/sec
b.	5 ft	2 lb	_____	90 rev/min
c.	800 ft	4000 lb	60 mi/hr	_____
d.	900 ft	45 tons	30 mi/hr	_____
e.	1.5 m	5 kg	_____	240 rev/min

8. Calculate the angle of lean or angle of tilt for a track or inclined curve.

   Use tan $\theta = V^2/rg$ and $g = 32$ ft/sec^2 = 22 mi/hr^2.

	V	r	tan θ	θ
a.	20 ft/sec	20 ft	_____	_____
b.	80 ft/sec	100 ft	_____	_____
c.	5 mi/hr	0.1 mi	_____	_____
d.	5 mi/hr	0.4 mi	_____	_____
e.	44 ft/sec	50 ft	_____	_____

9. A tangent to a circle is a straight line that is (parallel with/perpendicular to)_____ the circle's radius.

10. Radial acceleration is related to the linear velocity squared as a (direct/inverse) _____ proportion.

11. Radial force has a relationship with the radius of rotation that would be called a/an (direct/inverse) _____ proportion.

12. State two principles concerning the angle of lean of a runner.

Kinetic analysis of variable angular motion

Body of lesson
Study guidelines
Self-evaluation

INTRODUCTION

We now take up the more usual case of angular motion, in which we have a change in the magnitude as well as the direction of a velocity.

First, we shall return for a moment to the case of a ball being whirled horizontally at the end of a string. By means of the string, we transmit a radial force ($F_r = mV^2/r$) to the ball, pulling it inward in a circular path. While we are pulling inward on the ball, however, by Newton's third law of motion we know that the ball is pulling *outward* on us (as in Fig. 21-1). This outward reaction force, called *centrifugal radial force* (Latin *fugere,* to flee), is measured by the same equation as is centripetal radial force; i.e., $F_r = mV^2/r$.

These radial forces may properly be thought of as manifestations of inertia in the sense that Newton's first law tells us the tendency of the ball is to move in a straight path rather than be accelerated in a curved path. The action centripetal force is the force required to make the ball deviate from its straight line path. If the string breaks, the ball flies off at a tangent to its former circular path. This happens simply because there is no longer any centripetal force acting on the ball. Since the tangential path followed by the ball is a straight line, neglecting gravity, we can use the term tangential velocity as being synonymous with linear velocity. Therefore the symbols V_t and V are interchangeable for tangential and linear velocity.

From the above discussion it is evident that we can distinguish between the tangential ac-

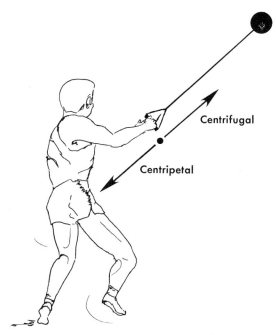

Fig. 21-1. For every applied centripetal force there is an equal and opposite reaction centrifugal force.

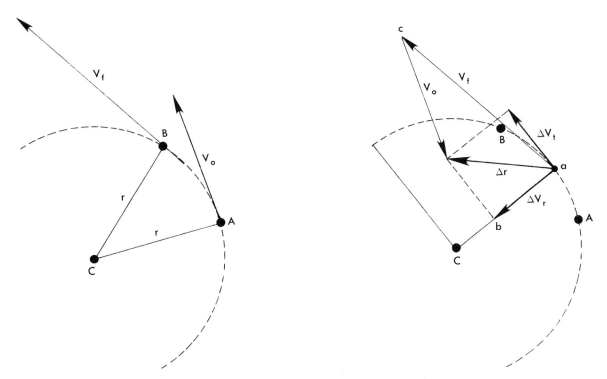

Fig. 21-2. Subtraction of the original velocity from the final velocity of an object in nonuniform rotation to obtain the total velocity change (ΔV) and its radial (ΔV_r) and tangential (ΔV_t) components.

celeration (a_t) of a particle in rotation, which results from a change in *magnitude* of the linear velocity, and the radial acceleration (a_r), which results from a change in the *direction* of motion. Thus *all* particles in circular motion have radial accelerations. Only particles whose speed changes in magnitude, however, have tangential accelerations.

ANALYSIS OF TANGENTIAL ACCELERATION AND FORCE

We now return to a consideration of a particle moving in a circle in which there is a change in the magnitude as well as the direction of its velocity. Fig. 21-2 shows a particle revolving in a circle. In moving from point *A* to point *B*, the velocity of the particle changes from V_o to V_f, where V_f has a greater magnitude than V_o as well as a different direction.

In the vector diagram of Fig. 21-2, *B*, the vector V_f has been transferred to point *a* (midway between *A* and *B*) and the vector change in velocity (ΔV) has been found as previously described in relation to Fig. 20-2. The ratio $\Delta V/t$ is the average acceleration between *A* and *B;* its direction is the same as that of ΔV, which, you will note, does not point toward the center of the circle.

As illustrated in Fig. 21-2, *B*, the vector ΔV can be resolved into two components: a radial component *(ΔV_r)* and a tangential component *(ΔV_t)*. Both of these are defined with reference to the radius; that is, the radial component is parallel with the radius, and the tangential component is perpendicular to the radius. Thus the two components are perpendicular to each other.

The ratio $\Delta V_r/t$ is the radial component (a_r) of the average acceleration of the particle in rotation, and the ratio $\Delta V_t/t$ is the tangential component of the particle's average acceleration.

The radial and tangential components of the instantaneous accelerations are the limits of these ratios as the time (t) approaches zero or as point *B* is taken closer and closer to point *A*. As *B* approaches *A,* the magnitude of the tangential component becomes more and more nearly equal to the difference between the magnitudes of V_f and V_o.

Since the tangential and radial accelerations are perpendicular to each other, a right triangle can be formed from them whose hypotenuse is the resultant instantaneous acceleration. Thus, using the Pythagorean theorem, we can find the resultant instantaneous acceleration (a):

$$a = \sqrt{a_t^2 + a_r^2}$$

(Equation 21-A)

Analysis of tangential acceleration

Since the tangential velocity is the same as the linear velocity discussed in Lessons 10 and 12, its magnitude is the product of the length of the radius of rotation and the magnitude of the angular velocity; i.e., $V_t = r\omega$. Likewise, the tangential acceleration of a body in rotation is the product of the length of the radius and the magnitude of the angular acceleration:

$$a_t = r\alpha$$

(Equation 21-B)

The tangential and radial components of acceleration of an arbitrary point on a rotating body are shown in Fig. 21-3, *A*.

Analysis of tangential force

The tangential force is obtained by substituting the tangential acceleration ($r\alpha$) for the linear acceleration *(a)* in Newton's second law (F = ma):

$$F_t = mr\alpha$$

(Equation 21-C)

Total forces involved in angular motion

The tangential and radial forces that cause the tangential and radial accelerations of a body in rotation are shown in Fig. 21-3, *B*. The angular displacement is measured from the positive x axis. Thus, if we know the total net force acting upon a body in rotation, this force can be resolved into its radial and tangential components through the trigonometric methods discussed in Lessons 8 and 13 (i.e., $F_t = F \sin \theta$ and $F_r = F \cos \theta$). If we do not know the total net force acting on a body in rotation, we can still determine the tangential and radial components through use of the following relationships: $F_t = mr\alpha$ and $F_r = mV^2/r = mr\omega^2$.

The total forces involved in an angular motion can be determined through the use of the Pythagorean theorem.

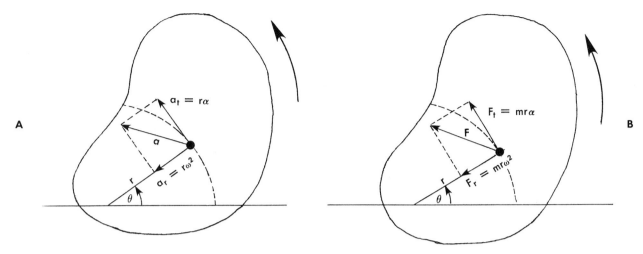

Fig. 21-3. The radial and tangential components of acceleration **(A)** and of force **(B)**.

$$F = \sqrt{F_t^2 + F_r^2}$$

(Equation 21-D)

Example

As an example of a motion situation in which the tangential and radial components can be determined, we can use the swing of a baseball batter as shown in Fig. 21-4. Assume that the angular velocity (ω) of the bat at point A is 10 rad/sec, that at point B ω is 15 rad/sec, and that it takes 1 sec to move from A to B. Assume also that the radius of the motion is 3 ft and that the mass of the parts involved is 1 slug. The force used to increase the speed of the bat can be determined through the use of Equation 21-C as follows:

$$\alpha = \frac{\omega_f - \omega_o}{t} = \frac{15 \text{ rad/sec} - 10 \text{ rad/sec}}{1 \text{ sec}} =$$

$$\frac{5 \text{ rad/sec}}{1 \text{ sec}} = 5 \text{ rad/sec}^2$$

$$F_t = mr\alpha = (1 \text{ slug}) (3 \text{ ft}) (5 \text{ rad/sec}^2) = 15 \text{ lb}$$

The force used to change the direction of the bat movement at point A can be determined through the use of Equation 21-D as follows:

$$F_r = mr\omega^2 = (1 \text{ slug}) (3 \text{ ft}) (10 \text{ rad/sec})^2 = 300 \text{ lb}$$

The radial force at point B is

$$F_r = mr\omega^2 = (1 \text{ slug}) (3 \text{ ft}) (15 \text{ rad/sec})^2 = 675 \text{ lb}$$

The total force that the batter is applying at A can be determined through the use of Equation 21-D as follows:

$$F = \sqrt{F_t^2 + F_r^2} = \sqrt{(15)^2 + (300)^2} = \sqrt{90,225} = 300.4$$

ANGULAR ANALOGS OF LINEAR MOTION QUANTITIES

Now that we know how to analyze the radial and tangential types of acceleration and the radial and tangential forces which cause these accelerations, we need to look at the application of Newton's laws in these motion situations. The principles of kinetics discussed in Lessons 17, 18, and 19 are as applicable to angular movements as they are to linear movements. However, the specific applications of these principles to angular motion do require the introduction of a few additional concepts and the identification of the angular analogs of linear motion quantities. The angular analogs of linear motion quantities are those concerned with inertia, momentum, impulse, kinetic energy, and work.

Inertia

In order to analyze the inertia of angular motion, we first need to change the force equation ($F = ma$) into a form applicable to angular motion. This is done by substituting the tangential force F_t for F and multiplying both sides of the equation by the radius of the circle (r) to obtain

Fig. 21-4. Angular motion of a baseball batter.

$F_t r$ = mar. The product on the left side represents the applied torque (T). Remembering from our discussion on kinematics that linear acceleration (a) equals the product of the radius (r) and the angular acceleration (α), we can replace the acceleration symbol *(a)* on the right side of the equation by its equal (rα):

$$F_t r = T = mar = mr^2\alpha$$

(Equation 21-E)

Since *m* and *r²* for a given body are both constants, they can be replaced by a single constant (labeled I):

$$T = mr^2\alpha = I\alpha$$

(Equation 21-F)

In this equation *I* is equal to *mr²* and is called the moment of inertia. The moment of inertia in angular motion is analogous to the mass (m) in linear motion, i.e., the property of a body which opposes acceleration.

Equation 21-E can also be obtained, of course, by multiplying both sides of Equation 21-C (F_t = mrα) by the radius *(r)*. In this case we obtain $F_t r = mr^2\alpha$.

Since the moment of inertia equals *mr²,* we can see that I has the same value regardless of the body's state of motion. The value of I depends upon the way in which the mass of a body is distributed relative to its axis of rotation. Quite possibly one body will have a greater moment of inertia than another even though its mass is much less. These relationships between inertia and the distribution of mass will be further discussed later in this lesson in conjunction with radius of gyration.

Momentum

Linear momentum (M) was defined in Lesson 18 as the product of an object's inertia or mass (m) and its velocity (V); i.e., M = mV. In angular motion the respective analogous terms are moment of inertia (I) and angular velocity (ω). Thus angular momentum (denoted by the symbol L), conceptually defined as the quantity of angular motion, is given as

$$L = I\omega = mr^2\omega$$

(Equation 21-G)

Impulse

The impulse equation describing the change-of-momentum situation for linear motion was net Ft = M_f – M_o. If torque (T) acting over a time (t) produces an angular momentum change (ΔL), then the analogous impulse equation for angular motion can be written

$$\text{net } Tt = L_f - L_o$$

(Equation 21-H)

Thus we can see that the angular change of momentum is affected by the amount of the causal torque (net T) and the duration of its application (t).

Kinetic energy

A rotating body possesses kinetic energy because its constituent particles are in motion, even though the body as a whole may remain in place. The linear speed of a particle that is a distance (r) from the axis of a body rotating with an angular velocity (ω) is, as we have already seen, V = rω. This expression of linear velocity can be substituted for *V* in the linear kinetic energy formula as follows:

$$\text{K.E.} = \tfrac{1}{2}mV^2 = \tfrac{1}{2}m(r\omega)^2 = \tfrac{1}{2}mr^2\omega^2 = \tfrac{1}{2}I\omega^2$$

(Equation 21-I)

Thus we can see that the farther a given particle is from the axis of rotation and the faster it moves the greater will be its contribution to the kinetic energy of the body.

Work

In Lesson 18, linear work was defined through the use of Newton's second law as Fd = $\tfrac{1}{2}mV^2$

+ mgh; or the work that can be done by a body is equal to its total mechanical energy (which is the sum of its kinetic and potential forms). This equation can be changed to angular form by substituting rθ for *d* on the left side and $\tfrac{1}{2}I\omega^2$ (the angular kinetic energy formula) for $\tfrac{1}{2}mV^2$ (the linear kinetic energy formula) on the right side to obtain Frθ = $\tfrac{1}{2}I\omega^2$ + mgh.

Substituting the torque symbol (T) for *Fr* on the left side gives us

$$T\theta = \tfrac{1}{2}I\omega^2 + mgh$$

(Equation 21-J)

ANGULAR ANALOGS OF NEWTON'S LAW OF MOTION

Newton's laws expressed in terms of angular quantities are as follows:

NEWTON I. An object either remains at rest or, if in angular motion, rotates at constant speed around a fixed axis unless acted upon by an unbalanced net torque.

NEWTON II. The angular acceleration (α) produced by an unbalanced torque (T) acting on a body is directly proportional to the net torque, in the same direction as the torque, and inversely proportional to the moment of inertia (I) of the body.

$$\alpha = \frac{T}{I} = \frac{T}{mr^2}$$

(Equation 21-K)

NEWTON III. Whenever one body acts upon a second body with a torque, the second body exerts a torque upon the first body. These torques are equal in magnitude and opposite in direction.

The applications of these laws in the analysis of angular motion will be discussed throughout most of the remaining lessons on kinetics.

MOMENT OF INERTIA AND THE RADIUS OF GYRATION

The analysis of linear motion is simplified through use of the concept of center of gravity, which reduces the distributed mass of an extended body to that of a particle. This concept, however, has limited application in the analysis of angular motion since, for the purpose of computing moment of inertia, the mass of a body, in general, cannot be considered as concentrated at its center of gravity. For example, a gymnast rotating around his/her center of gravity while in the air has zero distance from the axis to the center of gravity. However, the gymnast certainly does not have zero moment of inertia. In fact, the moment of inertia may be quite large.

When an object rotates around a fixed axis, the distribution of each particle of mass relative to the axis of rotation must be considered since each particle has its own radius of rotation and its own moment of inertia. Thus all these individual moments of inertia must be summed to provide the moment of inertia for the whole object. Such a procedure is obviously most difficult and requires specific information about the object's size, shape, density, etc. Much of this information about the segments of a given indi-

vidual is simply not available. Thus we can understand why the moment of inertia measured in terms of radius of rotation is calculated so rarely on a large scale. Much more commonly the procedure is done in terms of the *radius of gyration*.

The radius of gyration is a concept similar to the concept of center of gravity in linear motion analysis that has been introduced to help in the analysis of angular motion. Here the complex mass distribution of a body is visualized as being replaced by an infinitely thin circular "ring" of matter located around the center of mass of the body. This ring has not only the same quantity of mass as the body but also a radius about the

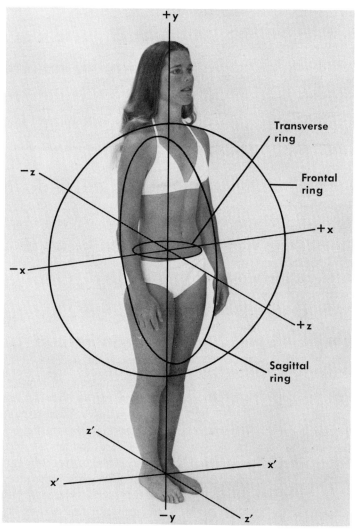

Fig. 21-5. The three cardinal axes, rings, and radii of gyration.

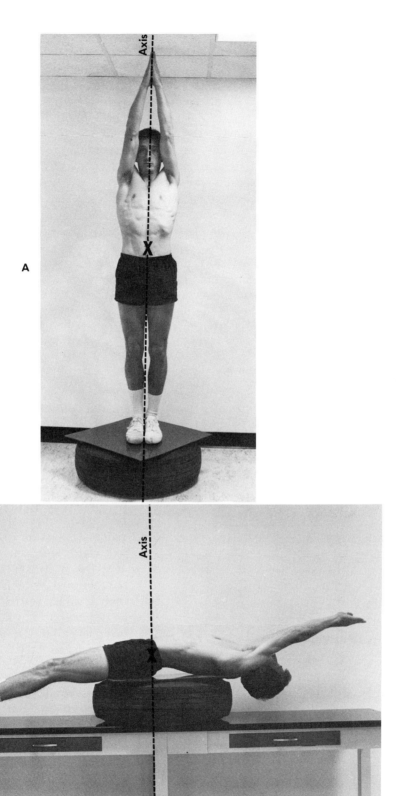

Fig. 21-6. Positions of minimum **(A)** and maximum **(B)** moment of inertia of the human body.

axis of rotation which gives it the same moment of inertia as the body. The radius of this ring is known as the *radius of gyration* and is symbolized by the letter k. The ring is known as the *ring of gyration*.

The three cardinal axes of motion of the human body are shown in Fig. 21-5. The ring of gyration associated with each axis is also shown with its approximate radius. These radii are nearly the same for the rings associated with the x and z axes; the radius of the ring associated with the y axis is much smaller. The lengths of these radii correspond to the magnitudes of the respective moments of inertia. In other words, I is much larger around the x and z axes than around the y axis. Secondary axes, rings, and radii of gyration exist for all other points of rotation of the body, such as for each of the joints around which the body segments rotate.

Positions of minimum and maximum moment of inertia

Relative to an axis passing through the center of gravity of the human body, the radius of gyra-

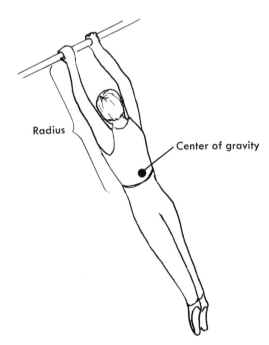

Fig. 21-7. When the axis of rotation does not pass through the center of gravity, the moment of inertia is greatest in a fully extended position with the hands as the axis.

tion and thus the moment of inertia is least about the longitudinal (y) axis when the arms are extended and close together above the head. It is greatest when the body is turning in a similar position, but with the arms parallel, about a horizontal axis. These positions of minimum and maximum moment of inertia as demonstrated on a turntable are illustrated in Fig. 21-6.

Relative to an axis not passing through the body's center of gravity, the moment of inertia is greatest in a fully extended position with the hands as the axis (Fig. 21-7). This is due to the fact that in such a situation the greatest distribution of mass is away from the center of rotation. Therefore the body's moment of inertia is always greater when the body rotates about an axis not passing through its center of gravity than when it rotates about an axis which does. For example, the moment of inertia of the individual shown in Fig. 21-5 is much greater about either of the horizontal axes, x' or z' (passing through her feet), than about either the x or the z axis (passing through her center of gravity).

Summary and definition equation

Whatever the shape of a body, a radial distance can always be found from any given axis where the mass of the body could be concentrated without altering the moment of inertia of the body about that axis. This distance is called the radius of gyration of the body about the given axis and is represented by k.

If the mass (m) of the body actually were concentrated at this distance, the moment of inertia would be that of a particle of mass *m* at *k* distance from the axis, or mk^2. Since this is equal to the actual moment of inertia (I), $mk^2 = I$ and

$$k = \sqrt{I/m} \qquad \text{(Equation 21-L)}$$

The use of this equation in estimating the radius of gyration and moment of inertia of body segments will be further discussed in Lesson 24. However, you should first understand that the moment of inertia is conceptualized as mr^2 though practically always measured as mk^2.

Examples

Fig. 21-8 shows four positions assumed by an individual on a turntable. Dyson (1971) has shown that the moments of inertia of these positions are approximately those given below each figure. For example, Fig. 21-8, *B,* shows a person standing rigidly upright on the turntable with arms extended horizontally. In this position the moment of inertia of his body about a vertical axis is approximately three times what it is when the arms are held at the sides (Fig. 21-8, *A*). In the arabesque skating and dancing position (Fig. 21-8, *C*) resistance to turning is about six times greater than in the position shown in *A*. Likewise, the position shown in Fig. 21-8, *D,* in which the individual is lying horizontally, has

a moment of inertia approximately fourteen times greater than does the position shown in *A*.

If a torque were actually applied to the person in the positions shown in Fig. 21-8, *A* and *B,* it could be demonstrated that he is three times more difficult to turn when his arms are held out than when his arms are held in. Likewise, it could be demonstrated that in order to produce an equal angular acceleration in these two situations three times as much angular impulse (Tt) would have to be applied when the arms were out. In other words, during a given time interval either a force three times as great is necessary or a cer-

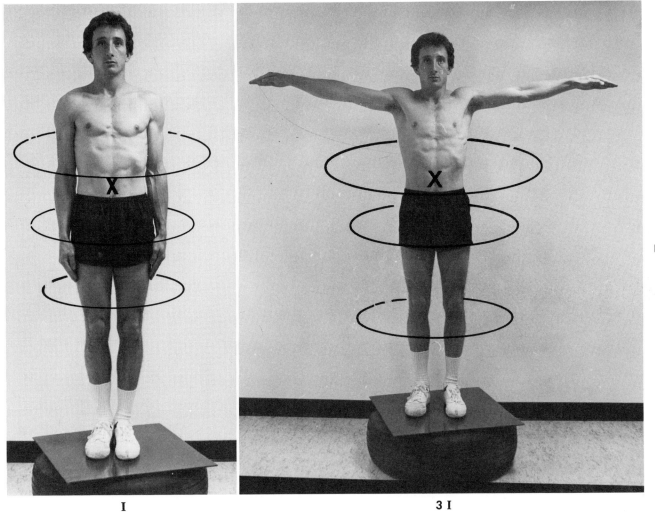

I

3 I

Continued.

Fig. 21-8. The moment of inertia of a body is affected by the distribution of its mass relative to its axis of rotation. The moment of inertia of each position is given below the figure.

C

6 I

D

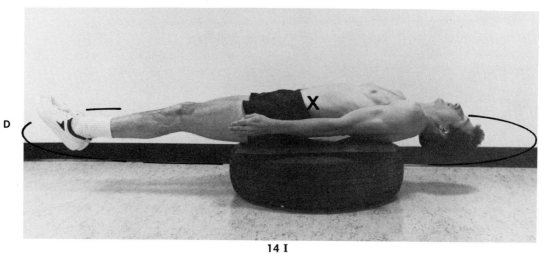

14 I

Fig. 21-8, cont'd. For legend see p. 343.

tain amount of force must be applied three times the distance from the axis when the arms are out as when they are in to produce the same acceleration.

If the person is rotating with equal speeds in the positions shown in Fig. 21-8, *C* and *D,* then it should be evident that position *D* will have more kinetic energy ($\frac{1}{2}I\omega^2$) and a greater amount of angular work ($T\theta$) must be performed to change his motion.

STUDY GUIDELINES
Introduction
Specific student objectives

1. Distinguish between centripetal and centrifugal forces.
2. Identify the formula for centrifugal force.
3. Differentiate between radial and tangential accelerations.

Summary

Newton's third law of motion (reaction) tells us that there must be an equal and opposite reaction force for every inwardly directed centripetal force. This outward force is called the centrifugal radial force and is a manifestation of the inertial tendency of all objects to travel in a straight line.

Radial forces (centripetal and centrifugal) produce radial accelerations and changes in an object's direction of motion. The force which changes a motion's speed is called a tangential force, and the resulting tangential acceleration is directed at a tangent to the circular path, i.e., at right angles to the radius and to the radial forces.

Performance tasks

1. Identify a specific angular motion task. Then identify and describe the following angular motion terms in regard to that task:
   a. Centripetal force and acceleration
   b. Centrifugal force and acceleration
   c. Tangential force and acceleration
2. All particles in angular motion have (radial/ tangential)_____accelerations.
3. An outward reaction force acting during angular motion is called (centripetal/centrifugal)_____.
4. Select two angular motion tasks and identify the acting centripetal/centrifugal forces.

5. A discus thrower in angular motion would be producing accelerations called:
   a. radial　　　　　　c. both a and b
   b. tangential　　　　d. neither a nor b

Analysis of tangential acceleration and force
Specific student objective

1. Describe the meaning of the following equations and each of the symbols used:
   a. $a = \sqrt{a_t^2 + a_r^2}$
   b. $F = \sqrt{F_t^2 + F_r^2}$
   c. $F_r = mV^2/r = mr\omega^2 = F\cos\theta$
   d. $F_t = mr\alpha = F\sin\theta$
   e. $a_t = r\,\alpha$

Summary

The tangential force (F_t) can be obtained from $F_t = mr\alpha$, where *m* is mass, *r* is the radius of rotation, and α is the total angular acceleration.

The total forces involved in angular motion can be determined through the use of $F = \sqrt{F_t^2 + F_r^2}$, while the total acceleration can be found from $a = \sqrt{a_t^2 + a_r^2}$.

Performance tasks

1. Identify and write out the meaning of the formulas used in determining tangential force and acceleration.
2. Using the angular motion information which follows, calculate F_t, a_t, F_r, and a_r.

	ω_0	ω_f	t	r	m
a.	20 rad/sec	30 rad/sec	2 sec	5 ft	2 slugs
b.	7 rad/sec	10 rad/sec	3 sec	4 ft	2 slugs
c.	10 rad/sec	50 rad/sec	4 sec	2 m	3 kg
d.	26 rad/sec	10 rad/sec	2 sec	1 m	5 kg

Angular analogs of linear motion quantities
Specific student objective

1. Identify and describe the angular analogs for
   a. mass　　　　　　d. impulse
   b. momentum　　　 e. kinetic energy
   c. causal agent (force)　f. work

Summary

Linear and angular motion analyses involve many similarities. In general, similar linear concepts are expanded and applied to circular motion. The following concepts (see next page) are essential to motion analysis.

	Linear	**Angular**
Causal agent	Force (F)	Torque (Ff) or (F_tr)
Inertia	Mass (m)	Moment of inertia ($I = mr^2$)
Momentum	$M = mV$	$L = I\omega = mr^2\omega$
Impulse	Net $Ft = M_f - M_o$	Net $Tt = L_f - L_o$
Kinetic energy	K.E. $= \frac{1}{2}mV^2$	K.E. $= \frac{1}{2}I\omega^2$
Work	$W = Fd$	$W = T\theta$

Performance tasks

1. The concept of inertia is expanded in angular motion to include a rotating mass plus the
   a. distance (squared) from the axis
   b. torque used
   c. both a and b
   d. neither a nor b
2. $I = mr^2$ is the angular analog of mass and is called
   a. kinetic energy   c. momentum
   b. impulse   d. moment of inertia
3. The causal agents for linear and angular motion are, respectively
   a. F and T   c. m and I
   b. m and L   d. all of the above
4. Develop a written table of analogous motion quantities for linear and angular motion.

Angular analogs of Newton's laws of motion
Specific student objectives

1. State Newton's three laws of angular motion.
2. Illustrate the angular analogs of Newton's laws of motion with specific motion examples.

Summary

The following are Newton's laws of angular motion:

I. Inertia. An object remains at rest or rotates at constant speed unless acted on by an unbalanced torque (net $T = 0$ implies $\alpha = 0$).

II. Acceleration. The angular acceleration (α) produced by a net torque (net $T = F_tr$) acting on a body is directly proportional to the net torque and inversely proportional to the object's moment of inertia (I). In formula form net $T = I\alpha$ or $\alpha = $ net T/I or net $F_tr = mr^2\,\alpha$.

III. Reaction. Whenever one body acts on a second body with a torque, the second body exerts a torque on the first body. These torques are equal in amount and opposite in direction.

Performance tasks

1. The tendency of a rotating object to maintain its existing state of motion is a characteristic called:
   a. acceleration   c. reaction
   b. inertia   d. torque
2. Angular acceleration is produced by a:
   a. net force   c. both a and b
   b. net torque   d. neither a nor b
3. Write out the angular analogs of Newton's laws of motion.
4. Illustrate each of Newton's laws of angular motion with specific motion examples.

Moment of inertia and the radius of gyration
Specific student objectives

1. Differentiate between the radius of rotation and radius of gyration.
2. Differentiate between the ring and radius of gyration.
3. Define the positions of minimum and maximum moment of inertia of the human body relative to an axis passing through its center of gravity.
4. Define the position of maximum moment of inertia of the human body relative to an axis not passing through its center of gravity.
5. Define radius of gyration (in both sentence and formula form).

Summary

The radius of rotation is the distance from the axis of rotation of each particle of mass making up a body. The radius of rotation of a total object is determined by vector adding the different radii of rotation of the different particles.

Whatever the shape of a body, it is always possible to find a radial distance from any given axis at which the mass of the body could be concentrated without altering the moment of inertia of the body about that axis. This distance is called the radius of gyration of the body about the given axis and is represented by k. In formula form the radius of gyration is $k = \sqrt{I/m}$.

The radius of gyration is a concept similar to the concept of center of gravity. Here it is conceptualized that the complex mass distribution of a body can be replaced by an infinitely thin circular "ring" of matter around the center of mass of the body with the same quantity of mass as the body itself about the axis of rotation. The radius of the ring is known as the radius of gyration, while the ring itself is known as the ring of gyration.

Relative to an axis passing through the body's center of gravity, the radius of gyration and thus the moment of inertia of the human body is least about its longitudinal (y) axis, with the arms extended and close together above the head, and greatest when the body is turning in a similar position, but with arms parallel, about a horizontal axis.

Relative to an axis not passing through the body's center of gravity, the moment of inertia is greatest in a fully extended position with the hands as the axis of rotation, because in this situation the greatest distribution of mass is away from the center of rotation.

Performance tasks

1. The radial distance from any given axis at which the mass of the body could be concentrated without altering the moment of inertia of the body about that axis is called
   a. radius of rotation
   b. radius of momentum
   c. radius of gyration
   d. radius of displacement
2. Write out descriptions of the ring and radius of gyration
3. Describe the position of minimum and maximum moment of inertia (a) relative to an axis passing through the center of gravity and (b) relative to an axis not passing through the center of gravity.
4. The formula for radius of gyration is
   a. $k = mr^2$ c. $k = \sqrt{I/m}$
   b. $k = I/m$ d. $k = \frac{1}{2}mV^2$

SELF-EVALUATION

Students should use no reference materials for this progress test, and they can check their answers by referring to Appendix A.

1. The causal agent for angular motion is
   a. force
   b. torque
   c. both a and b
   d. neither a nor b
2. State Newton's three laws for angular motion and illustrate each with a specific motion example.
3. The two types of radial forces in angular motion are
   a. torque and momentum
   b. mass and moment of inertia
   c. both a and b
   d. neither a nor b
4. $F_r = mV^2/r$ is the formula for
   a. centripetal force
   b. centrifugal force
   c. both a and b
   d. neither a nor b
5. Radial forces affect the motion of an object by changing its
   a. speed
   b. direction
   c. both a and b
   d. neither a nor b
6. Tangential acceleration of a rotating object means that the object is changing its
   a. speed
   b. direction
   c. both a and b
   d. neither a nor b
7. Throwing a ball with the angular motion of the arm results in an acceleration called
   a. radial
   b. tangential
   c. both a and b
   d. neither a nor b
8. Using $F_t = mr\alpha$ and $a_t = r\alpha$, calculate the respective tangential forces and accelerations in the motion situations that follow:

	m	r	α	F_t	$r\alpha$
a.	10 kg	1 m	15 rad/sec^2	_____	_____
b.	25 kg	500 ft	5 rad/sec^2	_____	_____
c.	5 slugs	5 ft	10 rev/min^2	_____	_____
d.	200 slugs	100 ft	10 rad/sec^2	_____	_____
e.	20 slugs	20 ft	1 rad/sec^2	_____	_____

9. Complete the motion analysis table below by identifying and describing the angular analog terms.

10. The vector sum of the different radii of rotation of the different particles making up an object is called the
    a. radius of rotation
    b. radius of gyration
    c. radius of displacement
    d. radius of momentum

		Linear	**Angular**
a.	Causal agent	Force (F)	
b.	Inertia	Mass (m)	
c.	Momentum	$M = mV$	
d.	Impulse	Net $Ft = M_t - M_o$	
e.	Kinetic energy	$K.E. = \frac{1}{2}mV^2$	
f.	Work	$W = Fd$	

Kinetic analysis of angular momentum

Body of lesson

ANGULAR MOMENTUM AND NEWTON'S LAWS

The angular momentum of a rotating body is the product of its moment of inertia and its angular velocity ($mr^2\omega$ or $I\omega$). Like linear momentum, angular momentum is a vector quantity possessing both magnitude and direction. Newton's laws can be expressed in terms of angular momentum as follows:

NEWTON I. A body in rotation will continue to turn about its axis with an angular momentum constant in both magnitude and direction unless an external net torque is exerted on it.

NEWTON II. The rate of change of angular momentum of a body is proportional to the torque causing the change and has the same direction as the torque.

$$T = \frac{L_f - L_o}{t}$$

NEWTON III. For every applied torque tending to produce angular momentum in one direction, there is an equal and opposite reaction torque tending to produce angular momentum in the opposite direction.

Some implications of Newton's first law

Newton's first law, also known as the law of the conservation of angular momentum, says that the total angular momentum of a system of particles remains constant when no net external torque acts upon it. More formally it states:

When the sum of the external torques acting upon a system of particles equals zero, the total angular momentum of the system remains constant in both magnitude and direction.

A skater or ballet dancer doing a spin capitalizes upon the conservation of angular momentum.

349

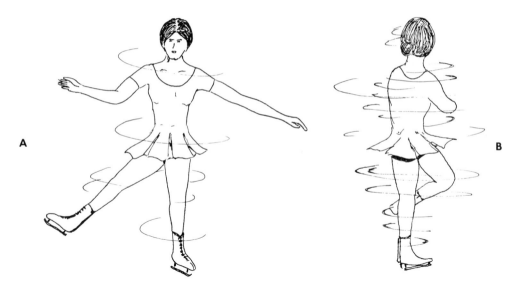

Fig. 22-1. An ice skater doing a spin capitalizes upon conservation of angular momentum. The skater in **A** already has an angular momentum with her arms and one leg outstretched. By bringing her arms and extended leg inward, **B**, she reduces her moment of inertia considerably and consequently spins faster. Thus her momentum ($mr^2\omega$) before and after the change in her moment of inertia has remained constant.

In Fig. 22-1, *A,* a skater is shown already with angular momentum starting her spin with her arms and one leg outstretched. By bringing her arms and extended leg inward (Fig. 22-1, *B*), she reduces her moment of inertia considerably and consequently spins faster; thus her momentum ($mr^2\omega$) before and after the change in her moment of inertia has remained constant. Likewise, when she wishes to decrease her spin, she increases her moment of inertia by re-extending her arms and leg outward. These changes in her rate of spinning (ω) are due solely to the changes in the length of her radius of gyration. Her mass is not changed but only redistributed.

Newton's third law is very much related to his first law, as illustrated in Fig. 22-2. A man sitting on a chair is free to rotate with a minimum amount of friction to simulate the rotation of a body in the air free of support. In Fig. 22-2, *A,* he is motionless so his angular momentum is zero. In Fig. 22-2, *B,* he swings his right arm to the left, which produces a reaction movement of his lower body to the right. These counterclockwise and clockwise angular momentums cancel one another out to leave a net angular momentum of zero.

Thus the original state of zero angular momentum is conserved.

Some implications of Newton's second law

According to Newton's second law, when a rotating body is acted upon by a net torque the resulting rate of change in angular momentum will be directly proportional to the net torque and be in the some direction as this torque; i.e., $T = (L_f - L_o)/t$. When we rearrange this equation by moving the time *(t)* to the left side, we have the equation for angular impulse. Thus the greater the torque and the greater the time during which it is applied, the greater will be the resulting change of momentum. This principle is especially important in eccentric thrust applications.

As was mentioned in Lesson 13, in order for angular motion to occur, a *lever arm* must be created by a force application which is not directed through the center of rotation. We called this "off-center" force an eccentric thrust. The magnitude of the eccentric thrust and the length of the lever arm determine the amount of torque acting. According to Newton's first law, angular

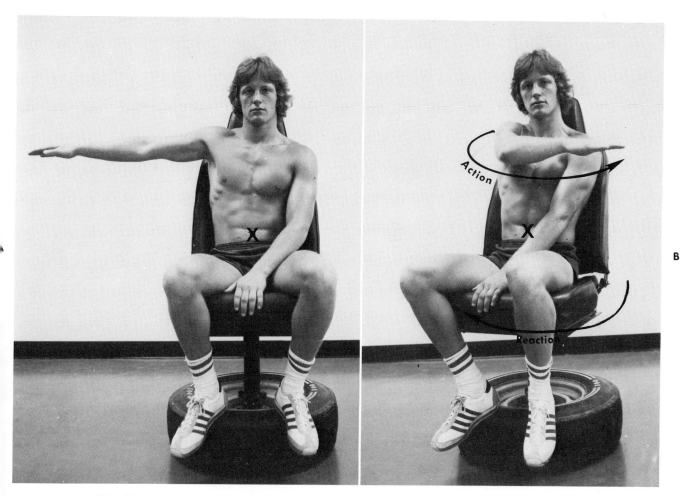

Fig. 22-2. Demonstration of Newton's first and third laws. In **A** a man is sitting on a chair that is free to rotate. The angular momentum of the man and chair is zero. In **B** he swings his right arm to the left, which produces a reaction movement of his lower body to the right. These counterclockwise and clockwise angular momentums cancel one another to leave a net angular momentum of zero. Thus the original state of zero angular momentum is conserved.

momentum will be established when the amount of applied torque is greater than the resistance torque. Newton's second law says that the *amount* of angular momentum which will be established by the unbalanced torque is directly dependent upon the amount of net torque and the time the torque acts; i.e., it is dependent upon the amount of angular impulse. Thus the factor responsible for giving a body angular momentum in the first place, and for increasing or decreasing the angular momentum once a body is in rotation, is angular impulse.

Angular impulse, for example, is created by a springboard diver leaning either forward (Fig. 22-3) or backward to get the center of gravity either ahead of or behind the feet in order to produce a torque. Springboard divers, as well as trampolinists, also learn that to obtain greater linear and angular impulse, they must "ride the board" or "ride the tramp" by staying in contact with it for a maximum length of time. The linear impulse is what causes the body to be projected into the air for the attainment of height, while the angular impulse is what causes the body to rotate (Fig. 22-3).

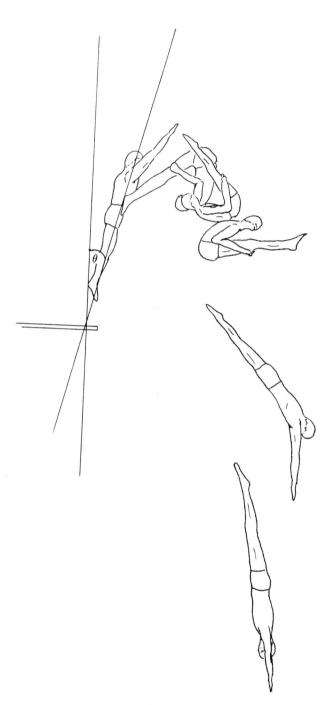

Fig. 22-3. The angle of forward lean *(θ)* is necessary to get the center of gravity ahead of the feet in order to produce torque. The linear impulse is what causes the body to be projected into the air for the attainment of height; the angular impulse causes the body to rotate.

Some implications of Newton's third law

Newton's third law says that whenever a torque tending to produce angular momentum in one direction is applied in a body there is an equal and opposite reaction torque created which tends to produce angular momentum in the opposite direction in another part of the same body. An example of this law applied to the free flight of a body would be a long jumper swinging his legs counterclockwise forward and up for landing, as shown in Fig. 22-4. This applied torque creates an equal but opposite torque which causes his upper body to rotate clockwise forward and down.

Action and reaction pairs are also created in the baseball batter in Fig. 22-5. The counterclockwise rotation of the batter's upper body and bat will create an equal and opposite torque which will act upon his lower body in a clockwise direction. If the feet are planted and not free to rotate, the reaction torque will be absorbed or transferred to the supporting surface. The magnitude of the clockwise reaction torque produced in the lower body, for example, can be measured through the use of a force platform as shown in Fig. 22-6, or it can be calculated through an analysis of the counterclockwise movements employing the principles discussed in the last lesson.

AXES OF ANGULAR MOMENTUM AND OF ANGULAR DISPLACEMENT

According to Newton's first law and the conservation of angular momentum, the total angular momentum of a rotating body will remain constant in both magnitude and direction when no net external torque acts upon the body. The magnitude of the momentum, of course, is equal to $mr^2\omega$. The direction of the angular momentum, however, is given by the spatial orientation (direction) of the axis around which the momentum is rotating. The spatial orientation or direction of this axis is the angle it makes with the +y and +z cardinal axes. This axis (called the axis of angular momentum) is determined by calculating the body's separate moments about the cardinal x, y, and z axes and then vectorally adding these moments to obtain a resultant momentum. Vector analysis of angular momentum is further discussed in a later section of this lesson.

The axis of momentum of a body moving through the air will remain fixed in direction

Fig. 22-4. Demonstration of Newton's third law, which says that whenever a torque tending to produce angular momentum in one direction is applied in a body an equal and opposite reaction torque tends to produce angular momentum in the opposite direction in another part of the same body. Thus, when the legs of the long jumper swing forward in a counterclockwise direction the upper body rotates forward in a clockwise direction.

Fig. 22-5. The counterclockwise rotation of a baseball batter's upper body and bat will create an equal and opposite torque which will act upon his lower body in a clockwise direction.

Fig. 22-6. The magnitude of the reaction torque produced in the lower body of a field hockey player or any athlete can be measured through the use of a force platform.

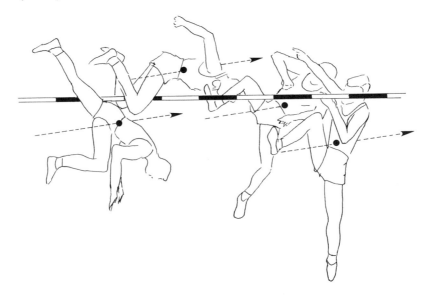

Fig. 22-7. The axis of momentum of a body moving through the air will remain fixed in direction when air resistance is disregarded until contact with a surface is regained. (Modified from Dyson, G. H. G.: The mechanics of athletics, London, 1971, University of London Press, p. 85.)

(when air resistance is disregarded) until contact with a surface is regained. The axis of angular momentum of a high jumper is shown in Fig. 22-7. This axis, which is directed at an angle of about 20° to the crossbar, remains fixed in direction throughout the jump.

Although the axis of momentum must be initiated when the body is in contact with the ground or other surface of support, rotations can be initiated in the air when the body is free of support. Such rotations occur around an axis of angular displacement that passes through the center of gravity but that in most cases is distinct from the total axis of angular momentum. The axis of angular displacement is the axis around which all action and reaction movements occur when rotations are initiated in the air. These movements, of course, maintain the state of zero total momentum change.

Fig. 22-8 shows a gymnast initiating a turn in the air while in contact with the ground. He first rotates his arms, shoulders, and head in a horizontal plane with his feet still on the ground. Then, as his legs drive him into the air, the angular momentum of his arm movements is transferred to his body. This transfer causes his whole body to turn in the air around an axis of momentum as shown in the illustration. Once in the

air, however, movements of the gymnast's body are initiated around axes of displacement which must be compensated for by equal and opposite reaction movements around these same axes.

Another example of a constant axis of momentum and variable axis of displacement is illustrated in Fig. 22-9. A diver is creating a horizontal axis of momentum while in contact with the board and then is creating an axis of displacement around which he initiates a half twist once he is in the air. His axis of momentum maintains its spatial orientation (direction) throughout the motion but his axis of displacement does not.

EFFECTS OF THE REDISTRIBUTION OF MASS ON THE RADIUS OF GYRATION AND THE SPEED OF ANGULAR MOTION

The illustrations and discussions presented so far in this lesson have demonstrated that the angular speed of a rotating body can be changed by the redistribution of the body's mass, which changes its radius of gyration and thus its moment of inertia. This concept is so important that this section of the lesson is devoted to a further discussion of it.

When a force is applied to a pivoted object, the tendency of the object to resist angular acceleration depends on its moment of inertia (mr^2).

Fig. 22-8. A gymnast must initiate a turn in the air while in contact with the ground.

Fig. 22-10. A runner shortens the radii of gyration of his upper and lower limbs by flexing his elbow and knee joints.

Fig. 22-9. Demonstration of the difference between the axis of momentum and the axis of displacement. The diver creates a horizontal axis of momentum while in contact with the board and then creates an axis of displacement once he is in the air. The axis of momentum maintains its spatial orientation (direction) throughout the motion, while the axis of displacement does not.

Since the mass of a body is a constant, the moment of inertia of the body is directly dependent upon the distribution of its mass round the axis of rotation, and this distribution determines the magnitude of the radius of gyration. Thus, in running, recovery movements of the arms and legs are often accomplished with the limbs in shortened positions.

Fig. 22-10 shows that a runner shortens the radii of gyration of his upper and lower limbs by flexing his elbow and knee joints. These flexions redistribute the masses of the limbs closer to the axis of rotation so they can be recovered more rapidly.

The effect of shortening the radius of gyration on angular speed can be seen quite vividly in the motion of a springboard diver rotating around an x axis as shown in Fig. 22-11. The angular speed of the diver's rotation can be increased by assuming a tucked position (*D*) and decreased by assuming a layout position (*G*).

A geometrical interpretation of angular momentum conservation as presented by Dyson (1971) is shown in Fig. 22-12. The motion of a diver's ankles around his center of gravity during a given time conserves angular momentum. If the angular momentums of positions *A* and *B* are equal, the areas swept out by the diver's ankles will also be equal. It should be noted that the straight position (*A*) has approximately three and one half times the moment of inertia as will be found in the tucked position (*B*) around a horizontal axis. Thus, if the diver leaves the springboard with just enough angular momentum for one complete somersault in the straight position, he will be able if he so chooses to perform from two to two and a half somersaults in the tucked position.

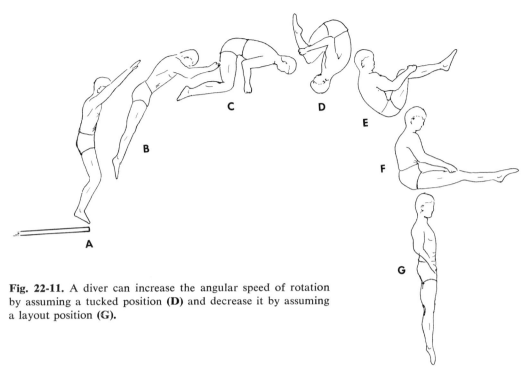

Fig. 22-11. A diver can increase the angular speed of rotation by assuming a tucked position (**D**) and decrease it by assuming a layout position (**G**).

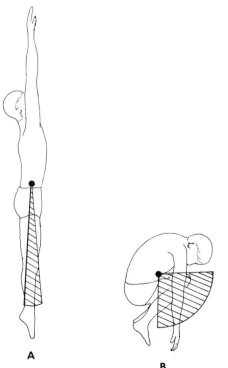

Fig. 22-12. Geometrical interpretation of angular momentum conservation. If the angular momentums of positions **A** and **B** are equal, the areas swept by the diver's ankles per unit of time will also be equal.

ANALYSIS OF ROTATIONS INITIATED IN THE AIR

Human beings are able to initiate rotations in the air using basically the same principles of dynamics as do cats, rabbits, and other animals which have the ability to right themselves after being released in the air upside down. As shown in Fig. 22-13, a cat is able to right itself by exploiting its relative moments of inertia. The movements of the cat follow the stages given below:

Stage 1. The cat "pikes" or bends in the middle and stretches out its hind legs almost perpendicular to an axis passing through its upper section and brings its front legs in close to its head.

Stage 2. After redistributing its mass in stage 1, the cat twists the forepart of its body through 180°—which makes the head, forelegs, and upper trunk ready for landing. During this time the hindparts are displaced through a much smaller angle, in the opposite direction, because of their much greater moment of inertia about this axis.

Stage 3. During the third stage of the animal's fall, it stretches out its front legs and brings in its hindlegs and its twisting takes place about an axis parallel with its hind section. The twist of the hind section occurs in the same direction as the twist of the head and trunk during the second phase. There-

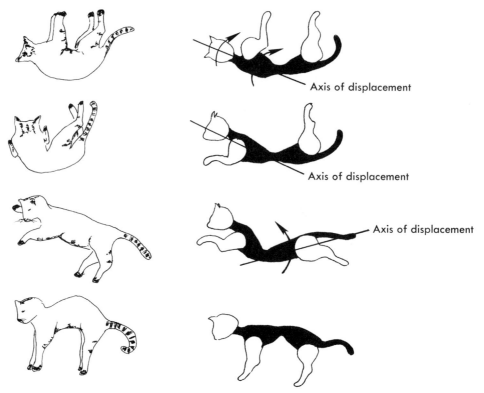

Fig. 22-13. A cat is able to right itself by exploiting its relative moments of inertia. (Modified from Hay, J. G.: The biomechanics of sports techniques, Englewood Cliffs, N.J., 1973, Prentice-Hall, Inc., p. 163.)

fore its hind section now turns through the larger angle, for the moment of inertia of its upper body about this new axis is much greater.

Stage 4. At the completion of this series of movements, the animal's whole body is ready for landing and has turned through 180°.

A diver or trampolinist is able to initiate rotations in the air by also exploiting relative moments of inertia. A good demonstration of this is to have a trampolinist bounce on a trampoline and initiate rotations in the air to either the left or the right upon a command shouted to him or her after the performance is started. A well-trained trampolinist can initiate the rotations by alternately increasing and decreasing the moments of inertia of the upper and lower sections. Fig. 22-14 shows a forward dive with one twist in the layout position being initiated in the air.

TRANSFER OF ANGULAR MOMENTUM

Not only can *mass* be redistributed in a body and thus change the length of the radius of gyra-

tion and the speed of a rotating body (as shown in Fig. 22-1) but there can also be a redistribution of *momentum* among the body's component parts. Momentum, in other words, can be traded or transferred from one part of the body to another, as well as from its linear to its angular form, or vice versa, without a change occurring in the total amount of momentum involved. Momentum can be transferred from the whole body to a part, from one part to another part, or from a part to the whole. During all of these transfers the total momentum is conserved in the absence of an external net force or torque.

Transfer of momentum to different parts within a system

The conservation of momentum law says that the momentum of the body as a whole can be transferred to a body part or to an external object such as a javelin. Likewise, the momentum developed in a body part—such as the arms in doing

Fig. 22-14. A diver is able to initiate rotations in the air by exploiting his relative moments of inertia.

a back flip in tumbling—can be transferred to the body as a whole. However, the javelin is considered to be an external object only in the sense that it is external to the body; in applications of the law of conservation of momentum, it is considered to be internal to and a component part of the rotating system.

The redistribution of momentum within a body (which we are calling the transfer of momentum) is illustrated in Fig. 22-15. When the diver is in the air, any increase in the angular momentum of one part of his body must be accompanied by an equal decrease in the angular momentum of some

other part to maintain the condition of zero momentum change.

When the diver is rotating in a pike position and begins to assume his position of entry, the angular momentum of his arms and trunk is reduced to approximately zero while the momentum of his legs is greatly increased during the time that they are moving into line with the rest of the body. This decrease in the momentum of the arms and trunk accompanied by an increase in the momentum of the legs maintains the constant angular momentum of the total body.

Fig. 22-15. Demonstration of the transfer of momentum to different parts within a system. The decrease in momentum of the arms and trunk and the increase in momentum of the legs during the time they are brought back into line with the trunk (as shown in the last three illustrations) maintain the constant angular momentum of the total body.

Transfer of momentum from one type to another

The conservation of momentum law also says that linear momentum can be transferred or changed into an equal amount of angular momentum and vice-versa as long as no external net force or torque is acting. The angular momentum of a ball being thrown, for example, is changed into an equal amount of linear momentum when

it is released. In other words, the total momentum of a body is the sum of its linear and angular components; i.e., $M_{total} = mV + mr^2\omega$.

Transfer of momentum when an external net force or torque is acting

Linear momentum can be transferred or changed into angular momentum and vice-versa when an external net force or torque is acting. The presence of the external net force or torque means that momentum will not be conserved exactly but will be transferred so a decrease in one type of momentum is accompanied by an increase in the other type.

When a body or body part is moving in a straight path and is suddenly checked at one end, the other end acquires an angular momentum and the angular momentum of the total body is related to the previous linear momentum. In vaulting over a horse (Fig. 22-16), for example, the gymnast, after a preliminary run-up, momentarily fixes both feet at takeoff. This fixation causes the rest of his body to rotate forward and to transfer (or change) part of its linear momentum into angular momentum. The clockwise turning thus started continues in the air, bringing the head and shoulders down and the feet up in relation to the center of gravity. When his hands finally strike the horse, they check both the linear and the angular momentums of the body and transfer these momentums into a reversed counterclockwise rotation of the body.

Linear momentum can be transferred into angular momentum and then back into linear momentum as seen in the javelin throw illustrated in Fig. 22-17. The thrower checks his forward linear momentum with his front foot, which transfers this momentum into angular momentum. At the same time he turns simultaneously about the horizontal x axis (passing through the point where his front foot strikes the ground) and a near vertical axis passing through the throwing base. These two turns impart considerable angular momentum (velocity) to the throwing shoulder and, thus, to the javelin. The angular momentum is transferred back to the linear momentum of the javelin at the time of release.

Since the linear velocity of a rotating body is directly proportional to its radius of rotation, body height can be advantageous to a thrower.

Fig. 22-16. Demonstration of the transfer of momentum when an external net force or torque is acting. In vaulting over the horse, the gymnast, after a preliminary run-up, momentarily fixes both feet at takeoff. This causes the rest of his body to rotate forward and to change part of his linear momentum into angular momentum, bringing his head and shoulders down and his feet up around his center of gravity (**A** to **C**). When his hands strike the horse (**C**), his linear and angular momentums are checked causing a reverse counterclockwise rotation of his body (**D**) and placement in position for landing (**E**).

Assuming for a moment that after checking with his front foot the javelin thrower continues his previous linear velocity, we can see that the shoulder above must move considerably faster. The greater the height of the shoulder above the center of gravity, the greater will be the multiplication of its linear velocity.

The linear velocity of a rotating body is also directly proportional to its angular speed. Thus the greater the speed at which the thrower turns around his horizontal and vertical axes just prior to releasing the javelin, the greater will be the linear speed transferred to the javelin.

Perhaps it should be pointed out here that the checking of momentum must cause a slight loss of speed due to the transfer of some of the momentum to the ground. For example, some of the linear momentum of a pole vaulter can be transferred to angular momentum when he plants his pole in the vaulting box (Fig. 22-18). The efficiency of the transfer, however, is directly proportional to the angle created between the ground and the pole at the first instant of contact. In other words, the smaller the angle between the

pole and the ground the greater the loss will be. This is due to the fact that only the tangential component of the linear momentum can be effectively transferred to angular momentum and, as we already know, the tangential component is directly proportional to the sine of the angle of application. The remainder of the momentum is radial and is transferred down the pole to the box and the ground.

VECTOR ANALYSIS OF ANGULAR MOMENTUM

Since the understanding of angular momentum is so fundamental to the understanding of angular motion, its vector nature must be thoroughly understood. This understanding is especially important in the analysis of gyratory motion which is the topic of Lesson 23.

Graphic representation of angular vectors

Any angular motion vector—angular velocity, angular acceleration, torque, etc.—can be graphically represented by a straight line whose length is the magnitude of the vector and whose direc-

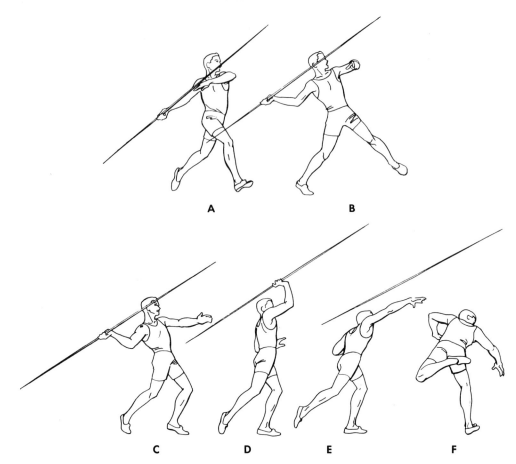

Fig. 22-17. Linear momentum is transferred into angular momentum when the javelin thrower checks his forward progress with his front foot (A). The angular momentum of the system is transferred back to the linear momentum of the javelin at the time of release. (From Cooper, J. M., and Glassow, R. B.: Kinesiology, ed. 4, St. Louis, 1976, The C. V. Mosby Co.)

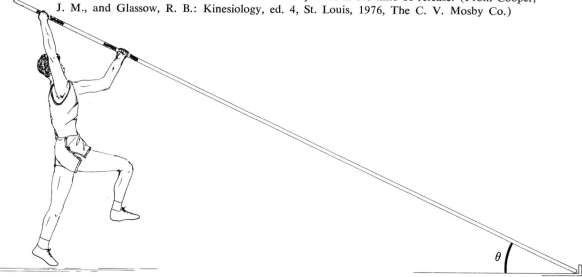

Fig. 22-18. The linear momentum of a pole vaulter is changed to angular momentum when he plants his pole in the vaulting box. The efficiency of the transfer, however, is directly proportional to the angle (θ) created between the ground and the pole at the first instant of contact.

Fig. 22-19. The right-hand thumb rule.

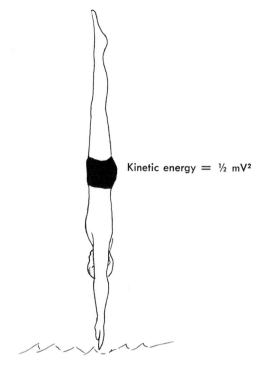

Kinetic energy $= \frac{1}{2}\, mV^2$

Fig. 22-20. A person diving into a swimming pool has kinetic energy that does work upon the water in the pool.

tion is parallel with the axis of rotation. The unique problem in graphical representation of angular motion vectors is the determination of direction, and in this regard the right-hand thumb rule has been found useful. If the curled fingers of a person's right hand point in the direction of the rotation, the extended thumb will point in the direction in which the motion vector is acting (Fig. 22-19). This direction is perpendicular to the motion plane and parallel with the axis of motion.

The fact that the direction of angular vectors is always perpendicular to the plane of the motion might seem strange. However, remember that angular quantities are usually derived as the products of other quantities. Torque, for instance, has been defined as the product of the perpendicular force (F_t) and the radius (r). Angular momentum is the product of $mr^2\omega$. Thus angular quantities result from the multiplication of vectors, and the direction of these vectors as well as their magnitudes must be considered in such an analysis.

The two distinct kinds of products obtained in the multiplication of vectors are the *scalar or dot product* and the *vector or cross product*.

Examples of scalar and vector products

The law of the conservation of energy and the law of the conservation of momentum can be used to demonstrate that energy is a scalar quantity and momentum is a vector quantity.

Energy as an example of a scalar quantity. That energy is a scalar quantity can be demonstrated by considering the various forms of energy involved when a person dives into a swimming pool (Fig. 22-20). The diver before entering the water has the energy of motion ($\frac{1}{2}mV^2 + \frac{1}{2}I\omega^2$). Upon entry his kinetic energy does work upon the water and produces some heat and sound. We know, however, from the conservation of energy law that the total energies produced by the dive must equal the total kinetic energy present before entry; but in what direction, for example, is the work done

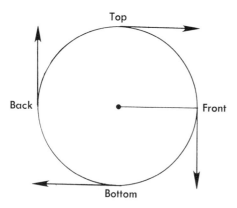

Fig. 22-21. Every particle on the surface of a rotating ball will exert a force in a direction parallel with its plane of motion and perpendicular to its radius. Because of opposite directions of these forces, their vector sum is zero.

upon the water? After a little thought you should be able to see that the water is moved by the diver in all possible directions: up, down, forward, backward, right, and left. Thus the total work done in this example is determined by simply adding the work done in all six directions without consideration of the direction factor.

That angular work is also a scalar quantity can be demonstrated by considering the work done by a rotating ball when it is also placed in a swimming pool. As shown in Fig. 22-21, every particle on the surface of the rotating ball exerts a force upon the water which is parallel with its plane of motion and perpendicular to its radius of rotation. However, the direction of this force changes with the rotation of the ball. The force produced by the particle when it is at the top of the ball rotating (let's say) with topspin is forward while the force produced when it is on the bottom is backward. Likewise, the force produced by the particle when it is located in front is downward and the force produced when it is located in back is upward. The fact that the work done by the ball is *not* zero demonstrates that work is a scalar quantity and that the total work done must be determined without consideration of direction.

The scalar nature of energy is also demonstrated when we think about the heat and sound produced by the diver upon entry into the pool. In what direction do the heat and sound act? Obviously they do not act in a single direction but in all possible directions. Thus work and energy

must be quantified without consideration of the direction factor.

Momentum as an example of a vector quantity. That angular momentum is a vector quantity was demonstrated earlier in the lesson (Fig. 22-2, *A*). A man sitting on a chair has an angular momentum of zero. When he moves his right arm to the left in a counterclockwise or positive direction, he produces a clockwise or negative rotation of his lower body to the right (Fig. 22-2, *B*). These positive and negative rotations cancel each other to leave a net angular momentum of zero. The zero momentum change demonstrates that angular momentum is a vector, with both magnitude and direction.

That linear momentum is also a vector quantity was demonstrated in Lesson 18, where a hand grenade of mass m initially at rest suddenly exploded into many particles (of masses m_1, m_2, m_3, . . . m_n) which flew apart. The forces acting on the hand grenade and causing it to break up were internal; there was no external force present. Since the hand grenade had an initial momentum of zero, the final momentums of the many smaller masses, when added together, must also have been zero. This zero momentum is possible when we consider the directions of the motion and subtract the momentums in the negative directions from the momentums in the positive directions. Linear momentum is thus a vector quantity (with both magnitude and direction).

Measurement of scalar and vector products

Linear work was defined in Lesson 14 as the product of force (F) and the distance (d) through which the force acts; i.e., W = Fd. Likewise, angular work was defined in the last lesson as the product of torque (T) and the angle (θ) through which the torque acts. Since both the F and the T are vectors, as are the displacements *(d* and *θ),* obviously both linear and angular types of work are the products of vectors. Furthermore, since work has magnitude but no direction, it is defined as a scalar quantity and identified as a *scalar* or *dot product.*

The theory of scalar or dot product measurement is presented in the boxed material.

We can take torque as an example of a *vector* or *cross* product. Torque was defined in Lesson 13 as the product obtained by multiplying

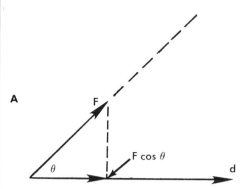

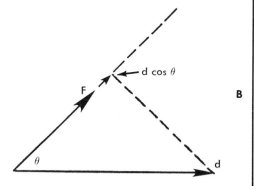

A graphical representation of two vectors (a displacement vector, *d*, and a force vector, *F*) whose directions make an angle *(θ)* with each other is shown above. In diagram **A** the component of *F* in the direction of *d* is given by *F cos θ* and is positive or negative depending on whether *θ* is smaller or greater than 90°. In diagram **B** the component of *d* in the direction of *F* is given by *d cos θ*. Since the two right triangles in these diagrams are similar, corresponding sides are proportional and we can write

$$F : F \cos \theta = d : d \cos \theta$$

Cross-multiplying to find the amount of mechanical work, we obtain

$$Fd \cos \theta = dF \cos \theta$$

Thus the order of the two vectors can be reversed without changing the value of the product.

$$\text{Work} = \mathbf{F} \times \mathbf{d} = \mathbf{d} \times \mathbf{F}$$

Summary

The scalar product of two vectors (A and B) whose directions make an angle *(θ)* with each other is equal to AB cos θ. The scalar product is written

$$\mathbf{A} \cdot \mathbf{B} = AB \cos \theta$$

(Equation 22-1)

If both vectors are in the same direction, cos θ = 1 and A · B = AB; if they are in opposite directions, then cos θ = −1 and A · B = −AB. If they are at right angles, cos θ = 0 and A · B = 0.

the tangential force (F_t) by the radius of rotation (r); i.e., $T = F_tr$. Since F_t and r are perpendicular to each other, their product is also a vector directed perpendicular to both of them. Whether the vector is pointing in the positive or negative directions, however, must be determined through use of the right-hand thumb rule.

The theory of vector or cross product measurement is presented in the boxed material which follows.

DETERMINING THE AXIS OF MOMENTUM

The axis of momentum, as already mentioned, can be determined by calculating the body's separate momentums about the x, y, and z axes while the body is still in contact with a supporting surface. The calculation of these momentums for a high jumper will be used as an example in this section.

Since angular momentum is a vector quantity,

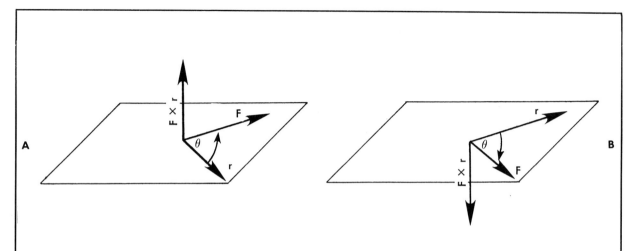

A graphical representation of two vectors (a force vector, *F*, and a radius vector, *r*) whose directions make an angle *(θ)* with each other is shown above. The vector product of these vectors is equal to torque, which is equal to *Fr sin θ* and whose direction is perpendicular to both *F* and *r*. The direction of the resultant vector *F* × *r* is determined through use of the right-hand thumb rule. Thus the vector product *F* × *r* has the same magnitude as *r* × *F* but in an opposite direction, as illustrated in diagrams **A** and **B** above. The direction is positive when the rotation of the radius is counterclockwise and negative when the rotation is clockwise.

Summary

The vector product of two vectors (A and

B) whose directions make an angle (θ) with each other is equal to a vector whose magnitude is AB sin θ and whose direction is perpendicular to both A and B, being positive relative to a rotation counterclockwise from A to B. A reversal of the order of the vectors reverses the sign of the product. The vector product is written

$$A \times B = AB \sin \theta$$

(Equation 22-2)

For two parallel vectors, sin θ = 0 and the vector product A × B = 0. If A and B are at right angles, sin θ = 1, A × B = AB, and all three vectors (A, B, and A × B) are mutually perpendicular.

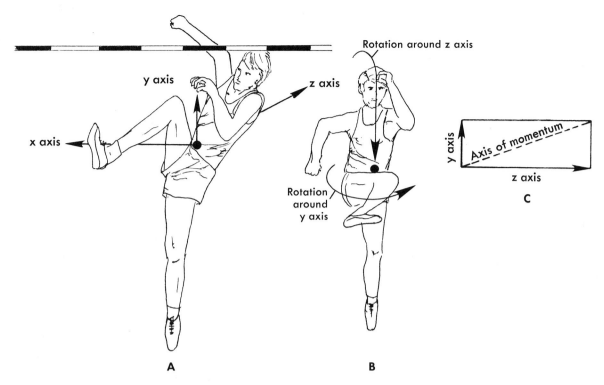

Fig. 22-22. The magnitude and direction of the axis of momentum are obtained by vector adding the momentums occurring around the x, y, and z axes. (Modified from Dyson, G. H. G.: The mechanics of athletics, London, 1971, University of London Press, pp. 85 and 118.)

the angular momentums around the x, y, and z axes of the high jumper can be represented by straight lines drawn as arrows whose lengths equal the magnitudes of the corresponding angular momentums. All three of the angular momentum vectors meet at the athlete's center of gravity. The direction of each of these vectors is determined through use of the right-hand thumb rule.

The x, y, and z axes of momentum of the high jumper are illustrated in Fig. 22-22. By using the three-dimensional parallelogram method we can establish the magnitude and direction of the total angular momentum. The resultant axis of momentum, of course, is directed parallel with the resultant angular momentum vector. This axis, as shown earlier in the lesson (Fig. 22-7), will remain fixed in direction throughout the jump.

It is assumed, for the particular jump illustrated in Fig. 22-22, that there is least angular momentum about the vertical (y) axis, most resulting from a sideward rotation about the z axis, and some backward rotation about the x axis. The

actual turning combinations, however, will vary from jumper to jumper and from style to style. The parallelogram in Fig. 22-22, *C,* shows the magnitude and direction of the resultant angular momentum obtained in two dimensions by the addition of momentums occurring around the *y* and *z* axes; but this resultant vector must be added to the momentum occurring around the *x* axis before the direction of the total axis can be obtained in three-dimensional space.

STUDY GUIDELINES
Angular momentum and Newton's laws
Specific student objectives

1. State Newton's three laws of angular motion in terms of angular momentum.
2. Identify two principles of human motion based on each of Newton's laws.

Summary

Angular momentum (L) is the product of a rotating body's moment of inertia and its angular velocity; i.e., $L = I\omega = (mr^2)\omega$. Angular momentum is a vector quantity.

Newton's laws for angular motion expressed in terms of angular momentum are as follows:

I. Inertia. A body motionless or in rotation with a given momentum will continue to be motionless or to turn about its axis of rotation with constant angular momentum unless an external net torque acts upon it.

II. Acceleration. The rate of change of angular momentum of a rotating object is directly proportional to the torque causing it and has the same direction as the causal torque:

$$\text{net } T = \frac{(L_f - L_o)}{t}$$

III. Reaction. For every applied torque tending to produce angular momentum in one direction, there is an equal and opposite torque tending to produce angular momentum in the opposite direction.

There are a number of other considerations relating to the aforementioned laws of angular motion:

1. Newton I is also known as the law of conservation for angular momentum.
2. In a closed system such as when the human body is in free flight, clockwise and counterclockwise angular momentums initiated in the air cancel each other.
3. $Ft = L_f - L_o$ is the angular impulse equation. It states that change in angular momentum depends directly upon the causal torque (T) and the time (t) during which it is applied.
4. Torque depends directly upon the magnitude of the eccentric thrust (force) and the length of the lever arm.
5. Action-reaction torques occur in all rotation situations involving the human body.

Performance tasks

1. Angular momentum is the product of
   a. mass and velocity
   b. mass and moment of inertia
   c. inertia and acceleration
   d. moment of inertia and velocity
2. $L = I\omega$ is the formula for angular
   a. acceleration   c. inertia
   b. length   d. none of the above

3. Write out Newton's three laws for angular motion in momentum terms.
4. Identify principles of human motion based on each of the three laws of angular motion and illustrate with motion examples.

Axes of angular momentum and of angular displacement
Specific student objectives

1. Differentiate between the axes of angular momentum and of angular displacement.
2. Describe what is meant by the spatial orientation (direction) of the axis of momentum.
3. Describe how the axis of momentum is determined.

Summary

The axis of angular momentum is the axis around which the angular momentum is conserved. Momentum around this axis must be imparted when the body is in contact with a supporting surface. Once free of support, the axis of momentum will remain fixed in direction (when air resistance is disregarded) until contact with a surface is regained. Rotations initiated in the air occur around axis of displacement, and these rotations must be compensated for by equal and opposite reaction rotations around the same axes. Axes of displacement usually do not remain fixed in direction.

The spatial orientation or direction of an axis is the angle it makes with the +y and either the +x or the +z cardinal axis.

The direction of the axis of momentum is determined by calculating the body's separate momentums about the cardinal x, y, and z axes and then vectorally adding these momentums to obtain a resultant momentum. The axis of momentum is directed parallel with the resultant momentum vector.

Performance tasks

1. The axis around which the angular momentum of a body is conserved is called
   a. axis of displacement
   b. axis of momentum
   c. axis of rotation
   d. axis of gyration
2. Write out descriptions of the axes of momentum and of displacement.

3. State how the magnitude and direction of angular momentum are described.

Effects of the redistribution of mass on the radius of gyration and the speed of angular motion
Specific student objectives

1. State how the radius of gyration of a body is affected by the distribution of the body's mass around the axis of rotation.
2. Describe the effect of shortening the radius of gyration on angular speed.

Summary

Since the mass of a body in rotation is a constant, its moment of inertia must be directly dependent upon the distribution of its mass around its axis of rotation. This distribution determines the magnitude of the radius of gyration. The speed of rotation is inversely proportional to the radius of gyration. Thus, when the radius of gyration of a rotating body decreases, the angular speed of the body will proportionately increase.

Performance tasks

1. When the radius of gyration of a body is increased, its angular speed is
   a. increased   b. decreased   c. maintained
2. The moment of inertia of a body is (directly/inversely) _____ dependent upon the distribution of its mass around its center of rotation.
3. Describe how the speed of limb movements of a runner is affected by the radius of gyration of the limbs.

Analysis of rotations initiated in the air
Specific student objectives

1. Given a specific angular motion task of the human body during free flight, describe the mechanics of the motions taking place.
2. State a principle concerning the initiation of angular motion in the human body during free projectile flight.

Summary

It is possible to initiate and carry out rotations of the human body during projectile flight. These movements are accomplished by manipulation of the moments of inertia of the body parts that interact when a rotation is initiated during free flight. It appears to be much easier to initiate rotations about the longitudinal axis than around either of the other two axes.

Performance tasks

1. Discuss the three principles concerning the initiation of rotational movements of the human body during free flight.
2. Select a specific diving or trampoline maneuver initiated in the air and describe the angular rotations which take place.

Transfer of angular momentum
Specific student objectives

1. State the law for conservation of angular momentum.
2. Illustrate the conservation law for angular momentum with specific motion examples of each of the following:
   a. Transfer of momentum between different parts within a system
   b. Change from linear momentum to angular momentum or vice-versa
   c. Transfer of momentum when an external causal agent is acting

Summary

Angular momentum in a closed system can be traded or transferred but is not lost or destroyed. During angular motion in a closed system, the total angular momentum of the system remains constant.

Angular momentum can be transferred to different parts within a closed system. Linear momentum can be transferred or changed to angular momentum and vice-versa. The same thing may occur when an external net force or torque is acting.

Performance tasks

1. Describe angular momentum.
2. Write out the law of conservation for angular momentum.
3. The example of the ice skater in Fig. 22-1 is an illustration of momentum conservation that occurs during
   a. transfers between types of momentum
   b. transfers of momentum when an external causal agent is acting

c. the redistribution of mass within a closed system
d. none of the above
4. Illustrate angular momentum transfer with two specific motion examples for each situation listed below:
   a. Transfer between different parts within a system
   b. Transfer from linear to angular or vice-versa
   c. Transfer when an external causal agent is acting

Vector analysis of angular momentum
Specific student objectives

1. Identify and describe the method of graphical representation of angular motion vectors.
2. Justify work as a scalar quantity and momentum as a vector quantity.
3. Describe the scalar or dot product of vectors.
4. Describe the vector or cross product of vectors.
5. Given two vectors, find their scalar and vector products.

Summary

The graphical representation of an angular motion vector is carried out by drawing:
1. a straight line (arrow) whose length represents the magnitude of the vector
2. a direction determined by the right-hand thumb rule (the curled fingers of the right hand pointing in the direction of the rotation and the extended thumb pointing in the direction of the angular motion vector)

The scalar or dot product of two vectors (A and B) whose directions make an angle (θ) with each other is equal to AB cos θ; i.e., $A \cdot B = AB \cos \theta$. (Remember: the dot product uses a dot but the cross product uses an X.)

The vector or cross product of two vectors (A and B) whose directions make an angle θ with each other is equal to a vector whose magnitude is AB sin θ, and whose direction is perpendicular to both A and B. A vector or cross product is written $A \times B = AB \sin \theta$.

Work and energy are examples of scalar products because they must be quantified without a consideration of direction. Momentum, on the other hand, is a vector product because the conservation of momentum law requires the consideration of momentum directions.

Performance tasks

1. The method of determining the direction of an angular motion vector is called
   a. scalar   c. both a and b
   b. vector product d. neither a nor b
2. Describe the method of graphically representing the magnitude and direction of an angular motion vector.
3. The scalar product of two vectors A and B would be
   a. AB sin θ c. A + B
   b. AB cos θ d. A − B
4. The cross product of two vectors has
   a. direction   c. both a and b
   b. magnitude   d. neither a nor b
5. Develop written description for the following:
   a. Graphical representation of angular motion vectors
   b. Scalar product
   c. Vector product
6. Find the scalar and vector products in the following situations:

	θ	A	B
a.____	30°	20 ft	80 lb
b.____	45°	45 m	120 kg
c.____	60°	152 cm	46 gm
d.____	120°	48 in	72 oz

Determining the axis of momentum
Specific student objective

1. Describe a method for determining the axis of momentum of a rotating body.

Summary

The position of the axis of momentum of a body rotating in the air can be calculated when the body's momentums about the x, y, and z axes are known at takeoff. By using the parallelogram method in three dimensions, we can establish first the magnitude of the total angular momentum and then the position of the axis of momentum (which gives the direction of the momentum vector).

Performance task

1. Write out a step-by-step procedure for determining the axis of momentum of a rotating object.

SELF-EVALUATION

Students should use no reference materials for this progress test, and they can check their answers by referring to Appendix A.

1. Angular momentum (L) equals
   - a. Fd
   - b. V/t
   - c. $I\omega$
   - d. ma
2. State Newton's three laws of motion in terms of angular momentum.
3. Identify two examples of human motion that demonstrate each of Newton's three laws.
4. State the law for conservation of angular momentum.
5. Illustrate the angular momentum conservation law with one specific example in each of the following areas:
   - a. Transfer between different parts within a system
   - b. Linear-angular momentum exchange
   - c. Transfer when an external force or torque is acting
6. State a principle concerning angular motion of the human body initiated during free flight.
7. Differentiate between axes of momentum and displacement.

8. The vector product magnitude of two vectors A and B is given by
   - a. AB sin θ
   - b. AB cos θ
   - c. A + B
   - d. A − B
9. Find the scalar products in the following situations:
   Use $A \cdot B = AB \cos \theta$.

a.＿＿＿	40°	10 ft	50 lb
b.＿＿＿	60°	45 lb	8 ft
c.＿＿＿	80°	22 m	32 kg
d.＿＿＿	150°	419 gm	206 cm

10. Describe the right-hand thumb rule.
11. Describe the application of Newton I for angular momentum with three motion examples.
12. Describe a method for determining the location of the axis of momentum.
13. Describe the effect of redistributing the mass of a body upon its radius of gyration.

LESSON 23

Kinetic analysis of gyratory motion

Body of lesson

CONCEPT OF GYRATORY MOTION

Gyratory motion is characterized by a spinning, whirling, or spiraling type of angular movement which occurs around an axis. An object undergoing this type of motion is called either a top or a gyroscope. A top is defined as an object rotating about an axis, one end of which is fixed. A common toy top set spinning like the one shown in Fig. 23-1 has its fixed point located at its lower tip. If the fixed point is at the center of gravity, as in a spiraling football, the object is called a gyroscope.

Characteristics of a top

The human body, as does any object, moves like a top when it is spinning or twisting while in contact with the ground or some other means of support. The rotation in this case always occurs around an axis, one end of which is fixed. The fixed point of rotation of a top can be either

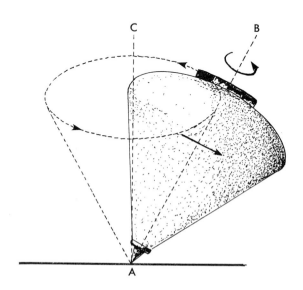

Fig. 23-1. A top rotates about an axis, one end of which is fixed.

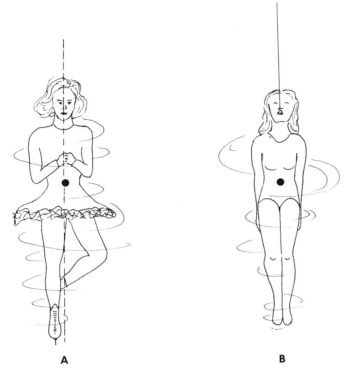

Fig. 23-2. A top can have its fixed point of rotation either below **(A)** or above **(B)** its center of gravity.

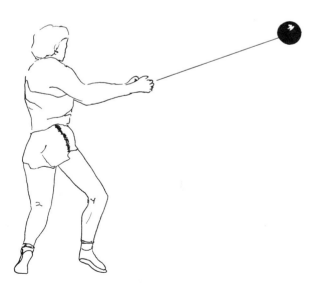

Fig. 23-3. A hammer thrower is an example of an asymmetrical top.

below or above its center of gravity. The pirouetting skater shown in Fig. 23-2, *A,* has the fixed end of her axis below her center of gravity at the point where her skates make contact with the ice. The spinning circus performer in Fig. 23-2, *B,* however, has the fixed end of her axis above her center of gravity at the point where she is gripping the cord with her teeth.

A top can also be either symmetrical or asymmetrical in its shape. It is symmetrical when one side is the mirror image of the other. In a symmetrical top the axis about which rotation occurs will always pass through the point of support and the common center of gravity as shown in the right-hand illustration above. When a top does not meet the requirement of symmetry, it is asymmetrical. In an asymmetrical top the axis around which rotation occurs will still pass through the base but may not pass through the center of gravity, and the top's position is said to be unstable. The rotation of an asymmetrical top as exemplified by a hammer thrower is illustrated in Fig. 23-3.

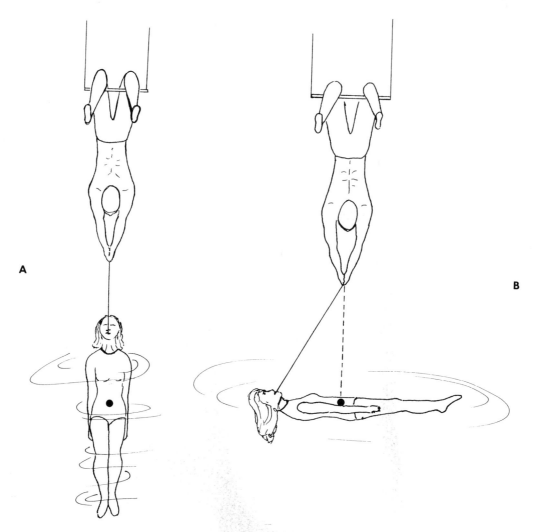

Fig. 23-4. A symmetrical top given sufficient time will finally settle down with stable equilibrium, selecting its axis of greatest moment of inertia (**B**).

In activities where some turning movements are made with both feet in contact with the ground (e.g., shot, hammer, javelin, and discus events), the support or base includes both feet and the intervening ground; therefore, in such cases, an axis can actually fall between the feet and move (as it must, in good throwing) from one point of the base to another.

A symmetrical top like that shown in Fig. 23-4, given sufficient time, will finally settle down to rotate with stable equilibrium, selecting its axis of greatest moment of inertia. This effect is illustrated in Fig. 23-4. A circus performer suspended from a cord gripped by the teeth is being turned by a partner above hanging upside down from a trapeze. According to Dyson (1971), at first the suspended person revolves rapidly about the body's long axis (Fig. 23-4, *A*) but quickly and automatically assumes a horizontal position turning about the body's axis of greatest moment of inertia (Fig. 23-4, *B*).

When a top is acted upon by a net external torque, the effects of the torque are to change either the magnitude or the direction or both the magnitude and direction of the angular momentum. The conical motion resulting from the

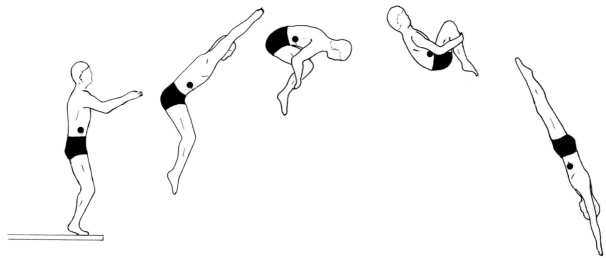

Fig. 23-5. A diver doing a forward somersault rotates about a horizontal axis of momentum passing through his center of gravity, and this axis maintains its direction relative to the ground or water until entry.

change in direction of the axis of momentum when a constant net external torque is acting is called *precession*. Precessional motion is illustrated by the conical axis of momentum of the toy top in Fig. 23-1.

Characteristics of a gyroscope

The human body, as does any object, moves like a gyroscope when it is rotating in the air free of support. The rotation always occurs around the center of gravity of the body. The movement also occurs around an *axis of momentum* which passes through the center of gravity. Rotation around this axis is imparted to the body prior to breaking contact with its supporting surface. In other words, rotation around the axis of momentum of a gyroscope is imparted by a reaction eccentric thrust or angular impulse imparted while the gyroscope is in contact with another object. This axis, unlike the body's cardinal axes, will remain fixed in direction (in the absence of an external net torque) throughout the movement until contact with a supporting surface is regained. This is the axis around which the conservation of angular momentum occurs.

Thus the diver in Fig. 23-5 will rotate about a horizontal axis of momentum passing through his center of gravity, and this axis will maintain its direction relative to the ground or water until entry.

A symmetrical gyroscope like that shown in Fig. 23-6, *A,* according to Hopper (1973), might rotate in the air around one of its cardinal axes, and in this case the cardinal axis would coincide with the axis of momentum. Any change in segmental positions, however, will make the gyroscope asymmetrical and will destroy the coincidence that may exist between the cardinal axis and the axis of momentum. Then the cardinal axis will be displaced and will no longer retain a constant direction in space. Instead it will trace out a conical surface about the axis of momentum.

Fig. 23-6 is a frontal view of an aerial pirouette. In *A* the person is rotating freely about his longitudinal or y axis. This cardinal axis, through the center of gravity, is also the axis of momentum, and the direction of the angular momentum vector is upward as indicated by the arrow. In *B* the asymmetrical positioning of the arms has changed the direction of the longitudinal or y axis in space so it no longer coincides with the axis of momentum, which, of course, is unaffected. The body now continues to rotate about the inclined y axis, but this y axis itself starts to rotate in a conical fashion about the axis of momentum, tracing out the surface of a cone as shown.

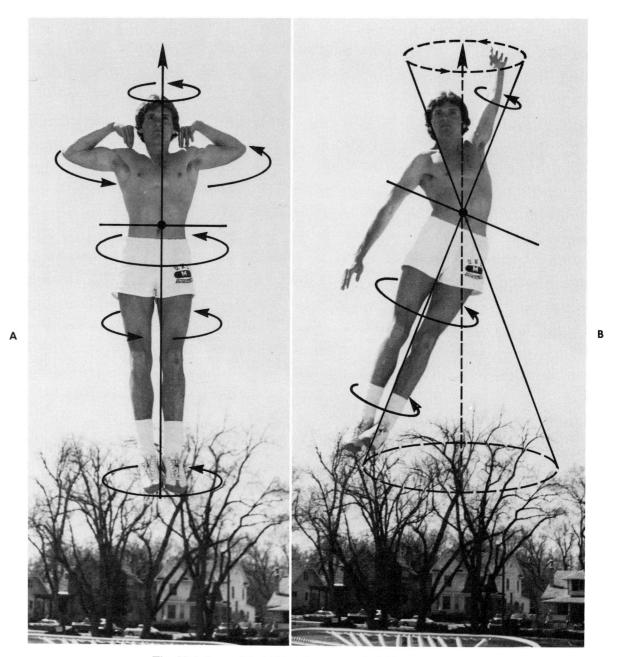

Fig. 23-6. Symmetrical **(A)** and asymmetrical **(B)** gyroscopes.

Such motion of a cardinal axis is called *nutation*. It results solely from the asymmetrical alignment of the body segments and not from the application of an external net torque. One possible effect of a net external torque application, however, is a change in the direction of the axis of momentum itself, and the resulting conical motion of this axis, as we already know, is called *precession*.

VECTOR ANALYSIS OF GYRATORY MOTION

The two effects that can be produced by an angular impulse, as mentioned before, are (1)

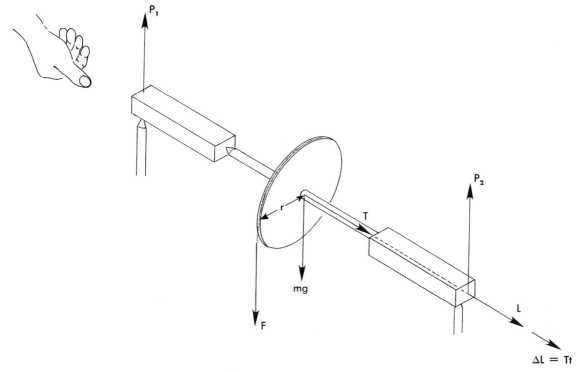

Fig. 23-7. A bicycle wheel mounted on a shaft through its center and rotating around its center of gravity is an example of a gyroscope. According to the right-hand thumb rule the angular momentum of the wheel points toward the right, along the axis of momentum. When a tangential force is applied to the wheel, the resulting torque acting during a time (t) produces a change in only the magnitude of the angular momentum $(\Delta L = Tt)$.

a change in the magnitude of the angular momentum and/or (2) a change in the direction of the angular momentum, i.e., a change in the direction of the axis of momentum.

Effect of angular impulse in producing a change in the magnitude of gyratory motion

Shown in Fig. 23-7 is a bicycle wheel mounted on a shaft through its center, with the shaft supported at each end so the weight of the whole is equally divided between the two supports. Since the wheel is rotating around its center of gravity, as indicated in the illustration, it is defined as a gyroscope. The angular momentum vector of the wheel *(L),* according to the right-hand thumb rule, points toward the right, along the axis of momentum.

Suppose a cord is wrapped around the rim of the wheel and a tangential force *(F)* is exerted on the wheel through the cord. The resultant

torque acting on the wheel is T = Fr, and the torque vector *(T)* also points to the right, along the axis of momentum. In a time (t) the torque produces an impulse and a vector change *(ΔL)* in the angular momentum. When this momentum change *(ΔL)* is added vectorally to the original angular momentum *(L),* the resultant is a vector of length $L + \Delta L$ which has the same direction as the original momentum *(L).* In other words, the magnitude of the angular momentum is increased but the direction remains the same. An increase in the magnitude of the angular momentum simply means that the body rotates more rapidly.

Effect of angular impulse in producing a change in the direction of gyratory motion

Fig. 23-8 shows the same rotating bicycle wheel we had before but with the support on the right side removed. The weight of the wheel and shaft

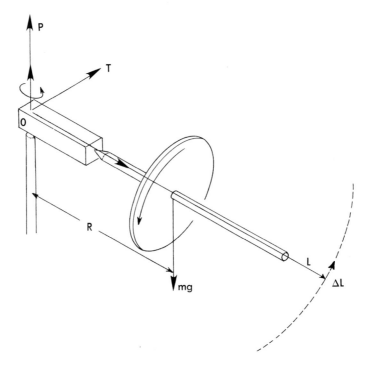

Fig. 23-8. When a gyroscope is supported only at one end of its axis, it becomes a top as well as a gyroscope. The force of gravity *(mg)* produces a torque *(T)* which during a time *(Δt)* produces a change in the momentum *(ΔL)*. The directions of both *T* and *ΔL* are horizontal and parallel with each other.

is no longer equally divided between two supports. If the support on the right is suddenly removed, the upward force at the support on the left in the first instant is still mg/2. The resultant vertical force is not zero, and the center of gravity of the system has an initial downward acceleration. With this downward rotation the bicycle wheel has now become a top with a fixed point of rotation at point *P* in addition to being a gyroscope rotating around its center of gravity.

The downward acceleration of the top is produced by the resultant torque, which is due to the weight of the wheel *(mg)* and is given by T = mgR. The direction of this torque is parallel with the axis of rotation of the top (at point *O*) and is perpendicular to the axis of rotation of the gyroscope in Fig. 23-8. In a time (t) this torque produces an impulse (Tt) and a change in angular momentum (ΔL). Since ΔL is perpendicular to *L* and is very small, the new angular momentum vector has practically the same magnitude as the old but in a different direction. The

tip of the angular momentum vector moves as shown, and as time goes on, it swings around a horizontal circle; but since the angular momentum vector lies along the gyroscope axis, the axis turns also, rotating in a horizontal plane about the point *P*. This motion of the axis of momentum is called precession.

Stabilizing effects of spin

Since angular momentum, like linear momentum, is a vector quantity possessing both magnitude and direction, we know from Newton I that an external torque is required to change either its magnitude or the direction of its axis. Since the direction of the angular momentum vector always points parallel with the axis of momentum, any change of direction of rotation requires a change in the orientation (direction) of the axis. The greater the value of the angular momentum, the greater is the torque needed to deviate it from its original direction.

This is the principle behind the spin stabilization of projectiles, such as footballs and rockets. Projectiles are set spinning about axes in their directions of motion so they will not tumble and thereby offer excessive air resistance. A toy top is another illustration of the vector nature of angular momentum. A stationary top set on its tip falls over at once, but a rotating top stays upright until its angular momentum is dissipated by friction between the tip and the ground. We also know that a moving bicycle is easier to keep upright than is a stationary one. The angular momentum of the bicycle wheels stabilize the bicycle in an upright position.

ANALYSIS OF PRECESSIONAL MOTION

In the last section, the statement was made that the *axis of momentum* of a top can itself rotate about a fixed point. This rotation was called precession. Thus a top or gyroscope can have two angular velocities: one, the angular velocity of the axis of momentum around a fixed point (precessional angular velocity), and the other, the angular velocity of the body rotating about the axis of momentum (rotational angular velocity).

Analysis of precessional angular displacement, velocity, and acceleration

The angle (θ) through which the axis turns in a time (t), called the precessional angular displacement, is given by tan $\theta = \Delta L/L = Tt/L$. The table of trigonometric functions in Appendix B shows that for small angles the tangent of θ is equal to the angle itself measured in radians. Thus tan $\theta = \theta$, and

$$\theta = \frac{\Delta L}{L} = \frac{Tt}{L}$$

(Equation 23-A)

The angular velocity of precession (Ω) (capital Greek letter omega) is θ/t, so

$$\Omega = \frac{T}{L}$$

(Equation 23-B)

The angular velocity of precession is therefore directly proportional to the acting torque and inversely proportional to the instantaneous angular momentum. Precessional angular acceleration can thus be produced by a change in either the acting torque or the angular momentum of the body. If the angular momentum is large, the precessional angular acceleration produced by an acting torque will be relatively small.

The axis of momentum of a gyroscope or top does not, of course, retain a fixed direction in space when acted upon by a net external torque. We can make a gyroscope or top turn in any direction we please by applying the appropriate torque. The magnitude of the torque acting on a top will determine the magnitude of the precessional angular acceleration, while the direction of the torque will determine the direction of the precessional angular acceleration.

Analysis of the precessional motion of a bicycle wheel

Returning to our example of gyroscopic motion illustrated in Fig. 23-8, we can see that the removal of the support on the right causes a downward acceleration of the bicycle wheel's center of gravity. At the same time precessional motion begins, though with a smaller angular velocity than in the final steady state. The result of this motion is to cause the end of the axle at point *P* to press down on the pivot with a greater force; thus the upward force at point *P* increases and eventually becomes greater than *mg*. The center of gravity then starts to accelerate upward. The process repeats itself, and the motion consists of a horizontal precession together with an up-and-down oscillation of the axis that is actually an alternating up-and-down vertical precession.

The top does not fall off the pivot because it was originally rotating as a gyroscope. The angular momentum produced by gravity when the support on the right was removed is added vectorally to the large angular momentum it already had. Since the change in momentum (ΔL) is horizontal and perpendicular to the original momentum, the result is a motion of precession with both the angular momentum vector and the axis remaining horizontal.

Summary example

As a summary example of precessional motion, we can again use the motion of a bicycle wheel mounted on an axle. The wheel has a moment of inertia (I) about its axis and a spin velocity (ω). If we place the left end of the spinning wheel and axle on the edge of the table as shown in

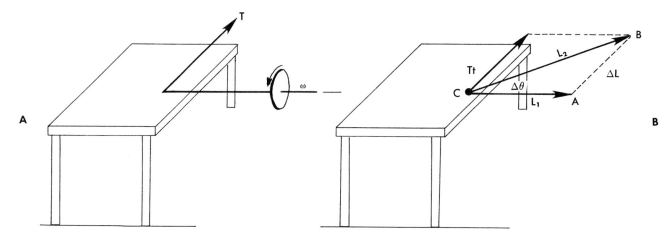

Fig. 23-9. The left end of a spinning wheel and axle is placed on the edge of a table **(A)**. Gravity produces a horizontal torque which during a time interval produces a horizontal impulse *(Tt)*. The resulting momentum change is also horizontal and parallel with the torque **(B)**.

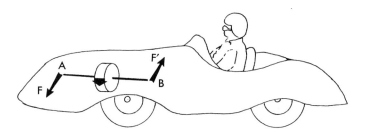

Fig. 23-10. The heavy flywheel of a racing car has an angular momentum. When the driver turns the car toward the left, a second angular momentum is created which produces the forces *F* and *F'* applied by the frame of the car. The rear end of the shaft, at *B*, moves upward and the front end, at *A*, moves downward. Thus the car digs in as it rounds the curve.

Fig. 23-9, the force of gravity will produce a torque which acts for a very short time interval (t), giving rise to an angular impulse *(Tt)*. By Newton's second law, we know that this impulse produces a change in the angular momentum of the wheel and axle which is directed horizontally to the right. If, while the wheel is spinning as shown in the figure, we attempt to increase the precessional motion by applying a horizontal force (F) to the axle, parallel with the direction of precession, the force we apply will give rise to a torque (T) whose vector is directed upward (right-hand thumb rule). Upward torque means that ΔL is upward and the axle rises at the unsupported end. Therefore a horizontal force gives rise to an upward motion.

Precessional motion of a race car

The peculiar reaction of a spinning wheel to a horizontal force application discussed above can be observed in the behavior of racing cars. The heavy flywheel of a racing car has a distinct precessional action of this sort (Fig. 23-10). When the driver turns the car toward the left, forces *F* and *F'* are applied by the frame of the car. The rear end of the shaft (at *B*) moves upward and the front end *(A)* moves downward. The car "digs in" as it rounds the curve. During a turn toward the right, the front end of the car "takes off" and tends to rise. To prevent this, speedway races are run with cars continually turning toward the left rather than toward the right. Of course, we are assuming that the flywheel

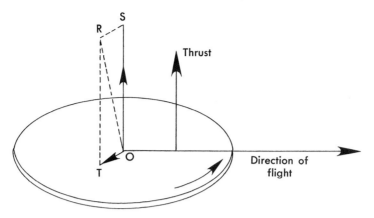

Fig. 23-11. Factors affecting the precessional motion of a projectile.

turns counterclockwise as viewed from the rear of the car. In fact, almost all race cars in the United States have counterclockwise-rotating flywheels.

Precessional motion of projectiles

Projectiles are constantly acted upon by the force of air resistance as they move along their trajectories. This force acts in front of the center of gravity and thereby produces a net torque which tends to cause precessional motion. Tricker and Tricker (1967) use a small cardboard disc to demonstrate the precessional motion of a projectile. They recommend that the disc be held between the index finger and thumb of the left hand and that it be projected forward with rapid spin by flicking it with the index finger of the right hand.

If the disc is projected forward with its plane horizontal or inclined slightly upward, the force of air resistance will tend to make it rotate backward. The disc will have two angular momentums: one imparted at the moment of projection and the other imparted by the air resistance. These two momentums are graphically represented as vectors in Fig. 23-11. The directions of the momentums are determined through use of the right-hand thumb rule. Vector *OS* is the angular momentum imparted at the moment of projection, and vector *OT* is the angular momentum imparted by the force of air resistance. The two vectors are combined through use of the parallelogram method to obtain the resultant *(OR)*. Thus the axis of momentum is displaced from *OS* to

OR by the eccentric thrust of the air and the disc precesses toward the right.

Because the eccentric thrust of air resistance produces the net external torque which causes the precessional motion of projectiles, this topic is important in the study of many sports. An interesting application of these concepts to the behavior of the boomerang is presented by Tricker and Tricker (1967).

ANALYSIS OF NUTATIONAL MOTION

It is possible to have gyratory movement around only one cardinal axis at a time, but this is rare and it is much more usual to have movement around two or three axes simultaneously (as exemplified by a wobbling football pass). For instance, the high jumper shown in Fig. 23-12 imparts a torque at takeoff around each of his cardinal axes. When these axes do not coincide with the axis of momentum, they will nutate around the axis *(M-M₁)*.

In the air a high jumper twists about his longitudinal axis, which at the same time commences to describe a conical path about his axis of momentum as shown in Fig. 23-12. If no rotations are originated in the air and no external torques are acting, then both the axis of momentum and the conical paths of the jumper's cardinal axes will not vary and there will be no interchange of angular momentums between these rotational motions.

Angular momentum, however, can be traded from one cardinal axis of displacement to another without changing the total angular momentum of

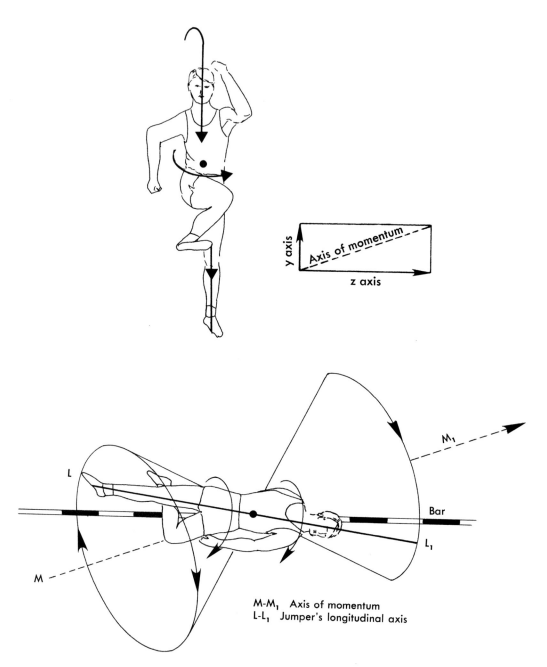

Fig. 23-12. A high jumper in the air will rotate around his longitudinal axis. The longitudinal axis, however, also nutates around the axis of momentum. (Modified from Dyson, G. H. G.: The mechanics of athletics, London, 1971, University of London Press, pp. 118 and 119.)

the system. This is accomplished by initiating movements in the air to bring one of the cardinal axes of displacement closer to the axis of momentum, and it causes the rotation to occur faster around the axis of displacement but to be reduced around the other two. In other words, the closer a given axis of displacement is to being in line with the axis of momentum the faster will be the spin of the body around that axis. The increase in spin around one axis, however, must be compensated for by an equal decrease in the spin around one or both of the other two axes.

The trading of angular momentum once an athlete is in the air is common in many types of motor tasks. For example, by originating movement in the air a straddle high jumper can bring his longitudinal axis more into line with his axis of momentum and thereby twist more rapidly away from the bar.

STUDY GUIDELINES
Concept of gyratory motion
Specific student objectives

1. Describe gyratory motion.
2. Distinguish between a gyroscope and a top.
3. Distinguish between symmetrical and asymmetrical tops and gyroscopes.
4. Distinguish between axes of momentum and displacement.
5. Distinguish between precession and nutation.

Summary

Gyratory motion is a special type of angular motion characterized by a spiraling type of rotation. If the fixed point of the rotational axis is at the center of gravity, the object is called a gyroscope.

A top is an object rotating about an axis, one end of which is fixed. The fixed point of rotation of a top can be either above or below the object's center of gravity. If one side is the mirror image of the other, the top is considered to be symmetrical. If there are differences, the top is asymmetrical.

A gyroscope posseses an axis of momentum, which passes through its center of gravity. Rotation occurs around the axis of momentum, but the axis remains fixed in direction if there is no external net torque throughout a movement. Motion about this axis is imparted to the object prior

to breaking contact with the supporting surface.

An axis of displacement is the axis around which all action and reaction movements occur when rotations are initiated in the air. This axis may change its spatial orientation during flight. Changes in the spatial orientation of the axis of momentum are called precession, while changes in the spatial orientation of the axis of displacement are called nutation.

Performance tasks

1. When the axis of momentum of a rotating body undergoes an angular displacement due to the influence of an external net torque, the motion is called
   a. gyratory   c. nutational
   b. precessional   d. angular
2. Describe the characteristics of nutational motion.
3. Identify two examples of gyratory motion.
4. A spiraling type of angular motion is called (gyration/nutation) _____.
5. An object in gyratory motion about is center of gravity is called a
   a. disc   c. both a and b
   b. top   d. neither a nor b
6. An object rotating about an axis, one end of which is fixed, is termed a
   a. gyroscope   c. both a and b
   b. top   d. neither a nor b
7. Gyratory motion initiated during free flight occurs around an axis of
   a. displacement   c. both a and b
   b. momentum   d. neither a nor b
8. An axis of motion passing through the center of gravity of a gyrating object when motion takes place on the support surface is called
   a. displacement
   b. momentum
   c. both a and b
   d. neither a nor b
9. Write out descriptions for the following:
   a. Gyratory motion
   b. Top
   c. Gyroscope
   d. Symmetrical tops and gyroscopes
   e. Axis of displacement
   f. Axis of momentum
10. Illustrate the motion terms of the preceding task with specific motion examples.

Vector analysis of gyratory motion
Specific student objectives

1. Describe the angular momentum resultant obtained through the vector addition of an original angular momentum (L) and an angular momentum change (ΔL) caused by an impulse applied in the same direction as the original rotation.
2. Describe the angular momentum resultant obtained when an original angular momentum (L) is added vectorally to an angular momentum change (ΔL) caused by an impulse applied so as to produce a moving axis of momentum.
3. Describe precession.
4. Describe the stabilizing effects of spin.

Summary

When an object is rotating with a given angular momentum and an impulse acts on the object in the same rotational direction, the result is an increase in the magnitude of the angular momentum and no change in the direction. The axis of momentum is unchanged.

An object rotating with a given angular momentum may also be acted upon by an impulse at a direction other than that of the rotation. The result is a movement of the axis of momentum which is called precession.

The inertial characteristic of a rotating object to resist changes in its angular momentum is used for spin stabilization of projectiles and other rotating objects.

Performance tasks

1. Set up an angular motion demonstration which shows a rotating object in each of the following situations:
   a. Constant axis of momentum but changing magnitude of momentum ◄
   b. Changes in both the angular momentum magnitude and the direction of the momentum axis
2. Describe precession.
3. State how the spiral of a football stabilizes its flight through the air.

Analysis of precessional motion
Specific student objectives

1. Describe the two angular velocities that a top or gyroscope can have.
2. Describe precessional angular displacement, velocity, and acceleration.
3. State why speedway races are run with cars continually turning toward the left.
4. Describe the precessional motion of a projectile.

Summary

Precessional movement occurs during angular motion when the axis of momentum of a top or gyroscope also rotates about a fixed point. The gyroscope can have two angular velocities: the angular velocity of the axis around the point (precessional angular velocity) and the angular velocity of the body about the axis (rotational angular velocity).

During precessional motion the angle (θ) through which the axis of momentum turns in a time (t) is given by $\theta = Tt/L$. The angular velocity of precession is given by $\Omega = T/L$. Thus precessional angular velocity can be increased (accelerated) by either increasing the applied net torque or decreasing the original angular momentum.

Speedway races are run with cars continually turning toward the left to prevent undesired precessions of the cars. When the cars turn toward the left, this angular momentum is added to the angular momentum of the flywheel to cause the front of the car to dig in as it rounds the curve. During a turn toward the right, the front end of the car tends to rise.

Projectiles are constantly acted upon by the force of air resistance. This force acts in front of the center of gravity and thereby produces a net torque which tends to produce precessional motion. The precessional motion that is produced is directly proportional to the torque produced by the air resistance and inversely proportional to the angular momentum of the projectile ($\Omega = T/L$).

Performance tasks

1. Many children have had the experience of rolling old automobile tires along the ground. As everyone probably knows, when the tire is rolling with relatively great speed, it easily maintains its upright orientation and continues its motion in a straight line. However, when it begins to tilt either to the right or to the left,

its motion tends to curve in that direction. This is an example of precessional motion produced by the eccentric pull of gravity.

    a. Explain why the greater the lean of the tire one way or the other the greater is the curving effect.

    b. Explain why a tire that leans toward the left will precess toward the left.

    c. Explain why a tire that is rotating forward slowly will precess faster than a fast rotating tire when both have the same angle of lean.

2. Differentiate between the angular velocity of precession and the angular velocity of rotation.

3. Describe the precessional tendencies of race cars and of projectiles.

Analysis of nutational motion

Specific student objective

1. Describe the angular motions involved in nutation.

Summary

    When the axis of displacement of a rotation does not coincide with the axis of momentum, the axis of displacement will nutate and describe a conical path about the axis of momentum. If no movements are originated in the air, the conical path of the axis of displacement will not vary and there will be no interchange of angular momentum between the two rotational motions.

    Angular momentum, however, can be traded from one cardinal axis of displacement to another without changing the total angular momentum of the system. This is accomplished by initiating movements in the air to bring one of the cardinal axes of displacement closer to the axis of momentum, and it causes the rotation to occur faster around that particular axis of displacement but to be reduced around the other two axes. In other words, the closer a given axis of displacement is to being in line with the axis of momentum the faster will be the spin of the body around that axis. The increase in spin around one axis, however, must be compensated for by an equal decrease in the spin around one or both of the other two axes.

Performance tasks

1. Identify three angular motion motor tasks that involve nutation.

2. Describe the axes of displacement and momentum in the above motion situations.

3. Describe the trading of angular momentum between the three cardinal axes of displacement.

4. Conduct an experiment involving a model of the human body or a cylinder by doing the following:

    a. Project the model or cylinder so it performs forward or backward somersaults.

    b. Project the model or cylinder so it performs forward somersaults with added twisting action.

    Describe the gyratory motions that result in these two situations.

SELF-EVALUATION

    Students should use no reference materials for this progress test, and they can check their answers by referring to Appendix A.

1. A spiraling type of angular motion is called
    a. gyration   c. both a and b
    b. linear   d. neither a nor b

2. Describe a gyroscope.

3. When the axis of momentum of a top or gyroscope also rotates about a fixed point, this motion is called
    a. angular motion   c. nutational motion
    b. precessional motion d. none of these

4. Identify and describe two examples of precessional motion as found in sports.

5. Identify and describe two examples of nutational motion as found in sports.

6. Precessional angular velocity is directly proportional to
    a. net torque
    b. original angular momentum
    c. both a and b
    d. neither a nor b

7. Describe the causes of precessional and nutational types of motion.

Kinetic analysis of pendular motion

Body of lesson
Study guidelines
Self-evaluation

CHARACTERISTICS OF PENDULAR MOTION

The primary type of swinging motion in human movement is that which is characteristic of a pendulum. The two types of pendulums are identified as simple and compound.

A *simple* pendulum consists of a small body (bob) suspended by a relatively long cord. Its total mass is contained within the bob. The cord is not considered to have mass.

A *compound* pendulum, on the other hand, is any pendulum such as the human body swinging by the hands from a horizontal bar. In this case the mass is distributed throughout the body.

The motion of simple pendulums is the topic of the present section of the lesson, while the motion of compound pendulums will be taken up in a later section.

When pulled to one side of its equilibrium position and released, a simple pendulum moves back and forth. If sufficient additional force is applied to the pendulum, it will move completely about its pivot point in a vertical circle. Once it is in motion, the primary forces acting are the angular components of gravity and the tension in the cord (as shown in Fig. 24-1). In this figure the instantaneous position of a small body of given mass swinging in a vertical arc at the end of a cord of length r is shown. The length of the cord is also the radius of rotation of the body.

Torque and angular components of gravity acting on a pendulum

The downward force acting on the body shown in Fig. 24-1 is the force of gravity *(mg)*, which can be resolved into its angular components $(F_r$

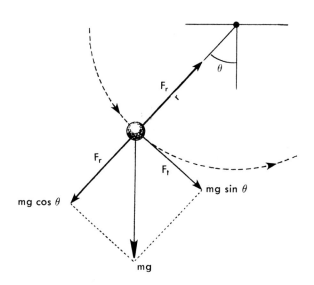

Fig. 24-1. The radial and tangential components of the force of gravity acting on a body in vertical circular motion.

and F_t). The direction of *mg* is vertically downward, while its respective components are directed parallel with and perpendicular to the supporting string *(r)*. The tangential force *(F_t)* acts to rotate the body to the midpoint or equilibrium position of its motion. Thus the torque causing the body to swing to the right is the product of the tangential component of the force of gravity *(F_t)* and the length of the cord *(r)*. Since the total force of gravity is represented by the weight of the body *(mg)* and since the tangential component of this force is equal to *mg sin θ,* then torque is given by the equation $T = (mg \sin \theta)r$.

As shown in Fig. 24-1, the force of gravity also has a radial centrifugal component, which is directed away from the center of rotation. This component is equal to *mg cos θ.*

Fig. 24-2 shows that the tangential component of the force of gravity acting on a swinging gymnast is greatest when his body is horizontal and diminishes progressively as he assumes a more vertical position. The loss of tangential force is accompanied by a corresponding increase in the radial component of the gravitational force.

Note also in Fig. 24-2 that when the gymnast is

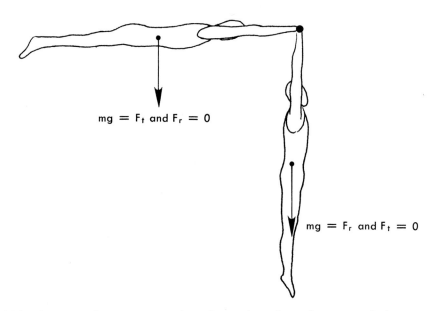

Fig. 24-2. The tangential component of the force of gravity acting on a swinging gymnast is greatest when his body is horizontal and diminishes progressively as he assumes a more vertical position.

horizontal the radius of his rotation is equal to the lever arm of the force of gravity; i.e., it is equal to the perpendicular distance from the axis of rotation to the line of action of gravity. The tangential component of gravity in this situation is also equal to the total force *(mg)*. In vertical angular motion, remember that the lever arm of gravity is always equal to the horizontal distance from the line of action to the axis. Thus the formula for torque in vertical angular motion situations can be written in either the symbols of force and lever arm (T = Ff) or the symbols of force and horizontal distance (T = Fd).

Newton's second law for the angular motion of a pendulum

Shown in Fig. 24-3 is the instantaneous position of a simple pendulum described in terms of its rectangular *(h,d)* and polar *(r)* dimensions with reference to the axis of rotation *(A)*. The vertical distance of the pendulum from the axis is given by *h,* while its horizontal distance from *A* is given by *d*. The straight line from *A* to the body is the radius *r*. Also shown in the figure is the force of gravity *(mg)* and its two angular components *(F_r* and *F_t)*.

The space triangle *hrd* and the force triangle *F_r mg F_t* shown in Fig. 24-3 are similar, since each contains a right angle and the two sides of one are parallel with the corresponding sides of the other. Angle θ in the force triangle is also equal to angle ϕ in the space triangle. Thus, with the aid of geometry, it can be shown that the ratio of the side opposite *(F_t)* of the force triangle to the side opposite *(d)* of the space triangle *(F_t/d)* is equal to the ratio of the hypotenuse *(mg)* of the force triangle to the hypotenuse *(r)* of the space triangle *(mg/r)*; i.e., $F_t/d = mg/r$.

By rearranging this equation, we can obtain $F_t r = mgd$. These are the two formulas that were developed in Lesson 13 for the calculation of torque; i.e., $T = F_t r$ and $T = Ff = mgd$. Since both $F_t r$ and *mgd* are equal to torque, they also

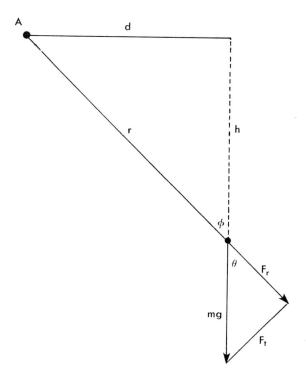

Fig. 24-3. Instantaneous position of a simple pendulum showing its spatial *(hrd)* and force *(F_r mg F_t)* components.

equal each other; i.e., $F_t r$ = mgd. Rearranging this equation we obtain

$$F_t = \frac{mgd}{r}$$

(Equation 24-A)

Since this is the force which accelerates a pendulum, it can be substituted for F in Newton II (a = F/m) to obtain

$$a = F/m = \frac{mgd/r}{m} = gd/r$$

(Equation 24-B)

This is Newton's second law for the angular motion of a pendulum. It means that the acceleration of a pendulum is directly proportional to the product of the gravitational constant (g) and the displacement (d) and inversely proportional to the pendular length (r).

ANALYSIS OF MECHANICAL ENERGY TRANSFORMATION IN SWINGING ACTIVITIES

In order to set a pendulum in motion, we must pull it aside from its vertical position (Fig. 24-4). This sideward pull also lifts it, and vertical work is done on the body. The amount of work done is the product of the weight *(mg)* of the body and the vertical distance *(h)* it is elevated *(mgh)*. The product mgh is also the potential energy of the body.

Releasing the elevated body subjects it to the accelerating force of gravity. Its velocity increases as long as it continues to fall. At the lowest point of its swing, it is moving rapidly and its inertia keeps it moving and it rises with decreasing velocity as gravity opposes its motion on the upswing. Finally it comes to rest again on the far side, and then the cycle is repeated.

At the upper ends of its swing, the body has no kinetic energy because it is momentarily at rest and its velocity is zero. At the bottom of its swing, it has maximum kinetic energy $(\frac{1}{2}I\omega^2)$ because there its velocity is greatest.

The potential energy of a pendulum, then, is at a maximum when its kinetic energy is zero, and vice-versa. As the pendulum falls and loses potential energy, there is a corresponding increase in the amount of its kinetic energy. The potential energy at the top of its swing equals the kinetic energy at the bottom, and at any intermediate position the sum of the two kinds of energy is equal to either the maximum potential or the maximum kinetic energy.

The work done by a body in angular motion such as a pendulum was described in Lesson 21 by Equation 21-J as $T\theta = \frac{1}{2}I\omega^2 + mgh$; that is, the angular work $(T\theta)$ which can be done by a body in pendular motion is equal to the sum of its kinetic and potential energies.

ANALYSIS OF VERTICAL ANGULAR MOTION

Fig. 24-5 represents a small body attached to a cord of length *r* and whirling in a vertical circle about a fixed point *(O)*, to which the other end of the cord is attached. The motion, while circular, is not uniform since the body accelerates on the way down and decelerates on the way up due to the force of gravity.

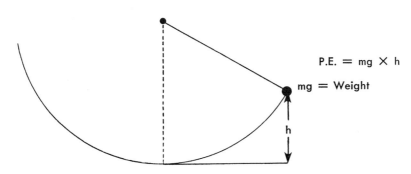

P.E. = mg × h

mg = Weight

h

Fig. 24-4. In order for a pendulum to be set in motion, it must be pulled aside. The potential energy imparted to the body in being pulled aside is equal to *mgh*.

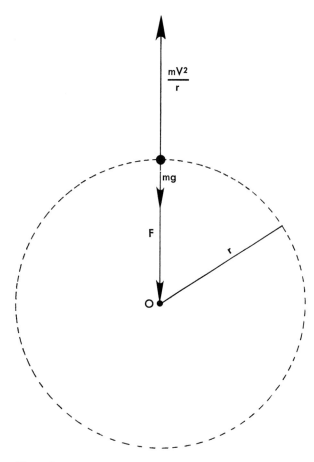

Fig. 24-5. A small body attached to a cord of length *r* is whirling in a vertical circle about a fixed point *(O),* to which the other end is attached. The tension in the cord *(F)* as the body passes the highest point is given by *mV²/r - mg.*

Determining the critical velocity required at the highest point for vertical circular motion

With vertical circular motion, we all know that there is a certain critical velocity below which the cord becomes slack at the highest point. We can analyze this critical velocity by letting *F* represent the tension in the cord as it passes the highest point. The total downward forces acting on the body at the highest point as shown in Fig. 24-5 are its weight *(mg)* and the tension *(F)* in the cord, so the resultant force is F + mg. Thus here the applied centripetal radial force *(Fr)* is equal to F + mg, which in turn is equal in magnitude but opposite in direction to the inertial radial

force, described in Lesson 20, mV²/r. In other words,

$$F_r = \frac{mV^2}{r} = F + mg$$

We can see therefore that at this point the radial force is provided partly by the body's weight and partly by the tension in the cord. The tension in the cord, by the way, is given by the equation

$$F = \frac{mV^2}{r} - mg$$

(Equation 24-C)

We can now use Equation 24-C to determine the critical velocity below which the cord becomes slack at the highest point. To find this velocity, we rearrange the equation F = mV²/r − mg after setting F = 0 as shown:

$$0 = \frac{mV^2}{r} - mg$$

and rearranging we obtain

$$V = \sqrt{gr}$$

(Equation 24-D)

This equation can be used to determine the critical velocity that a gymnast must obtain on top of a horizontal bar, for example, in order to execute a giant swing (Fig. 24-6).

Determining the greatest centripetal force required for vertical circular motion

If we now let *F* represent the tension in the cord shown in Fig. 24-7 at the lowest point of the circle, we can see that F = mV²/r + mg. The tension in the cord therefore is given by

$$F = \frac{mV^2}{r} + mg$$

(Equation 24-E)

These facts illustrate that the gymnast executing a giant swing on the horizontal bar must exert his strongest grip at the lowest point of his swing. The strength of his grip provides the centripetal force which prevents him from being pulled off the horizontal bar by the resultant centrifugal forces. The total centrifugal forces acting upon him would be the reaction centrifugal force produced by his angular motion (given by mV²/r) plus the centrifugal force produced by his body weight (mg). Thus the total centripetal force he must

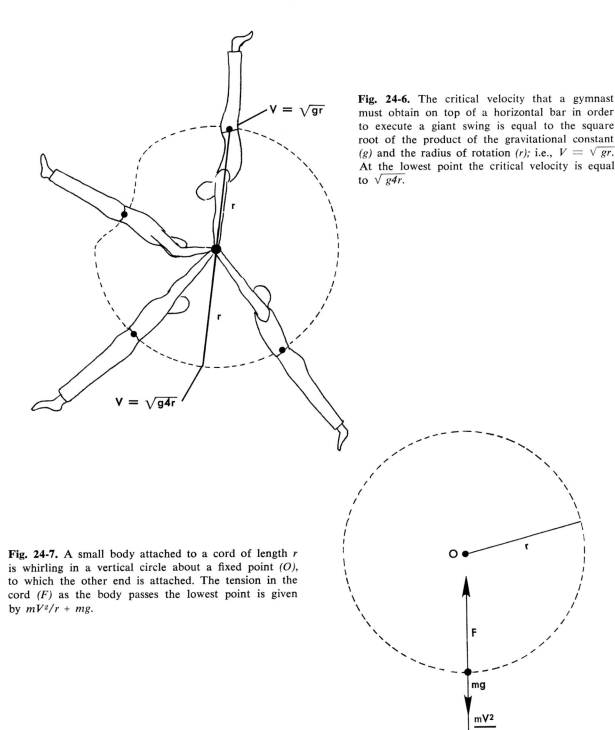

Fig. 24-6. The critical velocity that a gymnast must obtain on top of a horizontal bar in order to execute a giant swing is equal to the square root of the product of the gravitational constant (g) and the radius of rotation (r); i.e., $V = \sqrt{gr}$. At the lowest point the critical velocity is equal to $\sqrt{g4r}$.

Fig. 24-7. A small body attached to a cord of length r is whirling in a vertical circle about a fixed point (O), to which the other end is attached. The tension in the cord (F) as the body passes the lowest point is given by $mV^2/r + mg$.

apply at the lowest point of his swing is given by F = mV²/r + mg.

Determining the critical velocity required at the lowest point for vertical circular motion

As mentioned in the last section, the kinetic energy of a body at the lowest point of its swing equals its potential energy at its highest point. Thus ½ mV² at the lowest point is equal to mgh at the highest point. In circular motion the height (h) to which a weight (mg) must be raised is twice the length of the radius; i.e., h = 2r. Thus the equation ½ mV² = mgh becomes ½ mV² = mg2r. The critical velocity required at the lowest point in the swing in order to reach the top of the circle and therefore to have circular motion can be determined by rearranging the equation ½ mV² = mg2r to obtain

$$V = \sqrt{g4r} \qquad \text{(Equation 24-F)}$$

This equation can be used to determine the critical velocity that a gymnast must obtain on a horizontal bar, for example, in order to reach the top of a giant swing (Fig. 24-7); air resistance and friction between his hands and the bar, of course, would be ignored.

ANALYSIS OF ANGULAR HARMONIC MOTION

Pendular motion is the primary type of angular harmonic motion occurring in human movement. When pulled to one side of its equilibrium position and then released, a pendulum moves back and forth about this position. Any such repetitive back-and-forth motion, as defined in Lesson 19 is harmonic motion. Likewise, as defined in Lesson 19, any motion which repeats itself in equal intervals of time is periodic; and when the motion is back and forth over the same path, it is oscillatory. The movements of a simple pendulum therefore may be called harmonic, periodic, and oscillatory.

Restoring force and torque acting in angular harmonic motion

In Lesson 19 the concept was developed that the instantaneous restoring force acting to cause elastic harmonic motion is the product of the

force constant (K) and the amount of horizontal displacement (d); i.e., F = Kd. In our present discussion we have already seen that the tangential component of the force of gravity is what moves a pendulum toward the midpoint of its arc and when we multiply this component by the radius (r) we obtain the acting torque (T).

In Fig. 24-1 we saw that the tangential component is equal to mg sin θ. If θ is small, we may replace sin θ by θ so the equation for torque becomes T = mgrθ. Since T is also equal to Kθ, we can see that the force constant (K) equals *mgr*.

Thus we have the following relationships:

$$T = F_t r = mgr\theta = K\theta$$

Equation 24-G

Period of angular harmonic motion

A so-called compound pendulum, as already defined, is any real pendulum (in contrast to a simple pendulum, where all the mass is assumed to be concentrated at a point). Thus, in the analysis of compound pendular movements, the moment of inertia and the acting torque must always be considered.

The equation for the period of angular harmonic motion (T) was first introduced in Lesson 19. This equation is $T = 2\pi \sqrt{m/K}$. Since the moment of inertia (I) of a pivoted body corresponds to the mass of a body in linear motion and since the force constant (K) at small angles equals mgr, the quantities *I* and *mgr* can be substituted for *m* and *K* in the period formula to obtain

$$T = 2 I/K = 2 I/mgr$$

Equation 24-H

Analysis of compound pendular parameters

A compound pendulum is shown in Fig. 24-8 as a body of irregular shape pivoted about a horizontal frictionless axis and displaced from the vertical by an angle *(θ)*. The distance from the pivot to the center of gravity is *r*, the moment of inertia of the pendulum about an axis through the pivot is *I*, and the mass of the pendulum is *m*. According to Equation 24-H the period of oscillation of the pendulum is $T = 2\pi \sqrt{I/mgr}$. This equation can be used to measure the moment of

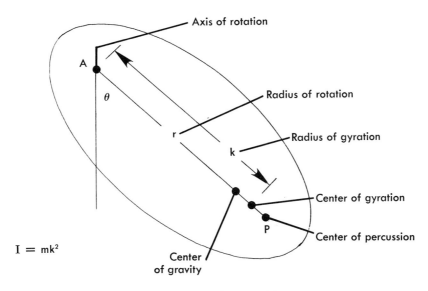

Fig. 24-8. A compound pendulum and its parameters.

inertia (I), the radius of gyration (k), and the radius of oscillation and percussion (q) of any compound pendulum. These parameters are illustrated in Fig. 24-8. The radius of rotation *(r)* is the distance from the axis to the center of gravity, the radius of gyration is the distance from the axis to the center of gyration, and the radius of oscillation and percussion is the distance from the axis to the center of oscillation and percussion.

Determining the moment of inertia. The moment of inertia of a compound pendulum can be found through the use of Equation 24-H, which can be rearranged to give

$$I = \frac{T^2 mgr}{4\pi^2}$$

(Equation 24-I)

The quantities on the right of the equation are all directly measurable. Hence the moment of inertia of a body of any complex shape (e.g., the human body) can be found by suspending the body as a compound pendulum and measuring its period of oscillation. The location of the center of gravity can be found through the approximation methods discussed in Lesson 15. Since *T, m, g,* and *r* are known, *I* can be computed. Equation 24-I is very useful in calculating the moment of inertia of the human body and its segments, as we will discuss later in this lesson.

Determining the radius of gyration. The radius of gyration of a particular body around a given axis of rotation can be determined through a knowledge of the moment of inertia of the body around the axis. The relationship between the radius of gyration (k) and the moment of inertia (I) was given by Equation 21-L as

$$k = \sqrt{I/m}$$

Thus if the moment of inertia *(I)* and the mass *(m)* of a body are known, the radius of gyration can be determined through the use of Equation 21-L.

Determining the center of oscillation and percussion. In Lesson 19 the center of percussion of an object was defined as the point that when struck will cause the object, suspended as a compound pendulum, to oscillate about its axis of rotation without producing pressure at this axis. Thus striking the center of percussion gives us the greatest conservation of force. The radius of percussion (q) can be determined through a knowledge of the distance from the pivot point to the center of gravity (r) and the radius of gyration (k). The relationship between these variables and the radius of percussion is

$$q = \frac{k^2}{r}$$

(Equation 24-J)

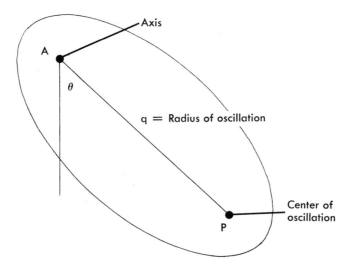

Fig. 24-9. The axis of rotation and center of oscillation are interchangeable.

The center of percussion is also called the center of oscillation because it is the point at which the mass of a compound pendulum may be considered to be concentrated as far as the period of oscillation is concerned. In other words, we can always find a *simple* pendulum whose period is equal to that of a given *compound* pendulum. The length of this simple pendulum is equal to the length of the radius of oscillation of the compound pendulum as given by Equation 24-J.

Fig. 24-9 shows a body pivoted about an axis through *A* and whose center of oscillation is at point *P*. The center of oscillation and the point of support are interchangeable; that is, if the pivot point of the pendulum is changed from *A* to *P*, its period is unchanged and *A* becomes the new center of oscillation.

Locating segmental parameters through the use of compound pendulums

The compound pendulum method has been used to locate the moments of inertia of selected segments of cadavers. Fischer (1906) dissected a cadaver, inserted parallel steel rods, one at each end of each segment, suspended each segment from the rod, and determined the period of oscillation. From these data he computed the moment of inertia and radius of gyration of each segment. Dempster (1955) followed a similar procedure, with the results shown in Table 24-1. The data in Table 24-1 can be used to estimate the location of the moment of inertia in a research subject.

Table 24-1. Moments of inertia of the major body segments about transverse axes through their centers of gravity*

Body segment	Moment of inertia (slug-ft^2)
Arm	0.016
Forearm	0.005
Hand	0.001
Forearm plus hand	0.022
Whole upper limb	0.077
Thigh	0.089
Leg	0.047
Foot	0.028
Leg plus foot	0.119
Whole lower limb	0.673
Trunk minus limbs	1.627

*Based on data presented in Dempster, W.: Space requirements of the seated operator, WADC technical report 55-159, Office of Technical Services, Department of Commerce, Washington, D. C., 1955, U. S. Government Printing Office, p. 198; as found in Hay, J. G.: The biomechanics of sports techniques, Englewood Cliffs, N. J., 1973, Prentice-Hall, Inc., p. 152.

ANALYSIS OF RESONANCE IN SWINGING ACTIVITIES

All simple or compound pendulums have natural periods or frequencies of oscillation depending upon their masses, their geometric characteristics, and the manner in which they are set into motion. Because of this fact the phenomenon of resonance

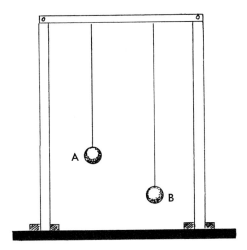

Fig. 24-10. Resonance in two pendulums.

is important in swinging activities just as it is in elastic harmonic motion activities. Resonance was defined in Lesson 19 as an increase in the amplitude of an oscillation.

A body is found to be set into oscillation very easily when a force acts upon it periodically with the body's natural frequency. For example, a child readily learns how to pump a swing; i.e., he learns that he can give the swing an oscillation of considerable amplitude if he properly times his impulse with the natural period of the swing. As mentioned in Lesson 19, the diver on the springboard also learns how to take advantage of the natural frequency of the diving board to increase his velocity of projection. Two oscillations with the same frequency are said to be *in phase* with each other if both start out together. Resonance occurs when two similar oscillations are in phase, but it does not occur when they are out of phase.

A common example of mechanical resonance is provided by a father pushing his child on a swing. The swing is a pendulum with a single natural frequency. If a series of regularly spaced pushes is given to the swing with a frequency equal to that of the swing, the motion of the swing may be made quite large. If the frequency of pushes differs from the natural frequency of the swing or if the pushes occur at irregular intervals, the swing will hardly execute a vibration at all; i.e., it will slow down.

Demonstration of resonance in a pendulum

An experimental demonstration of resonance is easily set up as shown in Fig. 24-10. Two simple pendulums are suspended from a flexible support. One *(A)* has a heavy bob made of metal, and the other *(B)*, a lighter bob made of wood.

The heavy metal bob *(A)* is pulled to one side and released to swing freely. The horizontal support responds to the motion and, in swaying back and forth with *A*, acts to set the other pendulum swinging. The response of pendulum *B* to this forced oscillation depends upon the relative lengths of the two pendulums. If there is considerable difference in the lengths, the response will be ever so slight; if the lengths are more closely equal, the response will be greater.

When *A* and *B* have the same length, their natural periods become equal and *B* responds to the swaying support and swings with a large amplitude. It responds in sympathy, or resonance, to the driving pendulum *A*.

ANALYSIS OF PENDULAR MOTION IN GYMNASTICS

All swinging-type activities in gymnastics are examples of pendular motions. In these activities the force of gravity is the primary propelling as well as resistive force. The torque produced by the force of gravity, as we already know, is directly proportional to the length of the radius of rotation. Thus the torque produced by gravity can be either increased or decreased by changing the distance (r) of the body's center of gravity from its axis of rotation. When swinging from a horizontal bar, for example, with the attainment of maximum momentum as the objective, the gymnast increases his radius of rotation on the downswing and decreases it on the upswing. These changes in the radius of rotation and thus of the acting gravitational torques result in greater accelerations on the downswing and lesser decelerations on the upswing.

The principle relating gravitational torque to the length of the radius is involved in all swinging-type activities. Even a child playing on playground swings knows he can acquire greater speed on his downswing by increasing his gravitational torque. He does this by crouching down during each downswing (to move his weight farther away from the axis of rotation) and thus producing

Fig. 24-11. A gymnast can shorten his radius of rotation by flexing the elbow, hip, and knee joints.

greater torque and greater acceleration. During the upswing he stands up straight (to move his body weight closer to the axis of rotation) and thereby reduces the gravitational torque acting to decelerate him.

In gymnastics, an athlete performing on a horizontal bar, parallel bars, or rings shortens his radius of upswing by flexing his elbows, hips, and knees and thus pulling toward the axis of rotation (Fig. 24-11). These flexions are performed as his body passes a point directly below the axis. He lengthens his radius by extending his elbows, hips, and knees at the height of the swing. His body is then held rigid and permitted to fall freely. Such

changes in the radii of rotation are repeated until the desired arc of swing is attained.

In order to eliminate the centrifugal forces acting on a gymnast during mounting activities that tend to pull him away from the apparatus, he must bring his center of gravity as close to the axis of rotation as possible. The critical moment at which this movement occurs varies with the exercise. It occurs in a kip, however, when the body's center of gravity passes below the bar as shown in Fig. 24-12. This movement may be done on the rings, parallel bars, or horizontal bar in essentially the same way, and in each case the timing of the movement is critical for success.

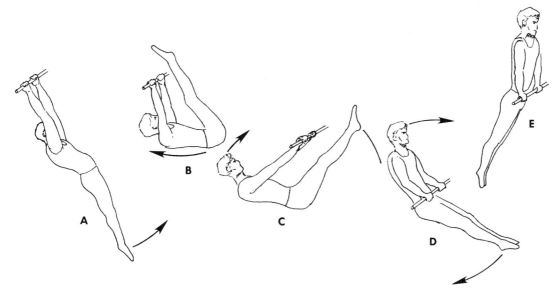

Fig. 24-12. A kip being performed on a horizontal bar.

In a kip on a horizontal bar, the athlete swings forward in an extended position (Fig. 24-12, *A*). At the end of the forward swing, forceful shoulder extension and trunk and hip flexion elevate the center of gravity to a level comparable to that in the support position (*B*). The extension of the hip at the bottom of the backswing shortens the radius of the angular motion and brings the center of gravity closer to the center of rotation (handgrip) as shown in position *C*. The force of hip extension produces angular motion of the total body at the time the center of rotation and center of gravity are at about the same height (position *D*). The hips are kept close to the handgrip; and if the force of momentum is great enough, the arms can shift from suspension to support so the performance will be successful (position *E*).

ANALYSIS OF THE PENDULAR MOTION OF BODY LIMBS

The rhythmic back-and-forth movements of the arms and legs of a runner are both periodic and oscillatory, similar to those of a compound pendulum. We can imagine, for example, the leg of a runner as being a pendulum. If set swinging, it will oscillate at a natural frequency which depends on its length and how its mass is distributed. Likewise, the arm of a bowler acts as a pendulum during back and forth movements.

Analysis of leg movements in running

At their natural frequency of oscillation the only energy needed to maintain the motion of body limbs is that required to overcome air resistance and the friction of the supporting joints. To move the limbs at other than their natural frequency, however, requires muscular effort and is tiring. This is so whether the limbs are being moved more slowly or more quickly than they would be when swinging freely. Thus the most economical pace for walking is at the same rate as the natural frequency of the limbs.

Any advantage that might exist for longer limbs in determining the speed of an animal diminishes and finally disappears when the animal forces its limbs to oscillate faster than their natural frequency as determined by their period of swing under gravity.

Tricker and Tricker (1967) have found that in animals of similar build, when the variation in size of their muscles is taken into account, the frequency of forced oscillation of a limb varies inversely with its length. This is consistent with Equation 24-B ($a = gd/r$), which shows that the acceleration of a pendulum is inversely proportional to its length (r). Thus, since running speed is the product of both the length and the frequency of stride, an animal's top speed is independent of size.

Surprising as this fact may seem, Tricker and Tricker (1967) have found that the top speeds of animals of similar shape but different sizes are not widely different. The fox, the hound, and the horse all have legs of much the same shape but markedly different lengths and can maintain similar top speeds of around 25 to 30 miles per hour.

Analysis of arm movements in bowling

The underhand pattern of throwing objects does not produce maximum force because of its restrictions on body rotation. However, it does allow for maximum use of gravity in the production of force. This advantageous use of gravity is important in bowling, in which a heavy ball is used.

In bowling, the ball is pushed forward away from the body. The arm and ball form a pendulum on which gravity acts. Since the force of gravity is straight downward, the velocity with which the ball is traveling at the bottom of the arc of the swing is dependent upon the height of the swing, i.e., $V = \sqrt{2gh}$.

The height of the backswing also depends on the height at which the ball is held in front and pushed away from the body because the last half of the backswing results from the momentum built up by gravity's pull on the arm and ball during the first half. The higher the push away, the greater will be the force which must be controlled, however, since gravity acts on the ball over a longer distance. Therefore more strength of trunk and shoulder muscle is required to stabilize the shoulder position when a higher push away is used.

In bowling, additional force is produced by utilizing the approach. Since the momentum of the body is transferred to any object held by the body, the forward momentum of the approach (if timed with the swing of the ball) is transferred to the ball and augments the momentum produced by the arm swing.

STUDY GUIDELINES
Characteristics of pendular motion
Specific student objectives

1. Differentiate between a simple and a compound pendulum.
2. Describe the torque and angular components of gravity acting on a pendulum.
3. Describe Newton's second law for the angular motion of a pendulum.
4. Describe the meaning of the following:
   a. Tangential component
   b. Radial component
5. Define a pendulum and describe the factors that affect its motion.

Summary

The two types of pendulums are identified as simple and compound. A simple pendulum consists of a small body suspended by a relatively long flexible cord. A so-called compound pendulum, on the other hand, is any real pendulum such as the human body as it might swing from a horizontal bar.

The torque acting in pendular motion is the product of the tangential component of the force of gravity (F_t) and the length of the cord (r). The total force of gravity (mg) can be resolved into its radial (F_r) and tangential (F_t) components. The tangential component decreases on the downswing of a pendulum starting from the horizontal and increases on the upswing to again be maximum when the pendulum is again at the horizontal.

For a pendulum of length r (which is the radius of rotation) and displacement d, the equations that can be used to describe pendular motion are $F_t = mgd/r$ and $a = gd/r$.

Performance tasks

1. Describe a pendulum.
2. Identify the following characteristics of pendular motion:
   a. Repetition in regular intervals
   b. Movement back and forth over same path
   c. Motion that repeats itself
3. The motion of the human body swinging from a horizontal bar is best described as that of a (simple/compound)_____ pendulum.
4. State Newton's second law for the motion of a pendulum.
5. Longer pendulums have (greater/lesser) _____ accelerations during motion.
6. The displacement of a pendulum is related to the resulting acceleration as a (direct/inverse) _____ proportion.
7. $a = gd/r$ is the formula for Newton's second

law. It tells us that the acceleration depends upon the
   a. length of the pendulum   c. both a and b
   b. displacement of   d. neither a nor b
      the pendulum
8. Give three examples of pendular motion as found in human motion situations.

Analysis of mechanical energy transformation in swinging activities
Specific student objectives

1. Identify the types of mechanical energy present during pendular action.
2. Describe the work done to develop the potential energy of a pendulum.
3. Analyze the mechanical energy considerations in pendulum-like actions by stating the following:
   a. Positions of minimum and maximum kinetic energy
   b. Positions of minimum and maximum potential energy
   c. Relationship of kinetic energy to potential energy during motion
   d. Application of the energy conservation law during pendular motion

Summary

During pendular motion the total mechanical energy, kinetic and potential, remains constant unless there is some transformation to heat or other types of energy.

The work done to begin motion and to produce potential energy in a pendulum is the product of the bob's height and its weight, i.e., P.E. = mgh. The position of maximum displacement is also the position where all the mechanical energy of the pendulum is potential. At the lowest point of the swing, all mechanical energy is in the form of kinetic energy. Thus the total M.E. of a pendulum equals K.E. plus P.E. at each point in the swing in between the two critical points (highest and lowest) already described.

Performance tasks

1. Kinetic and positional (potential) are the two types of
   a. gravitational energy   c. both a and b
   b. mechanical energy   d. neither a nor b

2. The work done to initiate the motion of a pendulum depends upon the
   a. elevation distance   c. both a and b
   b. pendulum weight   d. neither a nor b
3. Analyze mechanical energy considerations in pendular motion by identifying the following:
   a. Positions of minimum and maximum kinetic energy
   b. Positions of minimum and maximum potential energy
   c. Relationship of kinetic energy to potential energy along the motion path
   d. Application of the mechanical energy conservation law to pendular motion

Analysis of vertical angular motion
Specific student objectives

1. Given r in sample data, calculate the critical velocity required for vertical circular motion using $V = \sqrt{gr}$.
2. Given r, m, and V, determine the maximum centripetal force and critical minimum velocities at the highest and lowest points required for vertical circular motion. Use $V = \sqrt{g4r}$, $V = \sqrt{gr}$, and $F_c = mV^2/r + mg$.
3. Identify the forces acting on a body rotating in vertical angular motion.

Summary

Angular motion in a vertical plane is not uniform. At the highest point in motion, the tension is

$$F = \frac{mV^2}{r} - mg$$

and the critical minimum velocity is $V = \sqrt{gr}$; at the lowest point,

$$F = \frac{mV^2}{r} + mg$$

and $V = \sqrt{g4r}$.

The tension at the lowest point in the motion is the maximum centripetal force that must be generated during the swing. The torque producing angular motion due to gravitational force is shown by $T = (mg \sin \theta)r$ and reaches its maximum when the object is in a horizontal position.

Performance tasks

1. Calculate the critical minimum velocity for vertical angular motion required at the highest and lowest points ($V = \sqrt{gr}$ and $V = \sqrt{g4r}$)
   a. r = 3 ft, m = 4 slugs

b. $r = 2$ m, $m = 20$ kg
c. $r = 30$ in, $m = 2$ slugs
d. $r = 93$ cm, $m = -1$ kg

2. Determine the maximum centripetal force required during vertical angular motion at the lowest point of the swing ($F_c = mV^2/r + mg$).

	m	V	r	F_c
a.	3 slugs	10 ft/sec	3 ft	_____
b.	2 slugs	22 ft/sec	30 in	_____
c.	80 kg	15 m/sec	2 m	_____
d.	50 kg	20 m/sec	1.2 m	_____

3. The forces acting on a body in vertical angular motion are the components of the force of gravity called
   a. radial
   b. vertical and horizontal
   c. both a and b
   d. neither a nor b

Analysis of angular harmonic motion
Specific student objectives

1. Describe a compound pendulum.
2. Describe the meaning of the following equations for angular harmonic motion and define the pendulum quantities involved:
   a. Torque $(T) = K\theta = mg\theta r$
   b. Period $(T) = 2\pi \sqrt{I/mgr}$
   c. Moment of inertia $(I) = T^2mgr/4\pi^2$, where T is the period
   d. Radius of gyration $(k) = \sqrt{I/m}$
   e. Radius of oscillation and percussion $(q) = k^2/r$
3. Describe the meaning of ring of gyration and radius of gyration.
4. Describe center of percussion and radius of percussion.
5. Given the period (T), mass (m), and radius (r), calculate each of the pendulum quantities listed in Task 2 above.

Summary

A compound pendulum is any real pendulum (in contrast to a simple pendulum) where all the mass is assumed to be concentrated at a point. The equations for compound pendulum movements are as follows:

1. Torque $(T) = K\theta = mg\theta r$, where K is the force constant, m the mass, and r the radius of rotation
2. Period $(T) = 2\pi \sqrt{I/mgr}$, where I is the moment of inertia and r is the radius of rotation
3. Moment of inertia $(I) = T^2mgr/4\pi^2$, where T is the period
4. Radius of gyration $(k) = \sqrt{I/m}$
5. Radius of percussion $(q) = k^2/r$

For compound pendular analysis the complex mass distribution of a body can be replaced (as discussed in Lesson 21), by an infinitely thin circular ring of matter located around the center of gravity of the body and containing the same quantity of mass as the body itself. The ring is known as the ring of gyration, while the radius of the ring is the radius of gyration (k).

The center of percussion of an object is the point that when struck will result in the least elastic resonance and the greatest pendular resonance. The radius of percussion is the distance from the fulcrum to the center of percussion.

Performance tasks

1. A pendulum in which all mass is assumed to be concentrated at a point is said to be a (compound/simple)_____pendulum.
2. Write out descriptions of ring of gyration and radius of gyration.
3. The point at which the least amount of vibration occurs upon impact is known as the center of
   a. gyration
   b. percussion
   c. both a and b
   d. neither a nor b
4. Describe (in sentence form) the center of percussion and radius of percussion.
5. Given the data below, calculate the moment of inertia, radius of gyration, and radius of percussion using $I = T^2mgr/4\pi^2$, $k = \sqrt{I/m}$, and $q = k^2/r$.

	Period (T)	Mass (m)	Radius (r)
a. _____	4 sec	5 slugs	3 ft
b. _____	10 sec	11 slugs	8 ft
c. _____	3 sec	4 kg	1.4 m
d. _____	2 sec	3 kg	1 m

6. Conduct an experiment to gather the type data shown in Task 5 above for the pendular motion of (a) an individual's arm while holding a bowling ball, (b) the swing of a golf club, or (c) the swing of a clock pendulum.

Analysis of resonance in swinging activities
Specific student objectives

1. Define resonance.
2. Give the term which describes oscillations that have the same frequency and that occur together.
3. Identify an illustration of the use of resonance in a pendular motion activity.
4. State a principle of resonance for pendular motion.

Summary

All pendulums have natural periods or frequencies of oscillation depending upon their physical makeup.

Resonance is defined as an increase in the amplitude of an oscillation. A body can be set into pendular motion easily when a force synchronized with the body's natural frequency acts upon it.

Two oscillations with the same frequency are said to be in phase with each other if they start out together. Resonance will occur in pendular motion when two similar oscillations are in phase with each other.

Performance tasks

1. An increase in the amplitude of an oscillation is called
   a. frequency
   b. resonance
   c. both a and b
   d. neither a nor b
2. Oscillations with the same frequency and that occur together are said to be (in phase/out of phase)_____ .
3. Develop written descriptions for the following:
   a. Resonance
   b. In-phase and out-of-phase oscillations
   c. Two principles of resonance
4. Illustrate the two principles in Task 3c with a specific human motion example.

Analysis of pendular motion in gymnastics
Specific student objectives

1. State how the torque produced by gravity can be either increased or decreased during pendular motion in gymnastics.
2. Describe how the centrifugal forces acting on a gymnast during mounting activities can be reduced.

Summary

The primary propelling force, as well as the resistive force, acting during swinging-type activities in gymnastics is gravity. The torque produced by the force of gravity can be either increased or decreased by changing the distance (r) of the body's center of gravity from the axis of rotation.

In order to eliminate the centrifugal forces acting on a gymnast during mounting activities that tend to pull him away from the apparatus, it is important that the center of gravity be brought as close to the axis of rotation as possible.

Performance tasks

1. The torque produced by the force of gravity can be increased by (increasing/decreasing) _____ the radius of rotation.
2. Describe how a child playing on a swing can increase the arc of his swing by changing the length of his radius of rotation.
3. Describe how a gymnast can reduce the centrifugal forces tending to pull him away from the apparatus from which he is swinging.

Analysis of the pendular motion of body limbs
Specific student objectives

1. Describe the limbs of the body as compound pendulums.
2. Describe the pendular motion of a bowler's arm.
3. Describe the most economical pace of walking.
4. Describe the relation of leg length to running speed.
5. Describe the relation of height of swing of a bowling ball to the velocity of release.
6. Describe the components of the final momentum of a bowling ball at the time of release.

Summary

The body limbs at times undergo motions that are harmonic, periodic, and oscillatory. These movements are similar to those of a compound pendulum.

The most economical pace of walking is one that is equal to the natural frequency of the limbs. Any advantage which might exist for longer limbs in determining the speed of an animal diminishes and finally disappears when the animal forces its limbs to oscillate at a rate faster than their natural frequency. In fact, an animal's top speed is independent of its size.

In bowling, the height of the swing will determine the velocity of release at the bottom of the swing; i.e., $V = \sqrt{2gh}$. The final momentum of the ball will be the resultant of the angular momentum of the arm and ball plus the forward linear momentum of the body attained during the approach.

Performance tasks

1. Describe four motion tasks in which the body limbs are used in pendular fashion.
2. Describe the motor tasks of walking, running, and bowling by doing the following:
   a. State one mechanical principle of pendular motion related to the task.
   b. Apply the principle in a specific motion example.

SELF-EVALUATION

Students should use no reference materials for this progress test, and they can check their answers by referring to Appendix A.

1. Newton's second law of motion for a pendulum is $a = gd/r$. Identify and describe the variables which determine the acceleration of a pendulum.
2. Describe the formula for the period of a pendulum ($T = 2\pi \sqrt{I/mgr}$).
3. Define a compound pendulum.
4. The point at which the least amount of vibration increase occurs upon impact is known as the center of
   a. gyration   c. rotation
   b. percussion   d. none of these
5. Describe the ring of gyration and radius of gyration.
6. Given the data below, calculate the moment of inertia, radius of gyration, and radius of percussion:

	Period (T)	Mass (m)	Radius (r)
a. _____	5 sec	8 slugs	4 ft
b. _____	6 sec	9 slugs	5 ft
c. _____	4 sec	5 kg	2 m
d. _____	3 sec	2 kg	3 m

7. Identify two motion tasks in which the arms or legs of the human body are used in pendulum-like motion.

8. State a mechanical principle related to pendulums for each of the motor tasks below:
   a. Walking
   b. Doing a kip on a horizontal bar
   c. Bowling
9. Describe the work done to initiate a pendular motion.
10. Analyze mechanical energy considerations during pendular motion by stating principles related to each of the following:
    a. Positions of minimum and maximum kinetic energy
    b. Positions of minimum and maximum potential energy
    c. The relationships of kinetic energy to potential energy at all points in the range of motion
    d. The energy conservation law for pendular motion
11. An increase in the amplitude of an oscillation during harmonic motion is known as
    a. frequency   c. the period
    b. gyration   d. resonance
12. Describe in-phase oscillations.
13. State two mechanical principles for resonance and identify an application involving human motion for each principle.
14. For vertical angular motion, calculate the critical minimum velocity required at the top and bottom of the circle (using $V = \sqrt{gr}$ and $V = \sqrt{g4r}$) and the maximum centripetal force (using $F_c = mV^2/r + mg$) required at the bottom of the circle.
    a. Minimum velocity at the top of the swing
       (1) $r = 6$ ft   (3) $r = 20$ in
       (2) $r = 4$ m   (4) $r = 31$ cm
    b. Maximum F_c and minimum velocity at the bottom of the swing

	m	V	r
(1) _____	7 slugs	22 ft/sec	20 ft
(2) _____	27 slugs	50 ft/sec	4 ft
(3) _____	500 kg	58 m/sec	10 m
(4) _____	50 kg	15 m/sec	1.5 m

15. Describe the torque and angular components of gravity acting on a pendulum.
16. Describe two principles of pendular motion that affect the motion of a swinging gymnast.

Kinetic analysis of segmental motion

GENERAL APPROACH TO THE KINETIC ANALYSIS OF HUMAN MOTION

The three steps involved in a kinetic analysis are as follows: (1) define the mechanical system to be studied, (2) construct a free-body diagram of the system isolated from its surroundings showing all forces (vectorally) acting on it, and (3) select an analysis approach appropriate to the problem being investigated.

Definition of mechanical systems

A mechanical system is one or more components of a whole capable of performing some common motion function. These systems are actually idealized models of the human body in which certain assumptions are accepted. According to Miller and Nelson (1973) the human body can be treated in a kinetic analysis as a particle, a rigid body, a quasi-rigid body, or a linked system. The definitions of these four systems are:

1. *Particle system.* This approach is used when a body can be adequately represented by the motion of its mass center. Such is the case in predicting the trajectory of a diver or jumper.
2. *Rigid-body system.* This approach is used in the study of muscle and joint forces acting upon individual limb segments, but it is seldom used to portray the total body.
3. *Quasi–rigid-body system.* This approach is used to describe the body when its segments remain in the same positions with respect to one another. The quasi-rigid configuration is seen in a diver rotating in a layout position or in a skater maintaining a fixed attitude in a spin.
4. *Linked system.* This approach represents the human body as a series of interconnected rigid segments which demonstrate independent motion. This is the most accurate me-

403

chanical representation of the moving human body and it is the most often used in the kinetic analysis of human motion.

From these definitions it can be seen that a mechanical system may include the whole body, as in the case of a jumper moving through the air, or may be limited to a few segments, such as the trunk and the upper limb segments of a thrower. The focus of the study may even be limited to the motion of a sports implement (e.g., a tennis racket). Whatever the problem, the boundaries of the system must be defined and the forces acting upon it identified.

Construction of free-body diagrams

Because motion is governed by the action of forces, it is essential in any motion analysis to identify and describe the magnitudes, directions, and points of application of these forces. This process is facilitated through the use of free-body diagrams.

A free-body diagram is a simplified drawing of the mechanical system being studied isolated from its surroundings showing all the forces (vertorally) acting upon it. The two basic types of free-body diagrams are those associated with the particle and segmental types of mechanical systems.

The analysis of linear motion uses the particle approach. In this situation the athlete is treated mathematically as a particle, and it is assumed that his motion can be described by studying the kinematics of his mass center. Similarly, for the purpose of kinetic analysis, a sports implement in flight is often considered a particle. In these cases body positions, velocities, and accelerations are expressed as vectors with x, y, and z components. This particular form of Cartesian reference frame is convenient because all forces acting upon the particle can be resolved into a resultant vector whose magnitude and direction determine the motion of the body.

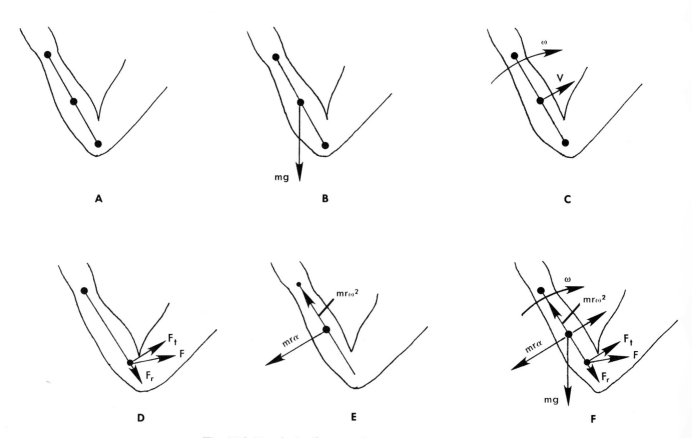

Fig. 25-1. Free-body diagram of a one-segment movement.

The analysis of angular segmental motion is based on the assumption that the human body can be represented mechanically as a linked system of rigid segments moving about axes of rotation through the joints. This system lends itself to a consideration of angular position, velocity, and acceleration of the limbs. The complete kinetic analysis of human motion, however, requires the incorporation of both linear and angular motions of the body segments. The incorporation of these motions in a free-body diagram is illustrated in Fig. 25-1.

One of the first steps in constructing a free-body diagram of body segmental motion is to identify the segmental end points, segmental lengths, and segmental centers of gravity as shown in Fig. 25-1, *A*. The weight of the segment is always vectorally represented by an arrow drawn vertically downward from the center of gravity (Fig. 25-1, *B*). The direction of angular motion is indicated through the use of a curved arrow pointing in the direction of the motion (ω), while the linear motion of a particle is indicated through the use of a straight arrow drawn to scale perpendicular to the radius of rotation at the location of the particle *(V)* (Fig. 25-1, *C*). Applied forces are drawn in as arrows from their points of application showing their magnitudes and directions. These forces can be resolved into either their angular (radial and tangential) (F_r and F_t) components (Fig. 25-1, *D*) or their Cartesian (horizontal and vertical) (F_x, F_y, and F_z) components. The inertial forces are drawn in the opposite direction of the motion to show that they act in opposition to the movement ($mr\alpha$ and $mr\omega^2$) (Fig. 25-1, *E*). Only reaction forces which result from forces being applied on external objects are drawn in since, according to Newton's third law, these are the forces that are acting on the system. A completed free-body diagram is shown in Fig. 25-1, *F*.

Selection of an analysis method

The three methods involved in the kinetic analysis of human motion are identified as (1) the force-mass-acceleration method, (2) the work-energy method, and (3) the impulse-momentum method. The *force-mass-acceleration* method is best suited for the instantaneous analysis of forces. This approach is used to examine the magnitudes and directions of fluctuating forces at particular instants during the course of a motion. The *work-energy* method is used to investigate forces acting over specific distances. The *impulse-momentum* method is used to analyze forces acting during given time periods.

KINETIC ANALYSIS OF ONE-SEGMENT MOTION

The kinetic analysis of one-segment motion involves (1) the application of d'Alembert's principle, which states that the force of inertia of a body is equal to and opposite an applied force used to move the body, resulting in a kinetic equilibrium (i.e., all free-body diagrams of a segment must be drawn to show the segment in static equilibrium), (2) the vector representation of inertial forces, and (3) the vector representation of applied forces.

D'Alembert's principle

D'Alembert's principle indicates that all inertial and applied forces acting on a segment should be drawn in opposite directions on a free-body diagram to show the system in equilibrium. The principle is named after the French mathematician Jean LeRond d'Alembert, who according to Plagenhoef (1971) was the first person to realize that all inertial and applied forces can be treated in terms of equal and opposing forces. Thus problems in dynamics are reduced to problems in statics which can be readily shown by means of free-body diagrams.

Vector representation of inertial forces

Plagenhoef (1971) uses an elevator problem to illustrate d'Alembert's principle as applied to linear motion. Everyone has felt the changing forces at the start and finish of an elevator ride. When the elevator starts upward you are pressed against the floor and you would weigh more if you were standing on a scale. Likewise you would weigh less when starting downward. Newton's second law specifies the relationship between the applied force (F) and the inertial force (ma) in this situation; i.e., F = ma.

As an example, we can assume that a 160 lb person is in an elevator moving downward with an acceleration of 2 ft/sec². The inertial force acting in this situation is determined by multiplying the person's mass (m = w/g = 160/32 = 5 slugs) by his acceleration (2 ft/sec^2), which gives a product of 10 lb (ma = (5) (2) = 10 lb). Since the inertial force is acting upward in this example, opposite in direction to the motion, its magnitude is subtracted from the person's weight as measured on the scale on which he is standing; that is, the scale would now show a weight of only 150 lb. Thus the inertial force due to the motion, in this example, is equal to and opposite the applied force causing the motion.

The two inertial forces acting in angular motion are the radial force (mrω^2) and the tangential force (mrα). Fig. 25-1 illustrates these inertial forces drawn on a free-body diagram. The radial inertial force *(mrω^2)* is drawn outward, parallel with the segment, equal to and opposite the applied radial force (which is acting toward the center of rotation). The tangential force *(mrα)* is drawn perpendicular to the segment from its center of gravity in the direction opposite the applied tangential force. This direction is opposite that of the movement when the acceleration is positive and the same as that of the movement when the acceleration is negative.

Vector representation of applied forces

The two types of applied forces acting on a body segment are identified as internal and external. The primary *internal* forces are the muscle-produced forces and the bone-on-bone frictional forces acting at the joint. The primary *external* forces are those of gravity, friction, and the pushes and pulls exerted either by other people or by external devices as in impact or continuous-contact situations.

Analysis of muscle forces

Muscles produce a pull upon the segments to which they are attached. According to MacConaill and Basmajian (1969) the laws of approximation and detorsion dictate their potential actions. These laws state that in a muscle contraction its bony attachments tend to be brought closer to one another and into the same plane. The actual forces and torques produced by individual muscles, however, cannot easily be predicted because of the indeterminate influence of a number of physiological and mechanical factors—including length-tension and force-velocity relationships as

well as the location of the muscle attachments with respect to the joint. In addition, the axis of rotation of the joint may not remain fixed during the course of the movement.

The analysis of muscle group–produced torques is also complicated by the fact that the number of muscles involved, the extent of their contractions, and the actual distance of their attachments from the joint centers are not known. Due to these unknowns, the direction of the resultant muscle force and the bone-on-bone joint forces cannot be determined satisfactorily. Valuable information results from the analysis, however, since the sum of all muscle forces times the sum of their resultant attachment distances from the joint center gives the torque, which is equal in magnitude and opposite in direction to the sum of the torques calculated from the external and inertial forces. In other words, the muscle-produced torque (T_m) is equal to the external torque produced by gravity (T_e) plus the inertial torque (T_I):

$$T_m = T_e + T_I = mg\ (\sin\theta)\ r + (mr\alpha)\ r$$

(Equation 25-1)

Since all torques act perpendicular to the segment, torque is represented in the free-body diagram by an arrow placed at an arbitrary point and indicating the direction of the tangential force with a magnitude necessary to maintain the system in equilibrium.

Analysis of external forces

In this lesson we are concerned only with the representation of external forces on free-body diagrams. The more detailed kinetic analysis of these forces is reserved for the next five lessons. The external forces which affect the movement of body segments are classified and symbolized as follows:

Gravitational forces (represented by weight and symbolized as w or mg)
Frictional forces (symbolized as f)
  Surface friction
    Starting friction (f_s)
    Moving (sliding) friction (f_m)
    Rolling friction (f_r)
  Fluid friction
    Air resistance (f_a)
    Water resistance (f_w)
Other externally applied forces (symbolized as F)

Gravitational forces

Gravitational forces are classified as distributed forces because their effects are distributed throughout a body. The resultant of these effects is represented by weight, which is considered to be concentrated at the center of mass of the body and directed toward the center of the earth.

Frictional forces

A sprinter running on a track experiences a surface reaction force when he pushes his foot against the track. This force can be resolved into vertical and horizontal components. The vertical component is called the radial or normal component (N), and the horizontal component is called the tangential or frictional component (f). The push of the foot against the track surface is backward and downward, while the reaction force of the surface against the foot is forward and upward. Only the forward and upward reaction forces are drawn on a free-body diagram since these are what cause the forward and upward motion of the sprinter. The points of application of these forces are indicated at the centers of the contact areas between the body segments and the surface.

Fluid friction, the friction between a body and the air or water through which it is moving, can be resolved into two components: parallel and perpendicular. The component parallel with the direction of motion but in the opposite direction is referred to as drag. This includes both skin drag caused by friction and form or pressure drag caused by forces normal to the surface. The second component acting perpendicular to the direction of the motion, is termed lift.

The force of buoyancy is a distributed force which acts vertically upward in opposition to the force of gravity. According to Archimedes' principle, the force of buoyancy of a body is equal in magnitude to the weight of the water that the body displaces. This force is considered to be concentrated at the center of buoyancy and coincides with the center of volume of the body.

KINETIC ANALYSIS OF TWO-SEGMENT MOTION

The kinetic analysis of two-segment motion involves (1) describing the basic data involved in the analysis, (2) constructing two-segment free-body diagrams, and (3) describing the Coriolis force.

Basic data involved in the analysis

The two types of data obtained in a mechanical analysis are termed absolute and relative. Data obtained about a segment's change of position in relation to the earth is *absolute*. If a change is measured in relation to the adjoining body segment it is *relative*. Thus the angular displacement of a segment measured from the right horizontal (+x) axis is called the absolute displacement, while the change in angle between adjacent segments is called relative displacement.

The two segments and two joints involved in a two-segment motion are identified as first or proximal and second or distal. The three radii are identified as primary (or segmental) and secondary. The two primary or segmental radii are symbolized as r_1 and r_2 for the first and second segments respectively. The secondary radius of the distal segment is symbolized by R. The radius of a segment is defined as the distance of the segmental center of gravity from its primary axis of rotation (adjacent joint); a secondary radius is the distance of the segmental center of gravity from its secondary axis of rotation (a remote joint). The Greek letter theta (θ) is used to describe the angular position of the segmental radius, while the Greek letter phi (ϕ) is used to describe the angular position of the secondary radius.

The data needed for a two-segment motion analysis are as follows:

R, Secondary radius of the second segment
ϕ, Absolute angular position of *R*
r_1 and r_2, Primary radii of the first and second segments
θ_1 and θ_2, Absolute angular positions of r_1 and r_2
w_1 and w_2, Weights of the first and second segments
m_1 and m_2, Masses of the first and second segments
l_1 and l_2, Lengths of the first and second segments
ω_1 and ω_2, Relative angular velocities of the first and second segments
α_1 and α_2, Relative angular accelerations of the first and second segments
V_2, Linear velocity of the center of gravity of the second segment

Constructing two-segment free-body diagrams

The free-body diagram (Fig. 25-2) is determined first for the second or distal segment. It

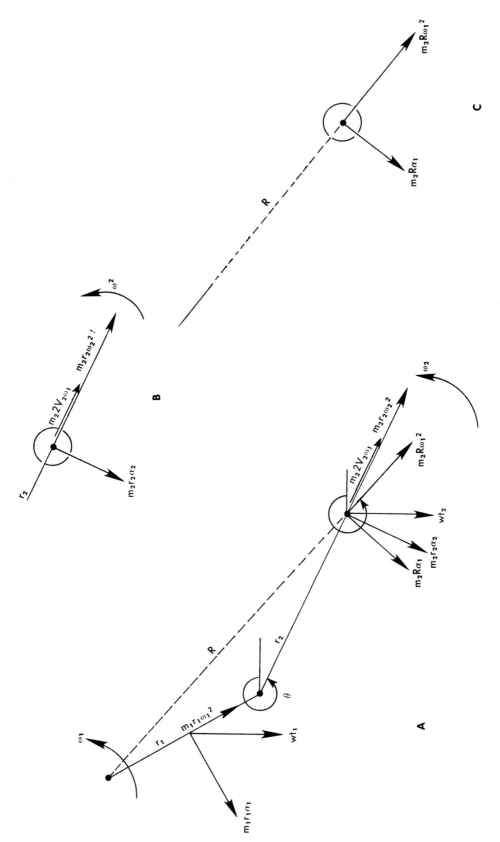

Fig. 25-2. Free-body diagrams of a two-segment movement.

must show both radii for this segment, r_2 and R, and all forces which act in opposition to the applied muscle forces. The applied muscle forces are represented by the perpendicular but arbitrary vector F_m.

The oppositional forces include the force of gravity (given by the weight of the segment) and five inertial forces. Fig. 25-2, *B*, shows the three inertial forces acting in reference to the segmental radius (r_2). Fig. 25-2, *C*, shows the two inertial forces which act in reference to the secondary radius (R). These inertial forces are defined and measured as follows:

1. Forces acting in reference to the segmental radius (r_2)
   a. The radial force $(mr\omega^2)$ acting parallel with r_2 and away from the primary joint or axis of rotation
   b. The tangential force $(mr\alpha)$ acting perpendicular to r_2 and in the direction of the resultant motion (F_m)
   c. The Coriolis force $(m_2 2V_2\omega_1)$ opposite direction and parallel with the radial inertial force

2. Forces acting in reference to the secondary radius (R)
   a. The radial force $(mR\omega^2)$ acting parallel with R and away from the proximal joint
   b. The tangential force $(mR\alpha)$ acting perpendicular to R and in the opposite direction of the resultant muscle force (F_m)

The forces acting on the first or proximal segment are (1) its weight, (2) its radial $(mr\omega^2)$ and tangential $(mr\alpha)$ inertial forces, which are described with reference to its segmental radius (r_1), (3) the sum of the forces of segment 2 applied to the distal end of segment 1, and (4) the moments

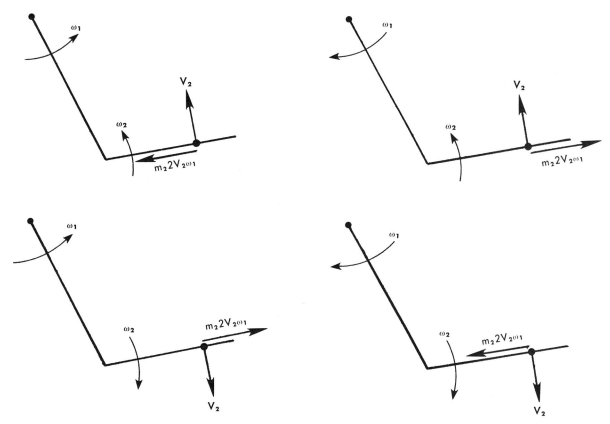

Fig. 25-3. The Coriolis force $(m_2 2V_2\omega_1)$ is a radial force pointing inward when the two segments rotate in the same direction and pointing outward when the segments rotate in opposite directions.

of force ...plied in the opposite direction ...n other words, the resultant ...see ...segment 2 also attempts to rotate ...opposite direction.

muscle ... now be familiar with all the inertial seg...ng on the two segments, with the exception ...of the Coriolis force (which acts upon the ...second segment). Thus there is a need for a further study of this force.

Coriolis force

The Coriolis force, named after the French scientist Gaspard Coriolis, is the force that exists when two segments are in motion—the first rotating about a fixed point, and the second rotating about the moving end of the first segment. This force is a radial force since it is directed parallel with the segmental radius. It points inward when the two segments rotate in the same direction and outward when the segments rotate in opposite directions as shown in Fig. 25-3.

Extension of the analysis technique to three-, four-, and five-segment motions

The technique presented above has been extended to the analysis of three-, four-, and five-segment motions by Plagenhoef (1971). You are encouraged to consult this source for additional information about these types of kinetic analyses.

STUDY GUIDELINES
General approach to the kinetic analysis of human motion
Specific student objectives

1. Identify and describe the three steps involved in a kinetic anlysis of human motion.
2. Differentiate between the particle, rigid-body, quasi–rigid-body, and linked systems of mechanical analysis.
3. Define a mechanical system.
4. Describe the representation of the following quantities on a free-body diagram of a body segment:
   a. Segmental end points, segmental lengths, and segmental centers of gravity
   b. Segmental weight
   c. Direction of angular motion of a segment and direction of linear motion of a particle
   d. Applied forces
   e. Inertial forces
   f. Reaction forces

5. Describe the following methods involved in the kinetic analysis of human motion:
   a. Force-mass-acceleration
   b. Impulse-momentum
   c. Work-energy

Summary

The three steps involved in a kinetic analysis are as follows: (1) define the mechanical system which is to be studied, (2) construct a free-body diagram of the system isolated from its surroundings showing all forces (vectorally) acting upon it, and (3) select an analysis approach appropriate to the problem being investigated.

A mechanical system is defined as the parts of a whole that are capable of performing some common motion function. The particle system is used to describe the motion of the center of gravity of a jumper, while the rigid-body and quasi–rigid-body systems are used to represent the static positions of the body and its parts. In the linked system the human body is represented as a series of interconnected rigid segments which demonstrate independent motion.

A free-body diagram is a simplified drawing of a mechanical system isolated from its surroundings showing all the force vectors acting upon it. The segmental end points and segmental centers of gravity are represented by points in space as described by either rectangular or polar coordinates. The straight line connecting the segmental end points of a segment is the segment length. Segmental weights as well as other applied and reaction forces are drawn in as vectors pointing in the direction in which they act, while inertial forces are drawn in the opposite direction of the motion to show that they act in opposition to the movement.

The force-mass-acceleration method is used to study the instantaneous effects of forces. The impulse-momentum method is used to analyze forces acting during time periods. The work-energy method is used to analyze forces acting over specific distances.

Performance tasks

1. List the three steps involved in a kinetic analysis of human motion.
2. The center of gravity of the body is best represented on a free-body diagram as a (particle/ linked) _____ system.

3. The parts of a whole capable of performing some common motion function constitute a
   a. mechanical system   c. rigid body system
   b. linked system   d. particle system
4. Describe the representation of the following quantities on a free-body diagram of a body segment:
   a. Segmental end points, segmental lengths, and segmental centers of gravity
   b. Segmental weight
   c. Direction of angular motion and direction of linear motion
   d. Applied force
   e. Inertial force
5. The analysis method best suited to the study of forces acting through a distance is
   a. force-mass-acceleration
   b. impulse-momentum
   c. work-energy
   d. all of these
6. The analysis method best suited to the study of forces acting through specific time intervals is
   a. force-mass-acceleration
   b. impulse-momentum
   c. work-energy
   d. none of these

Kinetic analysis of one-segment motion
Specific student objectives

1. Define d'Alembert's principle.
2. Describe the vector representation of inertial forces.
3. Describe the vector representation of applied forces.

Summary

D'Alembert's principle indicates that all applied and inertial forces acting on a segment should be drawn in opposite directions on a free-body diagram to show the system in equilibrium.

The two inertial forces acting in angular motion are the tangential force ($mr\alpha$) and the radial force ($mr\omega^2$). The two types of applied forces are identified as internal and external. The primary internal forces are the muscle-produced forces and the bone-on-bone friction forces acting at the joint. The primary external forces are those of gravity, friction, and the pushes and pulls exerted by other people or external devices as in impact and continuous-contact situations.

Performance tasks

1. Define d'Alembert's principle.
2. The principle which states that all problems of dynamics must be reduced to problems of statics was first proposed by
   a. Newton   c. Archimedes
   b. D'Alembert   d. none of these
3. Describe the vector representation of inertial and applied forces on free-body diagrams.
4. Construct a free-body diagram of the forearm showing all the inertial and applied forces normally acting on this segment during flexion of the forearm at the elbow.

Kinetic analysis of two-segment motion
Specific student objectives

1. Identify the basic data involved in the kinetic analysis of two-segment motion.
2. Differentiate between absolute and relative segmental positions.
3. Describe the construction of a two-segment free-body diagram.
4. Describe the Coriolis force acting on the second segment during a two-segment motion.

Summary

The basic data involved in the kinetic analysis of two-segment motion are listed as

R, Secondary radius of the second segment
ϕ, Absolute angular position of the secondary radius (R)
r_1, r_2, Primary radii of the first and second segments
θ_1, θ_2, Absolute angular positions of the primary radii of the first and second segments
w_1, w_2, Weights of the first and second segments
m_1, m_2, Masses of the first and second segments
1_1, 1_2, Lengths of the first and second segments
ω_1, ω_2, Relative angular velocities of the first and second segments
α_1, α_2, Relative angular accelerations of the first and second segments
V_2, Linear velocity of the center of gravity of second segment

Absolute angular quantities are measured from the right horizontal ($+x$), while relative angular quantities are measured in relation to the adjoining body segment.

The Coriolis force is the force that exists when two segments are in motion; the first is rotating about a fixed point and the second is rotating about the moving end of the first segment. In a two-segment motion this force is equal to $m_2 2 V_2 \omega_1$.

Performance tasks

1. List the basic data involved in the kinetic analysis of two-segment motion and explain the meaning of the symbols involved.
2. Angular position measured in relation to the adjoining body segment is termed
   a. absolute   c. both a and b
   b. relative   d. neither a nor b
3. Construct a two-segment free-body diagram of your arm and forearm showing the proper representation of the basic data involved in the analysis of the two-segment motion of shoulder joint flexion, and elbow flexion.
5. Describe the Coriolis force (in both sentence and formula form).

SELF-EVALUATION

Students should use no reference materials for this progress test, and they can check their answers by referring to Appendix A.

1. The three steps involved in a kinetic analysis are (1) define the mechanical system being studied, (2) construct a free-body diagram of the system, and (3) _____

   _____ .

2. State when each of the following mechanical systems is used in a kinetic analysis:
   a. Particle system
   b. Rigid-body system
   c. Quasi–rigid-body system
   d. Linked system
3. Describe the construction of a free-body diagram.
4. Identify and describe the three methods that can be used in a kinetic analysis.
5. What does d'Alembert's principle indicate?

   _____

   _____

   _____ .

6. Describe the vector representation of inertial and applied forces on free-body diagrams.
7. Angular position measured from the right horizontal axis is termed
   a. absolute   c. both a and b
   b. relative   d. neither a nor b
8. Describe the construction of two-segment free-body diagrams.

Kinetic analysis of external forces

Kinetic analysis of gravitational forces

NEWTON'S LAW OF GRAVITATION

Everyone has probably heard the story of how Isaac Newton was struck on the head by a falling apple while sitting under an apple tree. According to the story this incident started Newton thinking about falling bodies and led to his formulation of the law of gravity.

Actually Newton's law of gravitation concerns more than just the relationship between the earth and objects on or near the earth's surface. It applies to the interaction of all bodies and is, therefore, sometimes called the universal law of gravity or gravitation.

What Newton discovered was that "every object in the universe is attracted to every other object in the universe with a force which is directly proportional to the product of their masses and inversely proportional to the square of the distance between their centers of mass." We can reduce his statement to formula form by using F for the gravitational attraction, m_1 for the mass of one object, m_2 for the mass of the other object, d for the distance between the centers of mass of the two objects, and G for a gravitational constant (this is not g, or 32 ft/sec/sec) as follows:

$$F = \frac{Gm_1m_2}{d^2}$$

(Equation 26-A)

This equation is seldom used for calculation purposes in kinesiology because the attractions that exist between objects other than the earth are insignificantly small. However, it can be used to calculate the physical attraction between a man and a woman weighing 192 lb and 96 lb, respectively. The attraction force when they are 3 ft apart is about 0.00000007 lb—an extremely small value.

The chief benefit of Equation 26-A is in the expression of relationships between the affective variables that influence the force of gravity. For example, we can see that the gravitational force exerted upon a body by the earth, called its weight, will be directly proportional to the mass of the body. Thus we can say that a person weighing 200 lb has twice the attraction to the earth as one weighing 100 lb. The 200 lb person is also twice as massive as the one weighing 100 lb, i.e., 6.25 slugs compared to 3.125 slugs (200/32 = 6.25, and 100/32 = 3.125).

Equation 26-A also tells us that the force of gravity exerted by the earth upon a body is inversely proportional to the square of the distance between the center of gravity of the earth and the center of gravity of the body. Thus we can see that since we are farther away from the center of the earth when we are on top of a mountain we actually weigh less on the mountain than we do at sea level. For this reason we can usually jump farther and throw objects farther at higher altitudes than at lower altitudes; i.e., the applied muscle force is acting upon a smaller resistance force, which yields a greater net force that can be used to accelerate a given mass (Newton II). Since the earth is actually fatter at the equator than at its poles, a man weighing 190 lb in the Arctic weighs 189 lb in Ecuador. Therefore he should be able to long jump about 2 inches farther near the equator.

ANALYSIS OF BODY WEIGHT AS A PERFORMANCE VARIABLE

The importance of body weight as a variable in human movement can be explained by referring once again to Newton's second law, which states that a *net* force acting upon a body imparts to it an acceleration proportional to the net force in both magnitude and direction and inversely proportional to the mass of the body. Therefore we can see that one of the most important factors affecting the equilibrium and acceleration of a body is its *mass*. The mass of a body is given by its weight divided by the gravitational constant (32 ft/sec^2 or 9.8 m/sec^2); i.e., m = w/g. The greater the mass (weight of a body), the greater is its equilibrium as measured by the amount of force required to either positively or negatively accelerate it.

Whether body mass is to the advantage or disadvantage of a performer depends upon his movement objectives.

Should he desire to maintain his body in a state of *equilibrium,* either "at rest" or "in motion," a greater body mass, other things being equal, would be to his advantage. However, should he desire to produce an *acceleration* of his body, then a greater body mass, other things being equal, would be to his disadvantage. These relationships of body mass to performance are well known from everyday and sports experience. In matters of "at rest" equilibrium, for instance, we know that it is much more difficult to *move* a heavy object than it is to move a lighter one. Therefore, heavy football linemen and wrestlers usually have an advantage over lighter players. Also in matters of "in motion" equilibrium, we know that it is much more difficult to *stop* the movement of a heavy object than

it is to stop the movement of a lighter one. Therefore, the heavy running back in football usually has an advantage over the lighter players in this sport.

On the other hand, should the movement objective of a performer be to *accelerate* his body, then we know from Newton's second law (a = F/m) that the mass of the body as such would be an encumbrance. Therefore, performers in acceleration and endurance activities usually attempt to reduce to a minimum any *dead weight* they may be carrying in their body masses.

Dead weight is defined as the weight of all body parts not directly involved in the production of a movement. It is accounted for by the weight of skeleton, skin, viscera, blood vessels, and especially inactive muscles and body fat. Heavy inactive muscles are just as much dead weight in movement activities as is body fat. Therefore resistive exercises which develop muscle strength without a great increase in muscle bulk are preferred for performance in acceleration and endurance activities.

Analysis of body weight components

The weight of the body can be divided into two main components: a fat component and a fat-free component. The fat-free component is often referred to as the *lean body mass.* The relative proportions of body weight that are fat and fat-free can be determined through the use of either skinfold measures or body fluid measures.

Skinfold measures of body weight components. Instruments used to obtain skinfold measures of the amount of subcutaneous tissue and fat in different parts of the body are generally known as fat calipers. Pascale, Grossman, and Sloane (1955) reported the following regression equation for the estimation of body density from fat caliper measurements.

$$Y = 1.088468 - 0.007123X_1 - 0.004834X_2 - 0.005513X_3$$

(Equation 26-B)

Y is the estimated body density, X_1 the skinfold measurement on the chest at the midaxillary line (at the level of the xiphoid), X_2 the skinfold measurement for the chest at the nipple position, and X_3 the dorsum of the arm, midway between the tip of the acromion and the olecranon.

Keys and Brozek (1953) report that the following equation will give the proportion of the body weight which is fat from the measure of body density:

$$F = \frac{4.201}{D} - 3.813$$

(Equation 26-C)

F is the proportion of body weight which is fat, and *D* is the measure of body density.

Body fluid measures of body weight components. Morehouse and Miller (1963) report that a measure of the fat content of the body can be calculated from a knowledge of the total amount of water in the body. Methods of measuring the total body water volume are given by the authors. The body fluid measure of the fat content of the body is based on the fact that fat is simply added to the body without the addition of corresponding amount of water and that the fat-free portion of the body has a reasonably constant water content of about 72%. If the measured water content of an individual is 54% and if we let X equal percent fat, then the proportion of the body weight which is fat can be calculated as follows:

Measured water content equals nonfat water content plus water content in fat, or

$$(100)(0.54) = (100 - X)(0.72)$$
$$54 = 72 - 0.72X$$
$$18 = 0.72X$$
$$0.25 = X \text{ (i.e., 25\%)}$$

Importance of body weight components as performance variables

The fat content of the body, like body weight in general, can be to either the advantage or the disadvantage of the performer. Football players, especially linemen, employ the fat portion of their mass in achieving momentum (mass × velocity) and also as a cushion to absorb the shocks of repeated contact. In acceleration and endurance activities, however, body fat is simply dead weight and should be reduced to an absolute minimum.

The fat-free component of the body weight is, of course, mostly composed of the weight of the muscles and of the skeleton. Since muscle strength is roughly proportional to the amount of muscle an individual has, it should be easy to understand the importance of lean body mass as a variable in human movement.

According to McCloy and Young (1954), the muscles in an adult male of good muscular development constitute approximately 40% to 45% of his body weight. This percentage is 35% to 40% in an adult female. McCloy and Young give the following illustration of the importance of muscle weight and muscle development in motor performance: If a man weighing 150 lb has 40% of his weight comprised of muscle, he has 60 lb of muscle. If he were to increase his weight to 180 lb, and if all this increase were in muscular weight, then his increase in muscular weight would be 30 lb (i.e., 60 + 30) or an increase of 50%. The increase in the load would also be 30 lb (i.e., 150 to 180) or an increase of only 20%. Hence the increase in muscle strength would be proportionately greater than the extra strength called for by the increased load. Thus the undesirable effect of increasing the mass of the body is compensated for by the greater increase in the force available for the production of movements. For this reason athletes can increase their weight through a resistive exercise program and also increase their quickness because of the greater increases in strength relative to their total body mass resulting from these programs.

LOCATION OF THE CENTER OF GRAVITY IN HUMANS

The various methods of locating the common center of gravity in man are described in detail in most tests and measurements in physical education textbooks. There are also several methods for determining the masses and locating the centers of gravity of body segments. Fig. 26-1 illustrates a simple method of locating the common center of gravity of an individual in the transverse plane. A board long enough for the person to lie on is placed on two triangular blocks of wood (knife edges), with a block at each end. One end of the board rests on a platform scale, and the other on a block. The weight of the board can be read from the scale, and this reading must be subtracted from subsequent readings made during the measurement. In the illustration a man weighing 150 lb is shown lying on the board. The distance between the two knife edges is 60 inches, and the corrected reading on the scale is 80 lb.

Since the man in the illustration is in rotatory equilibrium, we can employ the law of the lever

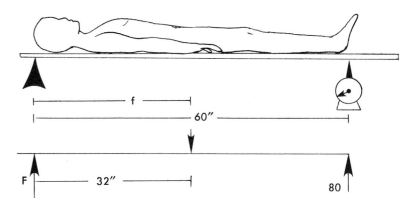

Fig. 26-1. Method of locating the body's center of gravity in the transverse plane.

$(F \times f = R \times r)$ to locate the center of gravity as follows:

$$150 \times f = 80 \times 60$$

thus

$$f = \frac{4800}{150} = 32.0 \text{ in}$$

The man's head is even with the knife edge, so his center of gravity in the transverse plane is 32 inches from the top of his head. This same method can be used to locate the center of gravity in the frontal and sagittal planes, except that the individual is standing erect and facing either straight ahead (frontal plane) or to one side (sagittal plane).

A simple method of determining the weight of body segments and locating the centers of each segment was devised by Cleaveland (1955). Marks are made on the body to indicate the limits of each segment, and the body is lowered into a tank of water to each mark in succession. The weight of each segment is calculated by the amount of total body weight lost and water displaced at the various stages of submersion. The center of gravity of each segment is located at the point where half its weight is lost.

The importance of locating a performer's center of gravity was discussed in Lesson 15. The key concept presented in these discussions is that the location of the common center of gravity of an individual greatly affects his "at rest" and "in motion" equilibrium in the performance of motor

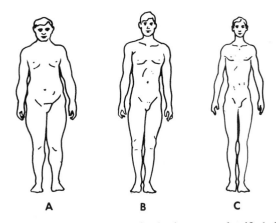

Fig. 26-2. Three types of physique as classified by Sheldon. **A,** Endomorph. **B,** Mesomorph. **C,** Ectomorph. (From Barham, J. N., and Wooten, E. L.: Structural kinesiology, New York, 1973, The Macmillan Co.)

activities. The segmental centers of gravity are also important in human movements because their locations are the points where the gravitational resistances to segmental movements are concentrated.

ANALYSIS OF BODY BUILD AS A PERFORMANCE VARIABLE

The distribution and relative proportions of the two components of body weight largely determine the type of body build an individual has. The most widely used method of classifying body builds, often referred to as somatotyping, is that associated with the work of W. H. Sheldon (1954) and his collaborators. According to Sheldon, body types can be classified as endomorphic, mesomorphic, and ectomorphic (Fig. 26-2). The following characteristics, determined through methods previously discussed in this lesson, are associated with each of the three body types:

Endomorphy is characterized by relative roundness and softness of the body. The head seems large and round, the bones small, the neck short, there is a central concentration of mass with a predominance of abdomen over thorax, and the shoulders are high and narrow. The chest is broad with fatty breasts, the arms are short, the buttocks heavy and fat, the legs short and heavy, and the hips broad. In this type of physique, the digestive system dominates the body economy.

Mesomorphy is characterized by a heavy, hard, rectangular outline with relatively prominent muscular development. The bones are large and heavy, the thorax is larger than the waist, the shoulders are broad, and the neck is usually fairly long. This type of physique is well adapted for most types of motor performances.

Ectomorphy is characterized by a slender frail body structure with small bones and thin segments. The limbs are relatively long, but the trunk is usually short (which prevents the individual from being tall). The neck is long and slender, the shoulders round, and the buttocks inconspicuous. This type of physique is associated with linearity, delicacy, and fragility of the body.

In Sheldon's technique, the somatotype of an individual consists of three numbers to represent the combination of the above characteristics. The first number represents the endomorphy component, the second number represents mesomorphy, and the third number represents ectomorphy. The numeral 1 represents the lowest amount of a component, while 7 represents the highest amount; a somatotype of 711 would be extreme endomorphy, 171 extreme mesomorphy, and 117 extreme ectomorphy.

The one component that seems to be involved in all physiques in which some degree of success is found in motor activities is mesomorphy. Extreme mesomorphs (because of their great lean body mass, muscular development, and streamlined build) are usually excellent in their general motor abilities. They make good football backs,

sprinters, middleweight wrestlers, boxers, etc. However, in activities requiring greater mass and greater cushions against repeated contact in sports, the best performers tend toward higher endomorphic components. For example, football linemen tend to be endomorphic mesomorphs. On the other hand, successful performers in speed and endurance activities—because of the need for height, long limbs, and less mass—tend toward higher ectomorphic components. For example, tennis players and endurance runners tend to be ectomorphic mesomorphs.

Since the weight and the linear girth dimensions of the body and its parts largely determine the somatotype, all the previous discussions of this lesson are directly applicable to the study of body builds and the roles they play in the successful performance of motor tasks. In addition, however, the aesthetic appeal of well-proportioned bodies to both men and women should not be overlooked as an important factor in physical education.

STUDY GUIDELINES
Newton's law of gravitation
Specific student objectives

1. State and describe the universal law of gravitation (in formula and sentence form).
2. State three principles derived from the universal law of gravitation.
3. Identify examples of the above principles applied to human motion.

Summary

Newton's law of universal attraction applies to the interaction of all material objects. It is stated as "every object in the universe attracts every other object in the universe with a force which is directly proportional to the product of their masses and inversely proportional to the square of the distance between their centers of mass." In formula form it is $F = Gm_1m_2/d^2$, where *F* is the attraction force between the two objects, m_1 and m_2 are the masses of the objects, *G* is a gravitational constant, and *d* is the distance between the centers of mass of the two bodies.

Several principles are derived from the universal law of gravitation:

1. The law applies to any pair of material objects.

2. It determines the force of attraction between any two bodies.
3. The attraction force is directly dependent upon the product of the two masses.
4. The attraction force is inversely proportional to the square of the distance between the objects' centers of mass.
5. The attraction force is a vector quantity whose direction is through the objects' centers of mass.

Performance tasks

1. The law governing the interaction between any two material objects is called Newton's law of
   a. acceleration   c. inertia
   b. gravitation   d. reaction
2. Write the universal law of gravitation (in sentence form).
3. Identify the symbols and explain the relationships in $F = Gm_1m_2/d^2$.
4. The attraction force between two material objects depends upon the
   a. product of their masses
   b. distance between them
   c. both a and b
   d. neither a nor b
5. The attraction forces F and m_1m_2 are (directly/inversely) —————— proportional.
6. F and d^2 are (directly/inversely) —————— proportional.
7. Increasing the distance between the centers of mass of two objects would (decrease/increase)—————— the attraction force between them.
8. The direction of the attraction force of gravity is
   a. horizontal   c. both a and b
   b. vertical   d. neither a nor b
9. Weight is an example of the application of an attraction force governed by the law of ——————.
10. List five principles relative to the universal law of gravitation.

Analysis of body weight as a performance variable

Specific student objectives

1. State a principle relating body mass to maintaining a state of equilibrium.
2. State a principle relating body mass to the production of acceleration.
3. Describe dead weight.
4. Identify and describe the two main components of body weight.
5. Identify an instrument for determining body density.
6. Determine body density and the fat component of body weight.
7. Describe the effect of the body weight components on motor performance.

Summary

Body weight as a performance variable can be advantageous or disadvantageous, depending on movement objectives. If the objective is maintenance of equilibrium (at rest or in motion) a greater body mass (weight) is advantageous. However, if maximum acceleration is the objective, a greater body mass puts an individual at a disadvantage.

Dead weight is defined as the weight of all body parts not directly involved in the production of a movement.

The weight of the body can be divided into two main components: a fat component and a fat-free component. The fat-free component is referred to as lean body mass. The relative proportion of the body weight that is fat and fat-free can be determined through the use of either skinfold measure or body fluid measures.

Fat content in the body can be to either the advantage or the disadvantage of a motor performer. The fat-free component of the body weight is composed mostly of the weight of the muscles and the skeleton. Thus lean body mass is found to be an important variable in human movement. Increases in strength from resistive exercise result in greater proportionate increases in muscle force relative to the total mass increase.

Performance tasks

1. A football running back who is moving and carrying the ball finds it to his advantage to have (greater/lesser) —————— mass.
2. A change of direction in motion can best be carried out, with a given amount of force, by a (greater/lesser) —————— mass.
3. The weight of all body parts not directly in-

volved in the production of a movement is referred to as
- a. body fat component
- b. body weight
- c. dead weight
- d. total body weight
4. State two methods for determining body weight components.
5. The two main components of body weight are
- a. fat and muscle
- b. fat and fat-free
- c. both a and b
- d. neither a nor b
6. If a kinesiology laboratory is available, determine body weight components by the skinfold measure and body fluid measure methods.
7. Discuss the advantages and disadvantages of varying proportions of the fat and fat-free components of body weight and the effect on motor performance.
8. State a mechanical justification for the development of strength with resistive exercise programs.

Location of the center of gravity in humans
Specific student objectives

1. Describe a method for locating the body's center of gravity.
2. Identify one application for the center of gravity in human motion analysis.

Summary

Numerous methods exist for locating the center of gravity of the body and body segments. The center of gravity specification is essential in the kinematics and kinetics of human motion.

Performance tasks

1. Describe center of gravity.
2. Consult at least two other references or sources for locating the center of gravity of a body or body segments and write a summary of a method described in the references.
3. If a kinesiology laboratory is available, conduct an experiment to determine the center of gravity in the human body.

Analysis of body build as a performance variable
Specific student objectives

1. Identify and describe Sheldon's body types.
2. Describe the numerical system of classification used by Sheldon.

3. Identify the body type component which seems to be a favorable factor present in motor activities.

Summary

The distribution of body weight and the relative proportions of fat and fat-free mass largely determine the type of body build an individual possesses.

Sheldon's classification system for body types are as follows:
1. Endomorphy—characterized by roundness and softness of body
2. Mesomorphy—characterized by a heavy, hard, rectangular outline with prominent muscular development
3. Ectomorphy—typically a slender frail body structure with small bones and thin segments

The numbering system of this classification consists of three numbers to represent the combination and degree of the three characteristics. The first number represents endomorphy, the second mesomorphy, and the third ectomorphy. The numeral 1 represents the lowest degree of a component, while 7 represents the highest degree.

Body build is another factor that determines the potential for effective and efficient motor performance.

Performance tasks

1. A body type characterized by a heavy, hard, rectangular outline with prominent muscular development is known as
- a. ectomorphy
- b. endomorphy
- c. mesomorphy
- d. all of the above
2. A slender frail body structure would be called an (ectomorph/endomorph) —————.
3. Describe Sheldon's classification numerals of
- a. 711
- b. 417
- c. 135
- d. 117
4. The most important characteristics for effective and efficient motor activities are commonly identified as
- a. ectomorphy
- b. endomorphy
- c. mesomorphy

SELF-EVALUATION

Students should use no reference materials for this progress test, and they can check their answers by referring to Appendix A.

1. Describe the relationships involved in $F = Gm_1m_2/d^2$.

2. The law governing the interaction between any two material objects is called Newton's law of
   a. acceleration   c. inertia
   b. gravitation   d. reaction

3. The attraction force between two material objects is directly proportional to the
   a. product of the masses
   b. distance between the centers of gravity
   c. both a and b
   d. neither a nor b

4. The attraction force (F) and the distance factor (d^2) in Newton's law of gravitation are (directly/inversely) ——————— proportional.

5. An individual would weigh (less/more) ——————— on the moon than on the earth.

6. The direction of the attraction force of gravity is
   a. horizontal   c. oblique
   b. vertical   d. none of the above

7. State a principle relating body mass to the force producing acceleration of the human body in a free fall situation.

8. The two main components of body weight are
   a. fat and muscle   c. both a and b
   b. fat and fat-free   d. neither a nor b

9. A change in speed in human motion is best carried out, with a given amount of force, by a (greater/lesser) ——————— mass.

10. Identify two methods of determining body weight components.

11. State a mechanical justification for the development of strength with resistive exercise programs.

12. The weight center of the body is called the center of ——————— .

13. Describe a method for locating this weight center and its segments.

14. A body type characterized by roundness and softness would be (ectomorphic/endomorphic) ——————— .

15. Identify and describe Sheldon's classification of body builds.

16. The body type that has been identified as the most important characteristic for effective and efficient motor performance is (ectomorphy/mesomorphy/endomorphy)——————— .

LESSON 27

Kinetic analysis of surface frictional forces

Body of lesson

TYPES OF FRICTION

The general statement can be made that wherever there is motion or a tendency toward motion within the earth's atmosphere there is friction. All frictional forces can be classified as being produced either by surface friction or by fluid friction. Surface friction is the resistance to motion created by contact between two surfaces. Fluid friction is the resistance to motion created when an object moves or tends to move through a fluid environment.

Types of surface friction

A body can be moved across the ground or similar surface by being made to either slide or roll. Therefore the two types of friction between a body and the surface over which it is moving are identified as *sliding* and *rolling*.

To start a body sliding or rolling requires a greater force than is needed to keep it sliding or rolling. In other words, starting friction (f_s) is greater than moving friction (f_m). In kinesiology the sliding and rolling types of friction are also subclassified into starting friction and moving friction. Thus there is starting sliding friction, moving sliding friction, starting rolling friction, and moving rolling friction.

Types of fluid friction

Since fluids are subclassified as liquids and gases, we shall consider these to be the two primary fluid environments through which human movements occur. For all practical purposes, however, we can consider water to be the primary liquid and air to be the primary gas. Thus the two primary types of fluid friction will hereafter be referred to as water resistance and air resistance, respectively.

In the next lesson fluid friction is considered in relation to its respective mediums of air and water. In this lesson we are focusing upon the

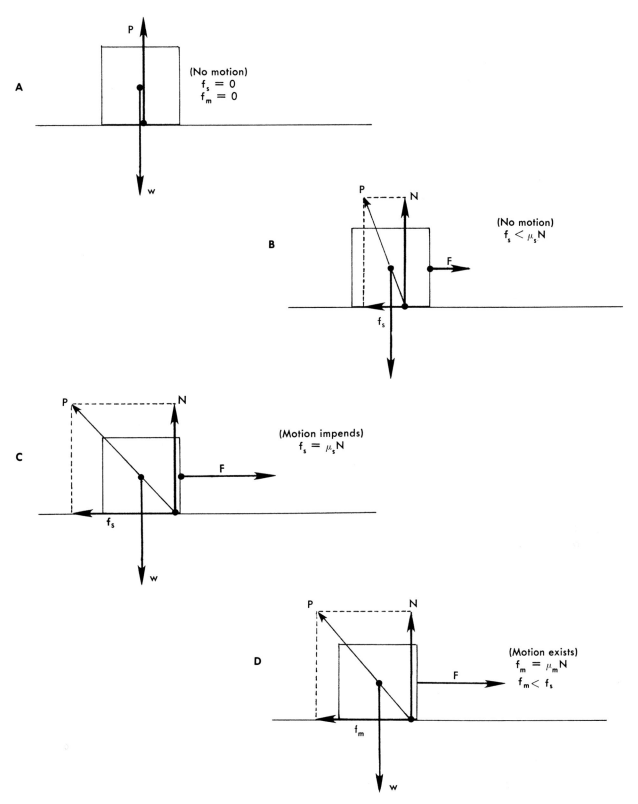

Fig. 27-1. The force of friction can be zero **(A)**, less than its maximum value **(B)**, or equal to its maximum value **(C)**. Once a body is in motion, surface friction decreases. In **D** moving friction (f_m) is less than starting friction (f_s).

frictional forces that exist between the surfaces of two solids.

MEASUREMENT OF SURFACE FRICTION

Surface frictional forces can exist in either the presence or the absence of motion. Whenever one body slides over another, for example, a frictional force is created between the surfaces of the bodies. This force is directed parallel with the surfaces and in the opposite direction of the motion. Thus a runner moving from left to right has a frictional force acting on the ground to the left, while an equal force acting on him is directed toward the right. A frictional force can even act when there is no relative motion. A horizontal force acting on a football defensive lineman, for example, may not be strong enough to set him in motion. In this equilibrium situation the applied force is balanced by an equal frictional force exerted by the ground so the vector sum of the forces is equal to zero.

Measurement of starting friction

Our quantitative treatment of starting friction will begin with a simple laboratory experiment. In Fig. 27-1, *A,* a box is at rest on a horizontal surface. It is in equilibrium under the action of its weight *(w)* and the upward force *(P)* exerted on it by the surface.

Suppose now a rope is attached to the box as shown in Fig. 27-1, *B,* and the tension *(F)* in the rope, as measured by a spring scale, is gradually increased. Provided the tension in the rope is not greater than the resistive frictional force, the box will remain at rest. The force *(P)* exerted on the box by the surface now has two components: one equal to and opposite the force of gravity and the other equal to and opposite the horizontal force being applied through the rope. The component of *P* parallel with the surface is called the force of starting friction *(f_s).* The other component, which is perpendicular to the surface, is called the normal force *(N).* From the conditions of equilibrium, we know that the force of starting friction *(f_s)* is equal to the applied force *(F)* and that the normal force *(N)* equals the weight *(w).*

As the force *F* is increased further, a limiting value is reached beyond which the box breaks away from the surface and starts to move. Thus there is evidently a maximum value that starting friction *(f_s)* can have, and motion occurs when this value is exceeded. Fig. 27-1, *C,* is the force diagram when *F* is just below the limiting value of starting friction and motion is about to begin. If *F* exceeds this limiting value, a net force exists and motion is imparted. Net force, of course, is the remainder obtained after the starting friction has been subtracted from the applied force (i.e., net $F = F - f_s$).

Experiments show that the force of starting friction (f_s) is due to the forces that press two surfaces together. These forces actually cause a welding of the surfaces at the areas of contact. Since only the forces acting perpendicular to the surfaces will have the effect of pressing the surfaces together and since these forces are called the normal forces *(N),* it can be seen that starting friction is proportional to the normal force. This proportionality is represented by the lower case Greek letter mu (μ), which is called the coefficient of friction. The coefficient of starting sliding friction (μ_s) is determined by dividing the weight of a body resting on a horizontal surface (Fig. 27-1) into the amount of force required to balance the maximum force of starting friction. In other words, $\mu_s = F/w$; or, since $w = N$ and $F = f_s$, in a horizontal equilibrium situation, we have

$$\mu_s = \frac{f_s}{N}$$

(Equation 27-A)

and rearranging we obtain

$$f_s = \mu_s N$$

(Equation 27-B)

Thus, if the box in Fig. 27-1 weighs 100 lb and 50 lb is required to balance the force of friction, the coefficient of starting friction is equal to 0.5.

$$\mu_s = \frac{f_s}{N} = \frac{50 \text{ lb}}{100 \text{ lb}} = 0.5$$

Knowing this coefficient should help us considerably in any attempt to determine the amount of force required to balance the maximum force of starting friction of other boxes made out of the same materials but with different weights. For example, the amount of force required to balance the maximum starting friction of a 500 lb box is determined through the use of Equation 27-B:

$f_s = (\mu_s) (N) = (0.5) (500 \text{ lb}) = 250 \text{ lb}$. Thus when a coefficient of starting friction of 0.5 exists between a 500 lb box and its horizontal surface, a force of 250 lb is required to balance the maximum force of starting sliding friction and any force greater than this will cause motion.

It should now be evident that for a given pair of surfaces, the maximum value of f_s is proportional to the normal force (N). The actual force of starting friction can therefore have any value between zero (when there is no applied force acting parallel with the surface) and a maximum value proportional to N or equal to $\mu_s N$. Thus $f_s \leq \mu_s N$. The equality sign holds only when the applied force (F), acting parallel with the surface, has such a magnitude that motion is about to start (Fig. 27-1, *C*). When F is less than this (Fig. 27-1, *B*), the inequality sign holds and the magnitude of the friction must be computed from the conditions of equilibrium.

Measurement of moving friction

As soon as sliding begins, the frictional force acting between the surfaces decreases. This new frictional force, the force of sliding friction, is also proportional to the normal force. The proportionality factor (μ) is now called the coefficient of moving friction. Thus when the box shown in Fig. 27-1 is in motion, the force of moving friction is giving by $f_m = \mu_m N$ (Fig. 27-1, *D*).

Measurement of rolling friction

A comparison of the force required to slide a heavy box along the ground with the force required to move the box on rollers shows that sliding friction is many times greater than rolling friction. For this reason wheels are used on vehicles instead of runners, and ball bearings are employed in some machines in place of sleeve bearings.

Rolling friction results from the deformation of two bodies where they make contact. The harder a rolling wheel or ball and the harder the surface over which it rolls, the less is the force of rolling friction. A better understanding of the origin of rolling friction can be gained by comparing the different kinds of wheels shown in Fig. 27-2. For a hard wheel on a soft dirt road as shown in Fig. 27-2, *A,* the applied force is continually pulling the wheel over a mound developed in the ground. For a soft wheel on a hard paved road (Fig. 27-2, *B*), the road is continually pushing the wheel out of shape. For a hard wheel on a hard road (Fig. 27-2, *C*), both wheel and road are distorted ever so little and the force of friction is exceedingly small.

The same equations which hold for the starting and moving types of sliding friction also hold for rolling friction, the only difference being that the coefficients for rolling friction are exceedingly small, $f_r = \mu_r N$.

Measuring friction through the use of an inclined plane

One method of measuring the coefficient of moving friction is to place a block on an inclined plane and then tilt the plane until the block slides down with constant velocity. In Fig. 27-3 a block has been placed on an inclined plane and the

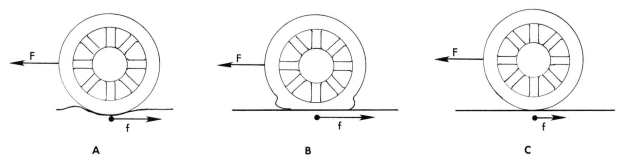

A　　　　　　　　**B**　　　　　　　　**C**

Fig. 27-2. Rolling friction between two surfaces is affected by their hardness. In **A,** a hard wheel is on a soft road. In **B,** a soft wheel is on a hard road. Both these situations produce relatively large amounts of friction. The least amount of friction exists when a hard wheel rolls on a hard road (**C**).

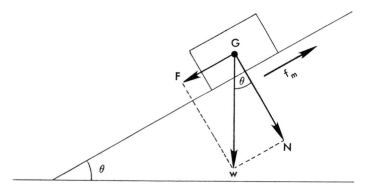

Fig. 27-3. The angle of uniform slip *(θ)* is determined by elevating the plane until the block slides down the plane at constant speed.

slope angle *(θ)* of the plane has been adjusted until the block slides down the plane at constant speed. The problem is to find the angle *θ*.

The forces acting on the block are its weight *(w)* and its normal and frictional components. Since motion exists, the friction force *(f_m)* is given by $\mu_m N$. If we take axes perpendicular to and parallel with the surface of the plane as shown in the figure, we can see that the normal force *(N)* is the "side adjacent" in the right triangle *GNw*. Thus the normal force is given by mg cos *θ*. The propelling force *(F)*, which causes the block to slide down the plane, is equal to the force of moving sliding friction when the sliding occurs at a constant speed. Thus both the propelling force and the force of sliding friction are given by mg sin *θ*.

Since f divided by N equals *μ* and since in this case we are also dividing the side adjacent *(N)* of a right triangle into the side opposite *(f)*, which gives the tan *θ*, we can see that

$$\mu = \tan \theta$$

(Equation 27-C)

Thus a block, regardless of its weight, will slide down an inclined plane with constant speed if the tangent of the slope angle of the plane equals the coefficient of moving friction. Measurement of this angle therefore provides a simple experimental method of determining the coefficient of moving friction.

We should now be able to see that if a block is placed on an inclined plane and the slope angle

of the plane is slowly increased the block will break away from the plane and start to slide when the tangent of the slope angle becomes equal to the coefficient of starting friction.

The inclined plane can also be used to measure the starting and moving types of rolling friction.

Relationship of the angle of uniform slip and the angle of lean

Using an inclined plane to measure friction illustrates that an increase in the coefficient of friction between the shoes of a performer and the surface on which the action is taking place directly increases the performer's potential angle of lean. Assume that surface conditions are found where the coefficient of friction *(μ)* between the surface and the soles of the shoes is 0.33. Since $\mu = \tan \theta$, the tan *θ* in our example also equals 0.33, which from Appendix B we can see is 18°. This is about the angle of inclination of a runner in the 100-yard dash. It can be seen therefore that very little lean would be possible here without slipping or falling. On the other hand, if the coefficient is found to be 1.5, which is not unusual for basketball shoes on a satisfactory playing surface, the angle of lean is 56°. This provides better gripping contact needed for even extreme maneuvers.

When the problem of slippage is too great and the surface cannot be changed, the performer will have to keep the center of gravity more nearly over the base of support. Short steps, avoidance of too much body lean, and dropping the center of gravity as low as possible by squatting will be necessary. In this way a playing posture can be maintained with reasonable efficiency.

Summary

The symbols used in the measurement of surface friction are as follows:

f, Friction force (parallel with the surfaces and acting in opposition to applied forces)
N, Normal force (perpendicular to the surfaces)
f_s, Starting friction force (before sliding or rolling motion begins)
f_r, Rolling friction force (during motion)
f_m, Moving sliding friction force (during motion)
μ, Coefficient of surface friction
μ_s, Coefficient of starting friction (before sliding or rolling motion begins)
μ_r, Coefficient of rolling friction
μ_m, Coefficient of moving sliding friction

The involved relationships can be stated as follows:

$$f_s = \mu_s N \text{ or } \mu_s = \frac{f_s}{N} \text{ for starting friction}$$

$$f_m = \mu_m N \text{ or } \mu_m = \frac{f_m}{N} \text{ for moving sliding friction}$$

$$f_r = \mu_r N \text{ or } \mu_r = \frac{f_r}{N} \text{ for rolling friction}$$

If in sliding or rolling situations the two contacting surfaces are continuously tilted at an increasing angle (θ) with the horizontal until motion starts, the limiting angle of starting sliding or rolling friction is termed θ_s and $\tan \theta_s = \mu_s$. Likewise if the surfaces are tilted to an angle (θ) with the horizontal until motion at a constant speed is reached, the angle of uniform slip is called θ_m or θ_r, $\tan \theta_m = \mu_m$, and $\tan \theta_r = \mu_r$.

FACTORS AFFECTING SURFACE FRICTION

Since surface friction (f) is equal to the normal force (N) multiplied by the coefficient of friction (μ), evidently it can be changed by changing either or both the normal force and the coefficient of friction. Thus we must review the factors that affect the magnitudes of the normal force and the coefficient of friction.

Factors affecting the normal force

The normal force (N) between two surfaces is the resultant of all forces acting perpendicular to the surfaces. When the surfaces are on the level and stationary, N can be considered to be the same as the gravitational force (called weight). The surface friction (f) between two solid objects is directly proportional to the magnitude of N pushing the two surfaces together.

A frictional force (f) therefore can be reduced by reducing the normal force (N). In football blocking, for example, this is done by a diagonally upward blocking force which introduces a vertical component *(F_y)* (Fig. 27-4). Since N is the resultant of all perpendicular forces acting on the contact surfaces, the vertical force (F_y) is subtracted from the opponent's weight so his normal force is reduced by the amount of F_y; i.e., $N = w - F_y$. This reduction in the normal force reduces f and makes it easier to move the opponent backwards.

Likewise, friction can be increased by any force which presses the surfaces more closely together. If we push a heavy piece of furniture with our hands against its top edge, the force applied is downward as well as forward (Fig. 27-5). Not only is the downward component *(F_y)* wasted as far as forward momentum is concerned, but the task is made more difficult by the friction to be overcome ($N + F_y$). Thus we can see that friction is dependent upon the total force pressing two surfaces together. This force includes both the weight of the body and all the additional forces pressing the object against the surface.

Relationship of friction to the actual contact area. The two types of contact areas are identified as actual and apparent. The difference between them is illustrated in Fig. 27-6. This figure also illustrates a method of measuring sliding friction. Fig. 27-6, *A*, shows that if the coefficient of starting friction is 0.25 a 125 gm force will be required to initiate the sliding of a 500 gm block. Likewise Fig. 27-6, *B*, shows that when the weight of the block is increased to 1000 gm by the addition of another 500 gm block the force required to start movement becomes 250 gm. Fig. 27-6, *C*, shows that when the two blocks in *B* are connected in tandem, one behind the other, the force of friction still equals 250 gm. Again, if the single block in *A* is turned on edge as in Fig. 27-6, *D,* the force of 125 gm is just enough to initiate sliding.

These observations may be explained largely in

Fig. 27-4. The frictional force (f) can be reduced by reducing the normal force (N). In football blocking, a diagonally upward force produces a vertical force component (F_y). When this component is subtracted from the defensive man's weight, it reduces his N and f components so he is easier to move.

terms of molecular attractive forces. The total contact area where molecular attraction is effective is generally small compared with the total apparent area. When a greater force is applied normal (perpendicular) to the surfaces, the contact areas increase in size and number and the following relationships are found to hold:

1. The total actual contact area is proportional to the total normal force.
2. The total actual contact area is independent of the total apparent area.
3. The force of sliding friction is proportional to the total actual contact area and independent of the total apparent area.

4. The force of sliding friction is proportional to the total normal force.

Apparent contact area is the total area of the contacting surfaces, whereas actual contact area is the smaller area within the apparent contact area that actually does the touching.

Racing cars use wide tires for the purpose of spreading the car's weight over a wider area, which thus allows for better wear and not for the attainment of greater friction.

Thus surface friction (f) is the same whether the force that presses the two surfaces together is concentrated over a large or a small apparent contact area. Furthermore, if the force is spread

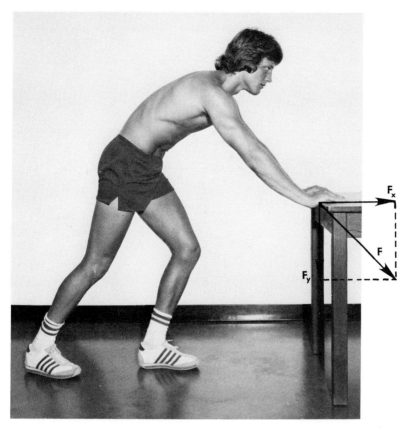

Fig. 27-5. Pushing forces directed downward as well as horizontally increase friction. The downward force (F_y) is added to the weight of the object to produce a greater normal force (N); i.e., $N = w + F_y$.

over a wide area, it will not be as great at any one point (even though the normal force remains the same) as it would if it were concentrated over a small area. The friction is the same regardless of the apparent contact area.

Factors affecting the coefficient of friction

The factors affecting the magnitude of the coefficient of friction are the nature and condition of the materials in contact as well as the relative motion and type of motion that exists between the surfaces.

Nature of the materials in contact. Because of greater molecular attraction, surfaces of the same material produce greater friction than do surfaces of different materials. The surface friction between rubber soled shoes and an artificial playing surface of rubberoid material is greater than between rubber soled shoes and a wood floor.

The coefficient of friction is affected by the roughness of the two surfaces. Basketball players attempting to play on a floor that has been covered with wax for a dance may have to make the soles of their shoes irregular by cutting them with a knife so there will be sufficient friction between them and the floor surface. Ping-Pong paddles with a round irregular surface enable the players to put more spin on the ball because of the increased friction. Surfaces such as the new unsmooth cement rings used for shot-putting and discus events keep the performer from slipping as he goes around and across the ring.

Condition of the materials in contact. The friction force depends not only on the material of the surfaces in contact but also on their condition. The friction of skis on snow changes drastically with different snow conditions. To offset this, the skier changes the condition of the

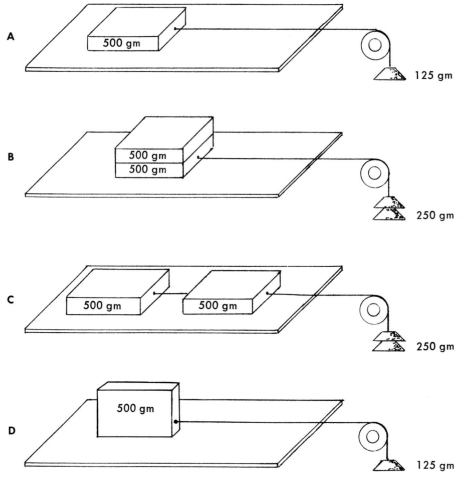

Fig. 27-6. Sliding friction between two surfaces is proportional to the normal force pushing the surfaces together and is independent of the apparent contact area.

ski surface by applying various types of wax. Firm-Grip, a liquid spray or paste, is applied to an athlete's hands to change the condition of the materials in contact—hands, bat, ball, or pole. This in turn increases friction.

Starting, stopping, changing speed, and changing direction are maneuvers that require a firm footing without slippage. A gripping effect is required between the shoes and the surface on which they are moving to produce a high coefficient of friction. In some sports such as baseball, football, and track, spiked shoes are used to accomplish this effect. In others, attention is directed to treating the surface played on and building shoes with soles having a high coefficient of friction. Treating wood playing surfaces

with finishes, using solvents on rubber soled shoes, and putting sticky substances in the soles of shoes are some of the current practices.

Relative motion between the surfaces. As we have already seen, once an object is in motion the force of sliding friction decreases considerably; that is, to keep it in motion is easier than to start it moving. Once an object is moving and increasing speed, there are no further significant changes in the coefficient of sliding friction. The principles involved here are as follows:

1. Starting sliding friction is much greater than moving sliding friction.
2. Moving sliding friction is relatively independent of speed changes.
3. Until a limiting value for starting friction

is reached, the action force tending to cause sliding motion will equal the reaction force of friction and the body will remain at rest.

Type of motion involved between the surfaces. Since sliding friction is many times greater than rolling friction, heavy objects should be placed on wheels whenever possible. The harder a rolling wheel or ball and the harder the surface over which it rolls, as has already been mentioned, the less will be the force of rolling friction. This interrelationship is obvious in the longer roll of a golf ball on a dry hard fairway than on a fairway of thick grass. When we attempt to push a wheelbarrow through loose dirt or sand instead of over a hard surface, we also get an idea of this interrelationship.

STUDY GUIDELINES
Types of friction
Specific student objectives

1. Describe fluid and surface friction.
2. Identify the types of surface friction.
3. Identify the types of fluid friction of primary interest in the study of human motion.
4. State a principle related to surface friction.

Summary

Wherever there is motion or a tendency toward motion within the earth's atmosphere, there is friction. Friction, a resistance to motion, is classified as surface or fluid. Surface friction is the resistance to motion created by contact between the surfaces of two solid objects. Fluid friction is the resistance to motion created by movement through liquids and gases. In human motion this would be motion through water and air.

The two types of surface friction are sliding and rolling. Starting friction is greater than movement friction. Sliding friction and rolling friction are subdivided into starting friction and moving friction. The two primary types of fluid friction are identified as water resistance and air resistance.

Performance tasks

1. Write out descriptions for
   a. fluid and surface friction
   b. friction
2. Starting friction is (greater/less) _____ than moving friction.

3. The fact stated in the preceding task applies to (fluid/surface) _____ friction.
4. The two types of friction are
   a. sliding and rolling   c. both a and b
   b. starting and moving   d. neither a nor b
5. Illustrate the surface friction principle "Starting friction is greater than movement friction" with demonstrations set up in an experimental setting.

Measurement of surface friction
Specific student objectives

1. Describe a frictional force.
2. State the relation of starting frictional force (f_s) to normal force (N).
3. Identify and describe the coefficient of moving sliding friction.
4. Identify and describe the angle of uniform slip.
5. State the relationship of angle of uniform slip to angle of lean.
6. State a principle relating rolling friction and sliding friction.

Summary

Frictional forces act whenever there is a normal force (N) between two solid objects or when there is relative motion.

The force of starting friction (f_s) exists whenever there is a force (called a normal force, N) acting perpendicular to the surfaces of two solids. The maximum value of f_s is proportional to N. The force of starting friction is called the coefficient of starting friction and is given by $f_s = \mu_s N$.

For moving sliding friction the moving frictional force (f_m) is also proportional to N. Thus $f_m = \mu_m N$.

A method for measuring the coefficient of moving friction (μ_m) is to place an object on an inclined plane and tilt the plane. When the object moves down the plane at a constant speed, μ_m can be found by determining the tangent of the angle θ (the angle of uniform slip or the angle between the tilted plane and the horizontal). A similar method can be used to determine μ_s when motion begins on an inclined plane. Note also that the angle of uniform slip will be equal to the angle of lean of a performer using the same surface conditions.

Sliding friction is greater than rolling friction. The same equations which hold for starting and moving sliding friction also hold for rolling friction.

Performance tasks

1. Develop written descriptions for the following:
   a. Frictional forces
   b. Starting friction force
   c. Normal force (N)
   d. Moving friction force (f_m)
   e. Coefficient of friction (μ)
   f. μ_s and μ_m
   g. Angle of uniform slip (θ)
2. Sliding friction is (greater/less) _____ than rolling friction.
3. Determine the coefficients of starting and sliding surface frictions by the angle of uniform slip method. Use at least two different materials on an inclined plane. An example would be two different types of shoes on a hardwood inclined plane.
4. Relate surface friction principles to
   a. angle of uniform slip and angle of lean
   b. rolling and sliding friction

Factors affecting surface friction
Specific student objectives

1. Define a normal force (N).
2. State the relation between surface friction (f) and the normal force (N).
3. Identify the affective factors for f and describe their relationship to f.
4. State three principles of surface friction.
5. Identify four principles relating surface friction to actual and apparent contact areas.
6. State three principles relating surface friction to relative surface motion.

Summary

The normal force (N) between two surfaces is the resultant of all forces acting perpendicular to the surfaces. The surface friction (f) between two solid objects is directly proportional to the magnitude of N pushing the two surfaces together.

The nature and condition of the materials in contact also determine the coefficient of surface friction. Surfaces of the same materials generally produce greater surface friction than do surfaces of different materials. Surface friction is affected by the construction of the two surfaces. The condition of the surfaces also determine the coefficient of friction.

The actual contact area between two surfaces is the part of the surfaces actually touching. The principles relating to this are as follows:

1. The total actual contact area is proportional to the total normal force.
2. The total actual contact area is independent of the total apparent area.
3. The force of sliding friction is proportional to the total actual contact area and is independent of the total apparent area.
4. The force of sliding friction is proportional to the total normal force.

Once an object is in motion, the friction decreases considerably ($f_m < f_s$). The following principles are involved:

1. Starting sliding friction is much greater than moving sliding friction.
2. Moving sliding friction is relatively independent of speed changes.
3. Until a limiting value for starting friction is reached, the action force tending to cause sliding motion will equal the reaction force of friction at rest.

Performance tasks

1. The resultant of all forces acting perpendicular to the surfaces of two solid objects is called the
   a. force of gravity   c. sliding force
   b. force of friction   d. none of the above
2. The relationship between surface friction and the normal force (N) is a/an (direct/inverse) _____ proportion.
3. Apparent contact area is (greater/less) _____ than actual contact area.
4. The coefficient of surface friction depends upon the nature and condition of the _____.
5. Actual contact area is (directly/inversely) _____ proportional to the total normal force.
6. Who will have the greatest surface friction between shoes and floor?
   a. A person weighing 100 lb with size 12 shoes
   b. A person weighing 100 lb with size 9 shoes

c. Neither a nor b
7. Sliding friction is related to the apparent contact area as
   a. a direct proportion   c. both a and b
   b. an inverse proportion d. neither a nor b
8. Moving sliding friction is related to relative speed change as
   a. a direct proportion   c. both a and b
   b. an inverse proportion d. neither a nor b
9. Write out six principles of surface friction.

SELF-EVALUATION

Students should use no reference materials for this progress test, and they can check their answers by referring to Appendix A.

1. The resistance to motion created by surface contact between two solid objects is known as
   a. fluid friction   c. sliding friction
   b. solid friction   d. surface friction
2. The resistance to motion through liquids and gases is referred to as (fluid/surface) _____ friction.
3. The two types of surface friction are
   a. fluid and solid   c. both a and b
   b. liquid and gas   d. neither a nor b
4. The magnitude of the frictional force be-

tween two solid objects is called the coefficient of
   a. rolling friction   c. both a and b
   b. sliding friction   d. neither a nor b
5. Identify the affective factors for surface friction.
6. The coefficient of sliding friction and the angle of uniform slip are related
   a. directly
   b. inversely
7. State the relationship between the angle of uniform slip and the angle of lean between two solid objects.
8. Sliding friction is (greater/less) _____ than rolling friction.
9. Define a normal force (N).
10. State five principles of surface friction.
11. Differentiate between actual contact area and apparent contact area.
12. The force of sliding friction is (directly/inversely) _____ proportional to the total normal force.
13. Starting sliding friction is (greater/less) _____ than moving sliding friction.
14. State a principle relating actual contact area to N in a surface friction situation.

LESSON 28

Kinetic analysis of fluid frictional forces

Body of lesson
  Concept of fluid friction
  Effect of relative velocity upon fluid friction
    Terminal velocity
    Cavitation
  Effects of active surface area and surface texture on fluid resistance
    Streamlining
  Effects of fluid pressure and temperature on fluid friction
    Concepts of density and specific gravity
    Concepts of fluid pressure
    Pressure and depth
    Pressure and velocity: Bernoulli's principle
    Effects of temperature on fluid friction
  Relationship of propulsion to resistance
Study guidelines
Self-evaluation

CONCEPT OF FLUID FRICTION

A common force involved in both aerial and aquatic motion is fluid friction. Air (a gas) and water (a liquid) are each considered to be fluids and to offer fluid friction to objects moving through them. Fluid friction results in two types of fluid flow during motion:

laminar flow The flow of air or water around a solid object which is relatively smooth
turbulent flow The flow of air or water around a solid object which produces rough eddies and low-pressure areas (cavitations)

Fluid resistance is the friction which manifests itself when a gas or liquid is made to flow around a stationary object or an object is made to move through a previously stationary fluid. In any discussion or treatment of fluid resistance, whether the fluid is considered to be moving and the object standing still or vice-versa makes no difference. We need to know only that there is a relative motion between the two.

The resistance of a liquid to an object moving through it can be determined by actually drawing the object through the liquid (Fig. 28-1). The resistance of a gas can be determined through the use of a wind tunnel as illustrated in Fig. 28-2. With techniques such as these, the following factors have been found to affect fluid friction:

1. Relative velocity between the object and the fluid
2. Shape of the object and its active surface area (especially its cross-sectional area acting at right angles to the line of flow)
3. Surface texture of the object
4. Fluid pressure
5. Temperature of the fluid

The magnitude of fluid friction affecting a body can be expressed by

$$f = \frac{KPSV^2}{2}$$

(Equation 28-A)

434

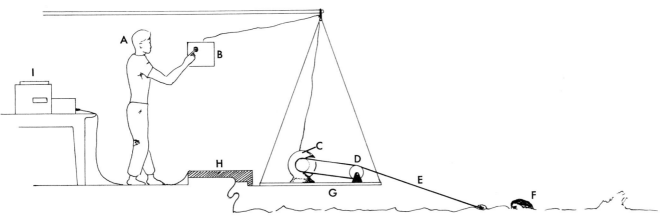

Fig. 28-1. The resistance of a liquid to an object moving through it can be determined by actually pulling the object through the liquid. Here a swimmer is being towed through the water. In this arrangement the operator *(A)* controls the on-off switch *(B)* to the motor *(C)*. The motor drives the shaft *(D)* at various speeds depending on the pulley arrangements. The towing rope *(E)* wraps around the shaft and pulls the swimmer *(F)* toward the platform *(G)*. The resistance caused by the swimmer exerts a force on the towing rope to pull the platform toward the swimmer. The platform is fixed to the side of the pool by the strain gauge beams *(H)*. The force on the towing rope is measured by the strain gauge beams and recorded on an electronic recorder *(I)*. (Modified from Counsilman, J. E.: The science of swimming, Englewood Cliffs, N.J., 1968, Prentice-Hall, Inc., p. 27.)

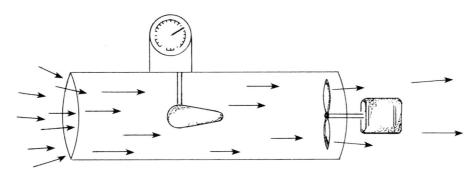

Fig. 28-2. A wind tunnel for measuring the air friction on an object. Air is pulled through the tunnel by the fan, and the force of air acting on the object is shown on the dial.

where f is the fluid resistance, K is a numerical nondimensional coefficient depending upon the texture and shape of the body and the temperature of the fluid, P is the pressure of the fluid (15 lb/in² for air at sea level), S is the active surface area of the body over which the fluid stream passes, and V is the velocity of the fluid flow with respect to the body. Thus it can be seen that the fluid resistance depends directly upon the fluid pressure, the active surface area, and the square of the relative velocity.

EFFECT OF RELATIVE VELOCITY UPON FLUID FRICTION

At relatively low speeds the flow of fluid around an object is smooth and regular and fluid friction is proportional to the velocity; $f = KV$, where K is a constant of proportionality (Fig. 28-3).

If initially $V = 0$, frictional resistance to motion is negligible and any applied force is entirely effective in producing acceleration. As the speed increases, however, friction increases proportionally so there is less and less net force avail-

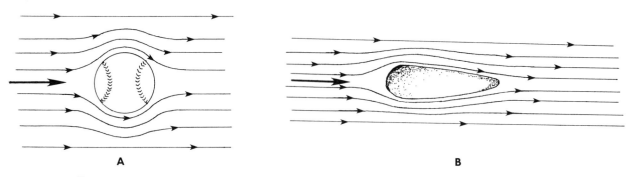

Fig. 28-3. Examples of laminar flow of a low-velocity fluid around solid objects. The baseball in **A** offers much greater resistance than does the streamlined object in **B**.

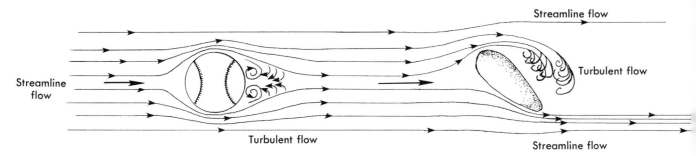

Fig. 28-4. Examples of turbulent flow of a high-velocity fluid around solid objects.

able for acceleration. Newton's second law, applied to motion through a fluid, therefore becomes

$$F = KV + ma$$

(Equation 28-B)

The above equation holds only for *laminar* flow, that is, for relatively low velocities. As the speed increases, a point is reached where *turbulence* sets in and the force of friction increases rapidly and becomes proportional to the square of the velocity.

$$f = KV^2$$

(Equation 28-C)

Turbulent flow is characterized by small eddy currents which form behind the object as shown in Fig. 28-4. Not only does the fluid have to move out and around the obstacle quickly, but considerable energy is taken up by greater friction. A baseball thrown with a relatively high velocity through the air will be resisted by a fluid frictional force that is proportional to V^2.

Fig. 28-5. The eddies formed by air resistance at high velocities develop not symmetrically in pairs along the sides of the object but alternately leaving a great deal of turbulence. This "turbulence trail" is shown by the waving of a flag in the wind.

The eddies created by air resistance at high velocities do not form in pairs on the sides of the object but form alternately, leaving a great deal of turbulence. This turbulence trail is shown by the waving of a flag in the wind (Fig. 28-5). The ripples in the flag indicate the alternate formation of eddies and whirlwinds, which produces turbulence.

Terminal velocity

It is well known that raindrops fall with a speed that depends upon their size and not upon the height from which they fall. Starting from rest, a particle falling in a gas or a liquid increases in velocity until the retarding force of friction becomes as great as the downward force of gravity. When this condition is reached, the body is in equilibrium and falls with a constant velocity called its terminal velocity.

The terminal velocity for small particles like fog droplets is so low that the airstream around them is one of laminar flow. George Stokes first discovered that the terminal velocity of small particles is proportional to their radius. This relation is known as Stokes' law.

For increasingly larger bodies the magnitude of the terminal velocity decreases; and as speed increases, turbulent flow sets in to eventually be the predominating part of the frictional resistance. Under these conditions resistance to both laminar flow and turbulent flow exists; thus equating downward force (weight) to the upward forces of friction leads to the following:

$$w = KV + KV^2$$

(Equation 28-D)

This equation applies not only to falling bodies but to airplanes in the air and ships in the water. The speed of planes and ships remains constant where the resistance is just equalized by the forward thrust of the propellers.

If a parachutist delays the opening of his chute long enough, he will attain a terminal velocity of from 130 to 150 mi/hr. At such speeds wind resistance pushes upward with a total force equal to his weight and the result is that he is no longer accelerating.

Terminal velocity is also a factor in swimming. When a force of given magnitude is applied to a body to move it through a liquid, the velocity of the body gradually increases until a certain limiting value is reached at which the resistance due to friction is equal to the applied force. After that value has been reached, the velocity remains constant as long as the force is applied.

Cavitation

Cavitation is caused when an object is pulled through the water so fast that a cavity is formed. When this happens, there is nothing but space for the propelling force to push against. According to Newton's third law there would be reduced forward movement. The action may be likened to slipping while walking or running.

The old paddle-wheel river boat experienced the effects of cavitation. If the wheels were rotated too fast, a cavity was created so there was no water for the wheel to push. This example and fact of cavitation suggests an optimum speed of movement of the arms and legs in swimming for the greatest efficiency. According to Bunn (1973) the optimal speed of limb movement is not now known on a scientific basis. Cavitation does not occur in swimming; however, it can play a major part in crew racing. The front oars of an eight-oared shell, for example, may be useless if they move slower than the rear oars.

EFFECTS OF ACTIVE SURFACE AREA AND SURFACE TEXTURE ON FLUID RESISTANCE

The magnitude of fluid friction in air and water also depends upon the shape and active surface area *(S)* of the body. A discus thrown as shown in Fig. 28-6, *A*, will have less fluid friction than one thrown as in Fig. 28-6, *B*. In movement through air or water, fluid friction is proportional to the greatest cross-sectional area of the body acting at right angles to the flow of the fluid. A swimmer with heavy gluteal muscles will tend to create greater fluid friction because of the greater active surface area.

A body moving through the water in a diagonal position encounters greater resistance than one moving in a horizontal position because it presents a greater surface area against which the water can act. The larger the surface area, the greater is the resistance to movement of the body through the water. Therefore the position of the body in swimming should be such that the smallest area

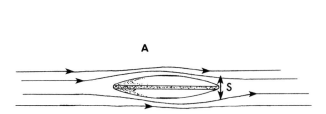

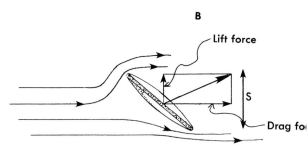

Fig. 28-6. In movement through air or water, the fluid friction is proportional to the greatest cross-sectional area of the body acting at right angles to the flow of the fluid (shown by the vector *S*).

possible is presented to the water in the direction of the desired movement.

Another factor in fluid friction is surface texture. The rougher the surface, the greater is the fluid friction. For example, Karpovich (1950) found that a wool suit offers more resistance to progress than does a silk suit. Also a loose suit, especially when relatively tight around the legs, adds to resistance. In addition, since the space between the suit and the body fills with water, the area of the body is, in effect, enlarged. We know furthermore that a softball with more seams and active surface than a baseball will encounter at identical speeds a larger amount of air resistance. The "dimpling" of a golfball, which creates a rougher texture, increases the fluid friction so essential for control and production of a spin effect on the ball. The smooth ball, used originally in the game, produced less fluid friction but was erratic and had little controllable spin effect.

Streamlining

When a body is shaped to the streamlines of the fluid through which it is moving, the retarding force of friction is greatly reduced. This is particularly true at high velocities when the conditions of turbulent flow would otherwise predominate. Streamlining is accomplished in human movements by reducing the active surface area and roughness of the body texture.

In swimming, for example, streamlining can be achieved by carrying the body parallel with the direction of motion and by keeping the legs close together so the width of the leg stroke is within the cross-sectional area formed by the body. When diving into water, the swimmer en-

counters less resistance when the head is kept between the arms; this presents a smaller front surface area.

A body can also be streamlined by reducing the roughness of its surface texture. A man, for example, who shaves his head and hairy body would have a tendency to decrease his fluid friction when swimming. This streamlining effect, plus that obtained by reducing the active body surface area, would reduce the tendency of the body to form eddies and the body could then slip through the water with minimum resistance.

EFFECTS OF FLUID PRESSURE AND TEMPERATURE ON FLUID FRICTION

In human motion the two primary mediums through which the body moves are air and water. At the same speed an object's fluid friction will be smaller in air than in water. The explanation for the greater fluid friction in water than in air is that water has greater density and fluid pressure than does air. An understanding of this fact requires development of the concepts of density, specific gravity, and fluid pressure, which are the topics of the following discussions.

Concepts of density and specific gravity

The mass of a body per unit of its volume is called its density (D); the weight of a body per unit of its volume is called its specific gravity (g_{spec}). Expressed algebraically,

$$D = \frac{m}{V}$$

(Equation 28-E)

$$g_{spec} = \frac{w}{V}$$

(Equation 28-F)

If we multiply both sides of equation 28-E by the acceleration of gravity (g), we obtain Dg = mg/V. Remembering that *mg* equals weight, we can see that we have just obtained Equation 28-F.

$$Dg = \frac{mg}{V} = \frac{w}{V} = g_{spec}$$

(Equation 28-G)

If the volume *(V)* and the density *(D)* of a body are known, they can be multiplied together to give mass as their product.

$$m = VD$$

(Equation 28-H)

When we multiply both sides of this equation by the acceleration of gravity (g), we obtain another interesting relationship:

$$mg = VDg = w$$

(Equation 28-I)

In other words, the weight of a fluid is the product of its volume and density times the gravitational constant.

Concept of fluid pressure

When a force (F) acts perpendicular to a surface whose area is A, the pressure (P) being exerted on the surface is defined as the ratio between the magnitude of the force and the area, P = F/A. Pressure is a useful quantity because fluids (gases and liquids) flow when forces are exerted upon them, and the flow is always from the high-pressure area toward the low-pressure area.

When we measure pressure, such as the pressure in a tire, we usually determine the difference between the unknown pressure and atmospheric pressure. This difference is called the gauge pressure, while the true pressure is called the absolute pressure.

Absolute pressure = Gauge pressure +
Atmospheric pressure

Thus a tire inflated at sea level to a gauge pressure of 24 lb/in² contains air at an absolute pressure of 38.17 lb/in² since sea level atmospheric pressure is 14.17 lb/in².

Pressure and depth

The pressure inside a volume of fluid depends upon the depth below the surface, since the deeper one descends the greater is the weight of the

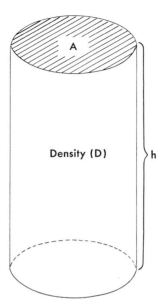

Fig. 28-7. A tank of height *h* and cross-sectional area *A* is filled with fluid of density *D*.

overlying fluid. Suppose we have a tank of height h and of cross-sectional area A which is filled with a fluid of density D (Fig. 28-7). With this information we can obtain the following:

1. Volume of the tank

$$V = Ah$$

(Equation 28-J)

2. Mass of the fluid contained within the tank

$$m = DV = DAh$$

(Equation 28-K)

3. Weight of the fluid in the tank

$$w = mg = DAhg$$

(Equation 28-L)

4. Pressure (P) exerted by the fluid on the bottom of the tank (i.e., weight of fluid divided by cross-sectional area of tank)

$$P = \frac{F}{A} = \frac{w}{A} = \frac{DAhg}{A} = Dhg$$

(Equation 28-M)

Since g is a constant, we can see that the pressure difference between the top and bottom of the tank is directly proportional to the height of the fluid column and to the fluid density. This result also applies to any depth (h) in a fluid,

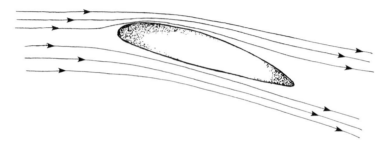

Fig. 28-8. A flow diagram of an airplane wing shows the crowding together of flow lines on top of the wing, which causes greater air speed at this location and results in a low-pressure area on top of the wing. Since motion takes place from high-pressure to low-pressure areas, the wing has a tendency to rise.

whether at the bottom or not, since the fluid beneath the specified depth does not contribute to the weight pressing down there.

The relationship of pressure and depth is important in understanding the force of buoyancy and Archimedes' principle, as will be discussed in the next lesson.

Pressure and velocity: Bernoulli's principle

In the 1700's the Swiss physicist Daniel Bernoulli found that the pressure in a fluid is greatest where the velocity is the least; that is, greater speeds mean less pressure. In the flow diagram of the airplane wing in Fig. 28-8, the upward tilt during climb causes a crowding together of flow lines on top of the wing which causes greater air speed. This results in a low-pressure area being developed on the top of the wing. Since motion takes place from high-pressure to low-pressure areas, the wing has a tendency to rise. Thus we can see why objects of low mass (e.g., paper) are drawn outside a rolled down window of a moving auto. The motion of the car creates high relative velocity—the air outside moving faster than the air inside. Thus according to Bernoulli's principle a low-pressure area is created outside the window and when the window is opened objects of low mass move from the higher pressure inside to the lower pressure outside.

Effects of temperature on fluid friction

Temperature has opposite effects on friction in air and water. It is directly related to fluid friction in air while being inversely related in water.

On a hot day there is increased air molecular motion, which produces more friction than on a cold day, thus *increasing* air resistance during motion. In air a temperature decrease causes a corresponding decrease in fluid friction.

An increase in water temperature seems to make the liquid thinner or less viscous, thus *decreasing* fluid friction. An object moving through warm water at the same speed as through cold water would experience less resistance.

RELATIONSHIP OF PROPULSION TO RESISTANCE

The fastest motion through a fluid results from the *reduction* of resistance in the direction of the motion and from an *increase* in resistance in the opposite direction. For example, the swimmer wishes to reduce resistance in the forward direction so his body will slip through the water with the least amount of friction. Forward movements, however, are the result of the backward propelling movements of body parts, and the greater the resistance they have to push against the greater will be the forward thrust. The factors which are important in increasing the propulsive force of any stroke are opposite many of those involved in the reduction of resistance to forward progress.

Maximum propulsive force is attained by presenting as large a surface as possible in the direction opposite the desired movement, by pushing through as great a distance in that direction as possible, and by moving the pushing members of the body as rapidly as possible. These movements will create maximum resistance to the backward displacement of the body part, and thus maximum forward propulsion will be produced.

STUDY GUIDELINES
Concept of fluid friction
Specific student objectives

1. Describe fluid friction.
2. List the two fluid mediums of primary importance in human motion.
3. List and describe the two types of fluid flow that occur during relative motion between a solid and a fluid.
4. List five factors for fluid friction.
5. Describe the factors involved in fluid friction (f) as shown in the equation $f = KPSV^2/2$.

Summary

Fluid friction is the resistance between a solid and a fluid (liquid or gas) during relative motion.

Fluid friction results in two types of fluid flow during motion: (1) laminar flow, the flow of air or water around a solid object which is relatively smooth, and (2) turbulent flow, the flow of air or water around a solid object which produces rough eddies and low pressure areas (cavitations). The two fluids of primary interest in human motion are air and water.

The affective factors for fluid friction are as follows:
1. Relative velocity between the object and the fluid
2. Shape of the object and its active surface area (the cross-sectional area acting at right angles to the line of flow)
3. Surface texture of the object
4. Fluid pressure
5. Temperature of the fluid

Fluid friction (f) can be expressed in the equation $f = KPSV^2/2$, where K is a numerical coefficient depending upon the shape and texture of the body and its attitude to the fluid flow, P is the fluid pressure, S is the active surface area of the body over which the fluid stream passes, and V is the relative velocity of the fluid to the body.

Performance tasks

1. List the two primary fluids affecting human motion.

2. The resistance between relative motion of a solid and a fluid is termed
   a. fluid friction
   b. rolling friction
   c. sliding friction
   d. solid friction
3. Fluid flow during motion between a solid and a fluid that is relatively smooth is known as (laminar/turbulent) _____ flow.
4. Make a list of the affective factors for fluid friction.
5. State the relationships shown in the equation $f = KPSV^2/2$.

Effect of relative velocity upon fluid friction
Specific student objectives

1. State the relationship of fluid friction to the relative velocity at low and high speeds.
2. Describe terminal velocity.
3. Describe cavitation.

Summary

At relatively low speeds fluid friction is proportional to the velocity. As relative velocity increases, a critical point is reached where turbulence begins and the force of friction becomes proportional to the square of the velocity.

When solid objects move through a fluid medium, they reach a constant terminal velocity when all forces are balanced. For example, objects falling through the air reach a terminal velocity when the downward force (weight) is equal to the upward forces of friction. Stokes' law states that the terminal velocity of such small particles is proportional to their radius.

Cavitation is caused when a solid object is moved through water at such a speed that a cavity is formed. When cavitation occurs, it results in a reduction of the potential for efficient motion.

Performance tasks

1. At low speeds, fluid friction and relative velocity are (directly/inversely) _____ proportional.
2. Write out descriptions of terminal velocity and cavitation.
3. Demonstrate cavitation by moving a solid object through water.

Effects of active surface area and surface texture on fluid resistance
Specific student objectives

1. State a principle relating fluid friction to active surface area.
2. Describe the relationship of surface texture to fluid friction.
3. Describe the task of streamlining the human body.

Summary

Fluid friction depends upon the shape and active surface area of the body; it is directly proportional to the greatest cross-sectional area of the body acting at right angles to the flow of the fluid.

Streamlining of the human body can be carried out by shaping to the streamlines of the fluid.

A body can also be streamlined by reducing the roughness of its surface texture. This is based on the principle that "the rougher the texture the greater is the fluid friction."

Performance tasks

1. Write out principles relating fluid friction to the following:
   a. Shape and active surface area of an object
   b. Surface texture of an object
2. Set up a demonstration in water of the above two principles. Develop written guidelines for the experiment.

Effects of fluid pressure and temperature on fluid friction
Specific student objectives

1. Differentiate between density and specific gravity.
2. Define fluid pressure.
3. State a principle relating fluid flow and fluid pressure.
4. State the relationship of depth to fluid pressure.
5. State the principles concerning fluid friction in air and water.
6. State Bernoulli's principle.

Summary

Density is the mass of a body per unit of volume; the weight per unit of volume is called specific gravity. Mass equals the product of volume and density for an object. The weight of a fluid is the product of its volume and density times the gravitational constant.

Fluid pressure is the ratio of force per unit area; i.e., $P = F/A$. Fluids flow from high to low pressure when forces are exerted upon them. Fluid pressure is directly proportional to the depth in a fluid.

Temperature affects fluid friction directly in air and inversely in water.

Bernoulli's principle states that the pressure in a fluid is greater where the velocity is less and vice-versa.

Performance tasks

1. Develop written descriptions for the following:
   a. Density
   b. Specific gravity
   c. Fluid pressure
   d. Bernoulli's principle
2. Specific gravity can be found by comparing the weight of an object to the weight of an equal volume of water.
   a. Describe an object with specific gravity of 1.
   b. Determine a method using the above principle to find the specific gravity of any object.
3. The mass of a body per unit of volume is known as (density/specific gravity) ———— .
4. Force per unit area is called
   a. density
   b. fluid pressure
   c. specific gravity
   d. none of the above
5. Write out principles concerning the following:
   a. Fluid friction
   b. Fluid pressure and fluid flow
   c. Fluid pressure and fluid depth
   d. Fluid friction and temperature
   e. Fluid velocity and fluid pressure
6. Illustrate the above principles with specific demonstrations.

Relationship of propulsion to resistance
Specific student objectives

1. State a principle relating fluid resistance to speed of movement. Consider the direction of

motion and the direction opposite the motion.

2. Identify the factors for maximum propulsive force in a fluid medium.

Summary

The highest speed in a fluid medium results from a reduction of resistance in the direction of motion and from an increase in resistance in the opposite direction.

Maximum propulsive force is attained in a fluid by presenting as large a surface as possible in the direction opposite that of the desired movement, by applying forces through as great a distance as possible in the direction opposite the intended motion, and by moving the force-producing members of the body as rapidly as possible.

Performance tasks

1. Maximum velocity in a fluid results from (increasing/reducing) _____ the resistance in the direction of motion.
2. Write out a principle relating maximum velocity in a fluid to resistance to motion in the opposite direction.
3. List the factors that produce maximum propulsive force in a fluid.

SELF-EVALUATION

Students should use no reference materials for this progress test, and they can check their answers by referring to Appendix A.

1. The two fluid mediums of primary importance in human motion are _____ and _____ .
2. Describe fluid friction.
3. List five factors that affect fluid friction.
4. Air resistance and water resistance are two types of
   a. fluid friction
   b. fluid pressure
   c. both a and b
   d. neither a nor b
5. Fluid flow during motion that produces disturbances and eddies is called (laminar/turbulent) _____ flow.
6. State the relationship of fluid friction to relative velocity at low and high speeds.
7. Objects falling through the air will reach a constant speed called
   a. original velocity
   b. terminal velocity
   c. both a and b
   d. neither a nor b
8. When a cavity is formed by an object moving through a fluid, this is called _____ .
9. State principles relating fluid friction to
   a. active surface area of a moving object
   b. surface texture
10. Weight per unit volume is known as
    a. density
    b. mass
    c. pressure
    d. specific gravity
11. Define fluid pressure.
12. Fluid depth and fluid pressure are (directly/inversely) _____ proportional.
13. State Bernoulli's principle.
14. Increases in water temperature will (decrease/increase) _____ fluid friction.
15. State principles relating fluid pressure and
    a. fluid flow
    b. fluid depth
    c. fluid velocity
16. Fluid pressure is greatest where fluid velocity is (greatest/least) _____ .
17. Identify the factors for maximum propulsive forces in a fluid medium.
18. Maximum velocity in a fluid results from (decreasing/increasing) _____ the resistance in the direction of motion.

Kinetic analysis of aerial and aquatic motion forces

Body of lesson

INTRODUCTION TO THE KINETIC ANALYSIS OF AERIAL AND AQUATIC MOTION FORCES

The study of forces affecting movement through air and water is termed aerodynamics and hydrodynamics, respectively. In the next section of this lesson, the movement of an airplane is used as an example of an aerodynamic situation. Then later in the lesson the movement of a swimmer is used to demonstrate the principles of hydrodynamics.

The three forces which affect aerial and aquatic motion, as developed in later sections of the lesson, are weight, thrust, and friction. The frictional force, however, can be resolved into an upward component (called lift) and a backward component (called drag). Thus the two pairs of opposing forces in horizontal aerial and aquatic motion are as follows:

1. *Vertical forces.* The two opposing vertical forces are the *upward* lift component of friction and the *downward* force of gravity.

An additional lifting force in aquatic motion is the force of buoyancy.

2. *Horizontal forces.* The two opposing horizontal forces are the *forward* propelling force that we can call thrust and the *backward* drag component of friction.

INTRODUCTION TO THE PRINCIPLES OF AERODYNAMICS

An airplane in a climb (Fig. 29-1) rises with constant velocity, and the conditions of equilibrium exist (i.e., all forces acting upon it form a closed polygon).

The external forces acting on the plane can be reduced to three: weight, thrust, and friction (Fig. 29-2). The *weight (w)* may be assumed to act vertically downward through the center of gravity of the plane. The *thrust (T)* is produced by the engine of the airplane and acts in the direction of the attitude of the longitudinal axis of the plane. The *friction (f)* is the resultant force of air on the plane and acts in a direction upward

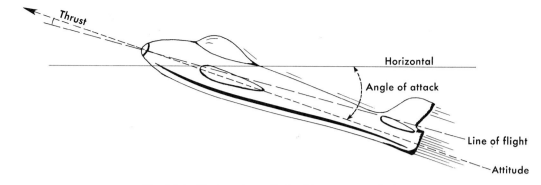

Fig. 29-1. Airplane in a climb with constant velocity.

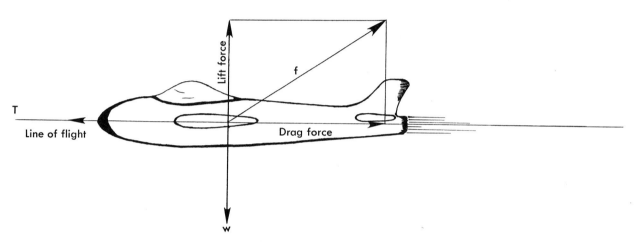

Fig. 29-2. The external forces acting on an airplane are its weight *(w)*, the thrust *(T)* produced by its engine, and friction *(f)*.

and back. The angle between the plane's attitude and the horizontal is called the *angle of attack* (Fig. 29-1). Note that the *line of flight,* the path along which the plane is flying, is not quite the same as the plane's *attitude,* which is the direction of its longitudinal axis.

It is customary to resolve the frictional forces acting on an airplane into two components: (1) a useful component perpendicular to the line of flight called *lift* and (2) a detrimental component parallel with the line of flight called *drag.* The conditions of equilibrium require that these combined forces form a closed polygon as shown in Figs. 29-2 and 29-3. This polygon is often used to determine certain factors in the performance of the plane. For example, if the forward thrust and weight are known and the line of flight determined, the force polygon can be used to find the lift *(L)* and the drag *(D).*

When an airplane is in level flight and has constant velocity, thrust *(T)* and drag will be practically horizontal, equal in magnitude, and opposite in direction while weight *(w)* and lift will be vertical, equal in magnitude, and opposite in direction. With the motor throttled down and the plane in a dive at constant velocity (Fig. 29-3), equilibrium conditions exist again and the forces form a closed polygon. Note that the drag component is taken parallel with the flight path and not with the plane's attitude.

Some of the applications of the principles of aerodynamics are found in the flight of sports objects through the air (e.g., a discus or a football), but they also apply to any movement

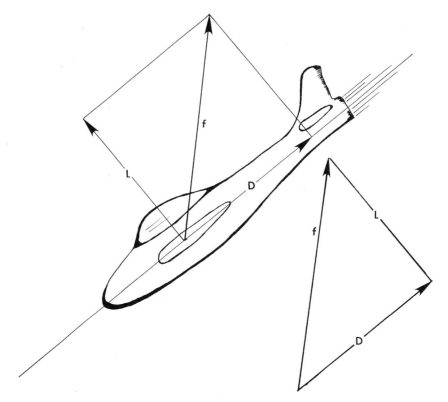

Fig. 29-3. When a plane is in a dive at constant velocity, equilibrium conditions exist and the forces acting on the plane form a closed polygon. *L,* Lift; *f,* frictional force; *D,* drag.

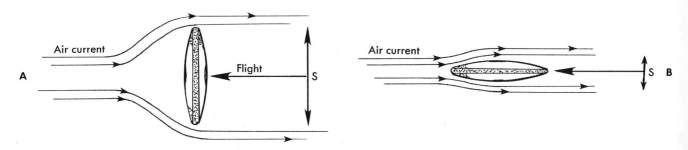

Fig. 29-4. Effect of air resistance on a body in flight. *S,* Cross-sectional area.

through an aerial environment. The forward lean of a runner, for example, reduces the active surface area over which the airstream passes and thereby reduces some of the retarding effects of air resistance. Our present discussion, however, will be confined to the examples afforded by the flights of a discus and a football.

Aerodynamic factors affecting the flight of a discus

For a particular example of the effect of the air on a body, the discus (because of its comparatively flat surface) is considered first. If it is presented with its surface at right angles to the airstream (Fig. 29-4, *A*), all the friction will

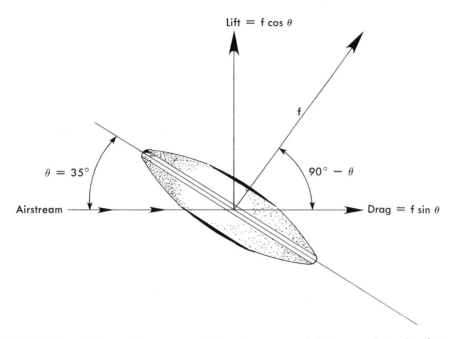

Fig. 29-5. Effect of air resistance on a rotating discus presented at an angle to the airstream.

be drag and there will be no lift. Thus the drag will be quite large per unit area because of the great change in direction (and therefore of momentum) of the air molecules involved in getting around the discus. On the other hand, if the edge of the discus is presented to the airstream (Fig. 29-4, *B*), the drag effect will be practically negligible because the change of direction of the air molecules is slight and comparatively few are involved. The surface resistance in this case is represented by the cross-sectional area through the short axis of the discus *(S)*. Friction is still wholly in the direction of the airstream with its y component (lift) being zero.

Finally, if the discus is presented at an angle to the airstream (Fig. 29-5), an intermediate value of the friction will be obtained whose magnitude depends upon the angle. Since air is considered as a fluid, the friction will be at right angles to the discus. If the leading edge is above the trailing edge, the reaction will produce a lifting effect (opposing the force of gravity). The y component of f is the lift and is directed upward in opposition to the force of gravity. The x component of f is the drag (parallel with the airstream) and is opposed to the movement of the discus. The ratio of lift to drag (L/D) is called

the *index of lifting efficiency* of a body which is acted on by an airstream.

Aerodynamic factors affecting the flight of a football

Because of the difference in mass between the discus and the football, the relative ability of the football to penetrate an airstream is much less. It should be pointed out, however, that the mass of the object has no effect on the amount of friction. Also, because of the shape and surface of the football, the direction of f is less than 90° to the axis of the ball (Fig. 29-6). Experimental data are not available to determine the best angle of inclination at which to pass a football under varying wind conditions; but in view of the difference in mass between the football and the discus and in view of the reduced lifting component, it may be concluded that under the same conditions as those faced by a discus the angle of inclination of the football should be less than that of the discus. In strong head winds the angle should be as near to zero as possible, with as much spin as possible for stabilization purposes. With strong tail winds the attitude of the football should be more nearly vertical.

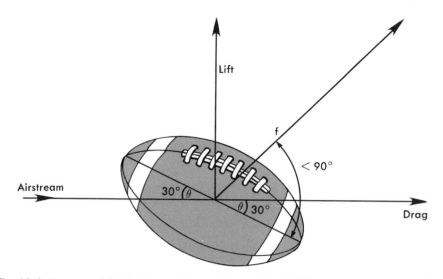

Fig. 29-6. Because of its lighter weight and thus reduced lifting component, a spiraling football needs less inclination than does a rotating discus. Compare Fig. 29-5.

ANALYSIS OF SPIN AS AN AERIAL FORCE

The movements of a body through the air are affected by its spin. Spin can produce a stabilizing effect that holds an object on course by resisting a change in the direction of the axis of momentum. Spin can also cause an object to curve through the air.

Stabilizing effect of spin

Motion pictures of the discus indicate that it rotates from eighteen to twenty-one times while in flight and tends to be steadied by the spin without its distance being affected because the forward force is so much greater than the spinning force. In the case of a football, if the spinning is not sufficient to keep the axis of the ball pointed on course the ball will float or tumble.

When spin or rotation is used for stabilizing purposes, as in the case of the discus and the football, the conditions for stable motion about a transverse axis can be expressed by

$$\frac{I^2\omega^2}{4T} > 1$$

where I is the axial moment of inertia, T is the torque about the transverse axis through the center of gravity, and ω is the spin in radians per second. Thus, when the ratio is greater than unity stable motion will exist.

The spin has a stabilizing effect that tends to hold an object on course by resisting a change in the direction of the axis of momentum. The spiral imparted to a football has the effect of stabilizing its flight through the air. If the spinning is not sufficient to keep the axis of the ball pointed on course, the ball will lose distance and forward speed by its erratic motion.

Curving effect of spin

In addition to its stabilizing effect, spin can have a curving effect upon a projectile. The curving effect is caused by an application of Bernoulli's principle that is called the *Magnus effect* (after the German physicist Heinrich Magnus).

When a ball moving at high velocity is spinning rapidly about an axis, pressure is built up on one side and reduced on the other because the surface of the ball tends to drag a little air along with it (Fig. 29-7). On the side where the air resistance to forward motion is in opposition to that of the air moving around the ball, a high-pressure area of turbulence is created. On the opposite side, where the two forces are in the same direction, the velocity of the moving air is increased and a low-pressure area results. We know that movement will take place from high pressure to low pressure. Therefore the ball tends to move toward the side where the pressure is less. Since with forward (top) spin the pressure is built up above the ball and reduced under the

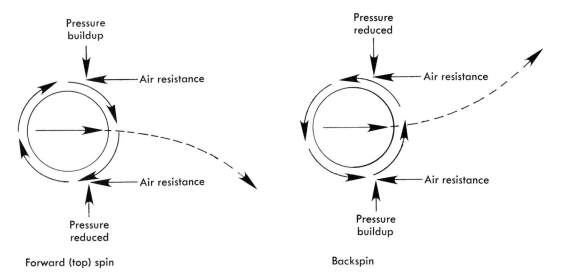

Forward (top) spin

Backspin

Fig. 29-7. Effect of forward spin and backspin on the path of a ball.

ball, there is a tendency for the ball to drop more rapidly than normal. With backspin the pressure is built up under the ball and reduced above the ball, and there is a tendency for the ball to remain in the air longer (or even to rise) since this pressure overcomes some of gravity's pull. When a ball is spinning to the right around a vertical axis, the pressure is built up on the left, curving the flight to the right (as in a golf "slice"); when the ball is spinning to the left, the pressure builds up on the right, curving the flight to the left (as in a "hook"). Because a head wind increases air resistance on the ball, it increases the rate of change of a spinning ball's path.

The effect of spin on a fast-moving ball is perceived late in the flight because the linear velocity of the ball is great in relation to the spin force. The effect of spin on a slower-moving ball is seen earlier in its flight because the linear velocity is less in relation to the force caused by the pressure resulting from the spin. Because the mass and the resulting greater momentum of a heavy ball make the heavy ball relatively less responsive to air pressure, spin has more effect in curving the flight of a light ball.

INTRODUCTION TO THE PRINCIPLES OF HYDRODYNAMICS

Aquatic locomotion, like walking and running, is a means of movement resulting from the body's pushing against a fluid and the fluid's exerting an

Fig. 29-8. The horizontal forces acting on a swimmer are the forward propelling force, labeled thrust *(T)*, and a backward resistive force, drag *(D)*. (Modified from Armbruster, D. A., Musker, F. F., and Mood, D.: Basic skills in sports for men and women, ed. 6, St. Louis, 1975, The C. V. Mosby Co.)

opposite force against the body. It differs from walking and running, however, in that the fluid medium against which the push is made is not as resistive as the usual walking surface; also the fluid medium offers a great deal more resistance to movements of the body.

In other words, the surrounding medium (water) both assists and hinders movement. It is essential to the propulsion of the body or boat, but at the same time it hinders the progress of the body or boat. In Fig. 29-8 a swimmer has two horizontal forces acting upon him: a forward horizontal propelling force (labeled thrust, *T*) and a backward horizontal resistive force (labeled drag, *D*). The propelling force is, of course, produced by the movements of the swimmer's

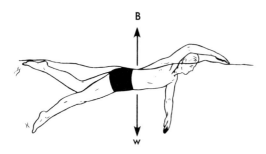

Fig. 29-9. The vertical forces acting on a swimmer are the downward force of gravity *(w)* and the upward force of buoyancy *(B).* (Modified from Armbruster, D. A., Musker, F. F., and Mood, D.: Basic skills in sports for men and women, ed. 6, St. Louis, 1975, The C. V. Mosby Co.)

body parts while the resistive force is due to the friction created between his body and the water.

The swimmer also has opposing vertical forces acting upon his body as shown in Fig. 29-9. In this figure the vertically downward force is the force of gravity, represented by the weight of the swimmer; the vertically upward force is the force of buoyancy. The force of gravity acts to cause the body to sink, while the force of buoyancy acts to cause it to rise. A body will float when these two forces are equal in magnitude.

ANALYSIS OF BUOYANCY AS AN AQUATIC FORCE

The famous and useful Archimedes' principle can be derived from Equation 28-M as shown in the boxed material below.

Archimedes' principle in equation form is

$$F = VDg$$

(Equation 29-A)

where F is the force of buoyancy, V is the volume of the body submerged in the fluid, D is the density of the fluid, and g is the gravitational constant. The force of buoyancy is equal to the weight of the displaced fluid.

Archimedes' principle enables us to determine only the buoyant force acting on a submerged object and not the resultant force. If the weight of the object is greater than the buoyant force, the ob-

---

Archimedes' principle may be derived from Equation 28-M ($P = Dgh$) and from the fact that the pressure at a point in a fluid is the same in all directions. Consider, for example, a block of height H and cross-sectional area A which is completely submerged in a fluid of a certain density (D) so its top is below the surface a distance h.

The downward pressure (P_1) on the top of the block is $P_1 = P_{atm} + Dgh$, and the upward pressure (P_2) on the bottom of the block is $P_2 = P_{atm} + Dg (h + H)$.

Since the downward force exerted by the fluid on the block is $F_1 = AP_1$ while the upward force is $F_2 = AP_2$, the *net* force on the block is net $F = F_2 - F_1 = A (P_2 - P_1) = ADgH$. Because the pressure on the bottom of the block is greater than on the top, the net force is upward (and is known as the *buoyant force* acting on the block). Because the volume of the block is $V = AH$, we have the simple result that $F = VDg$, or buoyant force equals weight of displaced fluid. The buoyant force acting on a submerged object is equal to the weight of fluid displaced by the object.

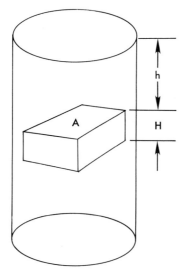

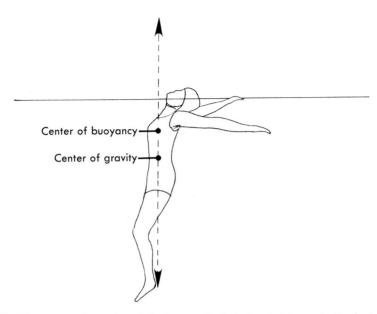

Fig. 29-10. The center of gravity of the human body is located lower in the body than is the center of buoyancy.

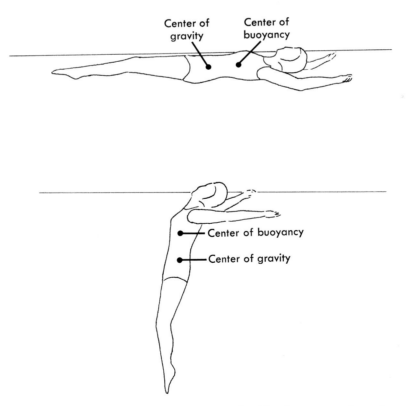

Fig. 29-11. The floating position of the body is determined by the amount of rotation required to bring the center of gravity under the center of buoyancy.

ject will sink; if the weight is less than the buoyant force, the object will rise.

Normally the human body floats because its specific gravity (weight per unit volume) is less than that of water. The degree of buoyancy is dependent upon body build. Bone and muscle have a higher specific gravity than does adipose (fatty) tissue. Bodies which are made up largely of bone and muscle are less buoyant than those containing a high percentage of fat. For this reason women generally are more buoyant than men; they have relatively lesser amounts of bone and muscle and more fat in their bodies than do men.

Center of buoyancy

When the human body is in the air, it will rotate around its center of gravity; but in the water its center of buoyancy becomes its center of rotation.

The center of buoyancy is located in the chest region because the chest area (containing the lungs, which are filled with air) is very light for its size and is the most buoyant segment of the body. The legs, on the other hand, are likely to have a high percentage of bone and muscle and therefore a high specific gravity. For these reasons the center of gravity is located lower in the body (hip region) than is the center of buoyancy (Fig. 29-10).

Floating position

Because the torque exerted by the legs around the center of buoyancy is greater than that exerted by the head (due to the greater weight and lever arm), the feet will sink lower into the water than will the head. They will sink into the water until the center of gravity of the body is directly below its center of buoyancy. Therefore the floating position of the body is determined by the amount of rotation required to bring the center of gravity under the center of buoyancy (Fig. 29-11).

STUDY GUIDELINES
Introduction to the kinetic analysis of aerial and aquatic motion forces
Specific student objectives

1. Define aerodynamics and hydrodynamics.
2. Identify the three forces that affect aerial and aquatic motion.

3. Describe the rectangular components acting upon an object in aerial or aquatic motion.

Summary

Aerodynamics is the study of forces affecting movement through air; hydrodynamics is the study of the forces involved in motion through water.

The three forces affecting aerial and aquatic motion are friction, thrust, and weight.

One pair of opposing forces in horizontal aerial and aquatic motion is the upward lift component of friction and the downward force of gravity. The lifting force in aquatic motion is referred to as buoyancy. The other pair of opposing horizontal forces is the forward propelling force (called thrust) and the backward or drag component (friction).

Performance tasks

1. Identify two human motion examples of aerodynamic and hydrodynamic situations.
2. The study of forces affecting motion through water is (aerodynamics/hydrodynamics) ____ _____ .
3. The forces of air resistance and water resistance are examples of _____ .
4. List the three forces affecting aerial and aquatic motion.
5. The vertical forces acting during horizontal aerial and aquatic motion are the
   a. lift component of friction and the downward force of gravity
   b. force of buoyancy and the force of gravity
   c. both a and b
   d. neither a nor b
6. Identify and describe the two opposing horizontal forces acting during horizontal aerial and aquatic motion.

Introduction to the principles of aerodynamics
Specific student objectives

1. Identify and describe the forces acting on an airplane in flight.
2. Describe flight attitude, line of flight, and angle of attack.
3. Describe the rectangular components of the air resistance during flight.
4. Define the lifting efficiency of an object in aerial motion.

5. State two principles of aerodynamics related to throwing or kicking a football.

Summary

The external forces acting on an airplane in flight are friction (air resistance), weight, and thrust. Weight (w) acts vertically downward, thrust (T) acts in the direction of the longitudinal axis of the power plant, and friction (f) acts in an upward and backward direction.

Attitude is the direction of the longitudinal axis of the plane, while angle of attack is the angle between horizontal and the attitude. The line of flight is the resultant flight path of the plane as all forces act upon it.

The frictional force (f) caused by air resistance can be divided into rectangular components: a lift component perpendicular to the line of flight and a drag component parallel with the line of flight.

During the flight of a discus, the amount of friction depends upon its angle of attack. The friction will be at right angles to the long axis of the discus. If the leading edge is above the trailing edge, the force of friction will produce a lifting (y component) and a drag (x component) effect on the discus.

The ratio of lift to drag (L/D) is known as the index of lifting efficiency of a body acted on by an airstream. It is symbolized by e.

A football is affected by an airstream to a greater degree than is a discus because of the differences in mass. The angle of attack or inclination of a football is usually less than that of a discus, especially into a head wind. Spin is used to stabilize the football's flight.

Performance tasks

1. Using a model or paper airplane, identify the following aerodynamic factors:
   a. Attitude
   b. Angle of attack
   c. Line of flight
   d. Weight
   e. Thrust
   f. Friction (including direction)
   g. Components of friction (including direction)
2. Write descriptions for the items in Task 1 above.

3. The ratio of lift to drag in aerodynamics is termed
   a. components of friction
   b. lifting efficiency
   c. both a and b
   d. neither a nor b
4. Describe the effects of friction upon a discus in flight if the leading edge is above the trailing edge.
5. Experiment with a discus throw to discover the effects of friction on a discus in the following situations:
   a. Attitude above horizontal with varying angle of inclination
   b. Attitude at horizontal
   c. Attitude below horizontal
6. Throw or kick a football to experimentally discover principles concerning the
   a. optimal angle of inclination
   b. effect of spin on the flight
7. State two principles related to the aerodynamics of
   a. kicking a football
   b. throwing a discus

Analysis of spin as an aerial force
Specific student objectives

1. Identify the two effects of spin upon objects in aerial motion.
2. Describe the Magnus effect on aerial projectiles.
3. State two principles relating spin to aerial motion.

Summary

The aerial motion of an object is affected by its spin. Spin can produce a stabilizing effect that holds an object on course by resisting a change in the direction of the axis of momentum, and it can cause an object to curve through the air.

The curving effect of a projectile results from an application of Bernoulli's principle called the Magnus effect. When an object is spinning during flight, pressure is built up on one side of the object and reduced on the other. This occurs when a turbulent high-pressure area exists on the side where there is opposition to the flow of air around the object. On the opposite side the spin force and the air resistance force are in the same direction, and thus a low-pressure area is

created. Since Bernoulli's principle states that
movement in fluids will take place from a high-
pressure to a lower-pressure area, the object will
move from the turbulent side to the opposite side.

The effects of spin on an object in aerial motion will be as follows:

1. Top spin causes a downward tendency.
2. Backspin produces lift.
3. Right spin results in motion to the right.
4. Left spin causes movement to the left.

Spin effect is also dependent upon the relative
magnitudes of the forward linear velocity and the
angular velocity of spin as well as upon the mass
of the moving object.

Performance tasks

1. Spinning of an object in aerial motion can
   have the effect of (dislocating/stabilizing)
   ——————————— the flight of the object.
2. The Magnus effect on a projectile results in
   (curvilinear/linear) ——————— motion.
3. The Magnus effect is an application of (Bernoulli's/Newton's) ——————— principle.
4. Write a description of the Magnus effect.
5. Use a football, baseball, softball, or basketball
   to demonstrate the two effects of spin on aerial
   motion.
6. State Bernoulli's principle.

**Introduction to the principles
of hydrodynamics**
Specific student objectives

1. Describe aquatic locomotion.
2. Identify and describe the two pairs of forces
   that determine the motion of a body in water.

Summary

Aquatic locomotion is a means of movement
resulting from the body's pushing against the
fluid and the fluid's exerting an opposite force
against the body to produce motion. During the
resulting motion progress is impeded by the
resistance forces of the fluid.

The forward horizontal propelling force (called
thrust, T) is produced through movement of the
body parts. The resistive force (known as drag,
D) is due to the friction created between a body
and the fluid medium.

An object in water also has opposing vertical
forces acting on it: the vertically downward force

of gravity and the vertically upward force of
buoyancy. A body will float when the force of
buoyancy is equal to or greater than the force
due to gravity.

Performance tasks

1. During aquatic locomotion the water
   a. assists motion
   b. hinders motion
   c. both a and b
   d. neither a nor b
2. The forward propelling force of a swimmer is
   called
   a. buoyancy
   b. friction
   c. gravity
   d. thrust
3. The vertically upward force of an object in
   water is known as
   a. buoyancy
   b. friction
   c. gravity
   d. thrust
4. Write out descriptions of the four forces acting
   upon a swimmer in water.

Analysis of buoyancy as an aquatic force
Specific student objectives

1. State Archimedes' principle.
2. Describe the resultant vertical force acting on
   an object in water.
3. State the conditions for flotation of an object
   in water.
4. Describe the center of buoyancy of an object.
5. Describe the relation of the center of buoyancy
   and the center of gravity when the human body
   is in a floating position.

Summary

Archimedes' principle states that the buoyant
force acting on a submerged object is equal to
the weight of fluid displaced by the object. This
principle describes the buoyant force acting upon
an object.

The relative amounts of the buoyant force and
the force of gravity acting upon an object in water
will determine its ability to float.

The human body will float if its specific gravity
is less than that of water. The degree of buoyancy depends upon body build, i.e., the relative

percentages of bone, muscle, and body fat content.

The center of buoyancy is the center of rotation of an object in water. The center of buoyancy is located higher in the human body than the center of gravity.

The floating position of the human body is attained in water when the center of buoyancy and the center of gravity are in a vertical line, i.e., when the former is directly above the latter. This usually occurs when the legs rotate around the center of buoyancy in a downward direction until the position described above is reached.

Performance tasks

1. Archimedes' principle states that the force of buoyancy is _____ the weight of displaced fluids.
   a. less than   c. greater than
   b. equal to   d. not equal to
2. The two vertical forces acting upon an object in water are _____ and _____ .
3. State the relation between the vertical forces when an object is floating in water.
4. Explain the statement, "I float like a rock."
5. Conduct an experiment in water demonstrating differing flotation capacities of objects in water.
6. Write a description of the center of buoyancy of an object.
7. Differentiate between center of buoyancy and center of gravity. Also describe their relative locations in the human body.
8. Conduct an experiment to determine the relative locations of the center of buoyancy and center of gravity of your body in water. Determine the floating position of your body in water.
9. Explain the statement, "I can't get my legs to float."

SELF-EVALUATION

Students should use no reference materials for this progress test, and they can check their answers by referring to Appendix A.

1. Define aerodynamics and hydrodynamics.
2. The forces for aerial and aquatic motion are gravity, thrust, and _____ .
3. The vertical forces acting during aquatic motion are the forces of _____ and _____ .
4. The horizontal component of friction during aerial and aquatic motion is termed
   a. drag   c. efficiency
   b. lift   d. none of the above
5. Describe the line of flight and angle of attack of an airplane in flight.
6. Define the lifting efficiency of an object in aerial motion.
7. State two principles of aerodynamics for throwing a football and a discus.
8. The two effects of spin upon objects in aerial motion are
   a. drag and lift
   b. attitude and inclination
   c. both a and b
   d. neither a nor b
9. The curving effect of spin is explained by an application of Bernoulli's principle, which is concerned with
   a. drag   c. efficiency
   b. lift   d. none of the above
10. Describe the Magnus effect of air resistance on an object in aerial motion.
11. The vertical upward force acting upon a body in water is known as
    a. buoyancy   c. gravity
    b. drag   d. lift
12. Identify and describe the two pairs of forces that determine the motion of a human body in water.
13. "The buoyant force acting on a submerged object is equal to the weight of the fluid displaced by the object" is known as
    a. Archimedes' principle
    b. Bernoulli's principle
    c. the Magnus effect
    d. Newton's law
14. Describe the resultant of the vertical forces acting on an object in water.
15. Under what conditions will an object float in water?
16. Describe the center of buoyancy of an object in water.
17. When the human body is floating the center of buoyancy and center of gravity are related as
    a. horizontal line   c. vertical line
    b. same position   d. none of the above

Kinetic analysis of impact and rebound forces

Body of lesson

TYPES OF COLLISIONS AND COLLISION FORCES

Collisions between objects can be classified as direct or indirect and as inelastic, partially elastic, or perfectly elastic. Since perfectly elastic collisions rarely occur in human movement, we shall disregard this classification and consider only the inelastic and partially elastic categories.

Direct and indirect types of collisions

In a direct collision two objects collide so the line of action of their impact forces is directed perpendicular to their surfaces and through their centers of gravity. Thus the force of direct impact of two colliding objects will tend to produce only linear motion. A head-on collision is called a direct impact. Indirect impacts therefore are described as any non–head-on collision that is at any angle other than a right angle. Indirect impact is also called oblique impact. A batter hitting the ball down the third base–line in softball would be producing an oblique or indirect bat-ball collision.

Inelastic and partially elastic collisions

An inelastic collision occurs when the elastic limit of a colliding object is exceeded and the object does not rebound to its original shape. An example of an inelastic collision would be an automobile crash that mashes in a fender. A partially elastic collision is one in which the deformation of an object is within its elastic limits and it rebounds to its original shape once the deforming force is removed. The collision and resulting deformation of a tennis ball and

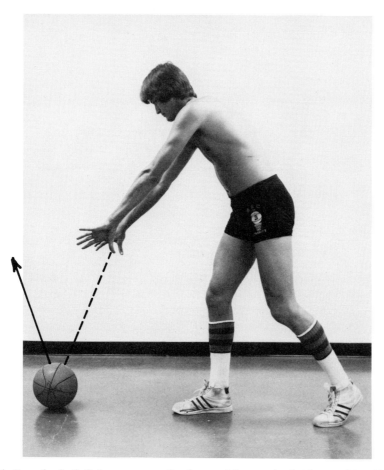

Fig. 30-1. In a basketball bounce pass the force of impact is the force with which the ball strikes the floor. The force that restores the ball to its original shape following deformation is called the force of rebound.

tennis racket on impact are an example of a partially elastic collision.

Definitions of impact and rebound forces

The basketball bounce pass is illustrated in Fig. 30-1. This activity can be used to demonstrate the concept of impact and rebound forces. The force of impact is, of course, the force with which the ball strikes the floor. The force that restores the ball to its original shape following deformation is called the force of rebound (also known as the force of restitution or the force of elasticity). It is called the force of restitution because of its shape-restoring effect, and the force of elasticity because it is produced by the elastic nature of the material. In Lesson 19 we called this force the elastic restoring force.

During a golf club–golf ball collision the impact and rebound forces can be identified as the action force of the club acting upon the ball (impact force) and the restoring reaction force of the ball acting upon the club (rebound force). In other words, the force of impact is the force with which one object strikes another while the force of rebound is the force with which a distorted object returns to its original shape following deformation and with which it pushes away from the body after collision.

FACTORS AFFECTING THE MAGNITUDE OF IMPACT FORCES AND STRESSES

The two factors affecting the force of impact of one object against another are the total kinetic energy of the two objects at the time of impact and the distance over which the forces are applied. The ratio of these impact forces to the area over which they are applied was defined in Lesson 19 as *stress*; or, in this case, the *stress of impact*. Thus the area over which forces are applied must be considered in the study of impact situations.

Total kinetic energy of the colliding bodies

The kinetic energy of a moving body was defined in Lesson 18 as the ability of the body to do work by virtue of its motion. The two types of kinetic energy are linear and angular. These can be quantified as K.E. linear $= \frac{1}{2}mV^2$, and K.E. angular $= \frac{1}{2}I\omega^2$.

We can increase the force of impact of a basketball with the floor by increasing its linear velocity of approach. We can also increase its impact force by imparting spin to it, which adds the effects of angular kinetic energy. Thus the force of impact is affected by the linear velocity (V) and the angular velocity (ω) components of kinetic energy. If the basketball were not of precise dimensions, we could also change its force of impact by changing its mass (m) and/or its moment of inertia ($I = mr^2$).

As a wheel rolls along a level road, it has both kinetic energy of rotation and kinetic energy of linear motion. In rotating about its geometrical center, it has a moment of inertia (I) and kinetic energy ($\frac{1}{2}I\omega^2$); its center of gravity, also moving along a straight line with velocity (V), has a kinetic energy of $\frac{1}{2}mV^2$. The total kinetic energy is therefore the sum of the two; i.e., total K.E. = rotational K.E. + linear K.E. $= \frac{1}{2}I\omega^2 + \frac{1}{2}mV^2$.

Distance over which the force is applied

The second factor affecting the force of impact is the distance over which the force is applied. A moving body (e.g., a basketball) has energy because in being brought to rest it must exert a force (F) on some other object and this force acting through a distance (d) does work. In other words, work can be done by a moving object. Conversely, by applying a constant horizontal force (F) on a body of mass m for a distance d, we can give it a kinetic energy ($\frac{1}{2}mV^2$), or $F \times d = \frac{1}{2}mV^2$. This is known as the *work equation*.

If we divide both sides of the work equation

by the distance d, we obtain $F = \frac{1}{2}mV^2/d$. Therefore a moving object is capable of exerting a force upon other bodies with which it impacts, and this force is directly proportional to the kinetic energy ($\frac{1}{2}mV^2$) of the object and inversely proportional to the distance over which the force is applied.

Thus a hard ball that does not greatly give upon impact and that is moving with great spin and linear velocity will impart more force upon the bodies with which it collides than will a softball that does give and that is moving with less linear velocity and less spin.

Area over which the force is applied

The third factor affecting impact situations is the area over which the force is applied. Force per unit of area is called mechanical stress. In other words, the stress placed upon a body during a collision is given by the force of impact (F) per unit of area (A):

$$\text{Stress} = F/A$$

(Equation 30-A)

Since $F = \frac{1}{2}mV^2/d$, the stress equation can be changed to Stress $= (\frac{1}{2}mV^2/d)/A$, which becomes

$$\text{Stress} = \frac{\frac{1}{2}mV^2}{dA}$$

(Equation 30-B)

Thus we can see that the stress applied to any part of a body is inversely proportional to the amount of give (d) in the body and the area over which the stress is applied (A). For these reasons baseball catchers use large catcher's mitts. The mitt gives and thus reduces the kinetic energy of the ball over a distance, but it also spreads the force of impact over the surface of the entire hand, i.e., a larger area.

Football shoulder pads also reduce the stress of impact acting on a given part of the body by spreading it over a larger area. Greater potential injury situations result when the impact force is concentrated on a small area.

The principle that "the impact stress is inversely proportional to the area of impact" can best be demonstrated by doing a shoulder roll when falling to the ground. This roll spreads the force over a broader area. Likewise, a softball player in order to reduce the stress of impact when sliding into base should land upon the ground by spreading out at flat as possible. In both these situations the stress of impact is spread over a broader area and thus the potential for injury is reduced.

Measurement of impact and rebound forces

Even though it is convenient to think of the forces of impact and rebound as being separate, they are both quantified through the use of the same formula, $F = \frac{1}{2}mV^2/d$. In other words, the magnitudes of the forces of impact and rebound are directly proportional to the magnitude of the total kinetic energy before impact and inversely proportional to the total deformation (give, force absorption, or force displacement) occurring during impact. Therefore, like centripetal and centrifugal force, the forces of impact and rebound are action and reaction forces. When we drop a ball upon a stationary surface, the force of impact is directed downward while the force of rebound is the equal and opposite reaction force directed upward.

FACTORS AFFECTING THE MAGNITUDE OF REBOUND FORCES

The factors affecting the magnitude of the rebound forces are exactly the same as those affecting the magnitude of the impact forces. However, special emphasis will be given here to the elastic restoring force, which is the force that causes objects to rebound from each other following impact. This elasticity factor is the same as the displacement factor (d) discussed in the last section for partially elastic collisions. When deformation occurs within the elastic limits of an object, the elastic restoring force is directly proportional to the amount of deformation (Hooke's law).

Effects of elasticity upon the force of rebound

Theoretically, in a perfectly elastic collision the kinetic energy of rebound is exactly equal to the kinetic energy of impact; but we know in actuality that in human movement situations we never have perfectly elastic collisions. Thus the energy of rebound is always less than the energy of impact because of the energy transfer that takes place in partially elastic collisions.

In other words, the kinetic energy of impacting objects does work during the collision by producing deformations and the work done during impact is ultimately converted into heat energy and is lost for the performance of work during rebound.

In Lesson 19 it was pointed out that the amount of give in a body upon impact is determined by the elasticity of the materials of which the body is made. Different kinds of materials behave differently at impact. They vary in the amount of deformation experienced during impact, and they move apart with different velocities during rebound.

In Lesson 19 the resistance of a body to deformation was also defined as elasticity. We might further identify this elastic and resistive force as the force of restitution because the greater the elasticity of a body the greater is its tendency to return to its original shape once the force deforming it is removed. When returning to its original shape, the material of the body exerts a force that according to Newton's second law ($F = ma$) determines the velocities with which the bodies separate following collision.

Thus we can say that the force of rebound is directly dependent upon the coefficient of elasticity or restitution as well as upon the magnitude of the kinetic energy present during impact.

The number that expresses the ratio of the velocity with which two bodies separate after collision to the velocity of their approach before collision was defined in Lesson 19 as the coefficient of elasticity or restitution. In formula form, r = velocity of separation/velocity of approach ($r = V_s/V_a$). In Lesson 19 the coefficient of elasticity or restitution was also given by the square root of the ratio of the height (h) to which a body will rebound after collision to the height (H) from which it is dropped; i.e., $r = \sqrt{h/H}$.

The elasticity of a basketball can be varied, of course, by increasing or decreasing the amount of air in it. Decreasing the amount of air in the ball will make it soft and easily deformed and thereby decrease the height to which it will rebound. Increasing the amount of air in a ball will, of course, have the opposite effect; that is, both the elasticity and the force of rebound will be increased, and therefore both the velocity of separation and the height of the rebound will experience an increase directly proportional to the increase in elasticity.

Effect of temperature and velocity of impact upon elasticity

In Lesson 19 the coefficient of elasticity (r) was shown to be affected by the temperature of the materials and the velocity of impact. Elasticity is directly affected by temperature. Thus a golfer who warms a golf ball by rubbing it with his hands or by placing it in his pocket is increasing its elasticity by increasing its temperature. The coefficient of restitution is also related to the velocity of impact, though not precisely since r is only slightly larger at lower speeds than at higher speeds. Thus there is a slight inverse relationship between elasticity and the velocity of impact.

Conservation and dissipation of impact forces

The total work done during an impact-rebound situation is found by adding the work done *within* the objects during impact and the work done *between* the objects during rebound. Work is done within the objects during impact when deformations and internal vibrations are produced. Thus the less the deformations and vibrations, the greater is the *conservation* of force and the greater will be the work done between the objects during rebound. Likewise the greater the deformations and vibrations produced during impact, the greater will be the *dissipation* of force and the less will be the work done between the objects during rebound. For this reason a baseball can be hit further when it has a high coefficient of elasticity (r) and is hit through its center of gravity at the center of percussion of the bat. Hitting the ball through its center of gravity minimizes the production of spin and maximizes the linear speed of projection.

FACTORS AFFECTING THE DIRECTION OF THE REBOUND FORCE

One of the most important factors determining the direction of the rebound force, which is defined as the angle of actual rebound, is the angle of impact. In a perfectly elastic collision between a nonspinning object and a horizontal surface, the angle of rebound will be exactly equal to the angle of impact. The angle of impact or entry describes the path of the object before impact, while the angle of rebound or departure describes the path of the object after impact. When an object is following a curved path at the time of impact, because of the curving effects of gravity the angle of impact is determined by the tangent to the arc of flight at the point of contact (Fig. 30-2).

The theoretical "angle of impact equals the angle of rebound" principle tells us that in direct impacts (head-on) the theoretical angles of impact and rebound are both equal to 90°. In an indirect or oblique impact with an angle of 50°, for example, the theoretical angle of rebound also equals 50°. This principle seldom is applicable in human movement, however, because of partially elastic collisions and frictional forces due to the effects of spin. Thus the angle of rebound is affected by the coefficient of elasticity and the frictional forces produced by spin.

Effects of elasticity on the angle of rebound

In partially elastic collisions the colliding objects are always deformed to some extent. The location of the deformation affects the angle of rebound.

In direct impacts all points at the sites of impact are equally deformed so the force of re-

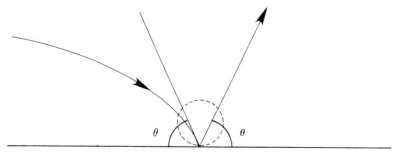

Fig. 30-2. The angle of impact (θ) is determined by the tangent to the arc of flight at the point of contact.

bound is directed back along the same line of action as the force of impact. For this reason the angle of rebound in direct collisions between nonspinning objects will always be along the line of impact (which is perpendicular to the impacting surfaces). The impacting surface of rounded implements, such as bats and balls, is a line tangent to the curve of the implement at the point of contact (Fig. 30-3).

During indirect or oblique collisions some parts of the colliding objects are more deformed than other parts. Thus the rebound force will not be directed back along the same line of action as the impact force. When a ball impacts at an angle to the floor, for example, the back of the bottom of the ball is deformed more than the front and the rebound force throws the ball forward and upward. How far back on the bottom of the ball the greatest depression is determines how low the resulting angle of rebound will be.

Elasticity affects partially elastic oblique collisions by lowering the angle of rebound. For ex-ample, the angle of rebound for a partially elastic object with an angle of impact of 75° will be less than 75°. In fact, the less elastic the object is the greater will be the deviation of the two angles. Thus the angle of rebound is directly affected by the elasticity of the colliding objects. A higher coefficient of restitution or elasticity (r) will cause a higher angle of rebound. Every tennis player knows that live tennis balls bounce at higher angles than do dead ones.

Effects of surface friction and spin on the angle of rebound

Surface frictional forces affect the angle and force of rebound of a spinning object during either a direct or an indirect collision. They have no effect on a nonspinning object in a direct impact because the rebound forces are equally distributed over the area of impact, but they do have an effect during an oblique impact.

The effect of forward spin is illustrated in Fig. 30-4, *A*. The spin force of the ball acting on the

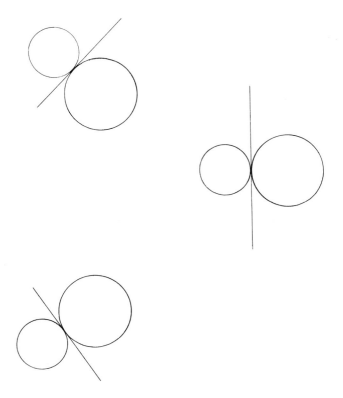

Fig. 30-3. The impacting surface of rounded implements, such as bats and balls, is a line tangent to the curve of the implement at the point of contact.

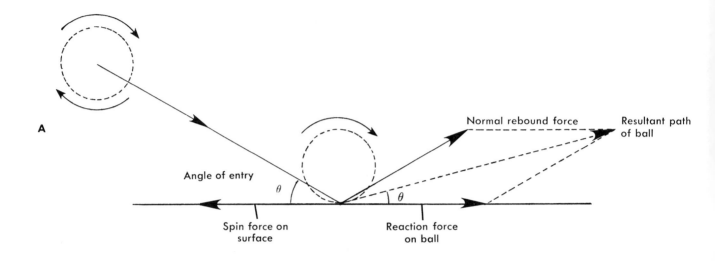

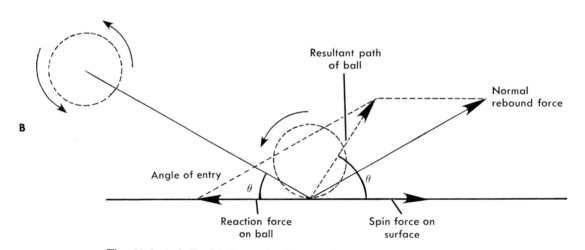

Fig. 30-4. A ball with forward spin upon impact with the floor **(A)** will exert a backward force on the floor. The reaction force acting on the ball will be directed forward. Thus the effect of forward spin is to increase the magnitude and decrease the angle of the resultant rebound force. Likewise, a ball with backspin **(B)** will rebound higher and slower than will a ball with no spin.

floor is directed backward while the reaction force acting on the ball is directed forward. Thus the effect of forward spin is to increase the magnitude and decrease the angle of the resultant rebound force. Therefore a ball with forward spin will rebound from a horizontal surface faster and lower than will a ball with no spin. Likewise, a ball with backspin will rebound higher and slower than will a ball with no spin (Fig. 30-4, *B*).

STUDY GUIDELINES
Types of collisions and collision forces
Specific student objectives

1. Differentiate between direct and indirect types of collision and give an example of each.
2. Differentiate between partially elastic and inelastic types of collisions and give an example of each.
3. Differentiate between impact and rebound forces and give an example of each.

Summary

The types of collisions between objects are classified as direct, indirect, partially elastic, and inelastic. Head-on collisions are termed direct, while non–head-on collisions are termed indirect. A basketball impacting at a 90° angle with the floor is an example of a direct collision; a ball impacting at an angle other than 90° is an example of an indirect impact. In an inelastic impact a body is deformed and does not return to its original shape (as when the fender of an automobile is bent). In a partially elastic collision the object does return to its original shape (as does a baseball after it collides with a bat). The force which causes the deformation is called the force of impact; the force which restores the original shape is the force of rebound. A baseball, for example, is deformed upon impact, but it regains its shape during rebound.

Performance tasks

1. Consult a dictionary for definitions of direct, indirect, elastic, impact, and rebound.
2. Give two examples each of the following:
   a. Direct collision
   b. Indirect collision
   c. Partially elastic collision
   d. Inelastic collision
   e. Impact force
   f. Rebound force

Factors affecting the magnitude of impact forces and stresses
Specific student objectives

1. List the three factors affecting the stress of impact of one object against another.
2. State how the kinetic energies of colliding bodies can be changed.
3. Describe the composition of the total kinetic energy of a rolling body.
4. Describe how the force of impact is affected by the distance over which the force is applied.
5. Describe how body injuries are affected by the area over which the force is applied.
6. Describe the measurement of impact and rebound forces.

Summary

The three factors affecting the stress of impact are the total kinetic energy of the two objects at the time of impact, the distance over which the forces are applied, and the area over which the forces are applied.

The kinetic energies of the colliding bodies can be changed by changing the linear velocities with which the bodies approach each other and/or the amount of angular velocity (amount of spin) they have. A rolling or spinning body has both kinetic energy of rotation ($\frac{1}{2}I\omega^2$) and kinetic energy of linear motion ($\frac{1}{2}mV^2$). The force of impact is directly proportional to the magnitude of the kinetic energies and inversely proportional to the distance through which the energies are absorbed.

The stress tending to produce body injury is inversely proportional to the area over which the force is applied. Thus the greater the area and the greater the distance over which the force is applied, the less will be the probability of injury. Both the force of impact and the force of rebound are measured by the same formula, $F = \frac{1}{2}mV^2/d$.

Performance tasks

1. State two ways that a baseball catcher can prevent injury to his hands during the act of catching a pitched ball.
2. Describe the best way to slide into base in softball. Justify your description in terms of the principles studied in this section.
3. State how the kinetic energy of a basketball can be changed.

Factors affecting the magnitude of rebound forces
Specific student objectives

1. State why objects rebound from each other following impact.
2. Describe the effects of elasticity upon the force of rebound.
3. State why the total kinetic energies of the colliding objects before impact are not equal to their total kinetic energies after impact.
4. Describe the effect of temperature and velocity of impact upon the force of rebound.
5. Differentiate between the conservation and the dissipation of impact forces and identify the factors which affect each.

Summary

Objects rebound from each other following impact because of the elastic restoring forces imparted to them during impact. The greater the elasticity of the objects the greater will be the restoring forces imparted to them. During impact part of the kinetic energy of the impacting objects is converted into heat and sound and thus dissipated and lost as energy for the production of rebound.

Elasticity is directly affected by the temperature of the materials and the velocity of impact. Thus the force of impact will vary directly with these factors.

The total work done during an impact-rebound situation is found by adding the work done *within* the objects during impact and the work done *between* the objects during rebound. Work done within the objects through the production of vibrations and deformation is dissipated. Thus the less the vibration and deformation during impact, the greater will be the conservation of force and the work done between the objects during rebound.

Performance tasks

1. A baseball can be hit further when it has a (high/low) _____ coefficient of elasticity.
2. The greater the deformation of an object on impact the greater will be its (conservation/dissipation) _____ of force.
3. The elasticity of a body is directly affected by its
   a. temperature   c. both a and b
   b. hardness   d. neither a nor b
4. Objects separate following impact because of their
   a. elasticity   c. both a and b
   b. restoring forces d. neither a nor b
5. During impact kinetic energy is (lost/gained) _____ by the impacting bodies.

Factors affecting the direction of the rebound force
Specific student objectives

1. Differentiate between the angles of impact and rebound.
2. Describe the relationship between the angle of impact and the tangent of the arc of flight of the object before impact.
3. Describe the theoretical relationship of the angle of rebound to the angle of impact.
4. Describe the effects of elasticity on the angle of rebound.
5. Describe the effects of surface friction and spin on the angle of rebound.

Summary

One of the most important factors determining the direction of the rebound force is the angle of impact. The angle of impact is the angle that the impact force makes with the striking surfaces; the angle of rebound is the angle the rebound force makes with the striking surfaces. When an object is following a curved path at the time of impact, because of the curving effects of gravity the angle of impact is determined by the tangent to the arc of flight at the point of contact. Theoretically the angle of impact equals the angle of rebound. The greater the elasticity of the impacting bodies, then the more closely will the actual angle of rebound approximate the angle of impact. The effect of spin on the angle of rebound is directly affected by surface friction. The greater the slippage due to the lack of friction, the greater will be the dissipation of force and the lower will be the angle of rebound.

Performance tasks

1. The direction in which two objects depart following impact is called their angle of (impact/rebound) _____.
2. When an object is following a curved path at the time of impact, the angle of impact is (directly/inversely) _____ propor-

tional to the tangent of the arc of flight.

3. Theoretically the angle of rebound (is/is not) _____ equal to the angle of impact.

4. The angle of rebound is (directly/inversely) _____ affected by the elasticity of the impacting bodies.

5. The effect of surface friction and spin on the angle of rebound is (direct/inverse) _____.

SELF-EVALUATION

Students should use no reference materials for this progress test, and they can check their answers by referring to Appendix A.

1. Differentiate between direct and indirect types of collision and give an example of each.

2. The collision of a bat and ball is an example of a/an (partially elastic/inelastic) _____ collision.

3. The force with which a golf club strikes a golf ball is called the force of
   a. rebound   c. both a and b
   b. impact   d. neither a nor b

4. The three factors affecting the stress of impact of one object against another are their kinetic energies, deformation, and _____ over which the force is spread.

5. Increasing the spin of a ball will have what effect on its kinetic energy?
   a. Increase it   c. Have no effect
   b. Decrease it   d. Dissipate it

6. The tendency of a force to do injury is called
   a. strain   c. both a and b
   b. stress   d. neither a nor b

7. Impact and rebound (are/are not) _____ measured in the same way and through use of the same formula.

8. A golf ball with a low coefficient of elasticity (can/cannot) _____ be hit further than one with a high coefficient.

9. In partially elastic collisions some kinetic energy (is/is not) _____ lost during impact.

10. The elasticity of a body is directly affected by its
    a. temperature   c. both a and b
    b. velocity of impact   d. neither a nor b

11. The greater the deformation of a body upon impact, the greater will be its (conservation/dissipation) _____ of force.

12. Differentiate between the angles of impact and rebound.

13. When an object is following a curved path at the time of impact, the direction of impact is given by a line on the curve called its
    a. sine   c. tangent
    b. cosine   d. cotangent

14. Theoretically the angle of rebound (is/is not) _____ equal to the angle of impact.

15. Elasticity and surface friction have a (direct/inverse) _____ effect on the angle of rebound.

16. The spin of a ball usually (does/does not) _____ have an effect on its angle of rebound following impact.

References

Adrian, M. J.: Electrogoniometry. In Cooper, J. M., editor: Selected topics on biomechanics, Proceedings of the C.I.C. Symposium on Biomechanics, Chicago, 1971, The Athletic Institute.

Adrian, M. J.: An introduction to electrogoniometry, Kinesiology Review 1968.

Adrian, M. J., Tipton, C. M., and Karpovich, P.: Electrogoniometry manual, Springfield, Mass., 1965, Springfield College.

Anderson, T. M.: Human kinetics and analyzing body movements, London, 1951, William Heinemann Medical Books, Ltd.

Ariel, G.: Computerized biomechanical analysis of human performance. In Bleustein, J. L., editor: Mechanics and sport, New York, 1973, The American Society of Mechanical Engineers.

Bade, E.: The mechanics of sport, Kingswood, England, 1952, Andrew George Elliot (Right Way Books).

Barham, J. N., and Thomas, W. L.: Anatomical kinesiology, New York, 1969, The Macmillan Co.

Barham, J. N., and Wooten, E. P.: Structural kinesiology, New York, 1973, The Macmillan Co.

Basmajian, J. V.: Electromyography of two-joint muscles, Anatomical Record **129:**371, 1957.

Basmajian, J. V.: Spurt and shunt muscles: an electromyographic confirmation, Journal of Anatomy **93:**551, 1959.

Basmajian, J. V.: Muscles alive—their functions revealed by electromyography, Baltimore, 1962, The Williams & Wilkins Co.

Bernstein, N. A.: Investigations on the biodynamics of walking, running, and jumping, Moscow, 1940, Central Scientific Institute of Physical Culture.

Bernstein, N.: The co-ordination and regulation of movement, Oxford, 1967, Pergamon Press.

Bierman, H. R., and Larsen, V. R.: Reaction of the human to impact forces revealed by high speed motion picture technique, Journal of Aviation Medicine **17:**407, 1946.

Bierman, W. A.: A new apparatus: a method for the measurement of minimal muscle force, Archives of Physical Medicine **38:**450, 1957.

Bierman, W., and Yamshon, L. J.: Electromyography in kinesiologic evaluations, Archives of Physical Medicine **29:**206, 1948.

Birdshistell, R. L.: Introduction to kinesics: an annotation system of analysis of body motion and gesture, Washington, D.C., 1952, U.S. Foreign Service.

Biomechanics Conference: Proceedings, Pennsylvania State University, 1971.

Broer, M. R.: An introduction to kinesiology, Englewood Cliffs, N.J., 1968, Prentice-Hall Inc.

Broer, M. R.: Efficiency of human movement, ed. 3, Philadelphia, 1973, W. B. Saunders Co.

Broer, M. R.: Laboratory experiences: exploring efficiency of human movement, Philadelphia, 1973, W. B. Saunders Co.

Broer, M. R., and Houtz, S. J.: Patterns of muscular activity in selected sport skills, Springfield, Ill., 1967, Charles C Thomas, Publisher.

Brunnstrom, S.: Clinical kinesiology, ed. 2, Philadelphia, 1966, F. A. Davis Co.

Bunn, J. W.: Scientific principles of coaching, ed. 2, Englewood Cliffs, N.J., 1973, Prentice-Hall, Inc.

Carlsoo, S.: How man moves, London, 1972, William Heinemann, Ltd.

Cerquiglini, S., Venerando, A., and Wartenweiler, J., editors: Biomechanics III, 1973.

Chaffin, D. B.: A computerized biomechanical model—development of and use in studying gross body actions, Journal of Biomechanics **2:**429, 1969.

Cleaveland, H. G.: The determination of the center of gravity of segments of the human body, Ed. D. dissertation, University of California–Los Angeles, 1955.

Cochran, A., and Stobbs, J.: The search for the perfect swing, Philadelphia, 1968, J. B. Lippincott Co.

Cooper, J. M., and Glassow, R. B.: Kinesiology, ed. 3, St. Louis, 1972, The C. V. Mosby Co.

Cooper, J. M.: Mechanics of human movement—throwing, 61st Annual Proceedings of the College of Physical Educators Association, 1958.

Cooper, J. M., editor: Selected topics on biomechanics, Proceedings of the C.I.C. Symposium on Biomechanics, Chicago, 1971, The Athletic Institute.

Courtney-Pratt, J. S.: A new method for the photographic study of fast transient phenomena, Research **2:**287, 1949.

Cureton, T. K.: Physics applied to health and physical education, Springfield, Mass., 1936, International YMCA College.

Cureton, T. K.: Elementary principles and techniques

of cinematographic analysis, Research Quarterly **10:** 3, 1939.

Daish, C. B.: The physics of ball games, London, 1972, The English Universities Press, Ltd.

Davis, J. F.: Manual of surface electromyography, WADC technical report 59-184, Office of Technical Services, Department of Commerce, Washington, D.C., 1955, U.S. Government Printing Office.

Dempster, W.: Space requirements of the seated operator, WADC technical report 55-159, Office of Technical Services, Department of Commerce, Washington, D.C., 1955, U.S. Government Printing Office.

Donskoi, D. D.: Biomechanik der Korperübungen, Berlin, 1961, Sportverlag.

Doolittle, T. L.: Errors in linear measurement with cinematographical analysis, Kinesiology Review 1971, p. 32.

Duvall, E. N.: Kinesiology: the anatomy of motion, Englewood Cliffs, N.J., 1959, Prentice-Hall, Inc.

Dyson, G. H. G.: The mechanics of athletics, London, 1971, University of London Press.

Edgerton, H. E., and Carlson, R. S.: The stroboscope as a light source for motion pictures, Society of Motion Picture and Television Engineers Journal **55:**88, 1950.

Edgerton, H. E., Germeshausen, J. K., and Grier, H. E.: High speed photographic methods of measurement, Journal of Applied Physiology **8:**2, 1937.

Evans, F. G., editor: Biomechanical studies of the musculo-skeletal system, Springfield, Ill., 1961, Charles C Thomas, Publisher.

Evans, F. G.: Selected bibliography on biomechanics. In Cooper, J. M., editor: Selected topics on biomechanics, Proceedings of the C.I.C. Symposium on Biomechanics, Chicago, 1971, The Athletic Institute.

Finley, F. R.: Kinesiological analysis of human locomotion, Eugene, 1961, University of Oregon Press.

Finley, F. R., and Karpovich, P. V.: Electrogoniometric analysis of normal and pathological gaits, Research Quarterly **35:**379, 1964.

Finley, F. R., and Wirta, R. W.: Myocoder-computer study of electromyographic patterns, Archives of Physical Medicine and Rehabilitation, vol. 48, Jan., 1967.

Fischer, O.: Theoretische Grundlagen für eine Mechanik der lebenden Körper mit speziellen Anwendungen auf den Menschen, sowie auf einige Bewegungs-vorgänge an Maschinen, Berlin, 1906, B. G. Teubner.

Frankel, V. H., and Burstein, A. H.: Orthopaedic biomechanics, Philadelphia, 1970, Lea & Febiger.

Frederick, A. B.: The analysis of gymnastics—a survey of the literature, Modern Gymnast, March, 1969.

Frost, H. M.: An introduction to biomechanics, Springfield, Ill., 1966, Charles C Thomas, Publisher.

Fung, Y., Perrone, N., and Anliker, M., editors: Biomechanics: its foundations and objectives, Englewood Cliffs, N.J., 1972, Prentice-Hall, Inc.

Garrett, R. E., Widule, C., and Garrett, G. E.: Computer-aided analysis of human motion, Kinesiology Review 1968.

Govaerts, A.: Biomechanics. A new method of analyzing motion, Brussels, 1962, Brussels University Press.

Groves, R., and Camaione, D. N.: Concepts in kinesiology, Philadelphia, 1975, W. B. Saunders Co.

Hainfeld, H.: Photography in sports, Journal of Health, Physical Education, and Recreation, vol. 38, 1967.

Handbook of high-speed photography, West Concord, Mass., 1963, General Radio Co.

Harris, R. W.: Kinesiology; workbook and laboratory manual, Boston, 1977, Houghton Mifflin Co.

Hawley, G.: An anatomical analysis of sport, Cranbury, N.J., 1940, A. S. Barnes.

Hay, J. G.: The biomechanics of sports techniques, Englewood Cliffs, N.J., 1973, Prentice-Hall, Inc.

Hay, J. G.: The center of gravity of the human body, Kinesiology III, p. 20, 1973.

Higgins, J. R.: Human movement: an integrated approach, St. Louis, 1977, The C. V. Mosby Co.

Hochmuth, G.: Biomechanik sportlicher Bewegungen, Frankfurt, 1967, Wilhelm Limpert-Verlag.

Hopper, B. J.: Notes on the dynamical basis of physical movement, Twickenham, Middlesex, England, 1959, St. Mary's College.

Hopper, B. J.: The mechanics of human movement, New York, 1973, American Elsevier Publishing Co., Inc.

Hughston, J. C., and Clarke, K. S., editors: Bibliography of sports medicine, Chicago, 1970, American Academy of Orthopaedic Surgeons.

Hyzer, W. G.: Engineering and scientific high speed photography, New York, 1963, The Macmillan Co.

Jensen, C. R., and Schultz, C. W.: Applied kinesiology, ed. 2, New York, 1977, McGraw-Hill Book Co.

Karpovich, P. V.: A frictional bicycle ergometer, Research Quarterly **21:**210, 1950.

Karpovich, P. V., and Karpovich, G. P.: Electrogoniometer: a new device for study of joints in action, Federation Proceedings **18:**79, 1959.

Karpovich, P. V., and Sinning, W. E.: Physiology of muscular activity, ed. 7, Philadelphia, 1971, W. B. Saunders Co.

Karpovich, P. V., and Wilklow, L. B.: A goniometric study of the human foot in standing and walking, U.S. Armed Forces Medical Journal **10:**885, 1959.

Kaufmann, D. A.: An annotated bibliography of electromyographic literature, unpublished paper, The University of Iowa, May, 1967.

Kelley, D. L.: Kinesiology—fundamentals of motion description, Englewood Cliffs, N.J., 1971, Prentice-Hall, Inc.

Kenedi, R. M., editor: Biomechanics and related bioengineering topics, New York, 1965, Pergamon Press, Inc.

Keys, A., and Brozek, J.: Body fat in adult man, Physiology Review **33:**245, 1953.

Kirby, R. F.: An introduction to sports photography, The Athletic Journal, Feb., 1972.

Komi, P. V., editor: Biomechanics V-A, 1975.

Komi, P. V., editor: Biomechanics V-B, 1975.

Krause, J. V., and Barham, J. N.: The mechanical foundations of human motion: a programmed text, St. Louis, 1975, The C. V. Mosby Co.

Lam, C. R.: An introduction to biomechanics, ed. 2, Springfield, Ill., 1967, Charles C Thomas, Publisher.

Lee, M., and Wagner, M. M.: Fundamentals of body mechanics and conditioning, Philadelphia, 1949, W. B. Saunders Co.

Lipovetz, F. J.: Basic kinesiology, Minneapolis, 1952, Burgess Publishing Co.

Lissner, H. R.: Biomechanics research, Journal of Engineering Education, vol. 52, no. 7, 1961.

Logan, G. A., and McKinney, W. C.: Anatomic kinesiology, ed. 2, Dubuque, 1977, William C. Brown Co., Publishers.

MacConaill, M. A., and Basmajian, J. V.: Muscles and movements—a basis for human kinesiology, Baltimore, 1969, The Williams & Wilkins Co.

Mathews, D. K., and Fox, E. L.: The physiological basis of physical education and athletics, Philadelphia, 1971, W. B. Saunders Co.

McCloy, C. H., and Young, N. D.: Tests and measurements in health and physical education, ed. 3, New York, 1954, Appleton-Century-Crofts.

Metheny, E.: Body dynamics, New York, 1952, McGraw-Hill Book Co.

Miller, C. E.: Handbook of high-speed photography, ed. 2, Concord, Mass., 1967, General Radio Co.

Miller, D. I., and Nelson, R. C.: Biomechanics of sport; a research approach, Philadelphia, 1973, Lea & Febiger.

Miller, D. I., and Petak, K. L.: Three-dimensional cinematography, Kinesiology III, p. 14, 1973.

Morehouse, L. E., and Miller, A. T.: Physiology of exercise, ed. 7, St. Louis, 1976, The C. V. Mosby Co.

Morton, D. J.: Human locomotion and body form, Baltimore, 1952, The Williams & Wilkins Co.

Nelson, R. C., and Morehouse, C. A., editors: Biomechanics IV, 1974.

Nelson, R. C., Petak, K. L., and Pechar, G. S.: Use of stroboscopic-photographic techniques in biomechanics research, Research Quarterly, vol. 40, 1969.

Noble, M. L.: Accuracy of tri-axial cinematographic analysis of determining parameters of curvilinear motion, M.A. thesis, University of Maryland, 1968.

Noble, M. L., and Kelley, D. L.: Accuracy of tri-axial cinematographic analysis in determining parameters of curvilinear motion, Research Quarterly, vol. 40, 1969.

Norris, F. H.: The EMG, a guide and atlas for practical electromyography, New York, 1963, Grune & Stratton, Inc.

Noss, J.: Control of photographic perspective in motion analysis, Journal of Health, Physical Education, and Recreation, vol. 38, 1967.

O'Connell, A. L., and Gardner, E. B.: The use of electromyography in kinesiological research, Research Quarterly, vol. 34, 1963.

O'Connell, A. L., and Gardner, E. B.: Understanding the scientific bases of human movement, Baltimore, 1972, The Williams & Wilkins Co.

Pascale, L. R., Grossman, M. I., and Sloane, H. S.: Correlation between skinfolds and body density in eighty-eight soldiers, Medical Nutrition Laboratory, vol. 162, 1955.

Perrott, J. W.: Anatomy for students and teachers of physical education, London, 1962, Edward Arnold, Ltd.

Plagenhoef, S. C.: Methods for obtaining kinetic data to analyze human motions, Research Quarterly **37:** 103, 1966.

Plagenhoef, S. C.: Computer programs for obtaining kinetic data on human movement, Journal of Biomechanics **1:**221, 1968.

Plagenhoef, S. C.: Gathering kinesiological data using modern measuring devices, Journal of Health, Physical Education, and Recreation, p. 81, Oct., 1968.

Plagenhoef, S. C.: Patterns of human movement: a cinematographic analysis, Englewood Cliffs, N.J., 1971, Prentice-Hall Inc.

Prior, T., and Cooper, J. M.: Light tracing used as a tool in analysis of human movement, Research Quarterly **39:**815, 1968.

Purdy, K. M.: Techniques of photography in physical education research, Ed.D. dissertation, Louisiana State University, 1969.

Rasch, P. J., and Burke, R. K.: Kinesiology and applied anatomy, ed. 3, Philadelphia, 1967, Lea & Febiger.

Reid, S. E., Tarkington, J. A., and Healion, T. E.: Medical telemetry in sports, Proceedings of the Fifth National Conference on Medical Aspects of Sports, Portland, Ore., 1963.

Roberts, V. L.: Strain-gauge techniques in biomechanics, Experimental Mechanics, p. 1, March, 1966.

Scott, M. G.: Analysis of human motion, ed. 2, New York, 1963, Appleton Century Crofts.

Sharpley, F.: Biomechanics for beginners, Ardmore, New Zealand, 1968, Ardmore Teachers College.

Sheldon, W. H.: Atlas of man, New York, 1954, Harper & Brothers.

Slater-Hammel, A. T.: Two approaches to kinesiological analysis, Physical Education **9:**17, 1954.

Steindler, A.: Kinesiology of the human body, Springfield, Ill., 1955, Charles C Thomas, Publisher.

Sweigard, L. E.: Human movement potential: its ideokinetic facilitation, New York, 1974, Dodd, Mead & Co.

Thomas, V.: Science and sport: how to measure and improve athletic performance, London, 1971, International Sports Co., Ltd.

Thompson, C. W.: Manual of structural kinesiology, ed. 8, St. Louis, 1977, The C. V. Mosby Co.

Tricker, R. A. R., and Tricker, B. J. K.: The science of movement, New York, 1967, American Elsevier Publishing Co., Inc.

Van Veen, F.: Handbook of stroboscopy, Concord, Mass., 1966, General Radio Co.

Vorro, J. R., and Hobart, D. J.: Multi-image stroboscopic-photographic techniques for the classroom and for research, Journal of Health, Physical Education, and Recreation, p. 63, May, 1973.

Vredenbregt, J., and Wartenweiler, J., editors: Biomechanics II, Basel, 1971, S. Karger AG.

Waddell, J. H., and Waddell, J. W.: Photographic motion analysis, Chicago, 1955, Industrial Laboratory Publications.

Wartenweiler, J., Jokl, F., and Hebbelink, M., editors: Biomechanics: technique of drawings of movement and movement analysis, Basel, 1963, S. Karger AG.

Waterland, J. C., and Shambes, G. M.: Electromyography: one link in the experimental chain of kinesiological research, Journal of American Physical Therapy Association **49:**1351, 1969.

Wells, K. F., and Luttgens, K.: Kinesiology, Philadelphia, 1976, W. B. Saunders Co.

Wickstrom, R. L.: Fundamental motor patterns, ed. 2, Philadelphia, 1976, Lea & Febiger.

Widule, C. J., and Gossard, D. C.: Data modeling techniques in cinematographic research, Research Quarterly **42:**103, 1971.

Williams, M., and Lissner, H. R.: Biomechanics of human motion, ed. 2, Philadelphia, 1967, W. B. Saunders Co.

Wilmore, J. H.: Exercise and sport sciences reviews, New York, 1973, Academic Press.

Wilt, F.: Mechanics without tears, Tucson, 1970, United States Track and Field Federation.

Yamshon, L. F., and Bierman, W.: Kinesiologic electromyography. II. The trapezius, Archives of Physical Medicine **29:**647, 1948.

Yamshon, L. F., and Bierman, W.: Kinesiologic electromyography. III. The deltoid, Archives of Physical Medicine **30:**286, 1949.

Zankel, H. T.: Photogoniometry, Archives of Physical Medicine **32:**227, 1951.

Answers for self-evaluation sections

Lesson 1

1. the study of human motion, i.e., of human motor behavior
2. c
3. *mechanical* kinesiology: the mechanical analysis of motor behavior, i.e., the physical laws of mechanics applied to human movement; *physiological* kinesiology: the study of human body function applied to human movement; *psychological* kinesiology: the study of behavioral variables as they affect human movement
4. the change in position of the body or body segments in space and time through the application of varying force magnitudes
5. c 6. b 7. b 8. b 9. c 10. c 11. c 12. b
13. a
14. *kinematics:* descriptive analysis of human movement; *kinetics:* causal analysis of human movement
15. b 16. c 17. a

Lesson 2

1. a 2. analysis 3. d 4. description 5. kinematics 6. c 7. c 8. quality 9. counted
10. continuous
11. *kinetics:* the study of motion causes; *kinematics:* the study of motion descriptions
12. a 13. balanced 14. inductive 15. inductive
16. d
17. three approaches: observational, theoretical, and experimental. The *observational* approach, used in kinematics, attempts to clearly describe the motion qualities and quantities involved. The *theoretical* approach, that of

inductive and deductive logic, attempts to develop a theoretical hypothesis which can be used to explain the observations. The *experimental* approach tests the tenability of the theoretical hypothesis.
18. a
19. Kinetic analysis is based upon kinematic descriptions. Its validity and reliability are dependent upon the validity and reliability of the underlying kinematic observations. Qualitative evaluation of a performance is based upon the development and verification of kinetic models. Reliability and validity of the evaluation are dependent upon reliability and validity of the underlying kinetic models.

Lesson 3

1. b 2. skill 3. charts, rating scales, and tests
4. b
5. data *acquisition:* gathering, measuring, and recording of information; data *reduction:* transformation of the data into a more condensed, organized, or useful form; processes involved are digitizing and mathematical analysis
6. measurable quantities: distance, force, speed, time, accuracy, difficulty, and standings
7. a 8. photographic 9. Still photography, motion pictures, and television recordings 10. a
11. data-acquisition system: transducers, a signal conditioner, and a recorder; data reduction carried out on digitizers and through mathematical analysis by electronic calculator or digital computer
12. is

Lesson 4

1. segmental
2. fundamental segmental movements of the body
3. a, b, c, b, a, c
4. c
5. a. phalanges, metatarsals, tarsals
   tibia, fibula
   patella, femur
   b. pelvis, lumbar vertebrae, thorax
   ribs, sternum
   cervical vertebrae, skull
   cranium, face
   c. scapula, clavicle
   radius, ulna
   carpals, metacarpals, phalanges
6. a. 14 phalanges: proximal and distal in the hallux and proximal, middle, and distal in each of the other four toes
   b. 5 bones numbered one through five from medial to lateral
   c. 7 bones: calcaneus, talus, navicular, cuboid, and three cuneiforms
   d. 2 bones: each composed of the fused ilium, pubis, and ischium
   e. 7 cervical, 12 thoracic, and 5 lumbar vertebrae, 1 sacrum, and 1 coccyx (the first two cervical vertebrae are termed atlas and axis)
   f. 2 scapulas and 2 clavicles
   g. 8 cranial bones and 12 bones of the face
   h. 8 bones: a proximal row (navicular, lunate, triquetral, and pisiform) and a distal row (trapezium, trapezoid, capitate, and hamate)
   i. 5 bones: numbered from the radial or thumb side to the ulnar or little finger side
   j. 14 bones: proximal and distal in the thumb (or pollex) and proximal, middle, and distal in each of the other four fingers
7. c, b, g, h, a, e, d, f
8. 12, 17, 10, 5, 4, 3, 1, 16, 15, 13, 11, 2, 14, 9, 8, 6, 7
9. 12, 9, 10, 3, 1, 6, 7, 8, 11, 4, 2, 5
10. 2-3, 9, 6, 4, 12, 5, 10, 11, 7, 8, 1
11. 17, 5, 3, 16, 15, 11, 14, 8, 7, 12, 10, 4, 1, 13, 2, 9, 6
12. 10, 2, 16, 11, 8, 6, 9, 13, 4, 12, 5, 15, 1, 7, 17, 14, 3

Lesson 5

1. articulation
2. the junction between two bones or a bone and cartilage
3. a
4. synarthrodial or immovable, amphiarthrodial or slightly movable, and diarthrodial or freely movable
5. b 6. d
7. sagittal (dividing the body into left and right sides), frontal (dividing the body into front and back areas), and transverse (dividing the body into upper and lower parts)
8. each perpendicular to the plane in which angular motion occurs: frontal (or x) axis passing horizontally from side to side perpendicular to the sagittal plane; longitudinal (or y) axis passing vertically perpendicular to the transverse plane; sagittal (or z) axis passing horizontally from front to back perpendicular to the frontal plane
9. a, b, c
10. c, d, e, f, b, a
11. gliding (carpals), hinge (elbow), pivot (radioulnar), condyloid (wrist), saddle (carpometacarpal of thumb), ball-and-socket (shoulder)
12. upright, feet together, arms at the sides, palms of the hands facing forward
13. linear 14. c 15. b 16. c
17. a. anterior or posterior surface of one segment approaching anterior or posterior surface of an adjacent segment while decreasing the angle between the longitudinal axes of the segments
    b. opposite of flexion, increasing the angle between the longitudinal axes of the segments
    c. movement beyond return to the anatomical position in extension
    d. movement of a segment away from the midline or part to which it is attached
    e. opposite of abduction, return toward the body or part
    f. movement about a longitudinal axis with no axis displacement
    g. movements at the shoulder in a transverse plane after the arm is abducted or flexed to a horizontal position

h. movement describing a cone
i. abduction in the trunk, i.e., sideward
j. movement at other than a right angle to one of the cardinal planes

18. a. flexion: at the knee, posterior leg toward posterior thigh

b. extension: at the knee, posterior leg away from posterior thigh

c. hyperextension: at the shoulder, arm backward from the anatomical position

d. abduction: at the shoulder, arm sideward from the anatomical position

e. adduction: at the shoulder, return of abducted arm to side of the body

f. rotation: at the radioulnar joint, pronation of forearm

g. horizontal flexion and extension: at the shoulder, arm horizontally forward and backward

h. circumduction: at the shoulder, arm swung in circle

i. lateral flexion: at the intervertebral joints, head-neck segment sideward

j. diagonal motion: at the shoulder, arm forward and out

Lesson 6

1. b
2. a. (1) flexion   (2) extension

b. (2) adduction   (3) abduction

c. tarsometatarsal

d. (1) eversion (or inward rotation)
   (2) inversion (or outward rotation)

e. ankle   (1) dorsal   (2) plantar

f. knee   (1) flexion   (2) extension
   (3) inward, outward

g. (2) hypertension   (5) rotation

3. a. dorsal flexion, plantar flexion, inversion, and eversion

b. at the ankle, plantar and dorsal flexion; at the knee, flexion and extension and outward and inward rotation

c. at the knee, flexion and extension; at the hip, flexion, extension, abduction, adduction, inward and outward rotation, and circumduction

4. b. breathing

c. (1) (a) flexion   (b) return from flexion

(c) hyperextension   (d) abduction (or lateral flexion)   (e) adduction (or lateral flexion in opposite direction)

(2) rotation

(3) (a) flexion-extension   (c) abduction-adduction

d. (1) forward tilt   (2) backward tilt

5. arm, forearm, hand

6. a. (1) elevation   (2) depression   (3) abduction   (4) adduction   (5) upward rotation   (6) downward rotation

b. humerus, scapula   (4) flexion, extension   (5) circumduction

c. elbow   (1) flexion-extension

d. (1) pronation   (2) supination

e. wrist   (3) abduction   (4) adduction

g. metacarpals, carpals   (c) opposition

i. metacarpophalangeal   (1) flexion-extension   (4) adduction   (5) flexion, extension

j. flexion, extension

Lesson 7

1. d
2. *time,* period when an action or process continues; *space,* area or volume occupied by a body and distance or angle through which it travels during movement; *matter,* anything having mass and occupying space (usually quantified by its weight or mass)

3. the process of assigning numbers and units to objects according to rules

4. b   5. a   6. c   7. d   8. c

9. Units of space, time, and weight are fundamental units (e.g., feet, seconds, and pounds in the British system); all other units are derived (e.g., mi/hr and ft/sec/sec).

10. a   11. mks and cgs

12. meter, foot; kilogram, pound; second, second

13. anything tending to change either the dimensions or the state of motion of an object

14. scalar   15. direction

16. the extension of the vector diagram in both forward and backward directions

17. d

18. the straight line connecting the center of gravity of the object with the center of the earth

19.

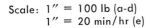

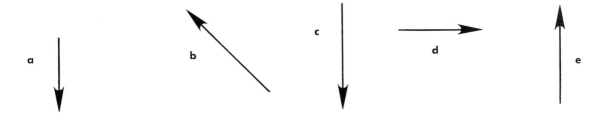

20. a 21. b
22. angle turned through, radius of rotation, direction of rotation about an axis, and arc of the circle cut by the angle
23. *radian:* unit of angular measure equal to approximately 57.3°; *goniometer:* instrument for measuring angles
24. a. 57.3 b. 0.01745
25. *chronoscope:* mechanical or electronic instrument for measuring time directly; *chronograph:* electronic instrument that records time on a chronogram
26. *Linear displacement* can be measured through the use of cinematographic techniques, but a multiplier is needed to convert projected lengths into real lengths. For measuring *angular displacement,* no such multiplier is needed. *Time* is measured by using a chronoscope in the field of view of the camera, recording directly on the frames of the film, or using the camera speed to calculate it from the frame rate.

Lesson 8

1. *Coplanar* vectors lie in the same plane; *concurrent* vectors have the same point of origin.
2. a 3. resultant 4. components
5. The process of combining several vectors to find one equivalent vector produces the same effect as the component vectors.
6. two methods: parallelogram and vector chain.

The *parallelogram* method is used to find the vector sum by forming a parallelogram with the two composite vectors. The resultant solution is the diagonal of the parallelogram originating at the point of application, i.e., at the intersection of the tails of the original vectors. The *vector chain* or triangle method is accomplished by chaining the composite vectors together (connecting the head of the first vector to the tail of the second). The resultant is found by drawing a line from the tail of the first vector (origin) to the head of the second vector (terminal point) and determining its magnitude and direction.

7. approximate (graphical) and precise (trigonometric). In the *graphical* method, draw and measure perpendicular lines from the head of the vector to the x and y axes. These lines, when multiplied by the scale value, are equal to the x and y components of the vector. In the *trigonometric* method, resolve a vector into its components by applying trigonometric functions as follows: $F_x = F \cos \theta$ and $F_y = F \sin \theta$.
8. For any right triangle the square of the hypotenuse equals the sum of the squares of the other two sides ($h^2 = o^2 + a^2$).
9. forming the negative of a vector (a vector of the same length but opposite direction); i.e., $B - A = B + (-A)$

10.

Scale: 2″ = 100 lb (a)
       1″ = 100 ft/sec (b)
       2″ = 100 lb (c)

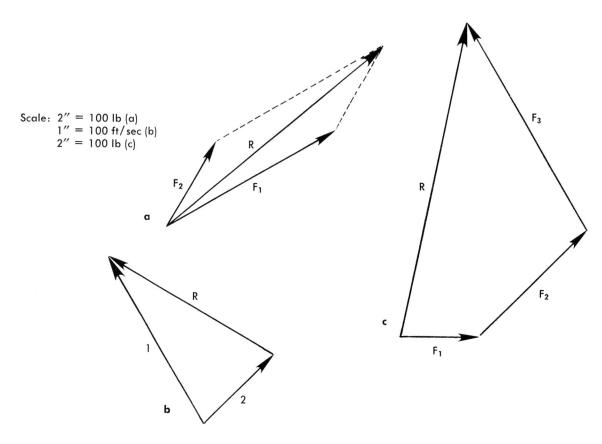

11.

Scale: 1″ = 100 mi/hr (a)
       1″ = 100 ft/sec (b)
       1″ = 200 lb (c)

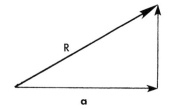
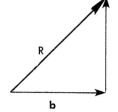

12.

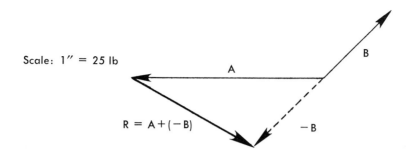

Scale: 1″ = 25 lb

A

B

R = A + (−B)

−B

13. *standing long jump:* vertical component projecting body into the air; horizontal component projecting body forward; *football lineman blocking:* upward force (F_y) lifting opponent and reducing his friction; horizontal force (F_x) moving opponent horizontally; *mile run:* downward component (F_y) and backward component (F_x) of force exerted by runner against the track surface

Lesson 9

1. a   2. secondary
3. a. plane passing through the body from front to back, dividing it into right and left sections
   b. plane passing through the body from side to side, dividing it into front and back sections
   c. plane passing through the body, dividing it into upper and lower sections
4. upright, feet together, arms at the sides, palms of the hands facing forward
5. a. right-to-left or left-to-right
   b. upward or downward
   c. forward or backward
6. Its distance from the center of gravity or some other point of reference is measured in the x, y, and z directions.
7. axis
8. a. perpendicular to the sagittal plane
   b. perpendicular to the transverse plane
   c. perpendicular to the frontal plane
9. a. forward or backward (e.g., a forward roll)
   b. to the right or left (e.g., a cartwheel)
   c. to the right or left (e.g., a full twist in tumbling)
10. a. x, sagittal
    b. z, frontal
    c. x, sagittal
    d. y, transverse
    e. x, sagittal
    f. y, transverse
    g. z, frontal

11. high diagonal of the shoulder; low diagonal of the shoulder; diagonal of the hip
12. The distance (r) of the point from the center of gravity or other reference is measured and the angle that r makes with the positive (x) axis is determined.
13. (a) Resolve all vectors into their x, y, and z components; (b) combine the horizontal, x and z, components to obtain a horizontal resultant; (c) combine the horizontal resultant with the vertical y component to obtain the total resultant (R).
14. The individual is visualized as standing inside a globe. Joint movement is recorded on the surface of the globe. Abduction and adduction are measured by the meridians; flexion and extension are measured by the parallels.
15. a. moving the camera as far away as possible and using a telephoto lens, photographing the object when it is located directly in front of the lens, aligning objects to be photographed in a plane perpendicular to the lens
    b. three cameras aligned perpendicular to each other and to the three cardinal planes of space parallel with the three cardinal axes or horizontally with their optical axes level, 120° apart, and intersecting in the middle of the visual field (Noss, 1967, and Miller and Petak, 1973)

Lesson 10

1. identical   2. a   3. scalar   4. c
5. a. rate of change of position
   b. rate of change of position in a given direction
   c. rate of change of velocity (i.e., change of speed and/or change of direction)
6. a. Speed or the magnitude of a velocity is

the distance traveled by an object per unit of time.

   b. Acceleration is the change of velocity $(V_f - V_o)$ per unit of time.

7. constant   8. variable

9. a. distance traveled per unit of time; mathematical average of all the instantaneous accelerations for a given interval

   b. velocity and acceleration of an object at a precise instant of time

   c. equal changes in velocity occurring in equal intervals of time

   d. unequal changes in velocity occurring in equal intervals of time

10. a. 22 ft/sec, 40 ft/sec, 56 ft/sec, 58 ft/sec

   b. 20 ft/sec^2, 30 ft/sec^2, 40 ft/sec^2, 30 ft/sec^2

   c. E to F

11.

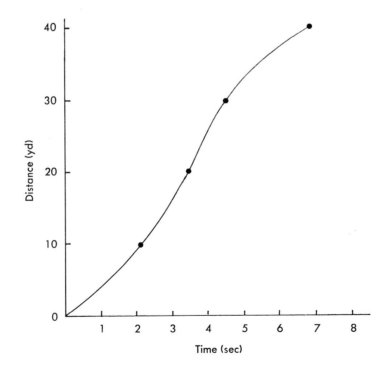

12. First second: 4 yd/sec

  Third second: 7 yd/sec

  Fifth second: 6 yd/sec

  Seventh second: 3 yd/sec

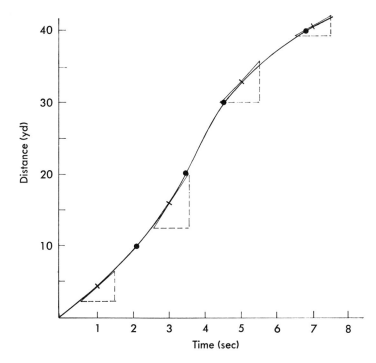

13.

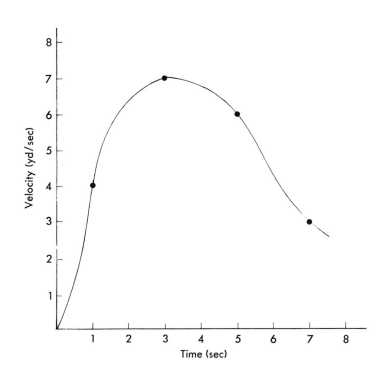

14.

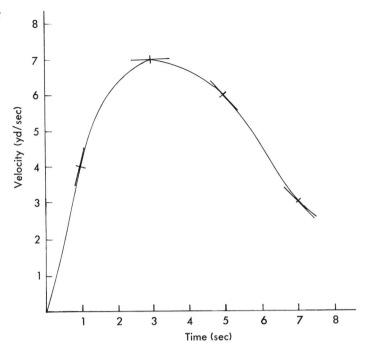

Velocity (yd/sec) vs Time (sec)

Lesson 11

1. a. theoretical path of a body moving through an air medium when the effects of air friction are neglected
   b. horizontal distance from the point of projection to the point where the projectile returns again to the projection plane
   c. time required for a projectile moving either obliquely or straight upward to rise and then return to the original level
   d. vertical (V_y) component
   e. horizontal (V_x) component
   f. a body moving through an air medium by its own momentum
   g. the constant acceleration produced by the force of gravity (32 ft/sec²)
2. d
3. The equations for free fall are used to solve for motion variables of projectiles when all but one of the quantities in an equation are known. V_f, Final velocity; V_o, original velocity; g, gravitational constant of 32 ft/sec²; t, time; d, distance
4. a. parallel with the surface of the earth
   b. straight upward or straight downward
   c. angled other than 90° to the horizontal and vertical
5. R = 1270.17 ft; t = 4.33 sec
6. a. 64.56 ft   b. 307.75 ft   c. 4.0175 sec

7. 187.94 ft, 196.96 ft, 200.00 ft, 196.96 ft, 187.94 ft
8. 2.95 sec, 193.5 ft
9. a. vertical jump, $\theta \approx 90°$
   b. 100-yard dash, $\theta \approx 0°$
   c. shot-put, $\theta \approx 40°$
10. b
11. (a) Takeoff height is equal to landing height, optimum angle 45°. (b) When takeoff height is above landing height, the optimum angle is less than 45° but increases with increasing speed of projection. (c) When takeoff height is above landing height, the optimum angle decreases with increasing height of release.

Lesson 12

1. d   2. a
3. the rate of change of angular position
4. The linear velocity of a tennis racket is directly proportional to the length of the radius of rotation and the angular speed of the racket.
5. a. a = –10 ft/sec²
   b. a = 24.44 ft/sec²
   c. α = –68 rad/sec²
   d. α = –0.33 rad/sec²

6. $\alpha = \dfrac{\omega_f - \omega_o}{t}$. Angular acceleration is the rate of change of angular velocity.

7. c 8. vector
9. direct variation
10. a. equal, equal
    b. unequal, equal
11. angular speed and direction of a body at a given moment in time
12. instantaneous
13.

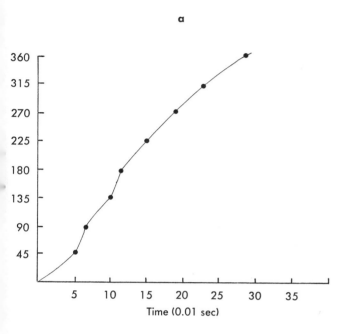

a

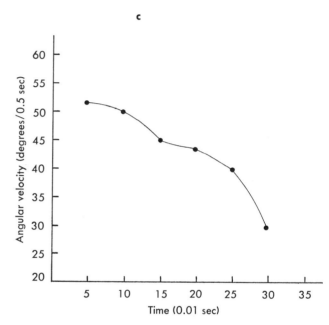

c

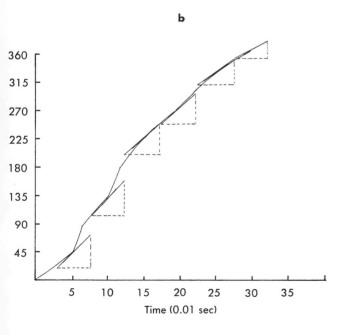

b

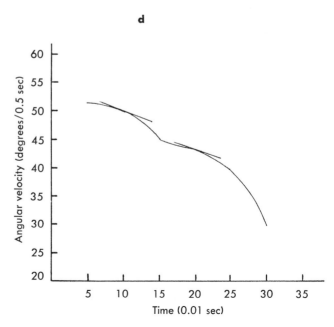

d

14. a. ratio of instantaneous angular velocity change to angular displacement
   b. perpendicular to the radius of rotation
   c. precise effect, because the projectile will travel in a direction that is 90° to the radius of rotation at the point of release
   d. the flatter the arc, the greater the control
15. c
16. a. extending the elbows in a golf swing just before hitting the ball
   b. flexing the knee so it can be brought forward more quickly in the recovery phase of running
   c. when a discus thrower decreases his radius of rotation during his turn but increases it immediately before releasing the discus

Lesson 13

1. a   2. a   3. b   4. internal
5. a. length of muscle unchanged
   b. length of muscle changed
      (1) muscle length decreased
      (2) muscle length increased
      (3) speed constant
6. a   7. dynamometer   8. transducer   9. c
10. a. perpendicular distance from the center of rotation to the line of action of a force
   b. line extending through the force vector both forward and backward
   c. producing counterclockwise rotation
   d. opposing the rotation tendency of an applied torque
   e. components acting, respectively, parallel with and perpendicular to the radius of rotation ($F_r = F \cos \theta$ and $F_t = F \sin \theta$)
11. centrifugal   12. eccentric   13. c   14. b
15. a. in discus throwing, muscular action and air resistance
   b. in bicycle racing, muscular action applied to the pedals and ground friction
16. c
17. a. 33 lb-ft
   b. 175 oz-in or 0.9115 lb-ft
   c. 150 lb-in or 12.5 lb-ft
   d. 600 lb-in or 50 lb-ft
   e. 21 lb-in or 1.75 lb-ft
      | | F_r | F_t |
18. a. 26.08   6.99
   b. 91.11   12.80

c. 86.60   50
d. 209.71   10.99
e. 39.76   130.06
19. a. 20.96 lb-in or 1.75 lb-ft
   b. 51.22 lb-in or 4.27 lb-ft
   c. 450 lb-in or 37.5 lb-ft
   d. 76.93 lb-ft
   e. 260.12 lb-ft

Lesson 14

1. energy
2. Work is the product of the resistance and the distance moved. W = fd
3. a
4. the ability to do work
5. heat, electrical, chemical, sound, mechanical, light
6. a bat striking a baseball (mechanical) and producing sound when it impacts
7. mechanical
8. *kinetic:* the ability of a moving object to do work, i.e., the energy of motion; *potential:* the ability of an object to do work by virtue of its position or state
9. a. foot-pound
   b. foot-pound per second and horsepower (1 hp = 550 ft-lb/sec)
10. Work is the ability to move a resistance through a distance, while power is the rate of doing work.
11. b   12. negative   13. $\overrightarrow{Fd} \cos \theta$, $\overrightarrow{Fd} \sin \theta$
14. a work-measuring device
15. a. Percent grade of a treadmill is equal to the tangent of the angle of inclination multiplied by 100. Work on a treadmill is measured through the use of W = wh = mg(V sin θ)t.
   b. Work on a bicycle ergometer is determined through the use of W = Fd = Frθ = F2πrn.
16. b
17. any device that aids in the performance of work
18. d
19. a. 342   ft-lb
   b. 4000   ft-lb
   c. 14   ft-lb
   d. 0.83 ft-lb
20. a. 110,382 ft-lb
   b. 115,098 ft-lb

c. 106,173 ft-lb

d. 33,537 ft-lb

21. a. 3.34 hp

b. 3.49 hp

c. 3.22 hp

d. 1.02 hp

22. a. 87,964.8 ft-lb

b. 150,796.8 ft-lb

c. 75,398.4 ft-lb

d. 140,743.7 ft-lb

Lesson 15

1. source of input energy (mover), machine for changing the energy into a more useful form, and resistor to serve as the load upon which the system acts

2. a. shoulder flexors

b. humerus

c. weight being lifted

3. points

4. a. axis between the force and resistance

b. resistance between the force and axis

c. force between the resistance and axis

5. a

6. *primary:* to balance forces, magnify forces, and magnify displacement and speed; *secondary:* to change the direction of the force application

7. *third class:* biceps brachii acting on the radius; quadriceps femoris acting on the tibia. Both have the primary machine function of magnifying displacement and speed. *first class:* triceps brachii acting on the ulna. This lever has the primary machine function of magnifying displacement and speed but the secondary function of changing the direction

of the force application.

8. d   9. b   10. two

11. a machine that functions to change the direction of the line of action of a force

12. d

13. type I: external oblique muscle pulling on the ribs to rotate the vertebral column (force magnification); type II: deep spinal muscles pulling on the vertebral column to rotate the trunk (speed magnification); pulley arrangement: patella changing the line of action of the quadriceps femoris muscle (direction change)

14. c

Lesson 16

1. d   2. a

3. a. applied force

b. force arm

c. resistance force

d. resistance arm

4. a   5. d   6. c

7. a. 4.0, 5.0, 0.8, to magnify forces

b. 1.02, 1.0, 1.02, to balance forces

c. 9.45, 5.0, 0.5, to magnify forces

d. 0.125, 0.17, 0.735, to magnify displacement

e. 0.07, 0.1, 0.7, to magnify displacement

8. c   9. a   10. b   11. b   12. d   13. c

14. metabolism   15. calorimetry   16. a   17. a

18. the ratio of external work accomplished to internal energy expended

19. a

20. a. treadmill   b. indirect calometry

21. a. 0.10, 0.25   b. 0.05, 0.125   c. 0.057, 0.2

22.

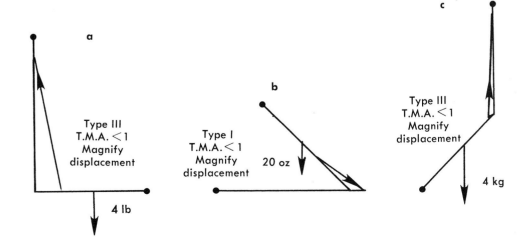

Lesson 17

1. b   2. a   3. mutually exclusive   4. b
5. inertia
6. (Newton I): gymnast doing a handstand; runner moving in a straight line with constant speed; football lineman in his starting stance

   (Newton II): greater difficulty in moving a heavy football lineman than a lighter one; greater propelling force and lesser resistive forces acting on a swimmer when proper techniques are used, thus more acceleration (i.e., greater net force); athlete working to increase his strength and reduce his weight

   (Newton III): force applied by a golf club to a golf ball equal to but opposite force of the golf ball on the club; reaction force of the ground on a runner greater than forces of water on a swimmer (thus faster movements of the runner); the greater ease one finds in running on a hard surface than on sand

7. b
8. The sum of all rectangular forces must equal zero, and the sum of all torques must equal zero.
9. d
10. (Newton I): An object at rest or in uniform motion continues in its state unless acted on by an external force.

    (Newton II): Acceleration produced by an external net force acting on a mass is directly proportional to and in the same direction as the force and inversely proportional to the mass.

    (Newton III): For every action there is an equal and opposite reaction.

11. scalar   12. b   13. changing
14. *resistance:* a force that acts in opposition to an applied force; *inertia:* a property whose magnitude is independent of the magnitude of an applied force
15. b   16. pound
17. the force equal to an applied force minus the resistive force; net $F = F - R$
18. a   19. a   20. a   21. direction
22. (Newton I): A body in motion will travel with constant speed unless acted upon by a net tangential force, and in a straight line unless acted upon by a net radial force.

    (Newton II): Rate of change of speed of a body in motion is directly proportional to the magnitude of the net tangential force, and inversely proportional to its mass.
    Rate of change of direction of a body in motion is directly proportional to the magnitude of the net radial force, and inversely proportional to its mass.

    (Newton III): For every applied tangential and radial force component there is an equal and opposite reactive tangential and radial force component.

23. *statics,* the study of balanced force situations; *dynamics,* the study of unbalanced force situations

Lesson 18

1. Work   2. d   3. b   4. d
5. ½ mV^2. Kinetic energy is directly proportional to the product of half the mass of an object and its velocity squared.
6. c   7. direct   8. a
9. G.P.E. = mgh. Gravitational potential energy is directly proportional to the product of an object's weight (mg) and its height above a surface (h).
10. a   11. b   12. b   13. causal
14. $Ft = mV_f - mV_o$. Impulse, the product of

the applied net force (F) and the duration of the force application (t), is equal to the momentum change it produces ($mV_f - mV_o$).

15. vector 16. d
17. newton, dyne, pound; kilogram, gram, slug; joule, erg, foot-pound; kilocalorie, calorie, BTU; joule/sec, erg/sec, ft-lb/sec

Lesson 19

1. vibratory 2. a 3. b 4. c 5. a 6. c
7. The deformation of a body is proportional to the applied force per unit of area (i.e., strain is proportional to stress).
8. a 9. c
10. r varies with the velocity of impact; r is highest at low velocities of collision; r has a range of values between 0 and 1; liquids and gases have high elasticity; air is used in inflated objects to increase r.
11. a. 0.95 b. 0.87 c. 0.96 d. 0.89 e. 0.88
12. Work done during elastic harmonic motion is directly proportional to half the force constant (K) and the amount of deformation squared.
13. c 14. a
15. (Newton I):   A body with a given shape will tend to retain that shape, and a body at rest will tend to remain at rest, unless a net distortion force acts upon it. If a body is in harmonic motion, it will remain in harmonic motion with equal displacements unless an unbalanced force acts upon it.
   (example) basketball at rest on the floor and one bouncing
   (Newton II):   Acceleration of an object returning to its equilibrium position or shape following distortion is directly proportional to the product of the force constant (K) of the object and its displacement (d) and is inversely proportional to its mass; i.e., a = Kd/m.
   (example) hard ball return-ing to its equilibrium position faster than a soft ball because of its greater force constant (K)
   (Newton III): For every force tending to distort a body within its elastic limits, there is an equal and opposite restoring force produced in the body that will tend to return the body to its original position or shape.
   (example) flattening of a basketball as it impacts with the floor and the resultant distortion imparting to it a restoring force which causes the ball to return to its original shape and thus rebound into the air

16. b 17. b 18. c

Lesson 20

1. a 2. toward
3. whenever there is a change in the speed, direction, or both the speed and the direction of a motion
4. a 5. b
6. *tangential force:* parallel with the linear path of the motion and perpendicular to the radius; *radial force:* parallel with the radius and perpendicular to the linear path

7.

	F_r	a_r
a.	2.75 lb	8 ft/sec^2
b.	27.76 lb	444.13 ft/sec^2
c.	1210 lb	9.68 ft/sec^2
d.	6046.88 lb	2.15 ft/sec^2
e.	483.41 kg	947.48 ft/sec^2

8.

	$\tan \theta$	θ
a.	0.625	32°
b.	2.0	63°
c.	11.36	85°
d.	2.84	70°
e.	1.21	50°

9. perpendicular to   10. direct   11. inverse
12. It is directly proportional to the linear velocity and inversely proportional to the radius of the turn.

Lesson 21

1. b
2. (Newton I): An object either remains at rest or, if in angular motion, rotates at constant speed around a fixed axis unless acted upon by an unbalanced net torque.

   (example) spiraling football continuing to spin at a constant speed until acted on by air resistance

   (Newton II): Angular acceleration (α) produced by an unbalanced torque (T) acting on a body is proportional to the net torque, in the same direction as the torque, and inversely proportional to the moment of inertia (I) of the body.

   $$\alpha = \frac{T}{I} = \frac{T}{mr^2}$$

   (example) spinning ice skater coming to a stop because of friction between skates and ice

   (Newton III): Whenever one body acts upon a second body with a torque, the second body exerts an equal but opposite torque on the first body.

   (example) dancer doing a pirouette because of the reaction force of the floor

3. d 4. c 5. b 6. a 7. c

8.

	F_t	$r\alpha$
a.	150 kg	15 m/sec²
b.	137,500 lb	2500 ft/sec²
c.	0.43633 lb	0.087 ft/sec²
d.	200,000 lb	1000 ft/sec²
e.	400 lb	20 ft/sec²

9. a torque (T)
   b. moment of inertia ($I = mr^2$)
   c. $L = I\omega = mr^2\omega$
   d. net $Tt = L_f - L_o$
   e. K.E. $= \frac{1}{2}I\omega^2 = \frac{1}{2}mr^2\omega^2$
   f. $W = T\theta = Fr\theta$
10. a

Lesson 22

1. c
2. (Newton I): A body in rotation will continue to turn about its axis with an angular momentum constant in both magnitude and direction unless an external net torque is exerted on it.

   (Newton II): The rate of change of angular momentum of a body is proportional to the torque causing it and has the same direction as the torque.

   $$T = \frac{L_f - L_o}{t}$$

   (Newton III): For every applied torque tending to produce angular momentum in one direction there is an equal and opposite reaction torque.

3. (Newton I): bowling ball tending to roll at a constant speed and in a straight line; springboard diver tending to twist with a constant speed around a fixed axis due to the lack of external net torque

   (Newton II): the greater the torque and the greater the time during which the torque acts, the greater the momentum change imparted to a gymnast; the greater the angular momentum of a ball, the greater must be the net torque to bring it to a halt

   (Newton III): the torque exerted on a ball by a bat equals torque exerted on the bat by the ball; the torque exerted by the floor on a gymnast equals torque exerted by the gymnast on the floor, but opposite in direction

4. The total angular momentum in a closed system remains constant without an external net torque.
5. a. angular momentum of the arms trans-

ferred to the body as a whole when doing a back flip

b. angular momentum of a ball being thrown changed to linear momentum at the time the ball is released

c. linear momentum of a pole vaulter partially transferred to angular momentum at the time of pole plant

6. Movements initiated in the air imparting momentum to body parts around an axis must be compensated by equal and opposite movements of other body parts around the same axis so the total momentum change will equal zero.

7. *axis of momentum:* around which the total angular momentum of the body is occurring; *axis of displacement:* around which movements initiated in the air and the resulting reaction movements occur

8. a

9. a. 383 ft-lb
   b. 180 ft-lb
   c. 122.25 m-kg
   d. 74,750.12 cm-gm

10. When the fingers of the right hand are curled in the direction of a rotation, the extended thumb will point in the direction of the axis of momentum of that motion.

11. An ice skater increases his/her rate of spin by bringing the arms in; a flexed knee increases the speed of forward leg movement in running; a diver in a layout position spins more slowly than one in a tucked position.

12. First, determine the separate momentums occurring around the x, y, and z axes. Then vector add them.

13. Moving the center of mass either away from or closer to the axis of rotation will increase or decrease the radius of gyration.

Lesson 23

1. a

2. an object that spins around its center of gravity

3. b

4. A moving bicycle has angular momentum due to the angular motion of its wheels; an external torque can be applied by the rider in leaning to either the right or the left, thereby producing precessional motion in that

direction; the bicycle will follow a path that curves depending upon the direction of the external torque. A spiraling football can be caused to wobble by the external torque produced by air resistance.

5. A high jumper will nutate around his/her longitudinal axis of momentum and will wobble even when not acted upon by the force of air resistance.

6. a

7. *precessional motion:* produced by an external torque; *nutational motion:* produced by an internal torque when the axis of displacement of a rotation does not coincide with the axis of momentum

Lesson 24

1. Acceleration is *directly* proportional to the gravitational constant (g) and horizontal displacement (d) and is *inversely* proportional to the radius (r).

2. The period is directly proportional to the product of 2π and the square root of the ratio of the moment of inertia (I) to the force constant (K).

3. any real pendulum such as the human body as it might swing from a horizontal bar

4. b

5. ring of gyration (of each of the three cardinal axes of angular motion): infinitely thin circular ring of matter located around the center of mass of a body and having the same quantity of mass as the body itself; radius of gyration: belonging to the ring of gyration and giving the ring the same moments of inertia as the body itself

		I	*k*	*q*
6.	a.	648.45	9	20.26
	b.	1313.12	12.07	16.21
	c.	39.72	2.82	3.98
	d.	13.40	2.59	2.23

7. bowling and running

8. a. The most economical pace is at the same rate as the natural frequency of the limbs.

   b. In order to reduce the centrifugal forces acting on a gymnast and tending to pull him away from the apparatus, the center of gravity should be brought as close to the axis of rotation as possible.

   c. The underhand pattern of projection allows for the maximum use of gravity in the production of force.

9. It is equal to the product of the pendulum's weight (mg) and the vertical distance (h) it is elevated (mgh).

10. a. The position of maximum displacement is the position of minimum kinetic energy; the lowest point of the swing is the position of maximum kinetic energy.

    b. The position of maximum displacement is the position of maximum potential energy; the lowest point of the swing is the position of minimum potential energy.

    c. On the downswing the loss of potential energy is equal to the gain in kinetic energy; on the upswing the gain in potential energy is equal to the loss of kinetic energy.

    d. The total mechanical energy, kinetic and potential, remains constant unless there is some transformation to heat or other types of energy.

11. d

12. oscillations that have the same frequency and start out together

13. (a) A force acting upon a body periodically with the body's natural frequency increases the total displacement of the motion, such as a child pumping a swing to increase its amplitude. (b) An impulse differing from the natural frequency of a body decreases the amplitude of the frequency, such as a diver stopping his/her bounce on a springboard.

14. a. (1) 13.86 ft/sec
       (2) 6.26 m/sec
       (3) 7.3 ft/sec
       (4) 1.74 m/sec

    b. (1) 393.4 lb and 50.6 ft/sec
       (2) 17,739 lb and 22.6 ft/sec
       (3) 129,900 kg and 19.8 m/sec
       (4) 7990 kg and 7.7 m/sec

15. the product of the tangential component of the force of gravity (F_t) and the length of the cord (r). The total force of gravity (mg) can be resolved into its radial (F_r) and tangential (F_t) components; the tangential component (F_t) decreases on downswing starting from horizontal and increases on upswing to again be maximum at horizontal.

16. (a) Torque produced by the force of gravity

acting on a gymnast can be either increased or decreased by changing the distance (r) of the body's center of gravity from the axis of rotation. (b) To reduce the centrifugal forces acting on a gymnast during mounting activities that tend to pull him/her away from the apparatus, the center of gravity must be brought as close to the axis of rotation as possible.

Lesson 25

1. Select an analysis approach appropriate to the problem being investigated.

2. a. when a body can be adequately represented by the motion of its mass center

    b. when the muscle and joint forces act upon individual limb segments

    c. when the body's segments remain in the same positions with respect to one another

    d. when the body's segments are moving

3. Segmental end points and segmental centers of gravity are represented by points in space as described by either rectangular or polar coordinates. The straight line connecting the segmental end points of a segment is the segment length. Segmental weights as well as other applied and reaction forces are drawn in as vectors pointing in the direction in which they act. Inertial forces are drawn in the opposite direction of the acceleration to show that they act in opposition to the movement.

4. *force-mass-acceleration:* to study the instantaneous effects of forces; *impulse-momentum:* to analyze forces acting during time periods; *work-energy:* to analyze forces acting over specific distances

5. that all applied and inertial forces acting on a segment must be drawn in opposite directions on a free-body diagram to show the system in equilibrium

6. Applied radial forces are drawn in the direction in which they act; inertial radial force is drawn outward; applied tangential forces are drawn in the direction in which they act; inertial tangential force is drawn in the opposite direction of the motion (when the motion is constant or positively accelerating) and in the same direction as the motion (when the motion is negatively accelerating).

7. a

8. (a) Describe the basic data involved in the analysis, (b) construct two-segment free-body diagrams, and (c) describe the Coriolis force.

Lesson 26

1. Every object in the universe is attracted to every other object in the universe with a force that is directly proportional to the product of their masses and inversely proportional to the square of the distance between their centers of mass.
2. b   3. a   4. inversely   5. less   6. b
7. The force of gravity acting upon a body is directly proportional to the mass of the body.
8. b   9. lesser
10. skinfold measures and body fluid measures
11. The increase in muscle strength resulting from resistive exercises is usually greater than the increase in mass.
12. gravity
13. reaction board method. A board long enough for the individual to lie on is placed on two triangular blocks of wood (knife edges), a block under each end. One end of the board rests on a platform scale, and the other on a block. The weight of the board is read from the scale, and this reading is subtracted from subsequent readings made during the measurement. If the weight of the person and the distance between the two knife edges are known, the center of gravity of the person can be determined through use of the law of the lever ($F \times f = R \times r$).
14. endomorphic
15. *endomorphy:* roundness and softness of body; *mesomorphy:* heavy, hard, rectangular outline with prominent muscular development; *ectomorphy:* typically a slender frail body structure with small bones and thin segments
16. mesomorphy

Lesson 27

1. d   2. fluid   3. d   4. c
5. normal force, nature and condition of materials in contact, actual contact area, relative motion between the surfaces, type of motion between the surfaces
6. a
7. The angle of uniform slip is equal to the angle of lean that is possible between two solid objects.
8. greater
9. the force acting perpendicular to two surfaces and pressing the two surfaces together
10. (a) It is directly proportional to the magnitude of the normal force, which is the resultant of all forces acting perpendicular to the surfaces. (b) Surfaces of the same materials generally produce greater surface friction than do surfaces of different materials. (c) Starting sliding friction is much greater than moving sliding friction. (d) Moving sliding friction is relatively independent of speed changes. (e) Until a limiting value for starting friction is reached, the action force tending to cause sliding motion will equal the reaction force of friction at rest.
11. *actual contact area:* that which is actually touching; *apparent contact area:* the two surfaces opposing each other, including vacuum spaces of no contact
12. directly   13. greater
14. The total actual contact area is proportional to the total normal force.

Lesson 28

1. air, water
2. the resistance manifesting itself when a gas or liquid is made to flow around a stationary object or an object is made to move through a previously stationary fluid
3. (a) relative velocity between the object and the fluid, (b) shape of the object and its active surface area, (c) surface texture, (d) fluid pressure, (e) temperature of the fluid
4. a   5. turbulent
6. directly proportional to the relative velocity at low speeds; directly proportional to the square of the velocity at high speeds
7. b   8. cavitation
9. a. It is directly proportional to the greatest cross-sectional area of the body acting at right angles to the flow of the fluid.
   b. The rougher the texture of a body, the greater is its fluid friction.
10. d
11. the ratio of force produced by a fluid to the area over which the force is applied

12. directly
13. Pressure in a fluid is inversely related to speed of movement.
14. decrease
15. a. Fluids flow from high-pressure to low-pressure areas.
    b. Fluid pressure is directly proportional to the depth of the fluid.
    c. Fluid pressure decreases with an increase in fluid velocity.
16. least
17. presenting as large a surface as possible in the direction opposite that of the desired movement; applying forces through as great a distance in the desired direction as possible; moving the force-producing members of the body as rapidly as possible
18. decreasing

Lesson 29

1. *aerodynamics:* study of forces affecting movement through air; *hydrodynamics:* study of forces affecting movement through water
2. resistance   3. gravity, buoyancy   4. a
5. *line of flight:* path along which the plane is actually flying; *angle of attack:* angle between the plane's attitude and the horizontal
6. the ratio of the lift to the drag components of air resistance
7. The angle of attack or inclination of a football will usually be less than that of a discus, especially into a headwind, because a football is lighter. Spin is used to stabilize the football's flight. During the flight of a discus, the amount of friction depends upon its angle of attack. The friction will be at right angles to the long axis of the discus. If the leading edge is above the trailing edge, the force of friction will produce a lifting (y component) and a drag (x component) effect on the discus.
   8. d   9. d
10. When an object is spinning during flight, pressure is built up on one side and reduced on the other. This occurs when a turbulent high-pressure area builds up on the side where there is opposition to air moving around the object. On the opposite side the spin force and air resistance force are in the same direction and a low-pressure area is created. The object will curve toward the low-pressure side.
11. a
12. *vertical pair:* gravity and buoyancy; *horizontal pair:* thrust and drag
13. a
14. downward force: gravity, represented by the weight of the swimmer; upward force: buoyancy
15. when the force of gravity and the force of buoyancy are equal in magnitude
16. the point around which a body will rotate while it is in water; also the point at which the force of buoyancy is concentrated
17. c

Lesson 30

1. *direct:* head-on; *indirect:* non-headon; basketball impacting at a 90° angle with the floor vs. ball impacting at an angle other than 90°
2. partially elastic
3. b   4. area   5. a   6. b   7. are   8. cannot   9. is   10. a
11. dissipation
12. *impact:* angle of entry; *rebound:* angle of departure
13. c   14. is   15. direct   16. does

Three-place values of trigonometric functions and degrees in radian measure

Radian	Degree	Sine	Tangent	Secant	Cosecant	Cotangent	Cosine	Degree	Radian
.000	0°	.000	.000	1.000	—	—	1.000	90°	1.571
.017	1	.017	.017	1.000	57.30	57.29	1.000	89	1.553
.035	2	.035	.035	1.001	28.65	28.64	0.999	88	1.536
.052	3	.052	.052	1.001	19.11	19.08	.999	87	1.518
.070	4	.070	.070	1.002	14.34	14.30	.998	86	1.501
.087	5	.087	.087	1.004	11.47	11.43	.996	85	1.484
.105	6	.105	.105	1.006	9.567	9.514	.995	84	1.466
.122	7	.122	.123	1.008	8.206	8.144	.993	83	1.449
.140	8	.139	.141	1.010	7.185	7.115	.990	82	1.431
.157	9	.156	.158	1.012	6.392	6.314	.988	81	1.414
.175	10	.174	.176	1.015	5.759	5.671	.985	80	1.396
.192	11	.191	.194	1.019	5.241	5.145	.982	79	1.379
.209	12	.208	.213	1.022	4.810	4.705	.978	78	1.361
.227	13	.225	.231	1.026	4.445	4.331	.974	77	1.344
.244	14	.242	.249	1.031	4.134	4.011	.970	76	1.326
.262	15	.259	.268	1.035	3.864	3.732	.966	75	1.309
.279	16	.276	.287	1.040	3.628	3.487	.961	74	1.292
.297	17	.292	.306	1.046	3.420	3.271	.956	73	1.274
.314	18	.309	.325	1.051	3.236	3.078	.951	72	1.257
.332	19	.326	.344	1.058	3.072	2.904	.946	71	1.239
.349	20	.342	.364	1.064	2.924	2.747	.940	70	1.222
.367	21	.358	.384	1.071	2.790	2.605	.934	69	1.204
.384	22	.375	.404	1.079	2.669	2.475	.927	68	1.187
.401	23	.391	.424	1.086	2.559	2.356	.921	67	1.169
.419	24	.407	.445	1.095	2.459	2.246	.914	66	1.152
.436	25	.423	.466	1.103	2.366	2.145	.906	65	1.134
.454	26	.438	.488	1.113	2.281	2.050	.899	64	1.117
.471	27	.454	.510	1.122	2.203	1.963	.891	63	1.100
.489	28	.470	.532	1.133	2.130	1.881	.883	62	1.082
.506	29	.485	.554	1.143	2.063	1.804	.875	61	1.065
.524	30	.500	.577	1.155	2.000	1.732	.866	60	1.047
Radian	Degree	Cosine	Cotangent	Cosecant	Secant	Tangent	Sine	Degree	Radian

$$\text{Sin } \theta = \frac{\text{Side opposite}}{\text{Hypotenuse}} \qquad \text{Cos } \theta = \frac{\text{Side adjacent}}{\text{Hypotenuse}} \qquad \text{Tan } \theta = \frac{\text{Side opposite}}{\text{Side adjacent}}$$

Radian	Degree	Sine	Tangent	Secant	Cosecant	Cotangent	Cosine	Degree	Radian
.541	31	.515	.601	1.167	1.942	1.664	.857	59	1.030
.559	32	.530	.625	1.179	1.887	1.600	.848	58	1.012
.576	33	.545	.649	1.192	1.836	1.540	.839	57	0.995
.593	34	.559	.675	1.206	1.788	1.483	.829	56	0.977
.611	35	.574	.700	1.221	1.743	1.428	.819	55	0.960
.628	36	.588	.727	1.236	1.701	1.376	.809	54	0.942
.646	37	.602	.754	1.252	1.662	1.327	.799	53	0.925
.663	38	.616	.781	1.269	1.624	1.280	.788	52	0.908
.681	39	.629	.810	1.287	1.589	1.235	.777	51	0.890
.698	40	.643	.839	1.305	1.556	1.192	.766	50	0.873
.716	41	.656	.869	1.325	1.524	1.150	.755	49	0.855
.733	42	.669	.900	1.346	1.494	1.111	.743	48	0.838
.751	43	.682	.933	1.367	1.466	1.072	.731	47	0.820
.768	44	.695	.966	1.390	1.440	1.036	.719	46	0.803
.785	45	.707	1.000	1.414	1.414	1.000	.707	45	0.785

Determination of square roots

Several motion formulas involve "squared" terms so it is often necessary to find the square root of a number in order to solve a motion problem.

Given a number N, then X is called the square root of N if $X^2 = N$.

Since $3^2 = 9$, then 3 is the square root of 9.
Since $0.4^2 = 0.16$, then 0.4 is the square root of 0.16.
Since $10^2 = 100$, then 10 is the square root of 100.

The symbol $\sqrt{}$ is used to indicate the positive square root of a number.

$\sqrt{16}\ =\ 4$ because 4^2 or $(4)(4)\ =\ 16$
$\sqrt{36}\ =\ 6$ because 6^2 or $(6)(6)\ =\ 36$
$\sqrt{0.04} = 0.2$ because 0.2^2 or $(0.2)(0.2)\ =\ 0.04$
$\sqrt{\frac{1}{4}} = \frac{1}{2}$ because $(\frac{1}{2})^2$ or $(\frac{1}{2})(\frac{1}{2})\ =\ \frac{1}{4}$

The square root of a number can be obtained through the use of slide rules, hand or desk calculators, or paper and pencil procedures as well as through the use of square root tables. Only the paper and pencil procedures and the square root tables, however, will be presented in this appendix.

PAPER AND PENCIL PROCEDURES

The two paper and pencil procedures are the simplified method of approximation and the arithmetical method.

Simplified method of approximating the square root of a number

1. By inspection, make a guess of the square root to two figures. For example, if the number is 685, inspection will show that it lies between $(20)^2 = 400$ and $(30)^2 = 900$ and that a reasonable guess might be 25.
2. Divide the original number by 25 to get 27.4.

3. The average of these two numbers, 26.2 is then the square root to three figures. If greater accuracy is desired, the averaged number may be assumed to be an original guess and the process repeated.

Example: Find $\sqrt{15}$ to the nearest hundredth; $\sqrt{15}$ is closer to 4 than to 3.

Thus 4 is the *first approximation* and $\sqrt{16}$ is approximately equal to ($\approx$) 4 to the nearest unit.

Divide the first approximation (4) into 15 and average the two factors:

$$\begin{array}{r} 3.7 \\ 4\overline{)15.0} \\ \underline{12} \\ 3\,0 \end{array}$$

Averaging 4 and 3.7 $\left(\begin{array}{r} 3.7 \\ +4 \\ \hline 7.7 \end{array} \text{ and } \dfrac{7.7}{2} \right) = 3.8$

Thus 3.8 is the *second approximation.*

Divide the second approximation into 15 and average the factors:

$$\begin{array}{r} 3.94 \\ 3.8\overline{)15.000} \\ \underline{11\,4} \\ 3\,60 \\ \underline{3\,42} \\ 180 \\ \underline{152} \\ 28 \end{array}$$

Averaging 3.8 and 3.94 $\left(\begin{array}{r} 3.8 \\ +3.94 \\ \hline 7.74 \end{array} \text{ and } \dfrac{7.74}{2} \right) = 3.87$

Thus 3.87 is the *third approximation.*

Divide the third approximation into 15 and average the factors:

$$\begin{array}{r} 3.875 \\ 3.87\,\overline{)15.0000} \\ 11\;61 \\ \hline 3\;390 \\ 3\;096 \\ \hline 2940 \\ 2709 \\ \hline 2310 \\ 1935 \\ \hline \end{array}$$

Averaging $\left(\dfrac{\begin{array}{r}3.875\\+3.87\\\hline 7.745\end{array}}{} \text{ and } \dfrac{7.745}{2}\right)= 3.872$

Thus 3.872 is the *fourth approximation.*

Note that each division is carried one decimal further than the preceding division. The fourth approximation (3.872) is closer to 3.87 than to 3.88, which means $\sqrt{15} \approx 3.87$ to the nearest hundredth.

Summary

1. Identify a *first approximation* by estimation or use of the table.

> For $\sqrt{29.4}$ the square root is closer to 5 than to 6; therefore 5 is the first approximation of 29.4.

2. Divide the first approximation into the original number to obtain two factors. Carry the division to one place further than the preceding approximation. Average the factors to obtain the next approximation.

$$\begin{array}{r} 5.8 \\ 5\,\overline{)29.4} \\ 25 \\ \hline 4\;4 \\ 4\;0 \\ \hline \end{array} \qquad \begin{array}{r} 5.8 \\ +5 \\ \hline 10.8 \end{array} \qquad \dfrac{10.8}{2} = 5.4$$

3. Repeat the process until you reach the result which is one decimal place past that desired.

$$\begin{array}{r} 5.44 \\ 5.4\,\overline{)29.400} \\ 27\;0 \\ \hline 2\;40 \\ 2\;16 \\ \hline 240 \\ 216 \\ \hline \end{array} \qquad \begin{array}{r} 5.44 \\ +5.4 \\ \hline 10.84 \end{array} \qquad \dfrac{10.84}{2} = 5.42$$

If the desired result is the square root to the nearest 0.1, then 5.42 is rounded to 5.4; i.e., $\sqrt{29.4} \approx 5.4$

Arithmetical method of obtaining the square root of a number

The arithmetical method is especially useful with numbers involving decimal values in which the first approximation is difficult to obtain for use of the simplified method (e.g., $\sqrt{0.8743}$).

Problem: What is the square root of 967.936?

1. Group the digits in pairs starting from the decimal point and going in both directions. Place little "tents" above each pair as shown.

$$\sqrt{\overset{\wedge\wedge}{9\,6}\overset{}{7}.\overset{\wedge\wedge}{9\,3}\overset{}{6}}$$

Move the decimal point vertically upward.
NOTE: The square of any single digit number will never be greater than two digits; for example, the square of 9 (the largest single digit) is 81. Conversely, the square root of a double digit number will always be a one digit number. This is the reason for separating the number into two-digit categories in each direction from the decimal point in Step 1. Thus, by inspection, any number of digits in a square root can be determined.

The tentlike pairings in our example tell us that the square root of this number will have two digits to the left of the decimal point and two digits to the right. Should we desire three digits to the right, we will have to add another tent as shown. This will give us an answer that has three digits to the right of the decimal to the nearest hundredth when we round off the last digit.

$$\sqrt{\overset{\wedge\wedge}{9\,6}\overset{}{7}.\overset{\wedge\wedge}{9\,3}\overset{\wedge}{6\,0}\overset{}{0\,0}}$$

2. Find the largest number whose square is less than or equal to ($\leq$) the digit or digits under the tent farthest to the left. In this case 3 is selected since $3^2 = 9$, which is equal to the digit under the farthest left tent.

$$\begin{array}{l}\boxed{3}\\ \sqrt{\overset{\wedge\wedge}{9\,6}\overset{}{7}.\overset{\wedge\wedge}{9\,3}\overset{\wedge}{6\,0}\overset{}{0\,0}}\\ \boxed{3}\quad 9\\ \qquad 0 \end{array}$$

Place this number in a box above the first tent and in another box to the left of and one space below the digit(s). Then, as in long division, multiply the numbers in the two boxes and place the product below the digit(s) under the farthest left tent to obtain a remainder (which in this case is 0).

NOTE: Finding a square root is a method of long division, and the following terms are important:

For example, in the equation $2\,\overline{)10}$ with quotient 5

> *2* is the *divisor*
> *10* is the *dividend*
> *5* is the *quotient*

3. Bring down the pair of digits under the next tent.

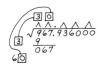

4. This step is the first in a series of *identical* repetitive steps carried out until the desired place is reached.

   Find each trial divisor from this point on by multiplying the quotient by 20 each time. Since our present quotient is 3, the new trial divisor is 20×3 (or 60). Place this trial divisor to the left of and one space below the new dividend. Put the last digit of this divisor in a small box and connect the box to another small box just above the second tent as shown.

5. Determine the number of times the new trial divisor will go into the new dividend.

   For example, 60 (new trial divisor) will go into 67 (new dividend) one time. Place this number in the two boxes as shown.

The number to be placed in the boxes is actually found by an estimation process; thus the product of the lower number (6 □) and the upper number (□) will be ≤ 67. (The digit 1 works nicely. Zero would be inappropriate because 6 ⓪ × ⓪ is nothing. The digit 2 would be too large because 6 ② × ② = 124.)

6. Now multiply the upper boxed number by the total (boxed and unboxed) lower numbers and place the product below the new dividend to obtain a remainder (which in this case is 6).

7. Repeat the process as follows:
   a. Bring down the digits under the next tent (in this case 93).
   b. Find a new trial divisor by multiplying the quotient by 20 (i.e., $31 \times 20 = 620$). Place this number to the left of and one space below the new dividend. Put the last digit of this divisor in a small box connected to another small box just above the third tent as shown.

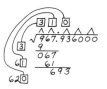

   c. By estimation, note that 620 multiplied by 1 must be the largest number ≤ 693. Therefore, place the number 1 in the two boxes.

   d. Now multiply the upper boxed number by the total lower numbers (i.e., 621) and place the product below the new dividend to obtain a remainder (in this case, 72).

The process is repeated until a quotient of 31.105 is obtained (which is rounded to 31.10).

Thus the square root of 967.936 is 31.10 to the nearest hundredth. A final check on our results can be found by squaring 31.10; it should be approximately 967.936.

$$
\begin{array}{r}
31.10 \\
\times 31.10 \\
\hline
3\,1100 \\
31\,10 \\
933\,0 \\
\hline
967.2100
\end{array}
$$

The check result, 967.21, is slightly smaller than the original number because of the rounding-off process. Note that 31.105 was rounded to 31.10 and not to 31.11. This is an arbitrary mathematical rule when the choice is halfway (i.e., 5). For example, if the digit before the 5 is an even digit (0, 2, 4, 6, or 8), the

5 is dropped and we "round down" as in our example. If the digit before the 5 is odd (1, 3, 5, 7, or 9), the number is "rounded up." For example, if the number had been 31.135 it would have been rounded up to 31.14.

Summary

Find $\sqrt{2718.2}$ to the nearest tenth.

1. Group all digits in pairs, beginning at the decimal point and moving in both directions;

$$\sqrt{\overset{\wedge\wedge}{27}\overset{.\,\wedge}{18}.\overset{\wedge}{20}\,\overset{\wedge}{00}}$$

this will reveal the number of digits in the answer. Move the decimal point vertically. Group one or more pairs to the right of the decimal point than the desired number of places.

2. Identify the largest number whose square is $\leq$ the one- or two-digit number beneath the

farthest tent to the left. Place this number in a box above that tent and also to the left of and below the digit(s) under the tent. Find the product of the numbers in the two boxes, place it below the tented digit(s), and subtract. Bring down the next pair of numbers to the right.

3. Multiply the quotient (5) by 20 and place the product to the left of and below the remainder

(i.e., 218). Position two single-digit squares, one above the next tent in the quotient and one around the last digit of the new trial divisor (i.e., 0). By estimation, select the single digit to place in both squares so the product of the single digit above and the total number below will be the remainder. The single digit 2 is selected because 102×2 (or 204) is the largest combination ≤ 218. (Note that $103 \times 3 = 309$.) Determine a new remainder and bring down the next pair of digits.

4. Repeat Step 3 until the result is carried one place further than desired. Round off the answer to the desired place.

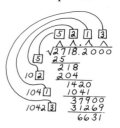

$\sqrt{2718.2} \approx 52.1$ to the nearest tenth

USE OF SQUARE ROOT TABLES

The square root tables included at the end of this appendix can be used to find the square root of any number from 1 to 10,000. The square roots of numbers greater than 10,000 can be obtained from the tables through experimentation. The rules for using these tables are as follows:

1. Finding the square root of a number *less than 1*
   a. It is usually best to use the arithmetical method described earlier when such a square root is desired.
2. Finding the square root of numbers *from 1 to 10.*
   a. Use the table for numbers from 100 to 1000.
   b. Place a decimal point between the boldface numbers in the left-hand column. This will be the number from which you wish to extract the root.
   c. Read the square root directly from the appropriate column, moving the decimal one place to the left (to reduce the answer by a factor of ten).

$$\sqrt{2.0} = 1.414 \qquad \sqrt{4.7} = 2.168$$
$$\sqrt{6.56} = 2.561 \qquad \sqrt{9.43} = 3.071$$

The table of proportional parts need not be used unless a result is desired to the ten-thousandths.

3. Finding the square root of a number *from 10 to 100*
   a. Use the table for numbers 1000 to 10,000.
   b. The square root is to be taken from the boldface numbers in the left-hand column.

c. Read the square root from the column directly across from the number, moving the decimal one place to the left (to reduce the answer by a factor of ten).

$$\sqrt{59} = 7.681 \qquad \sqrt{18} = 4.243$$
$$\sqrt{34.6} = 5.882 \qquad \sqrt{84.1} = 9.171$$

To find the square root of a number from 10 to 100 expressed to the nearest hundredth, use the *proportional parts* segment of the table.

$$\sqrt{72.62} = ?$$
$$\sqrt{72.62} = 8.521$$

Using the proportional parts segment, we see that

$$\sqrt{72.62} = 8.521 + 0.001* = 8.522$$

Likewise

$$\sqrt{19.47} = 4.405 + 0.008 = 4.413$$

and

$$\sqrt{48.76} = 6.979 + 0.004 = 6.983$$

4. Finding the square root of a number *from 100 to 10,000*
   a. Use either of the two tables, depending upon the size of the number.

───────

*Since the answer is in thousandths (8.521), the proportional parts will also be in thousandths.

b. Read the table directly: the first two digits in the left-hand boldface column, the third digit in the top row, and the fourth digit in the proportional parts segment.

$$\sqrt{1760} = 41.95$$
$$\sqrt{3500} = 59.16$$
$$\sqrt{212.5} = 14.56 + 0.02 = 14.58$$
$$\sqrt{7738} = 87.92 + 0.05 = 87.97$$

5. Finding the square root of a number *greater than 10,000*
   a. If the number of digits is *odd* (5, 7, 9, etc.), use the table from 100 to 1000. Locate the decimal point by experimentation and common sense.

   $$\sqrt{36,720} = 1916 + 1 = 1917$$

   By trial and error we locate the decimal point to get 191.7.

   $$\sqrt{92,500} = 304.1$$

   b. If the number of digits is *even* (6, 8, 10, etc.), use the table from 1000 to 10,000. Locate the decimal point by trial and error.

   $$\sqrt{360,000} = 600.0$$
   $$\sqrt{883,000} = 939.7$$

Square root tables

	100-1000										Proportional parts								
	0	**1**	**2**	**3**	**4**	**5**	**6**	**7**	**8**	**9**	**1**	**2**	**3**	**4**	**5**	**6**	**7**	**8**	**9**
10	10.00	10.05	10.10	10.15	10.20	10.25	10.30	10.34	10.39	10.44	0	1	1	2	2	3	3	4	4
11	10.49	10.54	10.58	10.63	10.68	10.72	10.77	10.82	10.86	10.91	0	1	1	2	2	3	3	4	4
12	10.95	11.00	11.05	11.09	11.14	11.18	11.22	11.27	11.31	11.36	0	1	1	2	2	3	3	4	4
13	11.40	11.45	11.49	11.53	11.58	11.62	11.66	11.70	11.75	11.79	0	1	1	2	2	3	3	3	4
14	11.83	11.87	11.92	11.96	12.00	12.04	12.08	12.12	12.17	12.21	0	1	1	2	2	2	3	3	4
15	12.25	12.29	12.33	12.37	12.41	12.45	12.49	12.53	12.57	12.61	0	1	1	2	2	2	3	3	4
16	12.65	12.69	12.73	12.77	12.81	12.85	12.88	12.92	12.96	13.00	0	1	1	2	2	2	3	3	4
17	13.04	13.08	13.11	13.15	13.19	13.23	13.27	13.30	13.34	13.38	0	1	1	2	2	2	3	3	3
18	13.42	13.45	13.49	13.53	13.56	13.60	13.64	13.67	13.71	13.75	0	1	1	1	2	2	3	3	3
19	13.78	13.82	13.86	13.89	13.93	13.96	14.00	14.04	14.07	14.11	0	1	1	1	2	2	3	3	3
20	14.14	14.18	14.21	14.25	14.28	14.32	14.35	14.39	14.42	14.46	0	1	1	1	2	2	2	3	3
21	14.49	14.53	14.56	14.59	14.63	14.66	14.70	14.73	14.76	14.80	0	1	1	1	2	2	2	3	3
22	14.83	14.87	14.90	14.93	14.97	15.00	15.03	15.07	15.10	15.13	0	1	1	1	2	2	2	3	3
23	15.17	15.20	15.23	15.26	15.30	15.33	15.36	15.39	15.43	15.46	0	1	1	1	2	2	2	3	3
24	15.49	15.52	15.56	15.59	15.62	15.65	15.68	15.72	15.75	15.78	0	1	1	1	2	2	2	3	3
25	15.81	15.84	15.87	15.91	15.94	15.97	16.00	16.03	16.06	16.09	0	1	1	1	2	2	2	3	3
26	16.12	16.16	16.19	16.22	16.25	16.28	16.31	16.34	16.37	16.40	0	1	1	1	2	2	2	2	3
27	16.43	16.46	16.49	16.52	16.55	16.58	16.61	16.64	16.67	16.70	0	1	1	1	2	2	2	2	3
28	16.73	16.76	16.79	16.82	16.85	16.88	16.91	16.94	16.97	17.00	0	1	1	1	1	2	2	2	3
29	17.03	17.06	17.09	17.12	17.15	17.18	17.20	17.23	17.26	17.29	0	1	1	1	1	2	2	2	3
30	17.32	17.35	17.38	17.41	17.44	17.46	17.49	17.52	17.55	17.58	0	1	1	1	1	2	2	2	3
31	17.61	17.64	17.66	17.69	17.72	17.75	17.78	17.80	17.83	17.86	0	1	1	1	1	2	2	2	3
32	17.89	17.92	17.94	17.97	18.00	18.03	18.06	18.08	18.11	18.14	0	1	1	1	1	2	2	2	2
33	18.17	18.19	18.22	18.25	18.28	18.30	18.33	18.36	18.38	18.41	0	1	1	1	1	2	2	2	2
34	18.44	18.47	18.49	18.52	18.55	18.57	18.60	18.63	18.65	18.68	0	1	1	1	1	2	2	2	2
35	18.71	18.73	18.76	18.79	18.81	18.84	18.87	18.89	18.92	18.95	0	1	1	1	1	2	2	2	2
36	18.97	19.00	19.03	19.05	19.08	19.10	19.13	19.16	19.18	19.21	0	1	1	1	1	2	2	2	2
37	19.24	19.26	19.29	19.31	19.34	19.36	19.39	19.42	19.44	19.47	0	1	1	1	1	2	2	2	2
38	19.49	19.52	19.54	19.57	19.60	19.62	19.65	19.67	19.70	19.72	0	1	1	1	1	2	2	2	2
39	19.75	19.77	19.80	19.82	19.85	19.87	19.90	19.92	19.95	19.97	0	1	1	1	1	2	2	2	2
40	20.00	20.02	20.05	20.07	20.10	20.12	20.15	20.17	20.20	20.22	0	0	1	1	1	1	2	2	2
41	20.25	20.27	20.30	20.32	20.35	20.37	20.40	20.42	20.45	20.47	0	0	1	1	1	1	2	2	2
42	20.49	20.52	20.54	20.57	20.59	20.62	20.64	20.66	20.69	20.71	0	0	1	1	1	1	2	2	2
43	20.74	20.76	20.78	20.81	20.83	20.86	20.88	20.90	20.93	20.95	0	0	1	1	1	1	2	2	2
44	20.98	21.00	21.02	21.05	21.07	21.10	21.12	21.14	21.17	21.19	0	0	1	1	1	1	2	2	2
45	21.21	21.24	21.26	21.28	21.31	21.33	21.35	21.38	21.40	21.42	0	0	1	1	1	1	2	2	2
46	21.45	21.47	21.49	21.52	21.54	21.56	21.59	21.61	21.63	21.66	0	0	1	1	1	1	2	2	2
47	21.68	21.70	21.73	21.75	21.77	21.79	21.82	21.84	21.86	21.89	0	0	1	1	1	1	2	2	2
48	21.91	21.93	21.95	21.98	22.00	22.02	22.05	22.07	22.09	22.11	0	0	1	1	1	1	2	2	2
49	22.14	22.16	22.18	22.20	22.23	22.25	22.27	22.29	22.32	22.34	0	0	1	1	1	1	2	2	2
50	22.36	22.38	22.41	22.43	22.45	22.47	22.49	22.52	22.54	22.56	0	0	1	1	1	1	2	2	2
51	22.58	22.61	22.63	22.65	22.67	22.69	22.72	22.74	22.76	22.78	0	0	1	1	1	1	2	2	2
52	22.80	22.83	22.85	22.87	22.89	22.91	22.93	22.96	22.98	23.00	0	0	1	1	1	1	2	2	2
53	23.02	23.04	23.07	23.09	23.11	23.13	23.15	23.17	23.19	23.22	0	0	1	1	1	1	2	2	2
54	23.24	23.26	23.28	23.30	23.32	23.35	23.37	23.39	23.41	23.43	0	0	1	1	1	1	1	2	2
	0	**1**	**2**	**3**	**4**	**5**	**6**	**7**	**8**	**9**	**1**	**2**	**3**	**4**	**5**	**6**	**7**	**8**	**9**

Square root tables

	100-1000										Proportional parts								
	0	**1**	**2**	**3**	**4**	**5**	**6**	**7**	**8**	**9**	**1**	**2**	**3**	**4**	**5**	**6**	**7**	**8**	**9**
55	23.45	23.47	23.49	23.52	23.54	23.56	23.58	23.60	23.62	23.64	0	0	1	1	1	1	1	2	2
56	23.66	23.69	23.71	23.73	23.75	23.77	23.79	23.81	23.83	23.85	0	0	1	1	1	1	1	2	2
57	23.87	23.90	23.92	23.94	23.96	23.98	24.00	24.02	24.04	24.06	0	0	1	1	1	1	1	2	2
58	24.08	24.10	24.12	24.15	24.17	24.19	24.21	24.23	24.25	24.27	0	0	1	1	1	1	1	2	2
59	24.29	24.31	24.33	24.35	24.37	24.39	24.41	24.43	24.45	24.47	0	0	1	1	1	1	1	2	2
60	24.49	24.52	24.54	24.56	24.58	24.60	24.62	24.64	24.66	24.68	0	0	1	1	1	1	1	2	2
61	24.70	24.72	24.74	24.76	24.78	24.80	24.82	24.84	24.86	24.88	0	0	1	1	1	1	1	2	2
62	24.90	24.92	24.94	24.96	24.98	25.00	25.02	25.04	25.06	25.08	0	0	1	1	1	1	1	2	2
63	25.10	25.12	25.14	25.16	25.18	25.20	25.22	25.24	25.26	25.28	0	0	1	1	1	1	1	2	2
64	25.30	25.32	25.34	25.36	25.38	25.40	25.42	25.44	25.46	25.48	0	0	1	1	1	1	1	2	2
65	25.50	25.51	25.53	25.55	25.57	25.59	25.61	25.63	25.65	25.67	0	0	1	1	1	1	1	2	2
66	25.69	25.71	25.73	25.75	25.77	25.79	25.81	25.83	25.85	25.87	0	0	1	1	1	1	1	2	2
67	25.88	25.90	25.92	25.94	25.96	25.98	26.00	26.02	26.04	26.06	0	0	1	1	1	1	1	2	2
68	26.08	26.10	26.12	26.13	26.15	26.17	26.19	26.21	26.23	26.25	0	0	1	1	1	1	1	2	2
69	26.27	26.29	26.31	26.32	26.34	26.36	26.38	26.40	26.42	26.44	0	0	1	1	1	1	1	2	2
70	26.46	26.48	26.50	26.51	26.53	26.55	26.57	26.59	26.61	26.63	0	0	1	1	1	1	1	2	2
71	26.65	26.66	26.68	26.70	26.72	26.74	26.76	26.78	26.80	26.81	0	0	1	1	1	1	1	1	2
72	26.83	26.85	26.87	26.89	26.91	26.93	26.94	26.96	26.98	27.00	0	0	1	1	1	1	1	1	2
73	27.02	27.04	27.06	27.07	27.09	27.11	27.13	27.15	27.17	27.18	0	0	1	1	1	1	1	1	2
74	27.20	27.22	27.24	27.26	27.28	27.29	27.31	27.33	27.35	27.37	0	0	1	1	1	1	1	1	2
75	27.39	27.40	27.42	27.44	27.46	27.48	27.50	27.51	27.53	27.55	0	0	1	1	1	1	1	1	2
76	27.57	27.59	27.60	27.62	27.64	27.66	27.68	27.69	27.71	27.73	0	0	1	1	1	1	1	1	2
77	27.75	27.77	27.78	27.80	27.82	27.84	27.86	27.87	27.89	27.91	0	0	1	1	1	1	1	1	2
78	27.93	27.95	27.96	27.98	28.00	28.02	28.04	28.05	28.07	28.09	0	0	1	1	1	1	1	1	2
79	28.11	28.12	28.14	28.16	28.18	28.20	28.21	28.23	28.25	28.27	0	0	1	1	1	1	1	1	2
80	28.28	28.30	28.32	28.34	28.35	28.37	28.39	28.41	28.43	28.44	0	0	1	1	1	1	1	1	2
81	28.46	28.48	28.50	28.51	28.53	28.55	28.57	28.58	28.60	28.62	0	0	1	1	1	1	1	1	2
82	28.64	28.65	28.67	28.69	28.71	28.72	28.74	28.76	28.77	28.79	0	0	1	1	1	1	1	1	2
83	28.81	28.83	28.84	28.86	28.88	28.90	28.91	28.93	28.95	28.97	0	0	1	1	1	1	1	1	2
84	28.98	29.00	29.02	29.03	29.05	29.07	29.09	29.10	29.12	29.14	0	0	1	1	1	1	1	1	2
85	29.15	29.17	29.19	29.21	29.22	29.24	29.26	29.27	29.29	29.31	0	0	1	1	1	1	1	1	2
86	29.33	29.34	29.36	29.38	29.39	29.41	29.43	29.44	29.46	29.48	0	0	1	1	1	1	1	1	2
87	29.50	29.51	29.53	29.55	29.56	29.58	29.60	29.61	29.63	29.65	0	0	1	1	1	1	1	1	2
88	29.66	29.68	29.70	29.72	29.73	29.75	29.77	29.78	29.80	29.82	0	0	1	1	1	1	1	1	2
89	29.83	29.85	29.87	29.88	29.90	29.92	29.93	29.95	29.97	29.98	0	0	1	1	1	1	1	1	2
90	30.00	30.02	30.03	30.05	30.07	30.08	30.10	30.12	30.13	30.15	0	0	0	1	1	1	1	1	1
91	30.17	30.18	30.20	30.22	30.23	30.25	30.27	30.28	30.30	30.32	0	0	0	1	1	1	1	1	1
92	30.33	30.35	30.36	30.38	30.40	30.41	30.43	30.45	30.46	30.48	0	0	0	1	1	1	1	1	1
93	30.50	30.51	30.53	30.55	30.56	30.58	30.59	30.61	30.63	30.64	0	0	0	1	1	1	1	1	1
94	30.66	30.68	30.69	30.71	30.72	30.74	30.76	30.77	30.79	30.81	0	0	0	1	1	1	1	1	1
95	30.82	30.84	30.85	30.87	30.89	30.90	30.92	30.94	30.95	30.97	0	0	0	1	1	1	1	1	1
96	30.98	31.00	31.02	31.03	31.05	31.06	31.08	31.10	31.11	31.13	0	0	0	1	1	1	1	1	1
97	31.14	31.16	31.18	31.19	31.21	31.22	31.24	31.26	31.27	31.29	0	0	0	1	1	1	1	1	1
98	31.30	31.32	31.34	31.35	31.37	31.38	31.40	31.42	31.43	31.45	0	0	0	1	1	1	1	1	1
99	31.46	31.48	31.50	31.51	31.53	31.54	31.56	31.58	31.59	31.61	0	0	0	1	1	1	1	1	1
	0	**1**	**2**	**3**	**4**	**5**	**6**	**7**	**8**	**9**	**1**	**2**	**3**	**4**	**5**	**6**	**7**	**8**	**9**

Square root tables

| | 1000-10,000 | | | | | | | | | | Proportional parts | | | | | | | | |
	0	1	2	3	4	5	6	7	8	9	1	2	3	4	5	6	7	8	9
10	31.62	31.78	31.94	32.09	32.25	32.40	32.56	32.71	32.86	33.02	2	3	5	6	8	9	11	12	14
11	33.17	33.32	33.47	33.62	33.76	33.91	34.06	34.21	34.35	34.50	1	3	4	6	7	9	10	12	13
12	34.64	34.79	34.93	35.07	35.21	35.36	35.50	35.64	35.78	35.92	1	3	4	6	7	8	10	11	13
13	36.06	36.19	36.33	36.47	36.61	36.74	36.88	37.01	37.15	37.28	1	3	4	5	7	8	10	11	12
14	37.42	37.55	37.68	37.82	37.95	38.08	38.21	38.34	38.47	38.60	1	3	4	5	7	8	9	11	12
15	38.73	38.86	38.99	39.12	39.24	39.37	39.50	39.62	39.75	39.87	1	3	4	5	6	8	9	10	11
16	40.00	40.12	40.25	40.37	40.50	40.62	40.74	40.87	40.99	41.11	1	2	4	5	6	7	9	10	11
17	41.23	41.35	41.47	41.59	41.71	41.83	41.95	42.07	42.19	42.31	1	2	4	5	6	7	8	10	11
18	42.43	42.54	42.66	42.78	42.90	43.01	43.13	43.24	43.36	43.47	1	2	3	5	6	7	8	9	10
19	43.59	43.70	43.82	43.93	44.05	44.16	44.27	44.38	44.50	44.61	1	2	3	5	6	7	8	9	10
20	44.72	44.83	44.94	45.06	45.17	45.28	45.39	45.50	45.61	45.72	1	2	3	4	6	7	8	9	10
21	45.83	45.93	46.04	46.15	46.26	46.37	46.48	46.58	46.69	46.80	1	2	3	4	5	6	8	9	10
22	46.90	47.01	47.12	47.22	47.33	47.43	47.54	47.64	47.75	47.85	1	2	3	4	5	6	7	8	9
23	47.96	48.06	48.17	48.27	48.37	48.48	48.58	48.68	48.79	48.89	1	2	3	4	5	6	7	8	9
24	48.99	49.09	49.19	49.30	49.40	49.50	49.60	49.70	49.80	49.90	1	2	3	4	5	6	7	8	9
25	50.00	50.10	50.20	50.30	50.40	50.50	50.60	50.70	50.79	50.89	1	2	3	4	5	6	7	8	9
26	50.99	51.09	51.10	51.28	51.38	51.48	51.58	51.67	51.77	51.87	1	2	3	4	5	6	7	8	9
27	51.96	52.06	52.15	52.25	52.35	52.44	52.54	52.63	52.73	52.82	1	2	3	4	5	6	7	8	9
28	52.92	53.01	53.10	53.20	53.29	53.39	53.48	53.57	53.67	53.76	1	2	3	4	5	6	7	7	8
29	53.85	53.94	54.04	54.13	54.22	54.31	54.41	54.50	54.59	54.68	1	2	3	4	5	6	6	7	8
30	54.77	54.86	54.95	55.05	55.14	55.23	55.32	55.41	55.50	55.59	1	2	3	4	5	5	6	7	8
31	55.68	55.77	55.86	55.95	56.04	56.12	56.21	56.30	56.39	56.48	1	2	3	4	4	5	6	7	8
32	56.57	56.66	56.75	56.83	56.92	57.01	57.10	57.18	57.27	57.36	1	2	3	4	4	5	6	7	8
33	57.45	57.53	57.62	57.71	57.79	57.88	57.97	58.05	58.14	58.22	1	2	3	3	4	5	6	7	8
34	58.31	58.40	58.48	58.57	58.65	58.74	58.82	58.91	58.99	59.08	1	2	3	3	4	5	6	7	8
35	59.16	59.25	59.33	59.41	59.50	59.58	59.67	59.75	59.83	59.92	1	2	3	3	4	5	6	7	8
36	60.00	60.08	60.17	60.25	60.33	60.42	60.50	60.58	60.66	60.75	1	2	2	3	4	5	6	7	7
37	60.83	60.91	60.99	61.07	61.16	61.24	61.32	61.40	61.48	61.56	1	2	2	3	4	5	6	7	7
38	61.64	61.73	61.81	61.89	61.97	62.05	62.13	62.21	62.29	62.37	1	2	2	3	4	5	6	6	7
39	62.45	62.53	62.61	62.69	62.77	62.85	62.93	63.01	63.09	63.17	1	2	2	3	4	5	6	6	7
40	63.25	63.32	63.40	63.48	63.56	63.64	63.72	63.80	63.87	63.95	1	2	2	3	4	5	5	6	7
41	64.03	64.11	64.19	64.27	64.34	64.42	64.50	64.58	64.65	64.73	1	2	2	3	4	5	5	6	7
42	64.81	64.88	64.96	65.04	65.12	65.19	65.27	65.35	65.42	65.50	1	2	2	3	4	5	5	6	7
43	65.57	65.65	65.73	65.80	65.88	65.95	66.03	66.11	66.18	66.26	1	2	2	3	4	5	5	6	7
44	66.33	66.41	66.48	66.56	66.63	66.71	66.78	66.86	66.93	67.01	1	1	2	3	4	4	5	6	7
45	67.08	67.16	67.23	67.31	67.38	67.45	67.53	67.60	67.68	67.75	1	1	2	3	4	4	5	6	7
46	67.82	67.90	67.97	68.04	68.12	68.19	68.26	68.34	68.41	68.48	1	1	2	3	4	4	5	6	7
47	68.56	68.63	68.70	68.77	68.85	68.92	68.99	69.07	69.14	69.21	1	1	2	3	4	4	5	6	7
48	69.28	69.35	69.43	69.50	69.57	69.64	69.71	69.79	69.86	69.93	1	1	2	3	4	4	5	6	6
49	70.00	70.07	70.14	70.21	70.29	70.36	70.43	70.50	70.57	70.64	1	1	2	3	4	4	5	6	6
50	70.71	70.78	70.85	70.92	70.99	71.06	71.13	71.20	71.27	71.34	1	1	2	3	4	4	5	6	6
51	71.41	71.48	71.55	71.62	71.69	71.76	71.83	71.90	71.97	72.04	1	1	2	3	3	4	5	6	6
52	72.11	72.18	72.25	72.32	72.39	72.46	72.53	72.59	72.66	72.73	1	1	2	3	3	4	5	6	6
53	72.80	72.87	72.94	73.01	73.08	73.14	73.21	73.28	73.35	73.42	1	1	2	3	3	4	5	5	6
54	73.48	73.55	73.62	73.69	73.76	73.82	73.89	73.96	74.03	74.09	1	1	2	3	3	4	5	5	6
	0	1	2	3	4	5	6	7	8	9	1	2	3	4	5	6	7	8	9

Square root tables

	1000-10,000										Proportional parts								
	0	**1**	**2**	**3**	**4**	**5**	**6**	**7**	**8**	**9**	**1**	**2**	**3**	**4**	**5**	**6**	**7**	**8**	**9**
55	74.16	74.23	74.30	74.36	74.43	74.50	74.57	74.63	74.70	74.77	1	1	2	3	3	4	5	5	6
56	74.83	74.90	74.97	75.03	75.10	75.17	75.23	75.30	75.37	75.43	1	1	2	3	3	4	5	5	6
57	75.50	75.56	75.63	75.70	75.76	75.83	75.89	75.96	76.03	76.09	1	1	2	3	3	4	5	5	6
58	76.16	76.22	76.29	76.35	76.42	76.49	76.55	76.62	76.68	76.75	1	1	2	3	3	4	5	5	6
59	76.81	76.88	76.94	77.01	77.07	77.14	77.20	77.27	77.33	77.40	1	1	2	3	3	4	5	5	6
60	77.46	77.52	77.59	77.65	77.72	77.78	77.85	77.91	77.97	78.04	1	1	2	3	3	4	4	5	6
61	78.10	78.17	78.23	78.29	78.36	78.42	78.49	78.55	78.61	78.68	1	1	2	3	3	4	4	5	6
62	78.74	78.80	78.87	78.93	78.99	79.06	79.12	79.18	79.25	79.31	1	1	2	3	3	4	4	5	6
63	79.37	79.44	79.50	79.56	79.62	79.69	79.75	79.81	79.87	79.94	1	1	2	3	3	4	4	5	6
64	80.00	80.06	80.12	80.19	80.25	80.31	80.37	80.44	80.50	80.56	1	1	2	2	3	4	4	5	6
65	80.62	80.68	80.75	80.81	80.87	80.93	80.99	81.06	81.12	81.18	1	1	2	2	3	4	4	5	6
66	81.24	81.30	81.36	81.42	81.49	81.55	81.61	81.67	81.73	81.79	1	1	2	2	3	4	4	5	6
67	81.85	81.91	81.98	82.04	82.10	82.16	82.22	82.28	82.34	82.40	1	1	2	2	3	4	4	5	5
68	82.46	82.52	82.58	82.64	82.70	82.76	82.83	82.89	82.95	83.01	1	1	2	2	3	4	4	5	5
69	83.07	83.13	83.19	83.25	83.31	83.37	83.43	83.49	83.55	83.61	1	1	2	2	3	4	4	5	5
70	83.67	83.73	83.79	83.85	83.90	83.96	84.02	84.08	84.14	84.20	1	1	2	2	3	4	4	5	5
71	84.26	84.32	84.38	84.44	84.50	84.56	84.62	84.68	84.73	84.79	1	1	2	2	3	4	4	5	5
72	84.85	84.91	84.97	85.03	85.09	85.15	85.21	85.26	85.32	85.38	1	1	2	2	3	4	4	5	5
73	85.44	85.50	85.56	85.62	85.67	85.73	85.79	85.85	85.91	85.97	1	1	2	2	3	3	4	5	5
74	86.02	86.08	86.14	86.20	86.26	86.31	86.37	86.43	86.49	86.54	1	1	2	2	3	3	4	5	5
75	86.60	86.66	86.72	86.78	86.83	86.89	86.95	87.01	87.06	87.12	1	1	2	2	3	3	4	5	5
76	87.18	87.24	87.29	87.35	87.41	87.46	87.52	87.58	87.64	87.69	1	1	2	2	3	3	4	5	5
77	87.75	87.81	87.86	87.92	87.98	88.03	88.09	88.15	88.20	88.26	1	1	2	2	3	3	4	5	5
78	88.32	88.37	88.43	88.49	88.54	88.60	88.66	88.71	88.77	88.83	1	1	2	2	3	3	4	5	5
79	88.88	88.94	88.99	89.05	89.11	89.16	89.22	89.27	89.33	89.39	1	1	2	2	3	3	4	4	5
80	89.44	89.50	89.55	89.61	89.67	89.72	89.78	89.83	89.89	89.94	1	1	2	2	3	3	4	4	5
81	90.00	90.06	90.11	90.17	90.22	90.28	90.33	90.39	90.44	90.50	1	1	2	2	3	3	4	4	5
82	90.55	90.61	90.66	90.72	90.77	90.83	90.88	90.94	90.99	91.05	1	1	2	2	3	3	4	4	5
83	91.10	91.16	91.21	91.27	91.32	91.38	91.43	91.49	91.54	91.60	1	1	2	2	3	3	4	4	5
84	91.65	91.71	91.76	91.82	91.87	91.92	91.98	92.03	92.09	92.14	1	1	2	2	3	3	4	4	5
85	92.20	92.25	92.30	92.36	92.41	92.47	92.52	92.57	92.63	92.68	1	1	2	2	3	3	4	4	5
86	92.74	92.79	92.84	92.90	92.95	93.01	93.06	93.11	93.17	93.22	1	1	2	2	3	3	4	4	5
87	93.27	93.33	93.38	93.43	93.49	93.54	93.59	93.65	93.70	93.75	1	1	2	2	3	3	4	4	5
88	93.81	93.86	93.91	93.97	94.02	94.07	94.13	94.18	94.23	94.29	1	1	2	2	3	3	4	4	5
89	94.34	94.39	94.45	94.50	94.55	94.60	94.66	94.71	94.76	94.82	1	1	2	2	3	3	4	4	5
90	94.87	94.92	94.97	95.03	95.08	95.13	95.18	95.24	95.29	95.34	1	1	2	2	3	3	4	4	5
91	95.39	95.45	95.50	95.55	95.60	95.66	95.71	95.76	95.81	95.86	1	1	2	2	3	3	4	4	5
92	95.92	95.97	96.02	96.07	96.12	96.18	96.23	96.28	96.33	96.38	1	1	2	2	3	3	4	4	5
93	96.44	96.49	96.54	96.59	96.64	96.70	96.75	96.80	96.85	96.90	1	1	2	2	3	3	4	4	5
94	96.95	97.01	97.06	97.11	97.16	97.21	97.26	97.31	97.37	97.42	1	1	2	2	3	3	4	4	5
95	97.47	97.52	97.57	97.62	97.67	97.72	97.78	97.83	97.88	97.93	1	1	2	2	3	3	4	4	5
96	97.98	98.03	98.08	98.13	98.18	98.23	98.29	98.34	98.39	98.44	1	1	2	2	3	3	4	4	5
97	98.49	98.54	98.59	98.64	98.69	98.74	98.79	98.84	98.89	98.94	1	1	2	2	3	3	4	4	5
98	98.99	99.05	99.10	99.15	99.20	99.25	99.30	99.35	99.40	99.45	1	1	2	2	3	3	4	4	5
99	99.50	99.55	99.60	99.65	99.70	99.75	99.80	99.85	99.90	99.95	1	1	2	2	3	3	4	4	5
	0	**1**	**2**	**3**	**4**	**5**	**6**	**7**	**8**	**9**	**1**	**2**	**3**	**4**	**5**	**6**	**7**	**8**	**9**

Mathematics review

The concepts covered in this appendix are formulas, the law of signs, the balance principle of formulas, formula transformations, proportionality, and changing units of measure.

CONCEPT OF FORMULAS

A formula is a shorthand statement of a mathematical law or relationship. Letters are used in formulas to represent numbers and to specify well-known facts. The letters have different values depending upon the problem under consideration, and their meaning must be clearly understood. They are frequently spoken of as *general numbers*. The study of mathematics is essentially concerned with the operations of, establishing laws for, and understanding the meaning of such general numbers. Examples of the use of general numbers in formulas are

$$\text{Newton's second law: } a = F/m$$
$$\text{Torque: } T = F \sin \theta \, r$$

Furthermore, we must be concerned with combinations of symbols. If each symbol in a combination represents a number, we shall call the combination an *expression*. Examples of expressions are

$$d/t \quad mV_f \quad - \quad mV_o \quad V_o t \quad + \quad \tfrac{1}{2} g t^2$$

If there are several parts connected by plus and minus signs, each part is called a *term*. Thus in the second of the above examples, mv_f is a term and mv_o is a term. Note that the sign is associated with the term, but if the sign is + it is usually not expressed unless for emphasis.

In discussing a term of an expression, we have additional terminology. For example, each of two or more quantities multiplied together is a factor of the product. Furthermore, any factor of

an expression is a coefficient of the remaining part. For example:

$$\tfrac{1}{2} 32 t^2$$
$$\tfrac{1}{2}, 32, \text{ and } t^2 \text{ are } \textit{factors}$$
$$\tfrac{1}{2} \text{ and } 32 \text{ are } \textit{numerical coefficients}$$
$$\tfrac{1}{2} 32 \text{ is the coefficient of } t^2$$
$$\tfrac{1}{2} t^2 \text{ is the coefficient of } 32$$

Names are also given to various expressions.

monomial An expression consisting of one term
polynomial An expression consisting of more than one term
binomial An expression consisting of two terms
trinomial An expression consisting of three terms

LAW OF SIGNS

There are four laws of signs.
1. To add two numbers having like signs, add their absolute values and prefix the common sign.

$$2 + 5 = 7$$
$$-2 + (-5) = -7$$

2. To add two numbers having unlike signs, take the difference of their absolute values and prefix to it the sign of the number having the larger absolute value.

$$8 + (-3) = 5$$
$$3 + (-8) = -5$$

3. To subtract one number from another, change the sign of the number to be subtracted and proceed as in addition.

$$9 - (+2) = 9 + (-2) = 7$$
$$9 - (-2) = 9 + (+2) = 11$$

4. The multiplication or division of two numbers with like signs yields a positive, while that of unlike signs yields a negative.

Positive by a Positive gives a Positive (+)
Positive by a Negative gives a Negative (−)
Negative by a Positive gives a Negative (−)
Negative by a Negative gives a Positive (+)

$$5 \times 2 = 10 \qquad 10 \div (-2) = -5$$
$$-2 \times 4 = -8 \qquad (-8) \div (-4) = 2$$

BALANCE PRINCIPLE

All formulas are in perfect balance; i.e., the left member is equal to the right member in *every* formula. Consider a formula as if it were a balance scale. In the formula $x + 2 = 6$, you might think of the equation as if each member were placed on the pan of a balance scale as shown in Fig. 1. Because of the balance, any formula can be re-arranged as long as the balance is not upset. This can be ensured by always doing the same thing (multiplying, dividing, adding, or subtracting) to both members of the formula so the scales never get out of balance.

To solve the formula for x in $x + 2 = 6$, you must eliminate the *+2* by subtracting 2 from this member (Fig. 2, *A*). The balance can be maintained as long as you also take 2 from the right member *(B)*. This produces the result shown in Fig. 2, *C*. Therefore $x = 4$.

FORMULA TRANSFORMATIONS

A formula transformation is the rearrangement of a formula carried out by applying the balance principle. Changing (or rearranging) a formula (sometimes called an equation) can be done mathematically by a process which retains the same idea relationship and values but transforms the equation into a handier form.

Formula transformations can be carried out by using the following steps:

1. Select the appropriate algebraic term and operation to isolate the desired quantity; i.e., select the unwanted quantity and the inverse of the present operation.

   Solve for d in the formula $V = d/t$. Here the unwanted quantity is t, which must be eliminated to isolate d. Note that t is related to d by the operation of division. Therefore t can be eliminated by selecting the inverse operation (i.e., multiplication).

2. Perform the inverse operation on both members of the formula.

   Using $V = d/t$, multiply both sides by t ($V \times t = dt/t$)

3. Simplify to obtain the solution.

   $$Vt = d$$
   $$d = tV$$

4. Repeat if necessary.

PROPORTIONALITY

A proportionality is the relationship that exists between two variables in a formula. The two types of proportionality are termed *direct* and *inverse*. Two variables are directly proportional if an exact increase in one causes an exact increase in the other (or vice-versa). Thus in $V = d/t$, if d is tripled (with t constant), V will have to also triple.

Direct variation or proportion has its counter-part in the idea of inverse proportion. In other

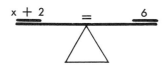

Fig. 1. (From Krause, J. V., and Barham, J. N.: The mechanical foundations of human motion: a programmed text, St. Louis, 1975, The C. V. Mosby Co.)

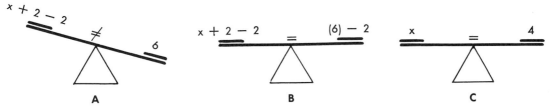

Fig. 2. (From Krause, J. V. and Barham, J. N.: The mechanical foundations of human motion: a programmed text, St. Louis, 1975, The C. V. Mosby Co.)

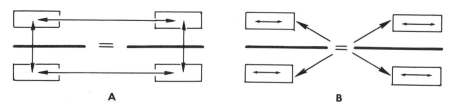

Fig. 3. (From Krause, J. V., and Barham, J. N.: The mechanical foundations of human motion: a programmed text, St. Louis, 1975, The C. V. Mosby Co.)

words, as one term increases the other one decreases (or vice-versa). Consider again the formula $V = d/t$. As t is increased five times (a change from 1 to 5), V is decreased by becoming five times smaller (a change from 5 to 1).

There is a shortcut method of determining the proportional relationship between two variables in a formula (Fig. 3).

1. If two variables are directly across or in vertical positions in a formula, they are in *direct* proportion. The arrows in the "mock" equation (Fig. 3, *A*) illustrate the potentially directly proportional pairs.

2. If two variables are diagonally across or both multiples are in the same position related by multiplication, they are in an *inverse* proportion. The arrows again indicate the potentially inversely proportional pairs (Fig. 3, *B*).

CHANGING UNITS OF MEASURE

The final topic in this review is the changing of units of measure, a common task in mechanical kinesiology. Most countries use the metric system of measurement; the United States temporarily is using the English system. Thus it is important to be able to convert metric units into English units and vice versa, as in changing yards to meters. Changing units must also be done many times within the same system as in changing yards to feet. These changes can be brought about by the following steps:

1. Obtain the conversion relationship between the given desired measurement units.

$$12 \text{ in} = 1 \text{ ft}$$

2. Express the relationship as a value of 1 so multiplication of the given measure by that quantity will eliminate the undesired units.

$$\frac{12 \text{ in}}{1 \text{ ft}} \times \frac{1 \text{ ft}}{12 \text{ in}} = 1.00$$

3. Select the relationship that when multiplied by the given quantity will result in the elimination of the undesired units.
Change 300 ft into yards.

$$300 \text{ ft} \times \frac{1 \text{ yd}}{3 \text{ ft}}$$

4. Simplify to obtain the solution.

$$300 \text{ ft} \times \frac{1 \text{ yd}}{3 \text{ ft}} = \frac{300 \text{ yd}}{3} = 100 \text{ yd}$$

Symbols used in text

Capital letters

A Area
   Amplitude
D Density
   Drag
E Efficiency
F Force

G Gravitational constant
H Height dropped from
I Moment of inertia
K Force constant
L Angular momentum
   Lift
M Linear momentum
N Normal force
O Origin
   Equilibrium point
P Surface force
   Pressure

R Range
   Resultant
   Resistance
   Secondary radius

S Surface area
T Torque
   Thrust
   Period
   Total time of flight
V Velocity
   Volumn

W Work

Points are described with
capital letters (A,B,C, etc)

lower case letters

a linear acceleration

d linear displacement

e index of lifting efficiency
f force arm of lever
   frequency
   frictional force
g acceleration of gravity
h height of projection or rebound

k radius of gyration
l length

m mass
n number

q radius of oscillation and percussion
r radius
   resistance arm
   restoring (as in restoring force F_r)
   radial (as in radial force F_r)
   coefficient of elasticity

t time
   tangential (as in tangential force F_t)

w weight
x,y,z axes of motion
Planes, sides, and lines are described
with lower case letters (a,b,c, etc)

Greek alphabet

A	α	*Alpha*		H	η	*Eta*		N	ν	*Nu*		T	τ	*Tau*
B	β	*Beta*		Θ	θ	*Theta*		Ξ	ξ	*Xi*		Υ	υ	*Upsilon*
Γ	γ	*Gamma*		I	ι	*Iota*		O	o	*Omicron*		Φ	ϕ	*Phi*
Δ	δ	*Delta*		K	κ	*Kappa*		Π	π	*Pi*		X	χ	*Chi*
E	ϵ	*Epsilon*		Λ	λ	*Lambda*		P	ρ	*Rho*		Ψ	ψ	*Psi*
Z	ζ	*Zeta*		M	μ	*Mu*		Σ	σ	*Sigma*		Ω	ω	*Omega*

θ, ϕ angles
Ω angular velocity of precession
ω angular velocity
α angular acceleration
μ coefficient of friction

Index